THEORISING TRANSITION

How do we understand the dramatic changes that are transforming the societies of Eastern and Central Europe and those 'peripheral' countries witnessing a shift to the market? *Theorising Transition* tries to answer this important question through a variety of studies of the complex political, economic and social transformations occurring in Eastern Europe and other parts of the non-western world that have moved away from strong state direction of their societies.

The authors in *Theorising Transition* challenge many of the comfortable assumptions unleashed by the euphoria of democratisation and the triumphalism of market capitalism in the first flush of 'post-Glasnost' openness. The transformations under way in post-communist societies, far from leading in a straightforward manner to a single model of economic and political life seven years after the events of 1989, now seem to have generated complex and highly differentiated regional systems of adjustment. *Theorising Transition* includes contributions by geographers, economists, sociologists, feminists, planners and scholars of international studies who attempt to unravel the complexity of change in post-communism through examining: theoretical perspectives on transition, industrial and regional restructuring, agrarian change, de-collectivisation and rural struggles, and ethnically and gendered transformations in a variety of Eastern and Central European contexts. The authors also examine the comparative experiences of de-collectivisation, and economic and state restructuring in 'peripheral' societies undergoing transformation.

Examining transformations using a variety of perspectives from critical political economy and cultural theory, *Theorising Transition* provides both a rich empirical map of the dimensions of post-communism and raises important theoretical issues about how we interpret these changes.

John Pickles is a Professor of Geography at the University of Kentucky, USA and **Adrian Smith** is a Lecturer in Human Geography at the University of Sussex, UK.

THEORISING TRANSITION

The Political Economy of Post-Communist
Transformations

edited by
John Pickles and Adrian Smith

London and New York

First published 1998
by Routledge
11 New Fetter Lane, London EC4P 4EE

Simultaneously published in the USA and Canada
by Routledge
29 West 35th Street, New York, NY 10001

Typeset in Garamond by Keystroke, Jacaranda Lodge, Wolverhampton
Printed and bound in Great Britain by Biddles Ltd, Guildford and King's Lynn

British Library Cataloguing in Publication Data
A catalogue record for this book is available from the British Library

Library of Congress Cataloging in Publication Data
Pickles, J. (John)
Theorising transition : the political economy of transition in
post-communist countries / John Pickles and Adrian Smith.
p.   cm.
Includes bibliographical references and index.
1. Post-communism—Europe, Eastern. 2. Post-communism—Former
Soviet republics. 3. Europe, Eastern—Economic conditions—1989–
4. Former Soviet republics—Economic conditions. 5. Social change—
Europe, Eastern. 6. Social change—Former Soviet republics.
7. Europe, Eastern—Social conditions—1989– 8. Former Soviet
republics—Social conditions. 9. Post-communism. I. Smith,
Adrian, 1966– . II. Title.
HC244.P4915 1998
338.947—dc21   97–30851

ISBN 0–415–16266–1 (hbk)
0–415–16267–X (pbk)

for Alex and Leon
who are experiencing their own transitions

CONTENTS

List of plates   x
List of figures   xi
List of tables   xii
List of contributors   xiv
Preface   xvii
Acknowledgements   xviii
List of abbreviations   xix

1   Introduction: theorising transition and the political economy of
    transformation   1
    Adrian Smith and John Pickles

Part I Theorising transition

2   Regulating and institutionalising capitalisms: the micro-
    foundations of transformation in Eastern and Central Europe   25
    Adrian Smith and Adam Swain

3   Organising diversity: evolutionary theory, network analysis and
    post-socialism   54
    Gernot Grabher and David Stark

4   Differential development, institutions, modes of regulation and
    comparative transitions to capitalism: Russia, the Commonwealth
    of Independent States and the former German Democratic
    Republic   76
    Michael Dunford

Part II Industrial restructuring, uneven development and networks in transition

5   Institutions, social networks and ethnicity in the cultures of

transition: industrial change, mass unemployment and regional
transformation in Bulgaria   115
Robert Begg and John Pickles

6 Economic restructuring and regional change in Russia   147
Michael Bradshaw, Alison Stenning and Douglas Sutherland

7 Restructuring state enterprises: industrial geography and Eastern
European transitions   172
John Pickles

8 Theorising trade unions in transition   197
Andrew Herod

9 Privatisation and the regional restructuring of coal mining in the
Czech Republic after the collapse of state socialism   218
Petr Pavlínek

**Part III Social and political movements and the politics of
agrarian transition**

10 Path dependence in Bulgarian agriculture   243
Mieke Meurs and Robert Begg

11 Defending class interests: Polish peasants in the first years of
transformation   262
Krzysztof Gorlach and Patrick H. Mooney

12 Rurality and the construction of nation in Estonia   284
Tim Unwin

Part IV Social transformation and the reconstruction of identities

13 'The political' and its meaning for women: transition politics in
Poland   309
Joanna Regulska

14 Imagined and imagining equality in East Central Europe:
gender and ethnic differences in the economic transformation
of Bulgaria   330
Mieke Meurs

15 Democratisation and the politics of water in Bulgaria: local
protest and the 1994–5 Sofia water crisis   347
Caedmon Staddon

16 Social exclusion and the Roma in transition   373
David Sibley

Part V From the developmental state to hybrid capitalisms: comparative transitions

17 Denationalisation of the Mauritanian state   389
   Mohameden Ould-Mey

18 'Sex and violence on the wild frontiers': the aftermath of state
   socialism in the periphery   408
   James Derrick Sidaway and Marcus Power

19 Economic transformation in China: property regimes and social
   relations   428
   Alan Smart

20 Recombinant capitalism: state, de-collectivisation and the
   agrarian question in Vietnam   450
   Michael J. Watts

   Index   506

PLATES

15.1 Barricades at Sapareva-Bania, 8 February 1995   349
(Source: Green Patrols Video, 1995, with permission)
15.2 Troops marching on protestors at Sapareva-Bania, 8 February
1995   350
(Source: Green Patrols Video, 1995, with permission)
15.3 Sofians filling water bottles at public taps, Central Mineral
Baths, Sofia   354
(Photograph by the author, February 1995)
15.4 Segment of Djerman–Skakavitsa diversion   363
(Photograph by the author, June 1995)

FIGURES

1.1 Central and Eastern Europe   3
2.1 Models of transition   40
2.2 Institutions and networks in transformation   41
4.1 GDP per capita by major regions, 1989 and 1991   77
4.2 Employment and productivity by country, 1992   79
4.3 Economic growth by major region, 1820–1992   81
5.1 *Oblast*-level unemployment, 1991   128
5.2 *Obstina*-level unemployment, 1994   129
5.3 Percentage of the vote for the MRF, 1991   135
6.1 Real GDP for CIS and Eastern Europe, 1989–97   148
6.2 Decline in the physical volume of industrial output between 1992 and 1994   152
6.3 Decline in industrial employment between 1991 and 1994   156
6.4 Share of labour force employed in SMEs at the end of first quarter 1995   159
7.1 The flow of power in a bureaucratic administrative command system   176
7.2 The structure of coercive bargaining between the state and enterprises   180
7.3 Embedded asymmetrical systems of social reproduction   180
9.1 Northern Bohemia coal mining region   219
9.2 Most District, Czech Republic   229
10.1 Percentage of the vote for the BSP, 1991   254
11.1 Selected Polish towns and cities   267
12.1 Percentage vote for KMU and Reform Party in the March 1995 parliamentary elections   296
15.1 Bulgaria study area map   348
17.1 Mauritania   390
18.1 Mozambique   412
19.1 China   436
20.1 Vietnam: growth rates, 1979–95   451
20.2 Vietnam   453

TABLES

4.1	Economic performance of the major regions, 1820–1992	80
4.2	Growth of the Soviet economy	84
4.3	Regional disparities in the former Soviet Union	88
4.4	Transition and GDP growth	94
4.5	Output and employment in the former GDR: the immediate impact of unification	101
4.6	Growth of income, prices and unit labour costs in East and West Germany, 1990–5	102
4.7	Labour market accounts for West and East Germany in thousands of persons	103
4.8	Public financial transfers to East Germany (DM billion)	106
5.1	Regional distribution of industrial output by *oblast*, 1990 and 1994	121
5.2	Per capita output by *oblast*, 1990 and 1994	122
5.3	Size distribution of Bulgarian manufacturing establishments, 1988	124
5.4	Geographic diversity by *oblast* by sector, 1990	125
5.5	World Bank estimates and actual employment in Bulgaria, 1991–4	127
5.6	Concentration of the Turkish vote with rising unemployment	136
5.7	Registered unemployment in the *obstini* with the ten highest rates, March 1994	136
5.8	Social characteristics of the regions with the highest unemployment (> 40 per cent), June 1993	137
6.1	Ranking of regions by key variables	153
6.2	Factors affecting regional change	155
7.1	Average utilisation capacities	182
9.1	Decline in coal production and coal mining employment in the Most District	225
10.1	Land claims under restitution, 1 July 1992	248
10.2	Bulgarian exports by selected trading partners, 1985, 1989–93 (per cent of exports)	251

10.3 Percentage output on personal or private plots, 1989–93   256
11.1 Prices of products sold by farms, 1989–92   271
11.2 Individual (peasant) farm income, 1989–92   271
11.3 Changes in real income (per capita) in families   272
11.4 Incomes in families   272
11.5 Peasant organisations   277
14.1 Sphere of employment by language competence and gender   333
14.2 Qualification level by language competence and gender   333
14.3 Firm leaders by language competence and gender   334
15.1 Bulgarian macroeconomic indicators, 1989–96   356
15.2 Social statistics, Kyustendil and Blagoevgrad *okrugs*, 1992   362
20.1 Vietnam: food production, 1976–95   461
20.2 The evolution of agrarian systems in the Mekong and Red River deltas, 1930–90   464
20.3 Peasant living standards and social equity in the Red River and Mekong deltas, 1930–90   465
20.4 The collective agricultural production system in 1986   472
20.5 Vinh Lac District in provincial and national context, 1993   477
20.6 Vinh Lac District: land, population and production, 1990–3   478
20.7 Dai Dông: village statistics, 1993   480
20.8 Tax loadings in Dai Dông village, Vinh Lac District, 1993   486

CONTRIBUTORS

Robert Begg is Professor of Geography and Planning at the Indiana University of Pennsylvania. He has been working on agricultural and manufacturing change in Eastern Europe, with a focus on Bulgaria.

Michael Bradshaw is Senior Lecturer in the School of Geography and an Associate Member of the Centre for Russian and East European Studies at the University of Birmingham. His research interests are in the regional dimensions of economic and trade restructuring in Russia.

Michael Dunford is Professor of Economic Geography in the School of European Studies at the University of Sussex, and is Editor of *Regional Studies*. His interests are in the contrasting geographies of economic and social modernisation in Europe, in the dynamics of urban and regional economies, and in regulation theory and historically and geographically oriented political economies of economic development.

Krzysztof Gorlach is Associate Professor of Sociology at Jagiellonian University in Krakow, Poland. His research focuses on agrarian transformation, the peasantry and the family farm in Poland.

Gernot Grabher is Professor of Regional Political Economy at the University of Constance in Germany and is working on regional networks and evolutionary economic approaches to regional development.

Andrew Herod is Assistant Professor of Geography at the University of Georgia. His research is concerned with the restructuring of, and new roles for, the Czech and Slovak metalworkers' trade unions.

Mieke Meurs is Associate Professor of Economics and Political Economy at the American University in Washington, DC. She has worked extensively on agricultural change in Cuba, Bulgaria and Romania, and gender issues in Bulgaria and is currently carrying out research on de-collectivisation in Hungary, Bulgaria and Mongolia.

Patrick H. Mooney is Associate Professor of Sociology at the University of Kentucky. He has been working on agricultural transformation in Poland since 1987 and before that researched agrarian transitions in the United States.

Mohameden Ould-Mey is Assistant Professor of Geography and Politics at Francis Marion University in South Carolina. He works on the recasting of the role of the state in response to globalisation and structural adjustment.

Petr Pavlínek is Assistant Professor of Geography at the University of Nebraska at Omaha. His research interests include the geographical and environmental implications of transition, and his most recent work has focused on foreign direct investment in the Czech Republic.

John Pickles is Professor of Geography and a member of the Committee on Social Theory at the University of Kentucky. His research deals with regional restructuring and industrial change, political economy, social theory, and the critical theory of geographic information systems.

Marcus Power is a Teaching Assistant in Human Geography at the University of Leeds. His research focuses on the restructuring of space and power in Mozambique, the formulation of British international development policy and the history of British development geography.

Joanna Regulska is Professor of Geography and Director of the Local Government Reform Project at the East European Research Institute at Rutgers University.

David Sibley is Reader in Geography at the University of Hull. He has written extensively on minority groups and social marginalisation and has recently been working on the Roma in Hungary.

James Derrick Sidaway is Lecturer in Geography at the University of Birmingham. He is interested in critical geopolitics and the transformation of the regional political economy of Mozambique. Recently he has been working on the changing geopolitics of regional integration in southern Africa in comparative perspective.

Alan Smart is Associate Professor at the Department of Anthropology, University of Calgary. He has been conducting research in Hong Kong and China on a wide variety of topics including the market in squatter dwellings, the political economy of squatter area clearance, foreign investment in China and its impact on local social organisation and labour relations, and the processes of social exchange across borders.

Adrian Smith is Lecturer in Human Geography in the School of Social Sciences and a member of the Centre on the Changing Political Economy of Europe at the University of Sussex. He works on the regional dimensions of transformation in Europe, industrial restructuring in Eastern Europe, and theorisations of transition. He is Deputy Editor of *Regional Studies*.

Caedmon Staddon is Lecturer at the University of the West of England. He has been carrying out research on the cultural politics of struggles over water in south-west Bulgaria.

David Stark is Arnold A. Saltzman Professor of Sociology and International Studies at Columbia University and has been conducting research on the network properties of East European economies.

Alison Stenning is Lecturer in the School of Geography and an Associate Member of the Centre for Russian and East European Studies at the University of Birmingham. She works on the changing politics of local economic development in Russia, and is beginning to undertake research in Poland.

Douglas Sutherland is a Research Fellow in the School of Geography and the Centre for Russian and East European Studies at the University of Birmingham.

Adam Swain is Lecturer in Geography at the University of Nottingham, where he works on the regional impacts of inward investment in the auto-mobile industry in East Germany and Hungary, and the regional dimensions of change in Ukraine.

Tim Unwin is Reader in Geography at Royal Holloway, University of London. He has been working on agricultural and rural restructuring in Estonia and has interests in geographic theory.

Michael J. Watts is Professor of Geography and Director of International Studies at the University of California, Berkeley. He has worked extensively on agrarian transition and geographical theory.

PREFACE

Theorising Transition has its origins in an attempt to understand the emerging triumphalism and universalism of transition theory following the events of 1989. In particular, we were interested in the limited theoretical explanation of those events and their consequences at the very time when western political economy, industrial geography and social theory were addressing questions of economic restructuring, globalisation and the changing geography of communities and regions, the recomposition of class identities, and the reworking of gender and ethnicity.

The book aims to bring together authors whose work is grounded in detailed analysis of transformation in reforming societies, and who are grappling with the challenges of contemporary debates in political economy and social theory. This project is part of a broader agenda in which the editors are attempting to bridge theoretical and methodological divides in our understanding of state socialist and post-communist societies. This project is being carried out within two specific contexts: first, that of the changing regional and industrial geographies of societies in transformation; and second, that of the reworking of democratisation, civil society and struggles over the recomposition of the state. These projects are in their infancy, but we hope that *Theorising Transition* has at least opened up a space in which the further development of these ideas will be possible.

John Pickles and Adrian Smith
Brighton, England

ACKNOWLEDGEMENTS

John Pickles acknowledges research support from the US National Science Foundation grants SBR 9515244 and INT-9021910, and the John D. and Catherine T. MacArthur Foundation award for 'The Environmental Consequences of Economic and Political Transition in Bulgaria'. He would also like to thank the participants of the John D. and Catherine T. MacArthur Foundation supported conferences on 'The Decollectivization of Socialist Agriculture' held in Hungary in 1993 and Cuba in 1995, and especially Michael Watts. Adrian Smith acknowledges research support from the 20th International Geographical Congress Fund of the Royal Society and the University of Sussex.

The editors would like to thank Franz Steiner Verlag Stuttgart for permission to reprint John Pickles 'Restructuring state enterprises: industrial geography and Eastern European transition' from *Geographische Zeitschrift* 83 Jg. Heft 2 (1995), and Carfax Publishing Company for permission to reprint Gernot Grabher and David Stark 'Organizing diversity: evolutionary theory, network analysis and post-socialism' from *Regional Studies* 31, 5 (1997).

We would also like to thank Dick Gilbraith of the Gyula Pauer Cartography Laboratory at the University of Kentucky for drawing some of the maps, and Sarah Lloyd, Rebecca Casey and Casey Mein at Routledge for steering the project through to the end.

Finally, our thanks go to Lynn Pickles and Angela Baxter for their love and support.

ABBREVIATIONS

AFL-CIO	American Federation of Labor-Congress of Industrial Organizations
APIE	State Property Agency (Mozambique)
BNM	National Bank of Mauritania
BSP	Bulgarian Socialist Party
CBOS	Centre for Public Opinion Research (Poland)
CDSS	Country Development Strategy Statement
CEC	Commission of the European Communities
CEE	Central and Eastern Europe
CIC	Civil Initiative Committee for the Protection of Rila Waters
CIS	Commonwealth of Independent States
CITUB	Confederation of Independent Trade Unions in Bulgaria
CMEA	Council for Mutual Economic Assistance
ČMKOS	Czech-Moravian Chamber of Trade Unions
CNUCED	Conférence des Nations Unies pour le Commerce et le Développement
COMECON	Council for Mutual Economic Assistance
CPV	Communist Party of Vietnam
CRCEES	Center for Russian, Central and East European Studies, Rutgers University
CSDR	Democratic Trade Union Confederation of Romania
ČSKOS	Czechoslovak Chamber of Trade Unions
DIW	Deutsches Institut für Wirtschaftsforschung, Berlin
DRV	Democratic Republic of Vietnam
EBRD	European Bank for Reconstruction and Development
EC	European Commission
ECE	Eastern and Central Europe
ECU	European Currency Unit
EIA	Environmental Impact Assessment
EIU	Economist Intelligence Unit
ENIP	Estonian National Independence Party

ESCBR	Economic and Social Council of the Basin Region (Czechoslovakia)
EU	European Union
FADM	Front for the Defence of Mozambique
FAO	Food and Agriculture Organisation
FBIS	Foreign Broadcast Information Service
FDI	Foreign direct investment
FEER	*Far Eastern Economic Review*
FTUI	Free Trade Unions Institute
G-7 (G-8)	Group of seven (eight) industrialised countries
GDP	Gross domestic product
GDR	German Democratic Republic
GPA	General Peace Accord (Mozambique)
GUS	Glowny urzad statystyczny (Main Statistical Office, Poland)
ICFTU	International Confederation of Free Trade Unions
IDS	Institute for Development Studies
ILO	International Labour Office
IMF	International Monetary Fund
IWC	Integrated world capitalism
KMU	Coalition Party and Rural Union (Estonia)
KSS 'KOR'	Self-Defence Committee (Poland)
KZKiOR	Agricultural Circles (Poland)
MCC	Most Coal Company
MIC	Military-industrial complex
MP	Member of Parliament
MRF	Movement for Rights and Freedoms (Bulgaria)
NATO	North Atlantic Treaty Organisation
NBBCM	North Bohemian Brown Coal Mines
NEK	National Electricity Company (Bulgaria)
NGO	Non-governmental organisation
NIC	Newly industrialising country
NPF	National Property Fund
NSI	National Statistics Institute (Bulgaria)
NSZZ RI 'Solidarnosc'	Independent Self-Governing Trade Union for Individual Farmers 'Solidarity'
OECD	Organisation of Economic Co-operation and Development
OMRI	Open Media Research Institute
OPZZ	National Agreement of Trade Unions (Poland)
OS KOVO	Czech Metalworkers' Federation
PFP	Policy framework paper
PHARE	Poland–Hungary Assistance for Economic Reconstruction

PRE	Programme of Economic Rehabilitation (Mozambique)
PRES	Programme of Economic and Social Rehabilitation (Mozambique)
PSL-PL	Polish Peasant Party 'People's Agreement'
PZPR	Polish United Workers' Party
R	Reform Party (Estonia)
R&D	Research and development
RAMS	Rural Assessment and Manpower Surveys
RF	Russian Federation
RFE/RL	Radio Free Europe/Radio Liberty
ROH	Revoluční odborové hnutí (Revolutionary Trade Union Movement) (Czech Republic)
SAL	Structural adjustment loan
SAP	Structural adjustment programme
SDR	Special drawing rights
SdRP	Social Democratic Party (Poland)
SL-Ch	Christian Peasant Party (Poland)
SME	Small and medium enterprise
TVE	Township and village enterprise
TKZS	Collective farms (Bulgaria)
UBD	Union of Development Banks (Mauritania)
UDF	Union of Democratic Forces (Bulgaria)
UFV	Independent Women's Association (German Democratic Republic)
UN	United Nations
US	United States of America
USAID	United States Agency for International Development
USSR	Union of Soviet Socialist Republics
VAT	Value added tax
ViK	Sofia Water Supply and Sewerage Company
ZSL	United Peasant Party (Poland)

1

INTRODUCTION

Theorising transition and the political economy of transformation

Adrian Smith and John Pickles

1989 and the geopolitics of transitions

In 1996, an OECD report on the transition from a state socialist, centrally planned economy in perhaps the most liberalised and westernised country in Eastern and Central Europe – the Czech Republic – raised the possibility that the 'transition' is at an end (OECD 1996). A year later, the western business press was full of reports about the failure of the Czech Republic to establish the conditions for competitive growth, largely as a result of the continued dominance of the economy by large, institutional investors (such as banks) with close ties to the state. Through mimicking the old relations of state socialism in a new guise the past had returned to haunt the post-communist scene. By contrast, in Bulgaria since 1989 almost annual changes of government, electoral reversals from communists to democrats to reformed communists and back again to democrats have typified the process of political democratisation. Industrial restructuring has been occurring in Bulgaria through loss of Eastern markets, plant closure and mass unemployment. Financial and fiscal instabilities permeate all institutions and had led, by 1996, to hyperinflation, as the value of the currency (the lev) fell precipitously against all other currencies: from around 60 leva to the US $ in June 1996 to around 3,500 leva to the US $ by February 1997. Over the same period, Vietnam, with an economic growth rate of around 9 per cent per year, was witnessing a partially agrarian-based economic restructuring centred around the reconfiguring of the role of the party-state, in complete contrast to the apparent liberalism of the Czech transition.

These diverse experiences of 'transition' raise two crucial issues for our understanding of the political economy of change in both Eastern and Central Europe (ECE) and the post-communisms and market-socialisms of the 'peripheral' states (Figure 1.1). First, the conventional, neo-liberal view

1

of transition wielded by western multilateral agencies and advisers to governments in ECE, that transition is a relatively unproblematic implementation of a set of policies involving economic liberalisation and marketisation alongside democratisation, enabling the creation of a market economy and a liberal polity, relies on an under-theorised understanding of change in post-communism. Such claims tend to reduce the complexity of political economic change in Eastern and Central Europe and fail to provide us with an adequate basis on which to move beyond policy prescriptions of transition as a set of end-points. The Czech experience, once held as the best practice solution to the problems of transition (low unemployment, rapid westernisation, the quick implementation of large-scale privatisation, etc.), with Prime Minister Vaclav Klaus presented on the international stage as the darling of East European neo-liberalism, suggests that even in Eastern Europe's most liberal polity 'best practice' transitional policies have not been implemented in the most effective way. In the Vietnamese case, however, with its parallels to China (Burawoy 1996; Smart in this volume), the continued role of the party-state has been a mechanism through which reform could occur. Yet in Bulgaria, where the party-state nominally was in control until 1991 (some would even argue until 1997, when the Bulgarian Socialist Party witnessed a resounding electoral defeat), the partial imposition of a market road to capitalism through price liberalisation but stalled privatisation has led to crisis. The challenge then is to negotiate ways in which we can understand the *diversity of forms* of transition.

Second, the causes of the 'Czech crisis', as identified by the western business press, point directly to the central role of the legacies of institutional frameworks and existing social relations derived from state socialism to an understanding of the multiple ways in which transition is playing itself out. Transition is not a one-way process of change from one hegemonic system to another. Rather, transition constitutes a complex reworking of old social relations in the light of processes distinct to one of the boldest projects in contemporary history – the attempt to construct a form of capitalism on and with the ruins of the communist system (Stark 1996; Smith 1997; see also Grabher and Stark in this volume).

Mainstream transition theory has, then, largely been written in terms of the discourses and practices of liberalisation. *Theorising Transition* attempts to move the ground away from such perspectives by directly engaging the criticisms identified above. Liberalisation can thus be thought of in terms of what Michel Foucault has called the 'technologics of the social body': as a series of techniques of transformation involving marketisation of economic relations, privatisation of property, and the democratisation of political life. Each seeks to de-monopolise the power of the state and separate the state from the economy and civil society. Marketisation seeks to free-up the economy. Privatisation aims to break up economic monopolies in the spheres of production, purchasing and distribution. Democratisation

Figure 1.1 Central and Eastern Europe

3

and de-communisation aim to break the hold of the Communist Party in political life and to enable a rejuvenated civil society to emerge. Each technique has, in turn, its own specific instruments: for example, the creation of markets and price reform for marketisation; restitution, voucher schemes and share ownership, and the selling off of state property for privatisation; multi-partyism and parliamentary democracy for democratisation.

Each technique of transformation, along with its specific instruments and policies, is bringing about a fundamental reorientation in the position of post-communist states in the global economy. The dismantling of the Council for Mutual Economic Assistance (CMEA) and the Warsaw Pact, and the rejection of links with former members by individual states, opened post-communist economies to international capital and institutions (Gowan 1995). The result was a fundamental reorganisation of material life within ECE and the former USSR, a transformation of geopolitical relations on a global scale, and a major ideological/discursive shift in the ways in which 'possibilities' and 'policies' were to be framed and acted upon in future years.

Early attempts by anti-Stalinist democratic socialists and social democrats within the democracy movements of Eastern and Central Europe to frame a transition 'with a human face' failed in the euphoria of 1989. Even as 'the Wall' fell, the Left was either swept away, or swept up into the machine of transition or western think-tanks, banks, and para-statal agencies who provided 'guidance'. For those who remained to articulate alternatives to 'shock therapy' and the 'three zatsias' (*privatizatsia, demokratizatsia, liberalizatsia*), anti-communist sentiment left little room for articulating alternative paths for transformation (a situation that applied equally to the reformed communists as well). There simply was no 'third way' available – a closure produced in part by the growing public imaginary of a West offering wealth, freedom and opportunity (Altvater 1993; see also Marshall Berman (1982) on the experience of the modern) and in part by the direct and indirect effects of western propaganda throughout the 1980s.

The possibility of a third way was also foreclosed by the immense power of the dominant discourse through which interpretations of transformation in ECE were projected: a reworked modernisation theory which recasts Sovietology and its Cold War ideology into what Michael Burawoy (1992) has called 'transitology'. In this view of transition, it was argued, socialist societies will undergo modernisation to market capitalism and democracy once the appropriate policies, institutions and mechanisms are in place. As central planning systems are dismantled, the bureaucratic economy will be replaced (through privatisation, new firm development and the emergence of an entrepreneurial class) with thriving examples of market economies and the emergence of social groups capable of cementing the market, liberal democratic practices and a re-emergent civil society.

In the past seven–eight years these techniques of transformation have contributed to the massive political and economic transformations and social dislocations attendant upon the collapse of Soviet power. The fall of the Berlin Wall in 1989 led to an immediate outpouring of optimism about the future possibilities opened to peoples of Eastern and Central Europe by the democracy movement and the overthrow of state socialist institutions and leadership groups. The subsequent turn to formally marketised and democratic systems has resulted in a deep and thoroughgoing modernisation of the structures and institutions of government, economy and social life. And, with the 'guidance' and pressure of western multilateral organisations (such as the IMF and World Bank), national and international assistance agencies, and NGOs and policy groups of one kind or another, the fabric of life has been reconfigured so as to construct capitalism and create the formal structures on which liberal democracy can be built. These changes have in turn brought with them a reworking of the geographies of the region in the forms of metropolitan growth, the economic collapse of peripheral regions, the polarisation of urban spaces as inequalities deepen, and the reworking of the territorial structure and democratic spaces of the state and civil society.

But the significance of the 'end of communism' runs far deeper than even these massive transformations. The geopolitical consequences of the end of a 'two-world' order and the consolidation of what George Bush called '*the* New World Order' have become increasingly clear in recent years. Super-power singularity supervises the emergence of regional power blocs and, once again, narrows the options open to individual state planners in all countries (see Ould-Mey in this volume). But precisely because of this geopolitical hegemony, we need to be aware of the *multiple* and *differentiated* strategies for transformation that are at work. From the boycott of Cuba and the blockading of Iraq to the 'openness' towards Vietnam and China, a variety of strategies are being deployed to sustain transitions with the goal of building post-communism, open markets, and stable relations for the flow of goods and capital.

Political economy, restructuring and theorising transition

At the end of the twentieth century there is a need, then, for an alternative set of conceptual frameworks on transition to challenge the neo-liberal hegemony and account for the variety of strategies, techniques and effects that constitute transition-in-process – actually existing transition. Unlike liberal transition theorists, however, we do not want to articulate a single, hegemonic perspective. Indeed, as we have seen from the cases of the Czech Republic, Vietnam and Bulgaria, the variety of actually existing transitions does not allow this. What we do need is a critical engagement with the *real* transformations in post-communism and some way of thinking through what our theorisations of transition should look like. And what we do have, even after the collapse of the one main legitimation for capitalism, state socialism, is a variety of starting points to enable us to begin to rebuild a critical political economy of transition. This book aims to make a modest contribution to this project – to begin to map, both empirically and conceptually, our understandings of post-communism; to produce alternatives to what Derrida in *Specters of Marx* (1994: 51–2) has described in the following terms:

> No one, it seems to me, can *contest* the fact that a dogmatics is attempting to install its worldwide hegemony in paradoxical and suspect conditions. There is today in the world a *dominant* discourse, or rather one that is on the way to becoming dominant, on the subject of Marx's work and thought, on the subject of Marxism (which is perhaps not the same thing), on the subject of the socialist International and the universal revolution, on the subject of the more or less slow destruction of the revolutionary model in its Marxist inspiration, on the subject of the rapid, precipitous, recent collapse of societies that attempted to put into effect at least what we will call for the moment, citing once again from the *Manifesto*, 'old Europe', and so forth. This dominating discourse often has the manic, jubilatory, and incantatory form that Freud assigned to the so-called triumphant phase of mourning work. The incantation repeats and ritualizes itself, it holds forth and holds to formulas, like any animistic magic. To the rhythm of a cadenced march, it proclaims: Marx is dead, communism is dead, very dead, and along with its hopes, its discourse, its theories, and its practices. It says: long live capitalism, long live the market, here's to the survival of economic and political liberalism!

Derrida's project is one of deconstruction, in which '[i]n order to analyze [the] . . . wars and the logic of [the] . . . antagonisms [of the post-Cold War world], a problematics coming from the Marxian tradition will be indispensable for a long time yet. For a long time and why not forever?' (Derrida 1994: 63–4). For Derrida (p. 75), deconstruction 'has never been Marxist, no more than it has ever been non-Marxist'. What is important is to act and work in the 'spirit of Marxism', although these spirits are many and heterogeneous.

This book remains open to this 'specter', in part because of the analytical purchase it gives to *some* of the authors, in part because of the symbolic function 'Marx' and 'Marxism' have played and continue to play in the geopolitics of communist and post-communist societies, and in part because of the fundamental challenges to hegemonic theories of transition and the destabilising role Marxian analyses of capitalism continues to perform. In this sense, the aim of this book is to explore some of the many ways in which a 'radical political economy' (see Sayer 1995) of post-communism can be constructed. This is not to argue that interrogating and 'deconstructing' the discourses that are wielded in thinking through and interpreting post-communism are unimportant. On the contrary, such a project should have a central place in our analyses. But the key is to think through how these discursive formations of post-communism relate to an understanding of transition in which the social and material worlds are transformed in conjunction with our ideas about them. In this introduction, we want to begin to identify starting points for such an alternative conceptual map. As editors, we argue that there is much to be learnt from an engagement between political economy and recent shifts in critical and cultural theory. Indeed, we have attempted to frame this introduction in terms of where our understandings of transition can be located in this engagement. However, we are also cognisant of the fact that our contributors may want to position themselves differently, some preferring to engage more closely with variants of political economy, others with a concern for pushing the boundaries of cultural theory. What we hope to achieve in *Theorising Transition* is an opening of our conceptual maps of transition to counter the closure of dominant discourses and the power of these ideas in shaping the policy agendas of transition in ECE and beyond.

Such a critical political economy is necessary now more than ever. In the face of the 'neo-evangelistic rhetoric' (Derrida 1994) of neo-liberalism, Eastern and Central Europe and other post-communist societies are being fundamentally transformed by globalised systems of capital accumulation. The experience of 'transition' after seven or eight years has been, in part, one of economic collapse, an onslaught on labour, and social and political disorientation (collapsing birth-rates and increasing death-rates suggesting a deep-seated social and psychological crisis), while at the same time enabling some to prosper while others fall into abject poverty. The result has been a profound increase in poverty and inequality. Milanovic (1994), for example, argues that poverty now affects some 58 million people in ECE, or 18 per cent of the region's population. Similarly, real wages have dropped dramatically and wage differentiation increased. Alongside increased inequality, homelessness has risen, health levels have declined and other social problems associated with polarisation have emerged.

Increasing poverty and inequality have proved to be fertile ground for two forms of coping strategies – and it is vital that any theory of transition understands the complex dialectical relationships between these: the first is the increased use of household survival strategies such as the exchange of household production including food and other basic items between friends and in networks established in the workplace and community which have led to a burgeoning of the informal economy. The second is the rise of illegal and semi-legal activities associated with, but not wholly confined to, the 'mafia' (Burawoy 1996). Both of these 'coping strategies' – and there are many more – can be read as ways in which individuals with differential power are able to mobilise existing social, political and economic resources to find a pathway through the maelstrom of transition. What is doubly interesting about these experiences is that they suggest that we have a great deal still to learn from the experiences of earlier (and uneven) transitions to capitalism.

Earlier transitions and the political economy of uneven development

Lenin, writing of Russia at the end of the last century, suggested that

> merchant's and userer's capital, on the one hand, and industrial capital . . . on the other, represent a single type of economic pheno-menon, which is covered by the general formula: the buying of commodities in order to sell at a profit. . . . Merchant's and userer's capital always historically precedes the formation of industrial capital and is logically the *necessary* premise of its formation . . . but in themselves neither merchant's capital nor userer's capital represent a *sufficient* premise for the rise of industrial capital (i.e., capitalist *production*); they do not always break up the old mode of production and replace it by the capitalist mode of production. . . . The inde-pendent development of merchant's capital is inversely proportional to the degree of development of capitalist *production* . . . the greater the development of merchant's and userer's capital, the smaller the development of industrial capital . . . and vice versa. . . . Conse-quently, as applied to Russia, the question to be answered is: Is merchant's and userer's capital being linked up with industrial capital? Are commerce and usury . . . leading to its [the old mode of production's] replacement by the capitalist mode of production, or by some other system?
>
> (Lenin 1967: 186–7; emphasis in original)

What is interesting about these debates a century ago is that they were embroiled in a working through and specification of a Marxist analysis of capitalism. Indeed, Lenin's project was one of demonstrating (*contra* the Narodniks) that capitalist production and social relations *were* transforming Russian society, yet they were doing so on the basis of a complex articula-tion of capitalist relations with mediating social formations such as merchant capital. What we have been in danger of losing in the face of the neo-liberal onslaught and the 'triumphalism' and 'neo-evangelism', as Derrida puts it, of the likes of Francis Fukuyama's thesis on the end of history, is a space for critical engagement with the transformative practices of capitalist modernities and the state socialist project. While many have decried the collapse of state socialism as the end of Marxism, we – as editors – would prefer to argue that the collapse has opened a space to rework a critical political economy, in part because the main justification for capitalism (the undemocratic and authoritarian forms of domination of state socialism) can no longer be wielded as proof that a Marxist project is doomed to failure.

What is also interesting about the debates taking place at the end of the last century is that they were taking place in the context of the uneven development of capitalism both at a European scale and at the global scale. In the contemporary European context, the post-socialist states are today seeing the continuation of their relatively peripheral position in Europe and its reworking through the project of future enhanced integration with the West. What this means in the long run is at present unclear. It is likely, though, that the forms of uneven development we are currently living with will at least continue in their intensity or at worst deepen, as the models of capitalist and geopolitical integration continue to be neo-liberal ones (Amin and Tomaney 1995). In earlier transitions, as the work of Gerschenkron (1962) recognised, the 'relative backwardness' of what became the Soviet system mattered to what unfolded there – the need for concerted state co-ordination of development in Tsarist Russia (in part to maintain control over large and diverse territories, in part to push resources into particular developmental projects) was partially replicated in the statist notion of Marxism-Leninism that emerged in the 1930s. Taking the argument further, however, it is clear that debates around capitalist transition in Russia at the end of the nineteenth century reflected the uneven development of the European space-economy. Lenin's concern for the uneven and contradictory emergence of capitalist relations in a feudal environment became central to his project of demonstrating politically that the emergent working class was in formation and was becoming a revolutionary vanguard. What is clear, however, is that there was only *partial* modernisation in the East, along with a high level of uneven development both between the underdeveloped East and more-developed West and within the East itself – an unevenness that in many ways, albeit under very different conditions, remains today (Dunford and Smith 1998). Kagarlitsky (1995a: 218), for example, has argued that '[t]he countries of the former Eastern bloc might be . . . described as the semiperiphery of the new world order established at the beginning of the 1990s'. In a similar vein, he (1995b: 92) has argued that the

> Russian capitalism [of the 1990s] was so closely and organically linked to barbarian and precapitalist structures that there could be no talk of modernization . . . the history of the early years of the century was repeating itself; modernization in Russia was impossible unless there was a break with the past but those in power were creatures of that past and incapable of a genuine new beginning.

Many of the contributions to *Theorising Transition* are also concerned with the important argument that treating post-communist Eastern Europe as a whole fails to recognise the ever-present diversity of some 27 states and 270 million people. Even at the end of the nineteenth century, such political-economic diversity was central to what was unfolding in the region. The uneven yet forward march of industrialisation in Bohemia and some parts of Poland, Hungary and the former East Germany, for example, contrasted markedly with other parts of the Austro-Hungarian Empire, the vast majority of the Balkans and most of Russia, in which only limited pockets of industrialisation occurred and where in many cases the 'second serfdom' led to a reinscribing of feudal relations in the East (Kochanowicz 1989). The diversity of historical experiences was replicated under state socialism, and while we would not argue for some form of historical determination, the state socialist economy in part relied upon these spatial divisions of labour and forms of social organisation and institutionalised practices, albeit that large-scale attempts at forced industrialisation were made to eradicate the legacies of 'peasant societies' and uneven capitalist development. Yet what is interesting is that this deep-seated historical context to understanding the diversity of Eastern and Central Europe remains true today.

What is even more compelling about this thesis is that the trajectories of change sought by the neo-liberal transition model are such that they have failed to recognise that diversity of experience, preferring instead to implement a stock set of policies to enable the supposed transition to capitalism at the end of the twentieth century to be achieved. Given the diversity of path-dependent trajectories in ECE, it is not surprising, then, that the neo-liberal project, in its universalisms and its neglect of particularity and complexity, results in quite divergent outcomes, from the 'economic involution' of Russia (Burawoy 1996), and the crisis state of Bulgaria, to the equally contradictory high growth (following massive collapse) of parts of the 'westerly' Czech Republic, Poland, Hungary and, to a certain extent, Slovakia and Slovenia.

The political economy of transition

Given these considerations concerning the geopolitical nature of transition, the contested and ideological discourses and technologies of transition, and the lessons of past transitions, *Theorising Transition* therefore deploys a number of conceptual claims with roots in political economy, feminism, evolutionary economics, and new cultural theory. Each of the contributors to *Theorising Transition*, however, takes a different stance on the balance between and relative merits of these conceptual approaches. These theories of political economy have themselves been greatly enhanced in recent years by a growing critical engagement with post-communist transitions and the experience of specific reform economies since 1989. The critique of neo-liberalism has been paralleled by internal debates within critical theory and political economy about different national modes of regulation and regimes of accumulation, questions of scale and levels of analysis, the importance of historical and geographical specificity in the path taken by reforming countries, and the role of different institutions, actors and social relations in the patterns of economic and political life that are emerging. These internal debates have similarly been paralleled by wider debates within critical social theory about the impacts of, and responses to, broader transformations in the international economy, debates about globalisation, the emergence of new systems and levels of governance within and beyond the nation-state, and the processes of identity formation and resource mobilisation that characterise different social groups and localities within reforming countries.

In these ways, the political economy of transition has challenged many of the taken-for-granted assumptions of neo-liberal perspectives and has, in interesting ways, begun a deepening of political economy itself. From arguments about path dependency to questions about regional regimes of accumulation, to the nature of the changing state form, to issues of locality, to institutional thickness, to the role of social networks in economic and political reform, political economists have begun to problematise the ways in which neo-liberalism has explained the path from state socialism to capitalism.

While the 'return to capitalism' may indeed be under way throughout the post-communist world, how we understand this return must now be rethought in the light of these critical perspectives. While the multilateralisation of state planning has clearly shifted much decision-making from state institutions to the G-7 (now G-8), World Bank and IMF, and while individual 'peripheral' states have had to respond to the winds of change brought about in the wake of Bretton Woods, the globalisation of capital, and the collapse of national development planning ideologies (what Guattari and Negri (1992) refer to as the emergence of Integrated World Capitalism (IWC)), not all states have responded in the same way, nor have they had the same capacities to respond to these international pressures. Nor – as we have seen more recently as 'flexibility' enters the lexicon of the World Bank – have all states been treated equally in the 'reform' strategies 'recommended' to them. A diversity of adjustments and responses to post-communism have emerged in local and regional circumstances, drawing on different local capacities, histories, conditions, and modes of insertion in the national and international economy.

Diversity and complexity in transition

It is precisely this diversity of conditions and responses to the 'package' of neo-liberal reform measures that is the focus of the authors in this book. Recognising that the reconfiguration of economic and political systems

11

occurs in concrete social and cultural milieux, and that these milieux provide the context within which institutional reform, economic liberalisation and political democratisation are reworked, has important implications for how each of the authors approaches the question of transition. Post-communist reform is, in this sense, about the reworking of modernity and the reconfiguration of the economic and political institutions and practices put in place (or adapted from pre-communist days) by state socialism. It is also, at the same time, a restructuring of the social relations of actors, the construction of new identities and the mobilisation of existing cultural resources to new ends. Reworking the social relations of communism and the building of post-communism thus also requires engagement with critical theories of subject formation, the role of discourse, and the nature of modern power in post-communist states and economies.

The first part of *Theorising Transition*, then, examines the complexity and diversity of transition through a series of theoretical debates in political economy. These debates are elaborated in the second part dealing with 'Industrial restructuring, uneven development and networks in transition', where the chapters focus on the economic constitution of change and the myriad responses to restructuring. The third part, 'Social and political movements and the politics of agrarian transition', examines the way in which agricultural change is contested and 'the rural' is mobilised in struggles in political life and the formation of national identity. Part IV, 'Social transformation and the reconstruction of identities', engages with ideas in cultural theory and feminism, and examines the way in which women, ethnic minorities and environmental groups are responding to transition and are mobilising the social and cultural resources at their disposal. The final part, 'From the developmental state to hybrid capitalisms: comparative transitions', raises at least two important issues: first, the comparative nature of political economic change in 'peripheral' post-communist and post-statist societies, and second, the ways in which hybrid capitalisms are being produced. All the chapters point to a multiplicity of possible paths to transition and the complexity of this process. Our starting point is Eastern and Central Europe, but the final part, which draws examples from 'peripheral' states undergoing reform, points to the broader significance of these paths and processes of transition.

Understanding transition

In attempting to forge the beginnings of a theoretical understanding of transition in the terms of a political economy of capitalist development, the contributions to *Theorising Transition* raise two sets of agendas. First, the authors' central attention is directed towards understanding how capitalism, in its variant forms, is constructed and how the definition and working of the central social relations of a political economy arise out of social struggles, the identities that compose those struggles, and the new social relations that result from them. Second, and in addition to a focus on the construction of capitalist materiality, we are also concerned with how, once having formed, the variant capitalisms that compose post-communism become regulated and reworked in the search for some form of more stabilised political economic model out of the crisis and collapse of the old system. The vibrancy of recent debates around economic and industrial restructuring is a central canvas on which we are suggesting that an understanding of transition, both in its ECE form and also in the context of 'peripheral' transitions, needs to be placed. Some of the most influential debates in western political economy have centred around the reworking of the industrial economies of advanced capitalism. From debates around globalisation to the restructuring of systems of enterprise organisation comes a set of useful analytical categories from which we can begin to work through our understanding of post-communist transition.

In addressing these central questions of transition, *Theorising Transition* revolves around five central issues: first, a set of questions concerning how political economies in transition are regulated; second, an exploration of the ways in which political economic transition is both evolutionary and path-dependent in the sense that it is based upon institutionalised forms of learning as well as struggles over pathways that emerge out of the intersection of old and new; third, a concern that particular trajectories of political economic development result from the ways in which social networks and social relations are transformed, struggled over and institutionalised in new forms; fourth, an exploration of the changing geographies of transition and a reworking of the scales of power in transitional contexts; and fifth, an examination of the limits and contested nature of democratisation.

Regulating political economies in transition

Theorising Transition addresses head-on debates over the mechanisms through which political economic life is regulated. Several of the chapters (Smith and Swain (chapter 2), Dunford (chapter 4) and Pavlínek (chapter 9)) draw upon regulation theory as a standpoint from which we can begin to understand how political economic transition occurs and is framed. As Smith and Swain point out, regulation theory is relatively underdeveloped in the context of transitional societies, but they argue that its conceptual understanding of how political economic forms are stabilised or enter into crisis through the coupling of forms of accumulation and mechanisms for their regulation is a valid lens through which one can *begin* to understand transition. Dunford, for example, takes the cases of Russia and the former East Germany as direct reference points to work through the ways in which the Soviet system was regulated and entered into crisis. He goes on to examine how the transition in these two countries is failing as a direct result of the inability of the transition programmes to appreciate how capitalist economies are regulated through complex sets of institutional structures that result from processes of institutional change and innovation over long time periods. The collapse and crisis of both the former East Germany and of Russia is a direct result, then, of the imposition of a set of regulatory frameworks 'alien' to the structures of accumulation in those two contexts. In a similar way, Pavlínek takes the case of brown coal surface-mining in northern Bohemia to examine how struggles over privatisation and environment are constituting new forms of regulation that may not provide a solution to the economic and environmental crisis in that particular locality.

The diversity of forms of regulation in transition suggests that one can identify a series of competing forms of transitional capitalism (market roads, mercantilist roads, involution) (see also Hodgson 1996). From sets of specifically local capitalisms identified in the chapters by Smith and Swain and by Smart (chapter 19), the chapter by Dunford suggests that different national roads to capitalism are evident in transitional countries.

Many of the chapters in *Theorising Transition* also push further our understanding of the political economy of transition by taking on specific debates about industrial restructuring in advanced capitalist economies, and how new industrial and economic activities are constituted and regulated. Whether forms of flexible specialisation and post-Fordism (vertically disintegrated production systems) are identifiable in the transitional contexts we examine after the end of what some (for example, Murray 1992) have called 'Soviet Fordism' is an open question. Many of our authors are concerned with the ways in which we understand the mechanisms of industrial restructuring in transitional societies/economies and the role of new sites of accumulation (speculative, small-firm, privatised large firms), new forms of organisational profiles of industrial economies (decentralised, de-monopolised enterprises and privatisation, locally embedded firms with dense networking arrangements), and new forms of labour processes resulting from the struggles over union rights to organise in the workplace. The chapters by Grabher and Stark (chapter 3), Begg and Pickles (chapter 5), Bradshaw, Stenning and Sutherland (chapter 6), Smart, Sidaway and Power (chapter 18), and Watts (chapter 20) all examine the diversity of ways in which new sites of accumulation are constituted in transition. Similarly, Pickles (chapter 7) and Pavlínek address, in varying contexts, the restructuring of existing (and often very large) state enterprises in response to pressures for privatisation and recomposition of organisational structures. Examinations of the reworking of labour processes and struggles over the definition of workplace organisation are directly addressed in chapters by Pickles, Pavlínek and Herod (chapter 8).

In the agrarian sphere, Watts explicitly examines Kornai's assertion that economic reform cannot occur without the complete dismantling of the party-state through the case of Vietnam, where the state remains a central regulatory agent. Watts uses the Vietnamese 'agrarian question' to demonstrate the ways in which the *doi moi* reforms initiated by the communist state seem to have enabled Vietnam to create a relatively successful economic restructuring (see also Burawoy 1996 on Russia and China). For Watts, the paradox of communist peasant entrepreneurs opens up the question of what we mean by capitalism and communism, and what the possibilities are for hybrid capitalisms. In a similar vein, Meurs and Begg (chapter 10) show how some Bulgarian farming co-operatives struggled hard to avoid being dissolved, and how this led to the striking paradox of co-operative managers deploying capitalist strategies and techniques to sustain and maintain the co-operatives during hard economic times. This is a theme also developed by Smart in his analysis of the growth of multiple 'local capitalisms' in China, several of which are located around the success-ful growth of township and village enterprises (TVEs) in which one sees the reinscribing of local elite and local state power to control accumulation. Smart, however, is careful not to ascribe primary significance to the role of state ownership in the promotion of successful restructuring and his central argument concerns the diversity of forms of local capitalism in the Chinese transition. Similarly, Dunford's analysis of the 'hollowing out' of the Russian state, and the failed implantation of West German state institutions into East Germany suggests that the neo-liberal claim that transition is most successful in situations where state organs wither away is highly problematic. The state, it seems, is required as a fundamental regulatory formation in transition.

Path dependency and evolutionary transitions

A second central theme running through many of the chapters examines how political economic transformation is an evolutionary and path-dependent process, based upon institutionalised forms of learning and struggles over pathways that emerge out of the intersection of old and new. This project has its roots in evolutionary political economy, and has been most forcefully pursued through David Stark's work (e.g., Stark 1996). As the chapter by Grabher and Stark claims, future forms of development arise out of the particular trajectories or paths that have been taken in the past. In this sense, 'legacies' are a central component of our understanding of both the possibilities and limits to transition. Smith and Swain take up this issue directly by suggesting that an understanding of the diversity of local responses to transition must arise out of the way in which previous sets of social relations in place are reworked within the limits of those relations. Similarly, the chapter by Dunford suggests that a failure to appreciate the distinctive nature of the Soviet systems in Russia and the former East Germany is a key factor in the application of inappropriate transitional strategies. In the agrarian sphere, Meurs and Begg argue that the weak performance of Bulgarian agriculture and, by implication, the Bulgarian

15

economy, is not the result of the uneven privatisation of collective and state farms, but that mass privatisation may be an inappropriate mechanism for agrarian restructuring in the context of the legacies of institutionalised forms of understanding, the economic structure of the country, and the technical characteristics of Bulgarian agriculture. Indeed, they show how some farmers in particular regions struggled to retain their co-operatives during the years of restitution and privatisation because this organisational form made sound economic sense in the changing political economy of food production.

Transformation of networks and social relations

A third major theme revolves around the claim that particular trajectories of political economic development result from the different ways in which networks and social relations are transformed and institutionalised in new forms. Again, with links into evolutionary theory (Grabher and Stark), these claims suggest that a fundamental concern of our understanding of social *and* economic change in transition must be the way in which networks of connectivity between social and economic actors are transformed and understood. Indeed, several of the chapters suggest that individual and household survival in times of economic crisis is strongly reliant upon the transformation of networks of social relations such as gender and ethnicity (Pickles, Regulska (chapter 13), Meurs (chapter 14), Sibley (chapter 16), Begg and Pickles). Pickles, for example, shows how social and economic networks in one large chemical enterprise have become central to the responses of managers and workers to changed external circumstances, and how these networks, social relations and work practices – forged under state socialist soft-budget constraints – have been deepened and extended in the lead up to privatisation and as hard budget constraints are imposed. Similarly, the reworking of the 'old' power relations through *nomenklatura* power and the rise of new sites for accumulation has been a major influence on the development of mobsters and mafia (Begg and Pickles, Sidaway and Power).

Regulska and Meurs each focus on the recomposition of gender relations in post-communism, and the chapters by Meurs, Sibley, and Begg and Pickles demonstrate how – as the economic crisis of transition impacts ethnic minorities to a much greater extent than other sectors of the population – ethnic groups have been forced to draw on the cultural resources of their communities. As well as documenting the social consequences of economic and political reform for women and ethnic minorities, Meurs, Regulska, and Begg and Pickles also focus on the ways in which a 'gendered politics' and cultural politics of defence and resistance may be emerging. Together, these chapters suggest that the political economy of gender (and ethnic) marginalisation demands that a social theory of the discourses of domination, masculinism and power be situated alongside the critical analysis of the actual subject positions women and minorities have been assigned and have assumed. As Regulska argues, the 'gendered political' must be rethought to take into account all the ways in which women do act politically, despite (and perhaps in part because of) their marginalisation from the formal institutions and practices of electoral politics. This reconfiguration of what constitutes political life, its relationship to the structure of household dynamics, and their articulation are important aspects of post-communist transformations.

Geographies of transition and the rescaling of power

While it may be possible to identify discrete 'national roads to transition', as we did at the start of our introduction, one of the central issues taken up in many of the chapters in *Theorising Transition* is the way in which uneven development is reworking the sub-national space economies of post-communism. We are witnessing the emergence of regionally differentiated transitions in which the dynamics of regional decline and growth play out in particular time–place contexts. Watts, for example, demonstrates how the two regional economies of the Mekong and Red River delta have responded differently to *doi moi* reforms in Vietnam. At another level, one can identify a geography of polarised transition around the expansion of new cores and peripheries. The chapters by Dunford and Bradshaw, Stenning and Sutherland, for example, echo the sentiments of a recent European Commission report (1996; see also Dunford and Smith 1998) that identified a set of regional economies experiencing or likely to experience dramatic restructuring and globalised growth usually based around the ability of capital city regions to mobilise increased global connectedness (in part through concentrations of foreign investment) and potential for institutional and industrial innovation. However, a set of marginalised regional economies is increasingly 'left behind' in the process of capitalist restructuring as enterprise decline and closure paves the way for the emergence of regions of mass unemployment. The chapter by Begg and Pickles takes these arguments further by claiming that our understandings of the extent of, and processes constituting, regional divergence and convergence are limited because of a fetishism of scale. Through the Bulgarian case, they argue that the structure of state socialist industries and the spatial allocation of specific enterprise forms are crucial to our understanding of the geography of regional economic change, specifically through identifying two types of industrial space – enterprise space (spaces of the core industries) and branch-plant space.

Vachudová and Snyder (1997: 3–4) have argued that one key element in understanding the diversity of national roads to transition between Hungary, the Czech Republic and Poland, on the one hand, and Bulgaria, Romania and Slovakia, on the other, is the importance of what they call 'ethnic geography'.[1] By this they mean the relative homogeneity, or otherwise, of a nation-state providing the resources to mobilise a non-ethnic national agenda in restructuring. The relative homogeneity of the Czech Republic, Hungary and Poland has meant that these countries do not 'confront the kind of pressing "national" problem that confounds domestic politics in Slovakia, Bulgaria and Romania' (Vachudová and Snyder 1997: 14–15). While this may be important in determining different national pathways, we would also point to the importance of 'ethnic geography' in the reconstruction and mobilisation of the geography of uneven development. Ethnicity has played two key roles in transition. First, it has been a fulcrum around which the break-up of the three communist federations revolved and in two cases (the former Yugoslavia and Soviet Union) resulted in war with devastating economic implications not only for these states, but also for neighbouring war-torn regions (Sussex European Institute 1995). War and conflict have been central moments in several post-communist transitions, and indeed one could begin to understand war in post-communist ECE as an extreme form of transition in which the contradictions of communist economic, political and social policies have resulted in almost complete devastation through fragmentation into competing war-zones and 'fiefdoms'. Yet, at the same time, this form of extreme fragmentation has reconfigured communist social powers and networks, if only as a survival strategy (see the chapter by Begg and Pickles for a less extreme, but no less important, case).

A final theme concerning the geographies of transition in *Theorising Transition* revolves around the relationship between globalisation and the changing nature of economic and political power. Just as state socialism had its own distinctive geographies (attempts to eradicate spatial inequality as a product of the legacies of earlier forms of uneven capitalist development), transition involves the uneven development of new relations of capitalism and the rescaling of accumulation and regulation from global to local. The so-called 'globalisation enthusiasts' such as Ohmae (1990) have claimed that changes within capitalism are resulting in the full-scale globalisation of economic life, largely through the mechanisms of global flows of capital mediated through the multinational company and through a globalisation of financial markets. The consequence of these processes is a wholesale 'hollowing out' of the nation-state as political and economic power is shifted, either up to the scale of global activity or downwards to enhance the role of local and regional forms of governance. Others (Hirst and Thompson 1996; Boyer and Drache 1996) have suggested that such universal claims about the all-encompassing power of globalised capital and the decline of the nation-state are overemphasised, if not (in the case of Hirst and Thompson) entirely misplaced. Rather, what is at work is a process in which internationalisation is in full swing, but that this process is mediated through nation-states, and in some cases is constituted through the very role of national governments reframing economic spaces of a globalised nature through regulation of financial markets, trade relations or sites for capital accumulation with global linkages through the 'planned' materiality of the export-processing zone and the free-trade zone. For understanding transition, these debates are of central importance. As several chapters show, there are many ways in which processes of globalisation are transforming economic and political life in transition, but at the same time we would not want to throw out a set of claims that states and other forms of institutionalised practice (networks, for example) are important in setting limits to and even constituting the mechanisms through which globalisation of 'transitional' nation-states and economies works.

The scope and limits to globalisation are taken up directly in the chapters by Ould-Mey, Sidaway and Power, and Unwin (chapter 12). Ould-Mey demonstrates how, in the case of Mauritania, there has been a thorough 'denationalisation' of the state through the imposition of a structural adjustment package by the Bretton Woods institutions and other agents of neo-liberalism. Sidaway and Power suggest, in their chapter, a similar set of relationships in peripheral post-communist Mozambique, where sets of 'transition fantasies' (the neo-liberal policy package) have led to the emergence of 'phantom states' in which the location of power is both contradictory and complex. In this sense, then, Sidaway and Power take Ould-Mey's argument that there is a rescaling of geopolitical power and economic control further by unpacking the contradictions and social pathologies ('sex and violence') of such development trajectories. While Sidaway and Power and Ould-Mey stress the uneven and contradictory 'hollowing out' of the national state, Unwin's chapter suggests that what is more at stake in the Baltic context of Estonia is a project of Europeanisation forged by an elite of urban-industrial interests, albeit with agendas set from outside by the European Union.

Contesting the limits of democratisation and transformation from below

The reworking of the political subject and our understanding of democracy constitute a fifth set of issues. Staddon's chapter (chapter 15), for example, highlights the ways in which social struggles over water resources in Bulgaria signify the fluid and contested nature of transition. Social relations are transformed as a result of the struggles between a diverse, locally based environmental mobilisation and the state. Democratisation, for Staddon, therefore becomes a process involving the empowerment of certain actors with knowledge about certain contexts. This is a theme taken up by Regulska, Meurs, and Gorlach and Mooney (chapter 11). Regulska calls for a nuanced reading of social transformation by the recognition of the important role played in restructuring by localities and gendered perspectives. In particular, she argues, as the 'gendered political' of transition emerges as a domain of action increasingly dominated by masculinist perspectives and practices, and as women are marginalised from the arenas of formal politics, we need to redefine the political to include those forms of social action that have until now been excluded or marginalised by the processes of formal electoral politics. Meurs demonstrates how women and minority ethnic groups in Bulgaria have – in practice – mobilised the resources of the household and community, often building upon the very experiences and conditions of marginalisation to create alternative opportunities for surviving the current crisis. In a similar fashion, Gorlach and Mooney show – *contra* Ost (1991) and Offe's (1992) arguments that the transition caught Eastern and Central Europeans without a theory or ideology to sustain their actions – how peasant farmers in Poland did indeed have a clear sense of their economic and political interests, and how they mobilised those interests behind one or another political party as their interests changed. Unwin also takes the contestation of identities as a way of examining the construction of nation in Estonia, from the reinscribing of old identities connected with forms of rurality to western-imposed notions of democratisation and political practice.

As struggles over particular sites and resources emerge in the transition, it therefore becomes important to assess how we conceptualise the nature of these struggles and the roles played by specific actors. However, these newly emergent subject-positions are politically ambiguous. In *Theorising Transition*, we want to avoid overemphasising any notion of an essentialised subject who emerges under transition as the 'truly democratic subject', 'the resistant democrat' and 'gendered bearer of social values' or to essentialise the local (or for that matter to prioritise any scalar category) as a privileged site for understanding the causal mechanisms, types of response or the nature of outcomes in the transformation process. Instead, we want to see the production and reconstitution of subjects, locales, institutions, discourses and practices as part of the broader political economy of change and as constituted through the process of change. To that end, we hope that *Theorising Transition* has opened as many agendas as it has closed.

Conclusion

The authors in *Theorising Transition* challenge many of the comfortable assumptions unleashed by the euphoria of democratisation and the triumphalism of market capitalism in the first flush of 'post-glasnost' openness. The transformations under way in post-communist societies, far from leading in a straightforward manner to a single model of economic and political life eight years after the events of 1989, now seem to have generated complex and highly differentiated regional systems of adjustment. The examples of the Czech Republic, Bulgaria and Vietnam with which we began our introduction serve to highlight the diversity of national roads to transition. Complicating the picture even more as we write in July 1997 is the turnover of Hong Kong – one of the icons of capitalist transformation in the 'periphery' – to China: amidst all the pomp and circumstance of the 'end of empire', intense discussions abound over what a transition from hyper-capitalism to Communist Party rule will mean. How we account for the diversity of transitions and for the increased divisions within the societies of Eastern and Central Europe and in the 'periphery' remains an open and contested question.

Note

1 Vachudová and Snyder (1997) also argue that two other factors, the nature of the regime change (the different way in which elites were able to forge capitalist and market-oriented agendas) and the state of the economy in 1989, are central to understanding the different pathways that these six countries have taken.

Bibliography

Altvater, E. (1993) *The Future of the Market*, London: Verso.

Amin, A. and Tomaney, J. (1995) *Behind the Myth of the European Union*, London: Routledge.

Berman, M. (1982) *All That Is Solid Melts into Air*, New York: Penguin.

Boyer, R. and Drache, D. (eds) (1996) *States Against Markets*, London: Routledge.

Burawoy, M. (1992) 'The end of Sovietology and the renaissance of modernization theory', *Contemporary Sociology* 21, 6: 744–85.

—— (1996) 'The state and economic involution: Russia through a China lens', *World Development* 24, 6: 1105–17.

Commission of the European Communities (1996) *The Impact of the Development of the Countries of Central and Eastern Europe on the Community Territory*, Luxembourg: CEC.

Derrida, J. (1994) *Specters of Marx: The State of the Debt, the Work of Mourning and the New International*, London: Routledge.

Dunford, M. and Smith, A. (1998) 'Uneven development in Europe', in D. Pinder (ed.) *The New Europe: Society and Environments*, Chichester: Wiley.

Gerschenkron, A. (1962) *Economic Backwardness in Historical Perspective*, Cambridge, MA: Harvard University Press.

Gowan, P. (1995) 'Neo-liberal theory and practice for Eastern Europe', *New Left Review* 213: 3–60.

Guattari, F. and Negri, A. (1992) *Communists Like Us*, New York: Semiotexte.

Hirst, P. and Thompson, G. (1996) *Globalization in Question*, Cambridge: Polity Press.

Hodgson, G. M. (1996) 'Varieties of capitalism and varieties of economic theory', *Review of International Political Economy* 3, 3: 380–433.

Kagarlitsky, B. (1995a) *The Mirage of Modernization*, London: Verso.

—— (1995b) *Restoration in Russia*, London: Verso.

Kochanowicz, J. (1989) 'The Polish economy and evolution of dependency', in D. Chirot (ed.) *The Origins of Backwardness in Eastern Europe: Economics and Politics from the Middle Ages until the Early Twentieth Century*, Berkeley: University of California Press.

Lenin, V. I. (1967) [1899] *The Development of Capitalism in Russia*, Moscow: Progress Publishers.

Milanovic, B. (1994) 'A cost of transition: 50 million new poor and growing inequality', *Transition* 5: 1–4.

Murray, R. (1992) 'Flexible specialisation and development strategy: the relevance for Eastern Europe', in H. Ernste and V. Meier (eds) *Regional Development and Contemporary Industrial Response: Extending Flexible Specialisation*, London: Belhaven.

Offe, C. (1992) 'Democratisation, privatisation, constitutionalisation', paper presented at Post-Socialism: Problems and Prospects Conference, Charlotte Mason College, Ambleside, Cumbria, July.

Ohmae, K. (1990) *The Borderless World*, London: Collins.

Organisation for Economic Co-operation and Development (OECD) (1996) *The Czech Republic*, Paris: OECD.

Ost, D. (1991) *Solidarity and the Politics of Antipolitics: Opposition and Reform in Poland since 1968*, Philadelphia: Temple.

Piirainen, T. (1994) 'Survival strategies in a transition economy: everyday life, subsistence and new inequalities in Russia', in T. Piirainen (ed.) *Change and Continuity in Eastern Europe*, Aldershot: Dartmouth.

Sayer, A. (1995) *Radical Political Economy: A Critique*, Oxford: Blackwell.

Smith, A. (1997) 'Breaking the old and constructing the new: geographies of uneven development in Central and Eastern Europe', in R. Lee and J. Wills (eds) *Geographies of Economies*, London: Arnold.

Stark, D. (1996) 'Recombinant property in East European capitalism', *American Journal of Sociology* 101, 4: 993–1027.

Sussex European Institute (1995) *Project on Post-War Reconstruction in the Balkans*, Brighton: Sussex European Institute.

Vachudová, M.A. and Snyder, T. (1997) 'Are transitions transitory? Two types of political change in Eastern Europe since 1989', *East European Politics and Societies* 11, 1: 1–35.

Part I

THEORISING TRANSITION

2

REGULATING AND INSTITUTIONALISING CAPITALISMS

The micro-foundations of transformation in Eastern and Central Europe

Adrian Smith and Adam Swain

Introduction

In the West, critiques of state intervention, market volatility and global economic crisis since the early 1970s contributed to the undermining of traditional state-led approaches to managing economic development (such as Fordist-Keynesianism). The result was the growth in the 1980s of a neo-liberal agenda stressing the importance of the self-regulating market above the state in the co-ordination of economic activity (Dunford 1990; Tickell and Peck 1995). The ascendancy of neo-liberalism represented the short-term victory of one side of a view of political economy, seeing a separation between state and markets as dominant co-ordinating mechanisms. However, despite the hegemony of neo-liberal economic approaches, the world economy has continued to lurch from one crisis to another (Peck and Tickell 1994). At the same time, the collapse of the Soviet system and the perceived crisis of centralised planning, together with the failure of neo-liberal marketisation to construct capitalism, further undermined the dominant view in political economy of the separation of states and markets. These changes have also challenged the argument that *either* the state *or* the market can effectively create dynamic institutions of economic governance.

The crisis of this dualistic reading of political economy has led to the growth of alternative claims surrounding the ways in which state, market and institutions intersect. In these challenges to the orthodoxy the emphasis has been placed upon the social construction and governance of socio-economic systems. As Piore (1992: 174; see also Stark 1992b) says:

[I]f one looks across the capitalist world, without the distraction of communism and its collapse, what is most striking is the variety of prevailing institutional arrangements. Economic structures seem to grow out of culture and history rather than nature; there is no obvious 'natural' core that all of these economies share that law and custom act to distort.

Broadly, this new political economy stresses three concepts. First, economic systems are discursively and socially constituted (Gibson-Graham 1996; Offe 1995), and governed by socio-political institutions (Painter and Goodwin 1995). Second, rather than co-ordination occurring through formal market relations and state intervention, economic systems have always depended on a multiplicity of (self-)organising principles and mechanisms, such as industrial networks (Grabher 1993; Sayer 1995; Jessop 1997). Third, these diverse organising principles of socio-economic systems mean that there can never exist one-way pathways to economic change, but that social systems evolve in a path-dependent manner in which past practices shape the options for future strategies and development (Hodgson 1988, 1993; Hausner *et al.* 1995; Jessop 1997).

This chapter draws from and seeks to contribute to several emerging debates in this new political economy through an examination of the micro-foundations of socio-economic regulation in Eastern and Central Europe (ECE). In doing so, it seeks to articulate claims derived from regulation theory with contemporary debates on the social embeddedness of economic action. The chapter argues that the concern of both these approaches to identify key social relations and institutions 'regulating' or 'governing' the economy can be brought closer together (Jessop 1995a; Mouleart 1997). While regulation theory has traditionally focused on macro-level processes of economic stabilisation and crisis resolution (however partial), we argue that drawing upon and integrating parts of the socio-economics literature enables a more precise specification of the nature and spatiality of micro-regulatory practices.

One of the paradoxes of post-Soviet development has been reliance on a neo-liberal agenda based on the traditional separation of state and market, with the emphasis on unleashing the power of the market. The roles played by Francis Fukuyama (1992), formerly of the US State Department, and Jeffrey Sachs (1990), as policy adviser to a number of ECE governments, translated this agenda into the all too familiar programme of so-called 'shock therapy'.[1] Shock therapy has been based on the view that capitalism could be designed for ECE and imposed by fiat (Stark 1995; Hausner 1995; Pejovich 1994) and that the unleashing of the power of capital will inevitably allow the institutions, regulations, habits and practices associated with the 'normal' functioning of a capitalist market economy to emerge.

However, the increasingly evident differences between the types of marketisation developing in ECE and the persistence of pre-existing rationalities have stimulated alternative conceptions of socio-economic transformations (Hausner *et al.* 1995; Kovács 1994; Grabher and Stark 1997; Offe 1995; Stark 1996a, 1996b; Sabel and Prokop 1996; Amin and Hausner 1997). *Contra* this view of smooth capitalist transition, a transformatory view has been suggested which examines changes in the social relations within which economic activity is embedded. This alternative view has six main claims. First, socio-economic change is driven by people (agents) in negotiation (interaction) with formal and informal institutions (structures) and change is the output of neither voluntarist design nor structural determinism. Second, change is mutually constitutive of and through both discursive representations and material practices which form the boundaries and nature of regional economies (Jessop 1997; Kosonen 1997). Third, change is path-dependent in the sense that a region's path of extrication from the Soviet system will shape its subsequent path of development. Fourth, path dependency constrains, but does not exclude, strategic choice so that change is not only path dependent but also path shaping. Fifth, there is a multiplicity of possible development paths in ECE and, moreover, such diversity and complexity are desirable for evolutionary change and in sustaining local capitalisms (see Grabher and Stark in this volume). Sixth, the development of local capitalisms requires local institutions to regularise conflicting accumulation strategies.

In this chapter we build upon these claims by seeking to demonstrate how regulationist and governance (specifically socio-economic) approaches can be combined to create a better understanding of both the uneven development of the Soviet system and the geography of post-Soviet transformations.[2] In the next section we elaborate on the contemporary debates in the new political economy introduced above and specify the articulation between regulation and governance approaches. Section three uses this framework to examine the regulatory mosaics, institutions and governance regimes of the Soviet system. The fourth section considers the diversity of post-Soviet systems of regulation and socio-economic governance.

Regulation and governance approaches

In recent years there have been a number of major shifts in political economy. First, in reaction to overly structuralist Marxist accounts of capitalism, more nuanced treatments of capitalist development emerged. Foremost among these were those associated with the regulation school (see, for example, Aglietta 1979; Lipietz 1987), in which a shift was made from structural determinism and the reproduction of capitalism to more nuanced claims concerning the fragility and cyclical nature of capital accumulation. Second, there have been attempts more recently to 'socialise' accounts of economic action further with the emphasising of the social embeddedness and governance of economic systems (see, for example, Granovetter 1985, 1992; Hodgson 1988, 1993; Smelser and Swedburg 1994). Third, the social and cultural mediation of economic systems has led to the identification of a diversity of production systems, which has resulted in the study of varieties of capitalism (Leborgne and Lipietz 1992; Esping-Anderson 1990; Hausner *et al.* 1993; Amin 1995).[3]

In this section we seek to elaborate a position utilising a combination of the insights provided by both regulation and governance approaches. Emerging as a challenge to both the structuralism of French Marxism and equilibrium theories of neo-classical economics, regulation theory's major contribution has been to identify institutional forces and social relations which guide, sustain and transform the process of accumulation in capitalist societies (Dunford 1990).[4] A central theme is the concern that capitalist economic change is not smooth, but cyclical and ridden with crises, contested and struggled over. Theoretical understandings of capitalism, it is argued, need to be able to cope conceptually with the challenge of dynamism and change in the productive forces of a political economy. Regulation theory has posited that capitalist growth, crisis and change are contingent upon the intersection of institutional formations in particular societal contexts (Lipietz 1987) – a strategic coupling of systems of accumulation and associated regulatory practices. Capital accumulation is not only dynamic and changing over time, but also varies between different national space-economies. The primary reason for these variations lies in the key concept of regulation – capitalist social relations do not automatically reproduce themselves on the basis of laws of change, but are actively constituted in particular times and places by social action and institutional structures which result from the playing out of strategic struggles.

The key theme in regulation theory has been the intersection of accumulation regimes with particular modes of regulation. Regimes of accumulation are understood as 'the stabilisation of relations between production and consumption through an efficient allocation of social product between reinvestment, profit, and consumption' (Goodwin, Duncan and Halford 1993: 70). In order to ensure medium- to long-term economic stabilisation and crisis avoidance, regimes of accumulation become coupled with systems of institutional regulation, or modes of regulation. These are institutional forms which are 'codifications of social relations' (Dunford 1990: 307), such as the way in which the wage relation is regulated (through collective-bargaining, for example), the particular form of the state and the way it provides collective consumption, and, we would argue, the systems of integration and embeddedness which structure the relations between labour and capital, and fractions of capital (see Smith 1995). A mode of regulation thus 'involves all the mechanisms which adjust the contradictory and conflictual behaviour of individuals to the collective principles of the regime of accumulation' (Lipietz 1992: 2). The power of regulationist analysis thus resides in its ability to understand the regulation of capitalist systems and crisis resolution as a complex intersection of forms of accumulation and institutional structures of regulation. It therefore provides a means of transecting the knife-edge between economic functionalism and empiricism.

However, two main criticisms can be placed at the door of regulation theory. First, regulation theory lacks a concern for the spatiality of political-economic life in two ways – in terms of the relationship between national regulatory systems and the global economy, and in terms of the sub-national uneven development of systems of accumulation and regulation (Tickell and Peck 1992). Second, considerations of what constitutes a mode of regulation have tended to be statist in their approach, thereby neglecting many of the important non-state dynamics which regulate the accumulation process (Mouleart 1997). For example, the clear link that is made between a Fordist system of mass production and mass consumption, on the one hand, and a Keynesian welfare state underwriting and supporting the reproduction of labour and regulating the wage relation through collective bargaining legislation results in a primacy being accorded to the state in regulation.

Paralleling the development of regulationist accounts of economic change, a diverse body of work has emerged which focuses on the 'governance' of socio-economic systems (see Jessop 1995a, 1997). Governance approaches share three common themes: first, 'a rejection of the conceptual trinity of market–state–civil society which has tended to dominate mainstream analyses of modern societies' (Jessop 1995a: 310); second, a concentration on processes of co-ordination implicit in the modes of interaction and forms of integration between different institutions; and third, they examine the regularisation and intersection of micro-processes of social practice. Two bodies of work which pursue a governance approach have been particularly influential. First, work on the organisation of the state has focused on the shift away from government towards governance, associated with the rise of non- and para-statal institutions in governing territorial development (Painter and Goodwin 1995). Second, work on the organisation of economic networks has examined the combination of different modes of co-ordination (from markets to hierarchies) between enterprises (Sayer and Walker 1992; Grabher 1993).

The growth of work on governance and the development of regulation theory have stimulated an emerging debate on the relationship between the two approaches and how they may be combined to specify the processes of regularisation (Hay and Jessop 1995; Painter and Goodwin 1995; Hay 1995; Tickell and Peck 1995). In the context of analysing the comparative dimensions of change in ECE, we argue that the relationship between governance and regulation is best understood along a continuum between abstract and concrete. At the most abstract level, all social systems require *governance*, in the sense that social systems are in themselves a process that is both structured and structuring (see Jessop 1997). This is achieved through governance mechanisms which act to reduce the complexity of the world, facilitate social learning, build methods for co-ordination and establish a system of stabilisations based on shared meanings (Jessop 1997). Thus, the *objects* of governance – interpersonal, inter-organisational and inter-systemic (between economic and political subsystems) relations – imply the co-existence of a requisite variety of *governance mechanisms*. The diversity of mechanisms for reducing real-world complexity, however, also suggests the impossibility of complete governance. Attempts at stabilising societal development are, in other words, always partial and open to contestation and transformation.

The great variety of economic systems gives rise to different organisa-tional principles and different sets of societal contradictions. Consequently, social systems develop different ways of *co-ordinating* or *managing* these contradictions. For example, capitalist systems are constituted by a set of contradictions around the wage relation requiring management through the regulation of markets (Aglietta 1979). In contrast, Soviet systems revolved around the contradiction between the production, appropriation and redistribution of the surplus product and its control (or lack of) by the state (Clarke *et al.* 1993; Konrád and Szelenyi 1979). This contradiction was co-ordinated (or, more accurately, denied) through its internalisation by the state, requiring the institutional fusion of the socio-political and economic spheres in which party/state apparatuses attempted to dominate all aspects of social, political and economic life. But, as recent events in ECE indicate, governance is always incomplete and sites of regulation mere 'moments' of co-ordination as all mechanisms of integration are prone to failure (Jessop 1997).

However, at the concrete level, we would argue that social systems comprise a myriad of social practices, the regulation of which – through interaction with other practices and institutions – may temporarily resolve or manage these contradictions in the form of *regulatory processes*; what Painter and Goodwin (1995: 350) call 'actually existing' regulation under-stood as 'processes constituted through social practices in particular historical and geographical contexts'.

One attempt to link these different levels of abstraction has been work which has focused on the time–space embeddedness of economic action and networks of socio-economic institutions (Amin and Thrift 1995; Granovetter 1985, 1992). These networks can be understood as constitutive of social practices through which 'more or less' effective regulatory processes (the conditions which allow the expanded reproduction of capitalist rela-tions) emerge. Socio-economics comprises three main claims (Granovetter 1985, 1992: 4): first, that pursuit of economic goals is normally accompanied by such non-economic ones as sociability, approval, status and power; second, that economic action (like all action) is socially situated, and cannot be explained by individual motives alone – it is embedded in ongoing networks of personal relations rather than carried out by atomised actors; and third, that economic institutions (like all institutions) do not arise automatically, but are socially constructed.

In this chapter we develop two claims. First, that networks play a role as social practices from which regulatory processes emerge by embodying and organising two types of properties: material and discursive. Material networks are networks of institutional relations which combine market, hierarchical and associative mechanisms. Discursive networks involve regulatory practices emerging from collective forms of understanding (Hodgson 1993). Together, these properties combine to create the normative 'rules of the game' which regularise economic action. Second, we claim that the organisation of space is not only an object of regulation – in the sense that uneven spatial development needs to be managed – but that it is also a subject of regulation; the organisation of space is thus an inherent part of the process of regulation. Thus, as Amin and Thrift argue (1994: 16–17), socio-economic practices are embedded in time and space:

> [T]he economic life of firms and markets is territorially embedded in social and cultural relations and dependent upon: processes of cognition (different forms of rationality); culture (different forms of shared understanding or collective consciousness); social structure (networks of interpersonal relationship); and politics (the way in which economic institutions are shaped by the state, class forces, etc.).

Moreover, social practices and regulatory processes are unevenly developed over space and time (Painter and Goodwin 1995). This helps to explain not only why economic development is geographically uneven, but also why regulatory processes are themselves uneven. Thus, we argue that regulatory processes are territorially institutionalised as 'bundles of social practices' or as 'network relations in place'. Indeed as Grabher and Stark (1997 and in this volume) argue, localities can be regarded as the prime means by which the complex diversity of network relations are organised or governed (cf. Jessop 1997). Yet, the way in which locally constituted social practices interact with each other to produce coherent modes of regulatory processes that hold regional economies together is never determined from the outset, but arises from the long-term adjustment of practices to discursively and materially constructed social conflicts and struggles at local, national and international levels. In other words, we may expect that in the relatively short time of ECE transformation a coherent set of regulatory practices superseding, yet building upon, those of the past is unlikely, if not impossible. Consequently, accounting for the nature of uneven development requires us to dissect the emergence of specifically localised social practices and emergent regulatory processes within the context of an external imposition of neo-liberal practices by institutions governing the world economy.

In the following section we begin to lay out those social practices and institutionalised mechanisms that attempted to 'hold together' the Soviet model of development in place. We seek to demonstrate that the absence of interaction and the tangled intersection of formal and informal social practices failed to stimulate the emergence of regulatory processes which could secure the expanded reproduction of the Soviet model. Following this, we move on to explore the complex ways in which these social practices have been reworked and restructured after 1989, and how these emergent practices are territorially institutionalised through new forms of local development. We conclude that the dominant practices in post-socialism are unlikely, under present conditions, to generate effective regulatory processes.

Regulation, institutions and the governance of the Soviet system

Regulation theory has hitherto had very little to say about the transformation of the political economies of ECE. Aside from the work of Kaldor (1990) on the political economy of the Cold War, Altvater (1993) on the collapse of 'state socialism' (see also Boyer 1995), and Smith (1994, 1995) on the restructuring of the armaments industry and regional uneven development in Slovakia, regulation theory remains underdeveloped in relation to understanding the transformation in ECE (see also Dunford in this volume). However, we would argue that regulation theory can provide a basis for the construction of a radical political economy of ECE transformation. We would argue that the coupling of forms of accumulation with regulatory institutions and structures has not only been a feature of advanced capitalism, but characterises every temporally and spatially specific political economy, including the Soviet system. What is required is the specification of the particular social relations structuring a political economy and how these relations work out, in terms of the conjunction of accumulation and regulation.

The model of development underpinning the Soviet system was primarily extensive, characterised by a growth model based upon the continued expansion of the means and forces of production (Altvater 1993; Smith 1994, 1998). This system was regulated through state and, to a lesser extent, non-state institutional forms which varied from country to country producing 'varieties of Soviet systems' (cf. Hodgson 1996). For example, planning mechanisms and bureaucratic regulation dominated in some of the more centralised political economies (the former Czechoslovakia, for example), while in Hungary a complex intersection of central planning and 'market' co-ordination emerged as the 'second economy' became increasingly central to the regulation of economic and social life (Stark 1992a).

However, where regulationist analysis has been extended to ECE it has transferred categories developed to analyse 'advanced' capitalism, such as 'Fordism' (Murray 1992; Altvater 1993). For example, in considering the relative failure of state socialism to compete with western capitalism, Altvater (1993) argues that state socialism failed to develop a full-blown form of Fordism. The conjoining of mass consumption with a system of rationalised mass production did not emerge, Altvater argues, in ECE as it did in the West. 'The missing half of a Fordist mode of regulation – mass production of consumer goods and mass demand for them – therefore undermined the half that was actually achieved, the rationalization and planning of labour in industry' (Altvater 1993: 32). Whilst state socialism failed to develop the consumer goods sector and expanded working-class consumption, as found under western Fordism, it is important to question the claim that the 'other half' of Fordism was to be found unproblematically in the East. It appears that Altvater is confusing the nature of the transformation of labour processes in the East with the aims and results of Fordism, as both a labour process and a regime of accumulation, in the West; that is, as the intensification of production. Fordism was based upon a particular regime of accumulation and instead of merely equating it with mass production we need seriously to address the form that accumulation took in ECE. Similarly, Murray's (1992) analysis also falls into the same trap as Altvater's. Murray argues that the Soviet Union represented a form of Soviet Fordism 'not simply because of the Soviet approach to the work process, scale and the structure of management, but also because the idea of central planning represents scientific management applied to the whole economy' (Murray 1992: 214). He argues that following the implementation of Leninist rationalities

> key features of Taylorism are all to be found here – the division of conception and execution, the control of technical knowledge by management, hierarchical authority, the weakness of horizontal ties, and the focus on the economy of time of individual 'atoms' of production.
>
> (Murray 1992: 214)

While important facets of Taylorism, as identified by Murray, were implemented in the Soviet Union and generalised to the rest of ECE after the Second World War, the intensification of the labour process was not the prime component of economic transformation in the Soviet system. What Murray fails to consider, like Altvater, is that these 'relations' of production were concerned with the expanded industrialisation of the Soviet economy based on *extensive* forms of accumulation – the continued expansion of the means of production, through what became the 'war economy' (Kaldor 1990). This confusion arises from a failure to consider the distinctive social relations of the Soviet system, the way in which these structured and were transformative of forms of accumulation, and the different form these have taken *vis-à-vis* western post-war development trajectories. In other words, we need to identify the levels at which we are using the concept of Fordism (Jessop 1992), and not conflate changes in the labour process with a particular system of accumulation (however important those labour process changes may be).

That said, however, the collapse of the Soviet system was closely connected to the exhaustion of this extensive model of development (Altvater 1993). The inability to intensify production, even with the increase of foreign loans to purchase more advanced western technologies in the 1980s, coupled with the ideological exhaustion of the state road to socialism, resulted in a fundamental erosion of the hegemonic power of the Communist Parties. The regulatory system failed to provide a transformatory framework – '[t]he actually existing socialist systems collapsed not because of shortcomings in material welfare, but because the institutions of society were insufficiently flexible in adjusting to crisis tendencies that had been concealed for too long' (Altvater 1993: 23).

Two insights provided by a socio-economic view help to explain the absence of a transformatory framework which lay behind the collapse of the Soviet system. First, in contrast to the conventional state-centred view of the Soviet system pursued by, amongst others, Altvater, the socio-economic approach emphasises the role of meso-, micro-, formal and informal institutions which combined to reproduce the planning system (Szelenyi, Beckette and King 1994). These institutions generated 'tangled hierarchies' (Jessop 1997), or more accurately 'tangled hierarchies within tangled networks' and 'tangled networks within tangled hierarchies', which undermined the ability of the state to regulate the economic system and thus secure the reproduction of the regime. Second, the decentralised nature of institutions in the Soviet system, and the role of tangled networks and hierarchies, suggest that both played a significant role in structuring space in a way which locked regions into a particular development trajectory and hindered the mobilisation of transformatory potential. We now look at these two insights in turn.

The so-called 'command economy' combined the 'scientific organisation of work' (Kössler and Muchie 1990) with centrally administered, vertically integrated industrial networks comprising a relatively small number of large enterprises which sought to mass-produce standardised products. The construction of tight, highly regulated networks between and within enterprises compounded the creation of an industrial system geared towards the *need for planning* (efficient allocation, through redistribution, of the surplus product and of production decisions), rather than a system in which the *production of social 'needs'* (use values) justified centrally planned decision-making (Heller, Feher and Markus 1983). Such 'allocative efficiency' (Grabher and Stark 1997) represented adaptation to the requirements of central planning at the expense of flexibility. Moreover, the industrial networks that were sanctioned by central planners contributed to undermine the *capacity* for evolution by eliminating other organisational forms, such as small and medium-sized enterprises, thereby reducing the 'genetic pool' inherent within systems of organisational diversity (see Grabher and Stark 1997 and in this volume). In this way, the system was deprived of the conditions necessary for the development of self-reflective networks capable of 'learning by monitoring' (Sabel 1994), and was thus rendered incapable of dynamic efficiency – the capability and capacity of adapting to changing environments (Grabher 1994b; Grabher and Stark 1997).

Despite the apparent domination of bureaucratic planners and hierarchical relations, reciprocal market-like social relations did operate inside and beyond the state sector (Stark and Nee 1989). The operation of the planning system generated chronic shortages owing to the existence of 'soft-budget constraints' which encouraged over-investment (the hoarding of the factors of production) and under-production (the diminution of the forces of production) (Kornai 1980). Consequently, redistribution by planners could never meet the allocative demands of state enterprises, resulting in the generation of 'plan bargaining' between enterprise directors and planners in which the former had an interest in underestimating capacity and over-estimating the need for investment (Pickles 1995). As a result, the planning system generated outcomes which could not be planned for; a process termed 'plan fetishism' by Burawoy (1985).

The smooth operation of the planning system (such as it was) therefore depended on a 'shadow plan' comprising three types of informal networks co-ordinated by reciprocal social relations which tangled with the formal planning system. First, the failure to separate the conception of work from its execution (which was related to the failure to shift from an extensive to an intensive mode of growth) generated a labour process which was at one and the same time 'despotic' (Burawoy 1985) and disorganised. Management was unable to secure sufficient control over the labour process, a factor which granted core workers considerable autonomy and power which led to informal bargaining (Stark 1992a). Second, poor co-ordination between and within enterprises generated informal plan bargaining amongst managers of production units and between them and enterprise directors. To ensure that plan targets were achieved managers entered into mutually convenient reciprocal or barter relations (Grabher 1994a). Third, to varying degrees states were unable to prevent the development of 'second economies' which in time assumed crucial roles in the reproduction of the system of central planning (Stark 1992a).

The increasing entangling of the informal bargaining networks and the planning system served only to undermine further the integrity and strategic capacity of the formal system, with the result that central planning generated ever more unpredictable outcomes. This in turn generated yet more informal bargaining, as enterprise officials sought to insulate themselves against the unpredictability of the planning system by *capturing* elements of it, in order to secure supplies, through shadow personal networks. 'Plan capture' merely organised a component of the system at the expense of undermining the coherence of the whole. In this way, the planning system became increasingly fragmented, contributing to its disintegration from the bottom up.

Moreover, as this behaviour spread, so the reproduction of the system became more dependent on action which was anti-systemic (and therefore anti-regulatory) in character. However, the political imperative of maintaining the Communist Party's formal political monopoly precluded the institutionalisation and democratisation of informal plan bargaining, which in turn blocked their development as vehicles for strategic change. As a result, interests were not represented or mediated, but were either ideologically denied or were internalised within the boundaries of the state apparatus. Informal networks were invariably ephemeral, arising in response to day-to-day problems, and were uncodified. Consequently, they did not constitute strategic forms of collective action, resulting in an absence of 'agency' which could have permitted the further development of the system and which could have resulted in the conditions necessary for evolution (Kosonen 1997). The combined effect of these trends was, first, to undermine the systemic coherence of the formal planning system generating increasingly uncoordinated social practices, and second, to stifle the development of informal networks which gradually became the domain of localised, illicit interests dedicated to plan *circumvention*. Consequently, a vicious circle developed in which the interaction of the formal and the shadow plan served only to undermine the integrity and strategic capability of both types of networks to affect strategic change.

This process of disintegration from the bottom up undermined the legitimacy of the system (see Szelenyi, Beckette and King 1994: 239). The planning system's legitimacy was derived from the combination of an 'instrumentalist rationality' which assumed that goals could be identified and met, and 'teleological rationality' in which planners assumed they alone could identify and guide the system to achieve its goals. The poor co-ordination of the system, particularly reflected in endemic shortage, and the resultant dependence on informal networks meant that an individual's personal experiences became increasingly dominated by the 'shadow plan'. This acted to block the internalisation of the social practices associated with the planning system. Furthermore, the complexity engendered by the intermingling of the formal and shadow plan resulted in 'opaque' lines of responsibility and accountability (Szelenyi *et al.* 1994), creating what might been termed a 'subject-less environment'. In such circumstances, it proved impossible to generate trust in either the formal or informal plan, the absence of which resulted in the dislocation of networks and institutions since they were increasingly isolated from one another and from the social formation of which they purportedly were a part. As the ineffectiveness of the planning system became apparent, so its already weakened legitimacy was eroded further and social actors confined themselves to merely 'painting socialism' (Burawoy 1992).

Crucially, the emergence of anti-regulatory processes from tangled social practices was bound up in the distinctive (dis)organisation and contradictory structuring of space under the Soviet system, which was implicated in the eventual disintegration of the system. Grabher and Stark (1997: 17–18; see also Grabher and Stark in this volume) identify two important features of the geography of the Soviet system. First, the requirements and ineffectiveness of the central planning system demanded the ever-increasing rationalisation and concentration of industrial production in a limited number of very large enterprises which not only reproduced the labour force through employment, but also through the provision of socialised welfare. This resulted in creating 'company towns' with a geography of localised specialisation. Second, the need to exploit new sources of labour as the system hit against the limits of extensive growth led enterprises to locate 'branch plants' in rural areas. This process generated a pattern of regional development marked by convergence (see also Smith 1996, 1998).

The central planning system generated a 'hub and spokes' network between the centre and outlying areas in which all significant decisions were taken outside the region irrespective of the resources, skills and identities of the regions concerned (van Zon 1992). As a result, formal regional institutions, enterprises and local authorities were merely 'transmission belts' conveying decisions taken beyond the boundaries of the region. The dependence of the regions on the centre was accentuated by the isolation of institutions in the same region, since all formal relations between them were articulated through the centre. The absence of local networks of interaction not only undermined the legitimacy of the system but also prevented the development of locally hegemonic socio-economic projects (Kosonen 1997).

Another consequence of the centralised structuring of space was the blurring of the boundary between territorial agencies, such as local and regional authorities, and state-owned enterprises (cf. Illner 1992). The ambiguity this created formed the basis for the development of local informal networks amongst proximate enterprises and between them and territorial authorities. As the informal networks became increasingly important in the process of securing the reproduction of the system at the local scale (at the expense of undermining the integrity of the system nationally), so they were 'sedimented' in particular places. These networks identified, represented and articulated local interests against the centre and as such were embryonic vehicles for local governance. However, the informal character of the networks, whose development, like those identified above, was blocked, meant they were guided by the goals of plan capture and subsequently plan circumvention and not by a strategic collective vision of local economic development. Consequently, notwithstanding the role of the informal networks, the combined effect of dependence on the centre (and often on one major state-owned combine), and the formal isolation of institutions located in the same region, contributed to a process of 'de-regionalisation' (van Zon 1992) and the consequent contradictory organisation of space.

The entangling of the formal governance mechanism, the central plan, and the informal forms of governance produced a regulatory crisis, or an 'under-governed' system, out of two contradictions. First, the diversity of forms of governance that actually developed to 'manage' the Soviet system created weaknesses in a system that was dedicated to the rationalised planning of the whole society. In particular, the absence of meta-governance, aimed at stabilising the interaction between the formal and informal co-ordination mechanisms, resulted in the complex entanglement of these mechanisms, which in turn served to undermine the efficacy of governance. Second, the organisation and structuring of space implicated in the centralised planning system prevented the formation of localised, hegemonic common projects which could have mobilised the resources necessary for strategic change. Together, then, the sorts of social practices which linked production and mass consumption that regulated western Fordism failed to develop in the Soviet system because the intersection of social practices was such that the regulatory dynamics of the Soviet system actually operated in an anti-systemic fashion by eroding the basis for continued evolution. As a result, there was also a failure to shift from extensive to intensive growth. In summary, the Soviet system collapsed for essentially two reasons: first, the ineffectiveness of institutionalised networks to act as a vehicle for strategic collective action; and second, because the networks and institutions lacked the capacity to evolve solutions to unexpected developments. In systems terms, this created complex, yet 'unintelligent', systems lacking the conditions (such as flexible feedback loops) essential for the long-term adaptability of social forms (see Grabher 1994b). In the following section we go on to discuss the path-dependent implications of this crisis for the diversity of forms of capitalism emergent in post-socialism.

Regulating and governing the transformation?

From a regulationist perspective, the demise of the Soviet system and the ensuing problems associated with it can be interpreted as a crisis of regulation and governance mechanisms to facilitate co-ordinated strategic action. As a result, the extent to which a new institutional coupling is occurring, and indeed achievable, in the relatively short time since the 'revolutions' in ECE must be questioned for a number of reasons. First, the legacy of the Soviet system bequeathed a paucity of social institutions from which collective strategic action could emerge. Second, high levels of emerging regional fragmentation and socio-economic dislocation, alongside the break-up and recasting of states (Czechoslovakia, Yugoslavia and the Soviet Union, for example) have led to great difficulty in establishing and identifying clear, discrete forms of accumulation and regulation. Third, as regulationist work on 'advanced' capitalism has argued, stabilised, long-term systems of accumulation governed by regulatory relations emerge over long time-frames (Aglietta 1979).

The current transformations in ECE can be seen as an attempt to institute and develop new systems of coupling between accumulation regimes and new regulatory processes. Yet, what is more in evidence is a reworking of past practices in the context of new forces engendering change. Hausner, Jessop and Nielsen (1993), for example, have argued that the intensification of accumulation is one of the primary goals of the transformation process. They argue that the transformation is currently concerned with the emergence of a set of 'search processes' to identify and institute new institutional structures of regulation and accumulation which are being crucially determined by the interpretation and implementation of 'western' development discourses (Figure 2.1). Search processes involve, first, a search for forms of market-isation in response to the breakdown of central planning – privatisation, deregulation, liberalisation and commercialisation; second, a search for forms of democratisation as one-party rule ends and new frameworks of political regulation need to be instituted – representation and mediation; and third, a search for economic diversification within the context of globalisation. The question is how these search processes are instituted. Hausner, Jessop and Nielsen (1993) argue that two processes are fundamental. First, decisions regarding forms of economic and political regulation are increasingly governed by western 'models' of development – the negotiated economy, neo-liberalism and the social market economy. Second, decisions and strategies are increasingly affected by the globalisation of economic and political process, resulting in a recasting of the nature of the state in ECE. Consequently, it is neo-liberal development discourse that is 'winning out' in ECE as a model for forging pathways from the Soviet system. This has been seen in the arguments over the need for privatisation, the deregulation of the state, and the decentralisation of economic decision-making. However, if we take seriously the claim that neo-liberalism is itself not only a response to but also a symptom of the crisis of global capitalism (Peck and Tickell 1994), the scope for its providing a solution to the institutionalisation of post-socialism needs to be questioned.

One of the key responses to the neo-liberal agenda in ECE is the emergence of a great diversity of unarticulated social practices which tend towards accentuating the crisis of governance in post-socialism, rather than providing solutions. Three types of processes affecting the network and institutional legacies of the Soviet system can be identified, suggesting that regulatory processes are unevenly developed as a result of the crystallisation of social struggles in particular contexts: first, the dissolution of networks and

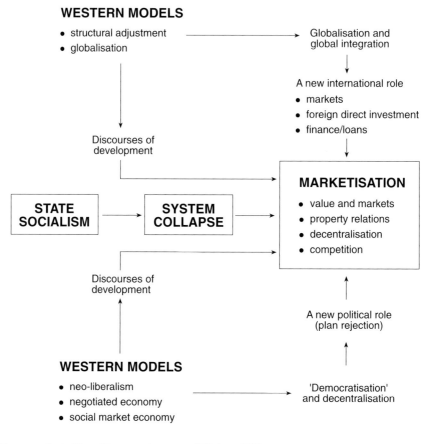

Source: adapted from Hausner, Jessop and Nielsen 1993

Figure 2.1 Models of transition

the isolation of institutions; second, the reconfiguration of networks in which institutions interact and learn new forms of action; and third, the endurance of pre-existing networks and the insulation of institutions (Figure 2.2).[5]

The first process of transformation we identify is the *dissolution of pre-existing networks* which resulted in the continued existence, but isolation, of the institutions which comprised the Soviet system. The late 1980s and early 1990s witnessed the acceleration of the process, which had been occurring almost imperceptibly for more than two decades, of fragmentation and disintegration of the system from the bottom up. The demise of the plan as a co-ordinating mechanism resulted in individual enterprises being 'cast adrift' (Burda, Singh and Török 1994). At the same time there was the attempt to enforce neo-liberal strategies designed to establish and disseminate new forms of economic rationality. These strategies, under the guise of 'state desertion',

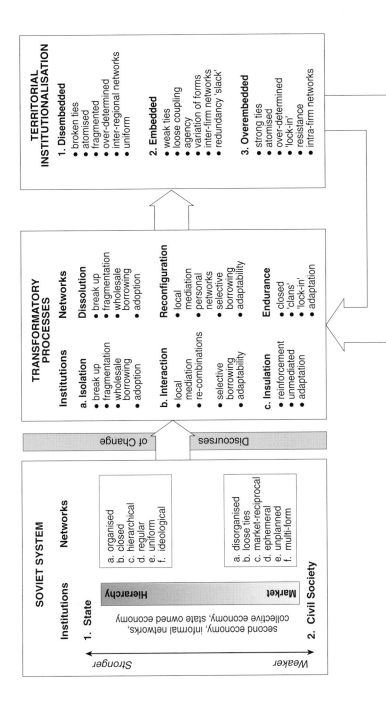

Figure 2.2 Institutions and networks in transformation

resulted in the uneven introduction of new forms of social relations generating the forced obsolescence and dissolution of the remnants of formal networks and institutions (such as ownership forms, business networks and government networks). The introduction of dollar accounting (thereby facilitating international trade), privatisation and inflows of foreign capital, for example, enabled some institutions to reorientate their strategies, but in doing so they disrupted any remaining integrity of the formal industrial networks they previously comprised. In such circumstances not only were the formal planning networks untied but also the informal networks were broken (Albach 1993; Balabanova 1994). Owing to a lack of interaction, through which institutions develop the shared expectations of behaviour, the *habitus* associated with the 'normal' functioning of a market economy was not internalised. In this way, both formal and informal 'network capital' leached away. Those institutions, having been cast adrift from the planning system and deserted by the state, and which were not able to integrate into new networks were left isolated, 'locked out' from attempts to enforce capital as a principal bearer of economic rationality. The isolation of these enterprises was accentuated by direct foreign investment, much of which was predicated on distancing itself from local crises of regulation, preferring instead integration into pan-European and global industrial networks (see Swain 1996, 1997).

The second process of transformation involved the *interactive reconfiguration of pre-existing networks* in which institutions remained part of functioning networks. Through interaction, dynamic institutional co-learning was possible which enabled the evolution of new networks which not only bore new economic rationality but also potentially the capacity of adaptability. Two types of such networks could be seen developing in the course of the 1990s. First, there were those enterprises (or parts of enterprises) which found a role in new networks through exporting to customers principally in Western Europe, together with a smaller number of firms which attracted direct foreign investment. In this way, enterprises were able to become acquainted with new socio-economic knowledge and practices associated with the 'normal' functioning of a market economy and selectively deploy them in the context of local conditions (Kogut 1996). Second, legal and semi-legal economic entities which had been previously tolerated were able to establish themselves on a more formal footing as new forms of company were legalised. This generated two types of networks of small firms: those which developed amongst collections of small firms and those which linked small firms and large former state-owned enterprises (of which the small firms had often formerly been a part). The latter type of network, loosely combining and bridging diverse organisational forms, created hybrid institutions which, faced with a regulatory crisis, employed 'hedging strategies' which created institutions ordered according to multiple forms of rationality, some market-led and others oriented towards the remnants of bureaucratic planning and bargaining. The result was the creation of what Stark (1996a, 1996b) has termed 'recombinant property' and 'recombinets' – networks of economic institutions which were neither wholly 'privately' owned nor wholly 'state' owned, but a complex combination of both.

Whilst at least the potential for learning was present in some networks, in others it was wholly absent. The third transformation process involves the *endurance of pre-existing networks* and the insulation of institutions as a means of resisting the enforcement of a new economic rationality. One response to the dissolution of networks and the enforcement of a new economic rationality has been the maintenance or reconstruction of networks dating from the period of central planning. The disintegration of the planning system from the bottom up, and the identification of the systemic problems with the centre led groups of enterprises, organised either geographically or sectorally, to combine in order to insulate themselves from the surrounding regulatory crisis (Burawoy and Krotov 1992, 1995). This resulted in the generation of networks which were tightly organised, hierarchical and closed. Accordingly, new institutions capable of articulating and representing interests for change were not constructed, nor were existing ones destroyed. In such circumstances, economic action remained ordered according to the rationality of the former system. Thus, instead of subordinating labour to the requirements of capital, partial insulation from external economic environments permitted the continued dominance of collectivist interests (Clarke *et al.* 1993). These 'clan-like' networks lacked the capacity to develop new ways of strategic action and were thus 'locked in' to existing paths of development.

Together, the three transformatory processes created a highly complex picture in which there was fragmentation not only in the nature of social practices, networks and institutions, but also of the ordering principles governing societies. Two trends were readily identifiable: first, the destruction of the very capacity to create new forms of collective action through the isolation and insulation of institutions,[6] and second, the formation of interactive complexity which might form the basis of resurgent socio-economic systems (Sabel and Prokop 1996). Together, the two trends resulted in the emergence of what Nuti (1996) calls 'post-communist mutations'.

Transformation and the reorganisation of institutionalised space

The transformatory processes were also deeply implicated in the reorganisation of space in response to the post-Soviet regulatory crisis. The process of 'de-regionalisation' under the Soviet system was gradually superseded by a countervailing tendency towards regionalisation (in part as a response to pressures towards globalisation) as a means of securing order at the local scale in the absence of order cascading down from the centre (Jessop 1995b; Kosonen 1997). To begin to examine territorial institutionalisation, the three transformatory processes can each be seen to have different implications for the organisation and structuring of space (Figure 2.2). There are two significant features of the post-Soviet organisation of space: first, the disintegration of the central planning system contributed to the fragmentation of post-Soviet space economies, and second (and mirroring the emergence of different types of networks), the variety of mechanisms for governing regionalised space economies has increased. Three dominant types of governance mechanism can be identified: *isolated, interactive* and *insulated.* These mechanisms are not necessarily mutually exclusive and are co-present in regions, but the relative significance of the three governance mechanisms has been crucial in shaping the development of regional economies.

The dominance of *isolated institutions* has created regional economies which are 'disembedded'. Disembedded regional economies are characterised by the dominance of two types of enterprises: 'tombs-in-the-desert', which no longer figure in networks and which are consequently left to demobilise gradually, and 'cathedrals-in-the desert', which form nodes in inter-regional industrial networks of business linkage. The result of the dominance of these institutions has been the failure to sediment any regional networks. The absence of such networks prevents inter-organisational interaction within the regional economy and has blocked the development and diffusion of the resources and skills needed for economic development. Characteristic of this form of regional economy is dependence on foreign know-how and foreign direct investment. Grabher, for example, concludes from his work on the former GDR, which received the most foreign know-how and capital in ECE, that (1994a: 182–3):

> the rupture of 'old ties' and the subsequent atomization of economic and social actors did not lead to the effects for which it had been justified, namely to unleash market forces. . . . The atomization of social forces and economic relations in eastern Germany has in no sense unleashed market forces. Instead, it has blocked the generation of indigenous economic activity.

Grabher argues that over-reliance on attracting direct foreign investment, which principally involved the creation of 'branch plants' integrated into West German industrial networks, resulted in the absence of local linkages. Moreover, certain investment strategies, particularly those undertaken by automotive firms, involved the development of new 'hyperefficient', so-called 'lean production', networks whose temporary competitiveness had the effect of 'crowding out' other types of industrial structure (Grabher 1994b, 1997; see also Swain 1996, 1997; Smith and Ferenčíková 1998). In this way, such regions not only lack inter-organisational interaction but also increasingly suffer from a reduced diversity of mobilised resources through which learning could take place. Moreover, the effect of this was to construct a 'truncated' form of 'capitalism without capitalists' in which capitalist rationality was not diffused locally, leading ultimately to the emergence of demobilised, disembedded regional economies.

Where *interactive learning networks* have been dominant, locally embedded regional economies have been formed. The presence of local networks has secured the mobilisation and organisation of local resources behind economic development projects. Interaction between institutions has resulted in the internalisation of new forms of economic rationality and behaviour which in turn further develops a locality's social infrastructure. In this way, 'powers of association' between economic actors have been established at the local level which serve to generate collective entrepreneurship (Amin and Thrift 1995). Embedded networks seem to be rare, but much emphasis is being placed on the transformation of the 'second economy' into a self-sustaining sector able to foster regional economic growth (see, for example, Kuczi and Makó 1997). To this end, state industrial and regional policies are particularly geared towards encouraging the development of small and medium-sized enterprises as a way of ensuring endogenous growth (Brezinski and Fritsch 1996; Smith 1997). In part, this is a reflection of the interest being paid to endogenous growth associated with the so-called 'Third Italy model', which sees flexible specialisation as a possible development model for ECE (Murray 1992; Bianchi 1992; see Smith 1997 for a critique).

However, more common has been the dominance of *insulated institutions* resisting marketisation, which has resulted in the development of over-embedded regional economies. In these cases, local networks are present but are over-determined successors to both formal and informal relations from the time of centralised planning. These networks are wedded to the formal or 'shadow' plan and the rationality of the old system. Despite this, the networks are not paralysed, but the restructuring that does occur is characteristic of defensive measures as institutions seek to insulate themselves from an increasingly threatening environment and are thus 'locked in' to the rationality of the old environment. Smith's (1994) study of a large, heavy engineering and armaments enterprise in Slovakia shows that the reformulation of the networks of integration which regulated what was one of the core enterprises of the Soviet system in Slovakia has resulted in the emergence of struggles over future pathways. The reformulation of the old networks has taken two forms. First, market integration was lost as CMEA markets collapsed and geopolitical alignments broke down in the 'Third World'. Second, the system of vertically and horizontally integrated industrial associations broke up as decentralisation of decision-making was introduced. The clear result has been that enterprises such as this were thrown into turmoil and faced high primary and secondary debt and bankruptcy. One strategy in particular which has emerged signifies 'over-embeddedness'. Variant survival strategies have been implemented by managers of the former shops (now divisions) in the enterprise to become independent and to forge a pathway alone. Yet the divisions remain highly integrated, reflecting the high degree of vertical integration in the original enterprise, and rely on reciprocity and barter exchange relations to secure internally produced products. Thus, regions in which these sorts of enterprise networks dominate are over-embedded in the sense that new forms of economic activity seem unable to develop despite marketisation. Indeed, in the case of the Slovak heavy engineering sector, recent evidence suggests the partial re-establishment of old systems of governance in the form of pseudo-industrial associations which are integrating enterprises previously made independent in the initial rush to decentralise after 1989 (Smith 1998). This is an argument also put forward by Burawoy and Krotov (1992, 1995) in their analyses of the regional dimensions of integration in the wood and coal mining industries in Russia. Recourse to para-statal organisations and old networks of barter to secure supplies has maintained the 'locked-in' social practices of the past as a survival mechanism for coping with an externally neo-liberal environment.

The disintegration of the Soviet system from below and the neo-liberal medicine that was poured on East and Central Europe from above generated variety and complexity in at least two senses. First, the repertoire of social practices increased as transformation processes came to bear on networks of socio-economic institutions. Second, regionalisation generated fragmented space-economies which were governed in a multiplicity of ways. Crucially, the repertoire of social practices and the fragmentation of regional economies to produce localised economic systems geared towards particular governance mechanisms meant that the quality of interaction and intersection, from which dynamic sustainable local capitalisms could emerge, was poor. This, in turn, indicated that the emergence of regulatory processes from the myriad of social practices depended on the organisation of space. In particular, there has been an absence of social practices bridging different spatial scales. There have emerged regions which are principally involved in inter-regional social practices (or institutionalised relations across regions), in which systems of cross-regional foreign investment represents the best example, and those principally engaged in intra-regional social practices, in which recourse is made to locked-in practices, reproducing old ways of working as a survival strategy in today's marketised environment. Accordingly, there has been an absence of regions engaged in both forms of regionalised social practices. In this sense, then, the three transformation processes have resulted in a replication of the organisation of space that characterised the regulatory failure of the Soviet system: de-regionalisation has occurred through the inter-regional social practices of inward investors replacing those of the old centrally planned system, and locked-in regional trajectories have emerged where social practices remain sedimented in particular places.

Conclusion

We have argued that regulation theory can provide a strong theoretical understanding of the economic dynamics and transformation of ECE political economies through highlighting the regulation of economic systems as a process of institutional coupling with forms of accumulation. We have also argued that the regulationist agenda needs to be extended to specify more clearly the nature of modes of regulation and regulatory practices in ECE transformation – how economies are governed and why governance is necessary. We have therefore suggested that economic and regional development in ECE is regulated in complex ways and that there are multiple methods through which systems of governance are constructed, stabilised and transformed. In order to move beyond a statist notion of regulatory relations, we have highlighted the centrality of networks and forms of integration in establishing particular social practices and regulatory processes. These regulatory processes develop through interaction which can temporarily stabilise the accumulation process. But clearly, no regulatory matrix of networks is static and without contradiction. Indeed, the current search in ECE for regulatory processes relevant to, and responding to, the imposition of a largely neo-liberal agenda has given rise to sets of uneven responses. These regulatory processes are themselves spatially constituted, providing part of the explanation of the complex geographies of development unfolding in ECE.

We have also attempted to show how many of these emergent regulatory processes fail to provide a sustainable and coherent governance system for ECE economies. Some proponents of the network paradigm have argued, for example, for institutional structures which can support the processes of participatory and democratic forms of regulation (Amin and Thrift 1995). In the context of ECE, a democratic and participatory system of institutional depth that supports and deepens development is all the more important in what are becoming, in an enlarged Europe, some of the 'least-favoured regions'. This is of further importance given the substantial social and geographical dislocation that is currently being produced through the transformation of the political economy of regional and economic development trajectories. While proponents of neo-liberal agendas argue that the embracing of a market capitalism will provide a recipe for economic rejuvenation, the fragmentation of regulatory practices that we have identified suggests instead that systems of market governance merely undermine the development of democratic and participatory economies in post-socialism. In this sense, then, neo-liberalism in ECE transformation remains a symptom of, rather than a cure for, the crisis of global capitalism and the Soviet system.

Acknowledgements

A previous version of this chapter was presented at the session on 'Theorizing Transition' at the Annual Conference of the Association of American Geographers, Charlotte, South Carolina, April 1996. We are very grateful to Ash Amin, Jamie Peck and John Pickles for their comments on earlier versions of the chapter.

Notes

1 For a critique of 'shock therapy' see Gowan (1995). Recently, Sachs has become adviser to the Bulgarian government and has argued strongly against the imposition of an IMF agenda on establishing a currency board, apparently in the interests of western creditors (*Transition* 1997: 24–5).
2 See also the recent paper by Mouleart (1997) which argues for an expanded concept of regulation comprising meta-regulation of subsystems which constitute the socio-economic structure of localities and which are themselves regulated by economic, non-economic, formal and informal practices.
3 Clearly, there are differences in the approach of these authors. Leborgne and Lipietz, for example, base their analysis of variants of post-Fordism and neo-Fordism around the centrality of the labour process while others, such as Esping-Anderson, see the role of welfare regimes as fundamental to the constitution of particular national capitalisms.
4 Regulation approaches are diverse. Jessop (1990), for example, has identified seven variations ranging from the early origins in Paris with the work of Aglietta (1979) and Lipietz (1987) to a branch of political economy in the United States focusing upon social structures of accumulation (Gordon *et al.* 1982; Kotz 1990).
5 Kosonen (1997) uses a similar framework of interpretation to examine the historical socio-economic development of Vyborg.
6 Burawoy (1996) has recently termed this 'involution'.

Bibliography

Aglietta, M. (1979) *A Theory of Capitalist Regulation*, London: New Left Books.
Albach, H. (1993) *Zerrissene Netz: Eine Netwerkanalyses ses ostdeutschen Transformationprozesses* [*Broken Networks: A Network Analysis of the East German Transformation Process*], Berlin: Edition Sigma.
Altvater, E. (1993) *The Future of the Market*, London: Verso.
Amin, A. (1995) 'Beyond associative democracy', mimeo, Department of Geography, University of Durham.
Amin, A. and Hausner, J. (eds) (1997) *Beyond Markets and Hierarchies: Interactive Governance and Social Complexity*, Cheltenham: Edward Elgar.
Amin, A. and Thrift, N. (1994) 'Living in the global', in A. Amin and N. Thrift (eds) *Globalization, Institutions, and Regional Development in Europe*, Oxford: Oxford University Press.
Amin, A. and Thrift, N. (1995) 'Institutional issues for the European regions: from

markets and plans to socioeconomics and powers of association', *Economy and Society*, 24, 1: 41–66.

Balabanova, T. (1994) 'An evolutionary institutional approach to systems transformation in eastern Europe', *European Association for Evolutionary Political Economy Newsletter*, 11 January.

Bianchi, G. (1992) 'Combining networks to promote integrated regional development', in T. Vaško (ed.) *Problems of Economic Transition: Regional Development in Central and Eastern Europe*, Aldershot: Avebury.

Boyer, R. (1995) 'The great transformation of eastern Europe: a regulationist perspective', *EMERGO*, 2, 4: 25–31.

Brezinski, H. and Fritsch, M. (eds) (1996) *The Economic Impact of New Firms in Post-Socialist Countries: Bottom-up Transformation in Eastern Europe*, Cheltenham: Edward Elgar.

Burawoy, M. (1985) *The Politics of Production*, London: Verso.

Burawoy, M. (1992) 'A view from production: the Hungarian transition to capitalism', in C. Smith and P. Thompson (eds) *Labour in Transition*, London: Routledge.

Burawoy, M. (1996) 'The state and economic involution: Russia through a China lens', *World Development*, 24, 6: 1105–17.

Burawoy, M. and Krotov, P. (1992) 'The Soviet transition from socialism to capitalism: worker control and economic bargaining in the wood industry', *American Sociological Review*, 57, 1: 16–38.

Burawoy, M. and Krotov, P. (1995) 'Russian miners bow to the angel of history', *Antipode*, 27, 2: 115–36.

Burda, J.C., Singh, I. and Török, A. (1994) *Firms Afloat and Firms Adrift: Hungarian Industry and the Economic Transition*, Armonk: M.E. Sharp.

Clarke, S., Fairbrother, P., Burawoy, M. and Krotov, P. (1993) *What About the Workers? Workers and the Transition to Capitalism in Russia*, London: Verso.

Dunford, M. (1990) 'Theories of regulation', *Environment and Planning D: Society and Space*, 8: 297–321.

Esping-Anderson, G. (1990) *Three Worlds of Welfare Capitalism*, Cambridge: Polity Press.

Fukuyama, F. (1992) *The End of History and the Last Man*, London: Hamish Hamilton.

Gibson-Graham, J.K. (1996) *The End of Capitalism (As We Knew It): A Feminist Critique of Political Economy*, Oxford: Blackwell.

Goodwin, M., Duncan, S. and Halford, S. (1993) 'Regulation theory, the local state, and the transition of urban politics', *Environment and Society D: Society and Space*, 11: 67–88.

Gordon, D., Edwards, R. and Reich, M. (1982) *Segmented Work, Divided Workers*, Cambridge: Cambridge University Press.

Gowan, P. (1995) 'Neo-liberal theory and practice for eastern Europe', *New Left Review*, 213: 3–60.

Grabher, G. (ed.) (1993) *The Embedded Firm: on the Socio-economics of Industrial Networks*, London: Routledge.

Grabher, G. (1994a) 'The disembedded regional economy: the transformation of East German industrial complexes into western enclaves', in A. Amin and

N. Thrift (eds) *Globalization, Institutions, and Regional Development in Europe*, Oxford: Oxford University Press.

Grabher, G. (1994b) *Lob der Verschwendung* [*In Praise of Waste*], Berlin: Edition Sigma.

Grabher, G. (1997) 'Adaptation at the cost of adaptability: restructuring the Eastern German regional economy', in G. Grabher and D. Stark (eds) *Restructuring Networks in Postsocialism: Legacies, Linkages and Localities*, Oxford: Oxford University Press.

Grabher, G. and Stark., D. (eds) (1997) *Restructuring Networks in Postsocialism: Legacies, Linkages and Localities*, Oxford: Oxford University Press.

Granovetter, M. (1985) 'Economic action and social structure: the problem of embeddedness', *American Journal of Sociology*, 91, 3: 481–510.

Granovetter, M. (1992) 'Economic institutions as social constructions: a framework for analysis', *Acta Sociologica*, 35: 3–11.

Hausner, J. (1995) 'Imperative vs. interactive strategy of systemic change in Central and Eastern Europe', *Review of International Political Economy*, 2, 2: 249–66.

Hausner, J., Jessop, B. and Nielsen, K. (eds) (1993) *Institutional Frameworks of Market Economies*, Aldershot: Avebury.

Hausner, J., Jessop, B. and Nielsen, K. (eds) (1995) *Strategic Choice and Path-Dependency in Post-Socialism*, Aldershot: Edward Elgar.

Hay, C. (1995) 'Re-stating the problem of regulation and re-regulating the local state', *Economy and Society*, 24, 3: 387–407.

Hay, C. and Jessop, B. (1995) 'Introduction: local political economy: regulation and governance', *Economy and Society*, 24, 3: 303–6.

Heller, A., Feher, F. and Markus, G. (1983) *Dictatorship Over Needs*, Oxford: Blackwell.

Hodgson, G.M. (1988) *Economics and Institutions*, Cambridge: Polity Press.

Hodgson, G.M. (1993) *Economics and Evolution: Bringing Life Back into Economics*, Cambridge: Polity Press.

Hodgson, G.M. (1996) 'Varieties of capitalism and varieties of economic theory', *Review of International Political Economy*, 3, 3: 380–433.

Illner, M. (1992) 'Municipalities and industrial paternalism in a "real socialist" society', in P. Dostal *et al.* (eds) *Changing Territorial Administration in Czechoslovakia: International Viewpoints*, Amsterdam: Instituut voor Sociale Geografie.

Jessop, B. (1990) 'Regulation theories in retrospect and prospect', *Economy and Society*, 19, 2: 153–216.

Jessop, B. (1992) 'Fordism and post-Fordism: a critical reformulation', in M. Storper and A. Scott (eds) *Pathways to Industrialization and Regional Development*, London: Routledge.

Jessop, B. (1995a) 'The regulation approach, governance and post-Fordism: alternative perspectives on economic and political change', *Economy and Society*, 24, 3: 307–33.

Jessop, B. (1995b) 'Regional economic blocs, cross border cooperation, and local economic strategies in postsocialism', *American Behavioral Scientist*, 38, 5: 674–715.

Jessop, B. (1997) 'The governance of complexity and complexity of governance: preliminary remarks on some problems and limits of economic guidance', in A.

Amin and J. Hausner (eds) *Beyond Markets and Hierarchy: Interactive Governance and Social Complexity*, Aldershot: Edward Elgar.

Kaldor, M. (1990) *The Imaginary War*, Oxford: Blackwell.

Kogut, B. (1996) 'Direct investment, experimentation, and corporate governance in transition economies', in R. Frydman, C. Gray and A. Rapaczynski (eds) *Corporate Governance in Central Europe and Russia*, Budapest: Central European University.

Konrád, G. and Szelenyi, I. (1979) *Intellectuals on the Road to Class Power*, New York: Harcourt.

Kornai, J. (1980) *Economics of Shortage*, Amsterdam: North Holland.

Kosonen, R. (1997) 'From patient to active agent: an institutionalist analysis of the Russian border town of Vyborg', in A. Amin and J. Hausner (eds) *Beyond Markets and Hierarchy: Interactive Governance and Social Complexity*, Cheltenham: Edward Elgar.

Kössler, R. and Muchie, M. (1990) 'American dreams and Soviet realities: socialism and Taylorism – a reply to Chris Nyland', *Capital and Class*, 40: 61–88.

Kotz, D. (1990) 'A comparative analysis of the theory of regulation and the social structure of accumulation theory', *Science and Society*, 54, 1: 5–28.

Kovács, J.M. (ed.) (1994) *Transition to Capitalism?*, New Brunswick: Transaction.

Kuczi, T. and Makó, C. (1997) 'Towards industrial districts? Small-firm networking in Hungary', in G. Grabher and D. Stark (eds) *Restructuring Networks in Post-Socialism: Legacies, Linkages and Localities*, Oxford: Oxford University Press.

Leborgne, D. and Lipietz, A. (1992) 'Conceptual fallacies and open questions on post-Fordism', in M. Storper and A. Scott (eds) *Pathways to Industrialization and Regional Development*, London: Routledge.

Lipietz, A. (1987) *Mirages and Miracles*, London: Verso.

Lipietz, A. (1992) *Towards a New Economic Order*, Cambridge: Polity Press.

Mouleart, F. (1997) 'Rediscovering spatial inequality in Europe: building blocks for an appropriate "regulationist" analytical framework', *Environment and Planning D: Society and Space* (forthcoming).

Murray, R. (1992) 'Flexible specialisation and development strategy: the relevance for Eastern Europe', in H. Ernste and V. Meier (eds) *Regional Development and Contemporary Industrial Response: Extending Flexible Specialisation*, London: Belhaven.

Nuti, D.M. (1996) 'Post-communist mutations', *EMERGO*, 3, 1: 7–13.

Offe, C. (1995) 'Designing institutions for East European transitions', in J. Hausner, B. Jessop and K. Nielsen (eds) *Strategic Choice and Path-Dependency in Post-Socialism*, Aldershot: Edward Elgar.

Painter, J. and Goodwin, M. (1995) 'Local governance and concrete research: investigating the uneven development of regulation', *Economy and Society*, 24, 3: 334–56.

Peck, J. and Tickell, A. (1994) 'Searching for the new institutional fix: the *after*-Fordist crisis and the global-local disorder', in A. Amin (ed.) *Post-Fordism: A Reader*, Oxford: Blackwell.

Pejovich, S. (1994) 'The market for institutions vs. capitalism by fiat: the case of Eastern Europe', *KYKLOS*, 47, 4: 519–29.

Piore, C. (1992) 'The limits of the market and the transformation of socialism', in

B. Silverman, R.Vogt and M. Yanowitch (eds) *Labour and Democracy in the Transition to the Market*, Armonk: M.E. Sharp.

Pickles, J. (1995) 'Restructuring state enterprises: industrial geography and Eastern European transitions', *Geographische Zeitschrift*, 83, 2: 114–31.

Sabel, C. (1994) 'Learning by monitoring: the institutions of economic development', in N.J. Smelser and R. Swedberg (eds) *Handbook of Economic Sociology*, Princeton: Princeton University Press.

Sabel, C. and Prokop, J.E. (1996) 'Stabilisation through reorganisation: some preliminary implications of Russia's entry into world markets in the age of discursive quality standards', in R. Frydman, C.W. Grey and A. Rapacynski (eds) *Corporate Governance in Central Europe and Russia: Insiders and the State* (vol. 2), Budapest: Central University Press.

Sachs, J. (1990) 'Eastern European economies: what is to be done?', *The Economist*, 23 January, 13–19.

Sayer, A. (1995) *Radical Political Economy*, Oxford: Blackwell.

Sayer, A. and Walker, R. (1992) *The New Social Economy: Reworking the Division of Labour*, Oxford: Blackwell.

Smelser, N.J. and Swedburg, R. (eds) (1994) *The Handbook of Economic Sociology*, Princeton: Princeton University Press.

Smith, A. (1994) 'Uneven development and the restructuring of the armaments industry in Slovakia', *Transactions of the Institute of British Geographers*, 19, 4: 404–24.

Smith, A. (1995) 'Regulation theory, strategies of enterprise integration and the political economy of regional economic restructuring in Central and Eastern Europe: the case of Slovakia', *Regional Studies*, 29, 8: 761–72.

Smith, A. (1996) 'From convergence to fragmentation: uneven regional development, industrial restructuring, and the "transition to capitalism" in Slovakia', *Environment and Planning A*, 28: 135–56.

Smith, A. (1997) 'Constructing capitalism? Small and medium enterprises, industrial districts and regional policy in Slovakia', *European Urban and Regional Studies*, 4, 1: 45–70.

Smith, A. (1998) *Reconstructing the Regional Economy: Industrial Transformation and Regional Development in Slovakia*, Cheltenham: Edward Elgar.

Smith, A. and Ferenčiková, S. (1998) 'Regional transformations, inward investments and uneven development in East-Central Europe', *European Urban and Regional Studies*, forthcoming.

Stark, D. (1992a) 'Bending the bars of the iron cage: bureaucratisation and informalisation in capitalism and socialism', in C. Smith and P. Thompson (eds) *Labour in Transition*, London: Routledge.

Stark, D. (1992b) 'Introduction', *East European Politics and Societies*, 6, 1: 1–2.

Stark, D. (1995) 'Not by design: the myth of designer capitalism in Eastern Europe', in J. Hausner, B. Jessop and K. Nielsen (eds) *Strategic Choice and Path-Dependency in Post-Socialism*, Aldershot: Edward Elgar.

Stark, D. (1996a) 'Recombinant property in East European capitalism', in G. Grabher and D. Stark (eds) *Restructuring Networks in Postsocialism: Legacies, Linkages and Localities*, Oxford: Oxford University Press.

Stark, D. (1996b) 'Networks of assets, chains of debt: recombinant property in Hungary', in R. Frydman, C.W. Grey and A. Rapacynski (eds) *Corporate*

Governance in Central Europe and Russia: Insiders and the State, Budapest: Central University Press.

Stark, D. and Nee, V. (1989) 'Towards an institutional analysis of state socialism', in V. Nee and D. Stark (eds) *Remaking the Economic Institutions of Socialism: China and Eastern Europe*, Stanford: Stanford University Press.

Swain, A. (1996) 'A geography of transformation: the restructuring of the automotive industry in Hungary and East Germany 1989–1994', unpublished Ph.D. thesis, University of Durham.

Swain, A. (1997) 'A tale of two enterprises: time–space embeddedness, industrial restructuring and self-transformation in Hungary and East Germany', *European Urban and Regional Studies*, 4 (forthcoming).

Szelenyi, I., Beckette, K. and King, L.P. (1994) 'The socialist economic system', in N.J. Smelser and R. Swedburg (eds) *Handbook of Economic Sociology*, Princeton: Princeton University Press.

Tickell, A. and Peck, J. (1992) 'Accumulation, regulation and the geographies of post-Fordism: missing links in regulationist research', *Progress in Human Geography*, 16: 190–218.

Tickell, A., and Peck, J. (1995) 'Social regulation after Fordism: regulation theory, neo-liberalism and the global-local nexus', *Economy and Society*, 24, 3: 357–86.

Transition (1997) 'Bulgaria', April 1997: 24–5.

van Zon, H. (1992) 'Towards regional innovation systems in central Europe', *FAST Dossier Continental Europe: Science, Technology and Community Cohesion*, Brussels: Commission of the European Communities.

3

ORGANISING DIVERSITY

Evolutionary theory, network analysis and post-socialism

Gernot Grabher and David Stark

Lessons from Labrador

Each evening during their hunting season, the Naskapi Indians of the Labrador Peninsula determined where they would look for game on the next day's hunt by holding a caribou shoulder-bone over the fire.[1] Examining the smoke deposits on the caribou bone, a shaman read for the hunting party the points of orientation of tomorrow's search. In this way, the Naskapi introduced a randomising element to confound a short-term rationality in which the one best way to find game would have been to look again tomorrow where they had found game today. By following the daily divergent map of smoke on the caribou bone, they avoided locking in to early successes that, while taking them to game in the short run, would have depleted the caribou stock in that quadrant and reduced the likelihood of successful hunting in the long run. By breaking the link between future courses and past successes, the tradition of shoulder-bone reading was an antidote to path dependence in the hunt.

Mainstream notions of the post-socialist 'transition' as the replacement of one set of economic institutions by another set of institutions of proven efficiency are plagued by similar problems of short-term rationality that the Naskapi traditional practices mitigate. As the economist's variant of 'hunt tomorrow where we found game today', neo-liberals recommend the adoption of a highly stylised version of the institutions of prices and property that have 'worked well in the West' (see, for example, Blanchard, Froot and Sachs 1994). Economic efficiency will be maximised only through the rapid and all-encompassing implementation of privatisation and marketisation. We argue, by contrast, from an evolutionary perspective, that although such institutional homogenisation might foster *adaptation* in the short run, the consequent loss of institutional diversity will impede *adaptability* in the long run. Limiting the search for effective institutions and organisational forms to the familiar western quadrant of tried and proven arrangements locks in the post-socialist economies to *exploiting* known territory at the cost of forgetting (or never learning) the skills of *exploring* new solutions.

With our Naskapi example we do not mean to suggest that policy makers in contemporary Eastern Europe should select institutions with a roll of the dice. For us, the lesson from Labrador is that institutional legacies that retard the quick pursuit of immediate successes can be important for keeping open alternative courses of action. Institutional friction preserves diversity; it sustains organisational routines that might later be recombined in new organisational forms. Resistance to change, in this sense, can foster change. Institutional legacies embody not only the persistence of the past but also resources for the future. Institutional friction that blocks transition to an already designated future keeps open a multiplicity of alternative paths to further exploration.

Our neo-liberal colleagues would be quick to argue that such exploration is costly, inefficient and unnecessary. In their view, the alternative, evolutionary course of search seems an indulgent squandering of resources, avoidable by exploiting institutions with proven returns. Given limited resources, the economies of Eastern Europe would do better to be quick to the chase, to learn from the leaders instead of the lessons of Labrador.

Recent studies in evolutionary economics and organisational analysis suggest, by contrast, that organisations that learn too quickly sacrifice efficiency. Allen and McGlade (1987), for example, use the behaviour of Nova Scotia fishermen to illustrate the possible trade-offs of exploiting old certainties and exploring new possibilities. Their model of these fishing fleets divides the fishermen into two classes: the rationalist 'Cartesians' who drop their nets only where the fish are known to be biting, and the risk-taking 'Stochasts' who discover the new schools of fish. In simulations where all the skippers are Stochasts the fleet is relatively unproductive – for knowledge of where the fish are biting is unutilised; but a purely Cartesian fleet locks in to the 'most likely' spot and quickly fishes it out. More efficient are the models that most closely mimic the actual behaviour of the Nova Scotia fishing fleets with their mix of Cartesian exploiters and Stochastic explorers. The purely Cartesian fleet in Allen and McGlade's (1987) study illustrates that organisations that learn too quickly exploit at the expense of exploration, thereby locking in to sub-optimal routines and strategies (see also March 1991).

This chapter counters the neo-classical prescriptions for the post-socialist economies with an alternative conception of development drawn from new insights in evolutionary theory and network analysis. These schools of analysis are not typically paired, and here we make the case that their combination provides fruitful insights for understanding the post-socialist transformations.

Our starting premise is that the proper analytic unit, because it is the actual economic unit, is not the isolated firm but networks that link firms and connect persons across them. Similarly, the unit of entrepreneurship is not the isolated individual but networks of actors. As such, our attention shifts from the attributes and motivations of individual personalities to the properties of the localities and networks in which entrepreneurial activity is reproduced. It follows that the economic unit to be restructured is not the isolated firm but networks of firms linking interdependent assets across formal organisational boundaries.[2] We shall also argue that networks are not only the units to be restructured but are also the agents to do the restructuring. That is, in place of the dichotomously forced choice of restructuring directed by state agencies versus restructuring via market processes we explore the possibilities of alternative co-ordinating mechanisms governed neither by hierarchy nor by markets (Grabher 1993; Powell 1990; Stark and Bruszt 1995; see also Smith and Swain in this volume).

The concepts of *Legacies, Linkages, and Localities* serve as the organising principles of this chapter. As we make the case for incongruence and explore the possibilities that ambiguity can be a resource for economic action, the reader should be prepared for some dissonance between the conventional meanings of these terms and their usage here. In developing these themes, we shall discover processes and logics quite different from notions that come first to mind. As we have already alluded, we shall see that legacies are not simple residues of the past but can serve as resources for the future. Similarly, the more systematically we pursue the logic of linkages, the more our analysis turns to the structural features produced by the absence of particular connections. And whereas 'localities' might evoke sites in which proximity shapes shared meanings, we examine localities as sites where the simultaneous presence of multiple logics (what we might think of as different 'species' of social action) yields complex ecologies of meaning.

Legacies

Fitness tests

In the neo-liberal prescription for the post-socialist transition, the persistence of organisational forms and social relationships of the old state-socialist system signals an incomplete change, a manifest symptom of a half-hearted implementation of the envisaged new social order. Accordingly, legacies indicate institutional pathologies contaminated with the deficiencies of the old regime obstructing the process of transformation: the future cannot be realised because the past cannot be overcome. The legacies of state socialism block the promising road to free markets. Free markets, the prominent advocates of neo-classical economics incessantly repeat, are a synonym for efficiency. Notoriously suppressed during state socialism, competition in free markets guarantees that more efficient organisational forms will survive and that inefficient ones perish.

Ironically, while economists can still embrace the crude Darwinism of Spencer's 'survival of the fittest', contemporary biologists (see, for example, Smith 1984; Gould and Lewontin 1984; and the essays in Dupré 1987) have challenged the received evolutionary model, arguing that evolution cannot simply be regarded as a one-dimensional process of optimisation, a beneficent and unilinear journey from the lower to the higher form of organisation, from the inferior to the superior. Natural selection does not yield the superlative fittest, only the comparatively and tolerably fit.

Evolution, in this sense, does not proceed along a single grand avenue towards perfection but along multiple paths which do not all lead to optimal change. That some developmental paths produce ineffective solutions and sub-optimal outcomes is not an indication of evolutionary failure but a precondition for evolutionary selection: no variety, no evolution. Hence, the evolutionary process necessarily entails development through failure: 'imperfections are the primary proofs that evolution has occurred, since optimal designs erase all signposts of history' (Gould 1987: 14). This critique of the 'survival of the fittest' paradigm offers an alternative evolutionary model for challenging the neo-classical assumptions of 'historical efficiency' in which survival implies efficiency and mere existence proves optimality (cf. Hodgson 1993). The lesson to be drawn from evolutionary theory is that competition in free markets does not necessarily favour the more fit and more efficient form of organisation: market competition is not an optimiser.

Fitness is not an absolute and invariant quantity. Rather, fitness depends on the environment, and the environment may change during the course of the selection process (Carroll and Harrison 1994). Thus, even if the selected characteristics of an organisational form were the 'fittest', they would be so only in regard to a particular economic, political and cultural context; they would not be the fittest for a changing or a different context. In fact, the very fitness of an organisational form might, through various mechanisms, induce environmental changes that undermine its efficiency. It follows that organisational forms that are most fit for the 'transition' are quite likely to be sub-optimal in the subsequently changed environment. In place of the search for the 'best' institutions to manage the transition, we might do better to reorient our analysis to identifying the types of organisational configurations that are better at search.

Evolutionary theory, moreover, turns our attention to how the future development of an economic system is affected by the path it has traced in the past. Once we reject the notion that 'from whatever starting point, the system will eventually gravitate to the same equilibrium', we are alerted to the possibilities that free markets might lock in economic development to a particular path that does not gravitate to the optimum (Hodgson 1993: 204). Positive feedback can have negative effects. Increasing returns from learning effects and network externalities yield real immediate benefits that can preclude selection in the long run of the most efficient organisational form (Arthur 1994; Carroll and Harrison 1994). Once an economy is locked into a particular trajectory, the costs of shifting strategies outweigh the benefits of alternatives. This approach to economic history stresses the possibility that the very mechanisms that foster allocative efficiency might eventually lock in economic development to a path which is inefficient when viewed dynamically. The mechanisms that are conducive for the synchronic adaptation of the economy to a specific environment may, at the same time, undermine an economy's diachronic adaptability (Grabher 1994).

The trade-off between allocative and dynamic efficiency constitutes a fundamental tension in the current transformation in Eastern Europe. Murrell (1991) argues from empirical data that state socialism was no less efficient in allocating resources than capitalist societies. Where it lagged was in dynamic efficiency, in its capacity to promote innovation. This imbalance has survived state socialism: current reform efforts seem preoccupied with removing institutional legacies for the sake of improving allocative efficiency. But a purging of organisational legacies to gain allocative efficiency can come at the cost of undermining dynamic efficiency (see also Hannan 1986).

We do not seek, of course, to reverse the evaluation of historical legacies from universally vicious to unequivocally virtuous. Instead, we aim to high-light the dual potential of legacies to block *and* to support transformation. It follows that instead of examining organisational forms in Eastern Europe according to the degree to which they conform to or depart from the ideal types of organising production in western-style capitalism, this chapter is concerned with variations and mutations emerging from the recombination of the inherited forms with emerging new ones (see Herod in this volume). Instead of simply conceiving these recombinations as accidental aberrations, we explore their evolutionary potentials.

Compartmentalisation: the organisation of diversity

We thus shift from the preoccupation with the efficiency of an individual organisational form to a concern for variety and diversity of forms central to the perspective of 'population thinking' (Mayr 1984). As we shall see, the recombination of old organisational forms in the reorganisation of the large state enterprises increases variety and diversity within the 'genetic pool' for the evolution of new organisational forms. For evolution to work there must always be a variety of forms from which to select: 'Selection is like a fire that consumes its own fuel . . . unless variation is renewed periodically, evolution would come to a stop almost at its inception' (Lewontin 1982: 151). Diversity and variety allow evolution to follow at the same time different paths which are associated with different sets of organisational forms. When selection starts off not simply from a single trajectory but from a broad and diverse range of evolutionary alternatives, the risk decreases that local maximisation results in an evolutionary dead end. Two or more evolutionary trajectories are thus able to cope with a broader array of unpredictable environmental changes than is the case with a single one.

In this perspective, different levels of efficiency associated with the different evolutionary paths are not symptoms of an inefficient selection mechanism. Rather, they are a precondition for improving overall efficiency since 'the rate of increase in fitness of any organism at any time is equal to its genetic variance in fitness at that time' (Fisher 1930: 35). The merciless competition evoked by the crude Darwinism of the 'survival of the fittest' is, according to neo-Darwinism, mitigated by the biological principle of *compartmentalisation.* Compartmentalisation buffers the various sub-populations from each other and, hence, allows less efficient ones to coexist with the currently most efficient ones without being exposed to selection immediately. Compartmentalisation allows for an increasing diversification of the evolutionary selection. In a compartmentalised genetic pool, rare genes have a greater chance to influence subsequent evolution than is the case with a non-compartmentalised genetic pool. Although compartmentalisation detracts from the fitness of the entire system, the sum of the subsystems keeps ready a broader spectrum of answers to environmental challenges and thus ultimately arrives at an even higher level of fitness (Weizsäcker and Weizsäcker 1984: 188).

The principle of compartmentalisation suggests that it is not simply the *diversity of organisations* but the *organisation of diversity* that is relevant for the recombination of organisational forms in Eastern Europe. The reproduction of diversity depends on the ability of different levels of efficiency to coexist. On the one hand, evolution comes to a stop in cases where less efficient forms are eliminated through selection immediately: too little diversity, no evolution. On the other hand, however, the absence of any evolutionary selective comparison might turn diversity into 'noise' in which none of the organisational forms would be able to influence the direction of any evolutionary trajectory: too much diversity, likewise, no evolution.

This tension between too little diversity (emerging from too low a degree of compartmentalisation) and too much diversity (resulting from too high a degree of compartmentalisation) is exemplified by the analysis of the restruc-turing of the large state-owned corporations in Eastern Germany (Grabher 1996) and in Hungary (Stark 1996). The resolute Eastern German approach led to a rapid dissolution of the old hegemonic form of the *combinat* and (through the establishment of western branch plants) to an increasing diversity of organisational forms. But, as Grabher (1996) argues, this diversity might yet shrink again in the medium-term future. The superior efficiency of the western branch plants could lead – from a lack of compartmentalisation – to a further crowding out of other organisational forms located mainly within the indigenous small firm sector. The great disparity between the invading front-runner and the indigenous laggards could produce a winner-takes-all situation that once again suppresses organisational diversity.

Seen from this perspective, the current Eastern German economy echoes the relative paucity of organisational forms of the old GDR economy whereas the transformation of the large enterprises in contemporary Hungary builds on the previous decade of organisational experimentation that allowed not only for competition among *firms* but also for competition of *forms* (Stark 1990). This competition of forms created a broad spectrum of variants in organising production that increasingly overlapped in terms of personnel, supplier relations and property rights. With this blurring of boundaries came greater organisational diversity. In contrast to the more recent experience in Eastern Germany, moreover, this diversity of forms has not been challenged by the emergence of a vastly more efficient form. That is, there is greater diversity of organisational forms in Hungary, but there is also much less obvious disparity of 'fitness' among them. Whereas in Eastern Germany a preponderant disparity runs the danger of suppressing diversity, in Hungary a 'noisy' diversity runs the danger of suppressing selection, with the result that less efficient forms might deprive more efficient forms of resources to an extent that blocks the evolution of the entire economy.

Legacies for entrepreneurial careers

The notion of compartmentalisation also figures implicitly in proposals for a 'two-track strategy' whereby resources are channelled into the indigenous small firm sector (the former second economy) while adopting more stringent administrative measures to harden the budget constraints of large firms remaining in the state sector. That strategy builds on the pioneering work of Gábor, who was among the first to perceive and analyse the significance of the second economy. Gábor (1979, 1986) demonstrated that the developmental potential of the second economy rested not in some spirit of individual entrepreneurship but in a *dynamic* tension between the twinned economies of late state socialism. Subsequent advocates of the two-track strategy such as Kornai (1990, 1992), Murrell (1992) and Poznanski (1993) argued that this dynamic tension would evaporate if privatisation and marketisation were to be attempted throughout the entire economy. That is, the transformative potential of the emerging marketised sector would dissipate if it was not buffered from the sphere of the large public enterprises. Attempts to 'privatise' everything at once would lead to privatising very little at all. A strategy of non-compartmentalised privatisation would yield firms that were private in name only. Similarly, expectations are not likely to change when those with new behaviours are scattered throughout the population. Actors are more likely to change their expectations when the probability of encountering a new behaviour trait is higher (Boyer and Orlean 1992). Buffering the sub-population of market-oriented actors increases this likelihood; and compartmentalisation (buffering that is not absolute but porous) increases the chances that the new patterns of behaviour can take hold in the broader population.

But the two-track strategy was nowhere adopted as official policy.[3] Nor can we assume, in any case, that a compartmentalised strategy would have selected behavioural traits of market orientation. What we can do is to examine actual behaviour in the emergent small firm sector. Doing so, we see first that the second economy has not necessarily promoted a dynamic capital-accumulating stratum, and second, that the second economy has not been the primary source of the new economic elite, as successful entrepreneurs are likely to come from the ranks of the socialist cadre. Each illustrates the ambiguous legacy of state socialism.

Gábor (1996), for example, observes that the small firm sector in postsocialist Hungary is marked by fragmentation and 'over-tertialisation'. Instead of finding small-scale proprietors growing into medium-sized employers, Gábor (1996) identifies an increasing tendency for small entrepreneurs to shun productive lines of business that involve higher investment intensity. He traces these features, at least partially, to economic preferences inherited from the second economy of the past regime including the income-maximising, consumption orientation of households; aversion to long-term business investment and risk-taking; the low appreciation of free time compared to income; and the poor tax morale.

Second, technocratic expertise acquired during state socialism provides an important source of entrepreneurship in the post-socialist period. As in advanced market economies, the elite in state socialism was an educated elite. It now appears, and not surprisingly so, that under post-socialism, education and entrepreneurship are closely linked. The legacy of socialism is that the former elite are well endowed to convert the cultural capital of the education and training acquired in the old order to advance to prominent positions in the new (Szelenyi and Szelenyi 1995). Empirical studies conducted in Hungary (Róna-Tas 1994), the Czech Republic (Benáček 1996) and Eastern Germany (Koch and Thomas 1996) are now providing evidence to support an argument that it is the common technocratic character of both party *and* entrepreneurial recruitment that is a main source of this continuity.

Taken together, these studies point to several legacies of state socialism in the field of entrepreneurial careers: whereas the old socialist hierarchies seem a launching pad for careers in the larger, legal firms of the emerging entrepreneurial sector, the heritage of the second economy pushes towards further fragmentation within the semi-legal sector of micro-firms.

Linkages

Loose coupling

In the predominant view, the implosion of state socialism has left behind an institutional vacuum and a social *tabula rasa* of atomised economic and political actors. Instead of atomisation and paralysis, however, this chapter examines the embeddedness of actors in social ties, whether official or informal (see also Smith and Swain in this volume). The relational approach adopted here starts not with the personal attributes of actors but with the networks of interaction that link actors (Emirbayer and Goodwin 1994). From this perspective, very strong and dense social networks facilitate the development of uniform subcultures and strong collective identities. But network analysis does not begin and end with social cohesion. A particularly dense and tightly coupled network (in its extreme form, where every actor in the network has a direct tie to every other) might promote cohesiveness while hindering the ability to gain information and mobilise resources from the environment. Recent trends in network analysis posit an inverse relationship, in general, between the density/intensity of the coupling of network ties on the one hand and their openness to the outside environment on the other. Similarly, in contrast to conventional cliquing models (e.g., 'who knows whom'), new research in the field is more likely to focus on absent ties in a network social space where actors lack direct connections. Research within this more robust relational analysis is now demonstrating that 'weak ties' (Granovetter 1973) indirectly connecting actors or bridging the 'structural holes' (Burt 1992) that become obligatory 'passage points' (Latour 1988) between relatively isolated groups of actors are crucial for the adaptability of networks.

The evolutionary advantages of loosely coupled networks were early appreciated and systematically differentiated by Weick (1976). First, a loosely coupled network is a good system for *localised* adaptation. If the elements in a system are loosely coupled, then any one element can adjust to and modify a local contingency without affecting the whole system. A second advantage is that loosely coupled networks preserve many independent sensing elements and therefore 'know' their environment better. Third, in loosely coupled networks where the identity and separateness of elements are preserved, the network can potentially retain a greater number of mutations and novel solutions than would be the case with a tightly coupled system.

Again, however, we are not claiming an unequivocally positive relationship between the loose coupling and the adaptability of a network. Although diversity and loose coupling might, on a structural level, support adaptability by allowing different levels of efficiency to coexist, they can also, on a cognitive level, result in a cacophony of orientations, perceptions, goals and world-views that confounds even minimal cohesiveness. Such is the danger noted by some observers of the Eastern European transformation who identify the 'chaos' resulting from the multiplicity and ambiguity of orientations and perceptions as a major obstacle to future-oriented economic action. Nevertheless, we invite a tolerance of ambiguity. That tolerance is not an unqualified embrace but an explicit ambivalence: it acknowledges that ambiguity can be an asset even while it recognises that these gains can come at the expense of accountability.

Aware that an excess of ambiguity can dissipate social cohesion, it is nevertheless alert to the possibilities that ambiguity can be a resource for credible commitments. Just as tolerance for ambiguity is regarded, on an individual level, as an attribute of a mature and robust personality, so here it is seen, on the system level, as a central cognitive precondition for adaptability. As in the ways that tolerance for different levels of efficiency enhances the evolutionary potentials of a network, so tolerance for ambiguous or even contradictory perceptions and goals facilitates the search for new answers to new questions. The communication of contradictions and conflicts, sparked by the ambiguity of goals, could act as a sort of 'immune system' for a network (Luhmann 1986: 185). In a sense, tolerance for ambiguity constitutes the 'intelligence' of a network, reducing the chance of contradictory signals being suppressed in favour of a singular but distorted knowledge and an internally consistent but mistaken interpretation.

Loose coupling in entrepreneurial networks

Rather than being extinguished for the sake of the logical principle of *tertium non datur* (there is no third case), ambiguity can be deliberately reproduced in particular situations by the *tertius gaudens* (the third who benefits). Taken from the work of Simmel (1923: 154 and 232), the *tertius* role is instructive in the Eastern European transformation because it points to an ambiguity from which 'the third who benefits' levers off a stable entrepreneurial position. In certain situations, emerging as the *tertius* depends on creating competition: 'Make simultaneous, contradictory demands explicit to the people posing them, and ask them to resolve their – now explicit – conflict' (Burt 1992: 76). Entrepreneurship, in this perspective, emerges from *tertius* brokering contradiction and ambiguity between others: no ambiguity, no *tertius*.

As Sedaitis's (1996) analysis of the emergence of new market organisations in Russia suggests, such a *tertius* strategy and the strategic utilisation of ambiguity seem more easily practised in loosely coupled networks than in tightly integrated ones. According to her study of the new commodity exchanges in Russia, exchanges organised around loosely coupled networks differ from tightly coupled networks in crucial aspects. Loosely coupled networks (with less density of direct ties among their founders) enjoy greater immediate returns on investment because of their greater manoeuvrability and more varied access to resources. They are able to serve market demand more directly and to exploit the lucrative opportunities in the disruption of established distribution patterns. With minimal constraints both internally and externally, they are relatively free to pursue *tertius* strategies. At the same time, however, their extraordinary diversity in turn provides little basis for social cohesion.

Commodity exchanges organised around the tightly knit networks grounded in the legacies of past institutional arrangements, by contrast, inherit institutional legitimacy yet they suffer a limited profitability. Sedaitis (1996) argues that the lower profitability of these tightly knit networks is due less to the constricted range of talent of their personnel than to the structural incapacity of their networks to pursue the aggressive *tertius* strategy favoured by the loosely coupled networks. Moreover, for the tightly knit networks, limited outside interaction inhibits processes of learning and unlearning: 'Shared past histories constrain the range of future possibilities . . . old ties limit organizational flexibility and maintain a "segmented" system of circumscribed action and responsibility that limits the potential of management to respond creatively to the new environment and the problems it poses' (Sedaitis 1996: 145).

Sedaitis's (1996) analysis of the Russian commodity exchanges thus marks an important departure from conventional approaches to entrepreneurship in two respects. It can be contrasted, first, to the research tradition that attributed entrepreneurship to the behavioural features of certain personality types, featured prominently, for example, in the early writings of Schumpeter (1912: 137), who provided a rich source of iconographic portraits of entrepreneurs as 'whole-hearted fellows' (*ganz Kerle*) combining the genius of creative discovery with the courage of 'creative destruction'. For Sedaitis (1996), entrepreneurship is not a function of an individual personality but of a social network. Second, her use of network concepts departs dramatically from a recent tendency to view network connections as the property of individuals. In that view, 'social capital' is a new individual-level variable that interacts with other assets ('human capital') in the process of status attainment or career mobility. Accordingly, researchers can now develop measures of the 'volume of network capital' in the possession of individual research subjects. However innovative in the field of mobility studies or the analysis of entrepreneurship, the addition of this new variable brings the notion of 'network' into the picture in a manner that neglects the relational dimension that is the fundamental insight of network analysis. In Sedaitis's (1996) study, by contrast, our attention shifts from *networks as property* to the *properties of networks* as she demonstrates that the shape, structure and characteristics of different kinds of networks make possible different economic activities.

Asset ambiguity

If the legacy of old networks and the structure of new ties are important for determining the types of entrepreneurial activity in post-socialism, might they also figure prominently in the restructuring of large corporations? This is the question posed in recent studies by McDermott (1996) and Stark (1996) on the Czech Republic and Hungary respectively.

In Czechoslovakia during the 1970s and 1980s, under the umbrella of meso-level 'Industrial Associations', constituent suppliers and customers, managers and workers, state bank branches, firms and local party members formed alliances to gain privileges from the centre and created informal compacts of economic co-ordination to limit and adjust to the uncertainties of an economy of shortage. McDermott (1996) argues that, over time, these informal networks became institutionalised, though not necessarily legally recognised, and became the frameworks to define and renegotiate claims to individual units of the large state-owned corporations. To the extent that these tightly coupled networks are also sources of mutual hold-up power among the actors, the discretion and the necessary knowledge to reorganise production are bound up in these relationships. Hence, the policy of the state to end-run the potential hold-up powers of firm actors – through rapid privatisation – would be 'one-legged' (McDermott 1996).

McDermott (1996) demonstrates that, despite its neo-liberal rhetoric, Václav Klaus's voucher privatisation programme did not eliminate the ties that bind so much as rearrange them (see also Brom and Orenstein 1994; Stark and Bruszt 1995). The outcome is a web of connections through which a multiplicity of actors are renegotiating not simply contractual ties but their mutual claims on interdependent assets. Through that web, firms, banks, investment companies, local governments and parts of the state bureaucracy identify firms that should be saved, devise strategies for restructuring assets, bargain about the allocation of resources, and renegotiate the very rules and governance institutions for resolving disputes among them.

The Janus face of networks also influences the Hungarian process of property transformation and corporatisation, driven by key actors in the old formal and informal networks who constituted the best organised social group in Hungary during the last decades. As Stark (1996) documents, managers of the large state-owned enterprises are breaking up their organisations – along divisional, plant, or even workshop lines – into numerous satellite corporations. Although these newly incorporated entities with legal identities were nominally independent, they combined private, semi-private and state property in a complex manner. Property shares in these satellite organisations are not limited to the founding enterprise but are also held by top and mid-level managers, professionals and other staff. In the typical pattern of this particular form of 'recombinant property', these private persons are joined in share ownership by other corporations and corporate satellites which are spinning around some other enterprises. At the same time, large enterprises are acquiring shares in one another, creating extensive inter-enterprise ownership networks. Like the ropes binding mountain-climbers on a treacherous face, these ties reduce risk, they buffer the networks from the uncertainty of the transformation shock, and they can facilitate innovation for some, even while retarding the selection process for many (Miner, Amburgey and Stearns 1990; Ickes and Ryterman 1994).

In contrast to the essentialist categories of private versus state property, these recombinant practices create networks of horizontal ties of cross-ownership intertwined with vertical ties of nested holdings in which the boundaries between state and private property are increasingly blurred. Recombinant property is not, however, a simple mixture of public and private: it is a hedging strategy that also blurs the boundaries of organisations themselves and blurs, as well, the boundedness of justificatory principles. In cases of extremely complex asset interdependence, it is not clear-cut property claims but an ambiguity of property claims that provide flexible adaptation. Such *asset ambiguity* should not be interpreted, however, as the simple polar opposite of Williamson's 'asset specificity' for it occurs in a volatile environment where the state's paternalistic efforts at the centralised management of liabilities create incentives for managers to employ a multi-plicity of justificatory principles to acquire resources. To survive in such an environment, managers become as skilled in the language of profitability for credit financing as in the syntax of eligibility for debt forgiveness. When they attempt to hold resources that can be justified by more than one legitimating principle, they make *assets of ambiguity*.

The same opportunistic blurring of boundaries that leads to a recombina-tion of assets and a decomposition of the large corporations also bears a social cost: it erodes (or, in the post-socialist case, retards) accountability. As Stark (1996) demonstrates, the problem with the peculiarly diversified portfolios in the 'polyphonic discourse of worth that is post-socialism' is that actors can all too often easily and almost imperceptibly switch among the various positions they hold simultaneously in the coexisting moral economies. To be accountable according to many different principles becomes a means to be accountable to none. Unless we are willing to posit 'flexibility' as an overriding value and a meta-legitimating principle, we cannot escape the challenge that post-socialism poses, not uniquely but acutely, for our epoch: if networks are viable economic agents of permanently ongoing restructuring, how can we make networks (as a new kind of moral actor) accountable?

Localities

Locality as ecology

In the dominant view, localities are irrelevant in constructing transition strategies. When not centred squarely at the level of the individual firm, analysis of the post-socialist transformations typically focuses on policies and institutions at the level of the national economy such as monetary policy, legal frameworks for corporate governance, or regulatory institutions for banking and finance. Place, the problem of localities, is out of place in these perspectives.

In arguing that localities should be brought into focus as sites of economic transformation (see also Gorzelak 1995; Smith 1995) we draw on the new economic sociology which demonstrates that globalisation does not displace the properties of localities but makes them all the more salient. As greater market volatility shifts strategic action from economies of scale to economies of scope and then to economies of time (Gereffi 1994), local knowledge, local culture and local networks give shape to the new organisational forms of flexible specialisation (see, for example, Amin and Thrift 1994; Scott 1996).

It was with the analysis of the Industrial Districts of northern Italy that the potential of localities to contribute to economic development most dramatically entered the research literature in the 1980s. The stories of regional production systems concentrated in the region Emilia Romagna have typically been written as success stories of a coherent system of economic institutions whose compatibility makes for the decisive transaction-cost efficiency of the regional co-operative networks.

But the story of the Italian Industrial Districts might also be read in a different light (Grabher 1994: 67–78). The Italian textiles and clothing districts in particular are composed of an extremely broad and heterogeneous spectrum of diverse institutions and organisational forms ranging from internationally renowned design ateliers and technologically highly advanced medium-sized firms at one pole to small artisanal firms and illegal home-workers at the other. Instead of regarding this spectrum as a coherent set whose efficiency is based on the transaction-cost savings gained through the compatibility of the various organisational forms, the evolutionary strengths of the Industrial District might be based on the very incompatibility of these forms. In this view, not systemic coherence but organisational discrepancy is the effective evolutionary antibody against hegemonic 'best practice solutions'. By preserving the richness of diverse organisational routines for the evolution of new organisational mutations, discrepancy increases the adaptability of the region.

The resistance against the economistic temptation to streamline,[4] at least in the Italian Industrial Districts, seems not to be an entirely intentional product of institutional design. In these districts, the spatial proximity of closely knit co-operative networks in small neighbourhoods is seen as a major source of their transaction-cost efficiency (see, for example, Pyke and Sengenberger 1992). From an evolutionary perspective, however, the transaction-cost effects are less important than the fact that spatial proximity allows for a continuous exchange of resources, information and personnel across these diverse, even incompatible, forms of production. Whether or not proximity economises on transaction costs, its long-term benefit is to facilitate a cross-fertilisation across disparate forms less likely if spatially dispersed. Like the Naskapi caribou ritual of our introduction, spatial proximity in the northern Italian districts acts as a sort of random generator disrupting the tendency towards transaction cost-efficient relations with compatible firms. In preventing hyper-efficient behaviour, spatial proximity does not dissolve incompatibility but enhances it.

Expressed in different terms, this view of Industrial Districts analyses localities as ecologies of diverse organisations (Grabher 1994: 70–8). Localities are sites of interdependence of even greater complexity than the proprietary ambiguities of complementary and co-specialised assets across the boundaries of enterprises. The interdependencies within localities are more complex because they entail ambiguities across different social logics, routines and practices involving not only business firms but political, religious, residential and family life. Because these logics cannot be reduced to each other, or expressed in the equivalents of a common currency, localities are not simply compartmentalising buffers separating sub-populations of the same species of organisation but are complex ecologies of diverse 'species' of social ordering principles.

Entrepreneurs in localities, entrepreneurial localities

In Eastern Europe, the emerging localised governance structure based on horizontal rather than on the hierarchical co-ordination of the past can contribute to the mobilisation of resources in the formation of new entrepreneurial units. In their study of a small community near Budapest, Kuczi and Makó (1996) indicate how local network ties reduce uncertainties and risks facing start-up ventures. That is, network linkages act as buffers retarding selection and reducing the 'liability of newness' – a problem facing new firms in any economy but particularly acute in the volatile uncertainties of post-socialist economic transformation. Kuczi and Makó (1996) point to trust-based relations where patterns of economic exchange are interwoven with ties of kinship and friendship. In that local community they studied, new contractual arrangements often follow informal relations among actors with shared experiences in the recent past, whether at the locally dominant state enterprise or through joint participation in the second economy. In such conditions, trust reduces the risks involved in the selection of suppliers, business partners and employees. Kuczi and Makó (1996) conclude that among these local networks, economic transactions are regulated by 'relational contracting' in which the stronger partner does not exploit situations where the weaker partner is vulnerable, and where maintenance of the tie itself is a value that regulates exchanges and moderates disputes.

The networks of small-scale proprietors in Kuczi and Makó's (1996) study bear some resemblance, at first glance, with the northern Italian Industrial Districts – for example, their preference for localised business contacts in the absence of a strong state, and the importance of traditional relations in contract enforcement. But the traditional elements are only a part of the success story of the northern Italian districts. And although Kuczi and Makó (1996) indicate that an entrepreneurs' club and a local foundation were in the planning stages at the time of their study, the community they examined showed few signs of the highly organised craft associations, trade unions and administratively competent local authorities so important in the northern Italian districts.

Moreover, there are reasons to question the causal connection between traditional ties, relations of trust and local development. For Gábor (1996), the liabilities of traditionalism are likely to outweigh the benefits. First, to the extent that second economy producers continue their old habits of making market transactions only where social relations have already preceded, they might be disadvantaged in establishing business ties where arm's length transactions are entirely appropriate (even to the point of forgoing advertising, for example). Second, in the absence of strong civic associations (blocked under communism, but thriving in Italy), Gábor (1996) is unwilling to assume that the legacy of the proximate ties of the second economy are relations of trust. It might just as well be that the most salient 'shared experiences' from the past are relations of mistrust and that new exchanges based on them will bear that stamp (Kemény 1996). In slightly different terms, instead of the northern Italian route to prosperity, for some post-socialist economies the Road to Europe might run through Sicily.

Finally, what if the direction of causality does not run from local identities to co-operative development strategies, but the reverse? Co-operative relations, Sabel (1992) argues, are not based on primordial loyalties but on 'studied trust'. One of the clues to these processes is that Sabel (1992) finds co-operative regional development projects in districts whose recent histories were marked by intense conflicts. Yet contemporary accounts by actors in these same localities repeatedly refer to harmonious pasts as history is reconstructed in line with the present. Thus, instead of shared identities giving rise to social relations of trust, this work suggests that co-operative configurations reshape identities that can then be shared. Although historically inaccurate, these identities are no less real in their effects as templates for current co-operative action.

In this alternative view, localities contribute to innovative and co-operative development strategies not because they are a locus of shared meanings but because they are sites of interdependence among different social groups and different social logics. Because localities cannot be indifferent to this interdependence, we can say that localities are means for organising diversity

– seen, for example, in the notion of actors manoeuvring not only through an ecology of organisations but through an ecology of ordering principles (Stark 1996) and in McDermott's (1996) analysis of how localities are the sites for complex negotiations among actors whose claims are not only competing but also very heterogeneous in character.

A similar conception of localities as ecologies of social logics informs the study of regional development in Poland by Hausner, Kudlacz and Szlachta (1996). Hausner and his colleagues examined economic development in nine provinces in south-eastern Poland in a study that takes the locality not only as the unit of observation but also as the unit of analysis. In seeking to explain why economic development takes off in some regions and not others, they turn from the properties of individuals to the properties (characteristics, qualities) of the localities themselves. In contrast to Kuczi and Makó (1996), who provide such a rich community study of *entrepreneurs in localities*, Hausner, Kudlacz and Szlachta (1996) might be seen to study *entrepreneurial localities*. Hausner, Kudlaez and Szlachta (1996) conclude that the best regional development strategies are not led by yet another administrative or quasi-governmental unit in the form of inter-mediate-level 'Regional Development Authorities'. Instead, a major factor explaining the differences in regional restructuring was the presence of networks linking diverse types of organisations.

Conclusion

In the opening pages of *The Economic Institutions of Capitalism*, Williamson (1985: 18–19) observes that 'Transaction costs are the economic equivalent of friction in physical systems.' Williamson's (1985, 1993) contribution to economics has been to develop an analytic strategy to understand 'friction' in economic transactions – with the aim of guiding policies and promoting institutions that minimise these transaction costs. This chapter can be seen as bringing the analysis of friction into the study of the transforming post-socialist economies. It differs from Williamson's project, however, in two fundamental ways. First, the friction it has examined is not that of economic exchanges *per se* but the friction of economic restructuring: that is, whereas Williamson turns our attention to *transaction* costs, we are concerned here with *transformation* costs. In fact, to the extent that institutionalisation is a kind of 'investment in forms' (Thévenot 1985) that reduces the costs of future transactions, such transformation costs might be conceptualised as sunk transaction costs. Second, unlike the Williamsonian tendency to assess as superior those forms that minimise friction, the chapter here sees a positive role for economic friction. To be sure, we are not advocating higher transformation costs or seeking to promote institutions with steep transaction costs; but it does seem to us useful to question the notion of a 'smooth' or frictionless 'transition'.

That position begins from the insight that some friction may be essential for the functioning of markets by undermining positive feedback loops that can lead to lock-in. Such was the lesson drawn by the Federal Securities and Exchange Commission in the aftermath of the 508-point crash of the New York Stock Exchange on 19 October 1987. As trading in some fields was approaching an almost frictionless character with advances in computerised 'program trading', the Securities and Exchange commissioners saw a danger that some markets could pass from volatility to chaos. To maintain orderly markets, the commissioners designed a set of 'collars' that trigger temporary halts in computerised index arbitrage when the Dow skips more than a certain number of points in either direction. Like the Naskapi caribou shoulder-bone that disrupts the negative effects of positive feedback, these so-called circuit breakers bring time, and hence friction, back into the Exchange (see, for example, Petruno 1994).

Our aim in this chapter has been to begin the analysis of the circuit breakers that bring friction to the post-socialist transformations. Institutional legacies produce the friction that grinds against a smooth transition but preserves diversity for future recombinant strategies. Inter-enterprise linkages buffer firms and retard selection, but the redundant relations of loosely coupled networks produce the friction of ambiguity that facilitates entrepreneurial strategies. And the multiple ordering principles of localities produce the friction that inhibits too-simple harmonisations but yields more complex ecologies that are the basis for regional development strategies. With the concepts of compartmentalisation, asset ambiguity and local ecologies of meaning we can proceed to analyse how actors reconfigure legacies, linkages and localities to forge pathways from state socialism.

Acknowledgements

This chapter is reprinted from *Regional Studies* Volume 31, Number 5, 1997. The authors and editors are grateful to the Carfax Publishing Company for granting permission to reprint it. The chapter is based on the authors' lead essay in their volume *Restructuring Networks in Post-Socialism: Legacies, Linkages, and Localities*, Oxford: Oxford University Press, 1997.

Notes

1 This account is drawn from Weick (1977: 45).
2 To be sure, shifting the level of analysis from isolated firms to networks does not imply that individual firms are conceived as homogeneous actors. On the contrary, as we shall demonstrate later, networks comprise a broad and heterogeneous spectrum of actors, all of whom exerting different levels of control and power. In this sense, we do not intend to characterise networks in terms of concord and harmonious collaboration among equals. Rather, in following

Håkansson and Johanson (1993: 48), we regard power as a functional element of networks: 'In contrast to the market model, in which power is seen as some kind of imperfection, the network model views power as a necessary ingredient in exploiting interdependencies.'

3 A plausible argument might be made that, despite official rhetoric, Poland's *de facto* policies came closest to the two-track strategy.

4 The economistic temptation to streamline grows stronger with increasing imbalances of power within the network. In other words, the more powerful individual firms are *vis-à-vis* other firms in the network, the more they tend to rationalise network ties according to their own needs. However, to the extent that this rationalisation increases the transaction-cost efficiency within the network, it might undermine the network's adaptive capacities by reducing the diversity of organisational forms. Such counter-productive impacts of pronounced imbalances of power within a network have, for example, been elaborated in research on Baden-Württemberg (see Herrigel 1993; Cooke, Morgan and Price 1993).

Bibliography

Allen, P. M. and McGlade, J. M. (1987) 'Modelling complex human sytems: a fisheries example', *European Journal of Operational Research* 24: 147–67.

Amin, A. and Thrift, N. (1994) *Globalization, Institutions, and Regional Development in Europe*, Oxford: Oxford University Press.

Arthur, W. B. (1994) *Increasing Returns and Path Dependence in the Economy*, Ann Arbor, MI: University of Michigan Press.

Benáček, V. (1996) 'Private entrepreneurship and small businesses in the transformation of the Czech Republic', in G. Grabher and D. Stark (eds) *Restructuring Networks in Post-Socialism: Legacies, Linkages, and Localities*, Oxford: Oxford University Press.

Blanchard, O. J., Froot, K. A. and Sachs, J. D. (1994) *The Transition in Eastern Europe*, Chicago, IL: University of Chicago Press.

Boyer, R. and Orlean, A. (1992) 'How do conventions evolve?', *Journal of Evolutionary Economics* 2: 165–77.

Brom, K. and Orenstein, M. (1994) 'The "privatized" sector in the Czech Republic: government and bank control in a transitional economy', *Europe–Asia Studies* 46: 893–928.

Burt, R. (1992) *Structural Holes: The Social Structure of Competition*, Cambridge, MA: Harvard University Press.

Carroll, G. R. and Harrison, J. R. (1994) 'On the historical efficiency of competition between organizational populations', *American Journal of Sociology* 100: 720–49.

Cooke, P., Morgan, K. and Price, A. (1993) 'The future of the Mittelstand. Collaboration versus competition', Regional Industrial Research Report No. 13, University of Wales, College of Cardiff.

Dupré, J. A. (ed.) (1987) *The Latest on the Best: Essays on Evolution and Optimality*, Cambridge, MA: MIT Press.

Emirbayer, M. and Goodwin, J. (1994) 'Network analysis, culture, and the problem of agency', *American Journal of Sociology* 99: 1411–54.

Fisher, R. A. (1930) *The Genetic Theory of Natural Selection*, Oxford: Clarendon Press.

Gábor, I. (1979) 'The second (secondary) economy', *Acta Oeconomica* 22: 91–111.

—— (1986) 'Reformok második gazdaság, államszocializmus. A 80-as évek tapasztalatainak feljödéstani és összehasonlító gazdaságtani tanulságairól' (Reforms, second economy, state socialism: speculation on the evolutionary and comparative economic lessons of the Hungarian eighties), *Valóság* 6: 32–48.

—- (1996) 'Too many, too small: small entrepreneurship in Hungary', in G. Grabher and D. Stark (eds) *Restructuring Networks in Post-Socialism: Legacies, Linkages, and Localities*, Oxford: Oxford University Press.

Gereffi, G. (1994) 'The organization of buyer-driven global commodity chains: how US retailers shape overseas production networks', in G. Gereffi and M. Kornzeniewicz (eds) *Commodity Chains and Global Capitalism*, Westport, CT: Praeger.

Gorzelak, G. (1995) *The Regional Dimension of Transformation in Central Europe*, London: Jessica Kingsley.

Gould, S. J. (1987) 'The panda's thumb of technology', *Natural History* 1: 14–23.

Gould, S. J. and Lewontin, R. C. (1984) 'The spandrels of San Marco and the Panglossian paradigm: a critique of the adaptationist programme', in E. Sober (ed.) *Conceptual Issues in Evolutionary Biology*, Cambridge, MA: MIT Press.

Grabher, G. (1993) 'Rediscovering the social in the economics of interfirm relations', in G. Grabher (ed.) *The Embedded Firm: On the Socioeconomics of Industrial Networks*, London: Routledge.

—— (1994) *Lob der Verschwendung. Redundanz in der Regionalentwicklung* (In praise of waste. Redundancy in regional development), Berlin: Edition Sigma.

—— (1996) 'Adaptation at the cost of adaptability? Restructuring the East German regional economy', in G. Grabher G. and D. Stark (eds) *Restructuring Networks in Post-Socialism: Legacies, Linkages, and Localities*, Oxford: Oxford University Press.

Granovetter, M. (1973) 'The strength of weak ties', *American Journal of Sociology* 78: 1360–80.

Håkansson, H. and Johanson, J. (1993) 'The network as a governance structure: interfirm cooperation beyond markets and hierarchies', in G. Grabher (ed.) *The Embedded Firm: On the Socioeconomics of Industrial Networks*, London: Routledge.

Hannan, M. T. (1986) 'Uncertainty, diversity, and organizational change', in *Behavioral and Social Sciences: Fifty Years of Discovery*, Washington, DC: National Academy Press.

Hannan, M. T. and Freeman, J. H. (1989) *Organizational Ecology*, Cambridge, MA: Harvard University Press.

Hausner, J., Kudlacz, T. and Szlachta, J. (1996) 'Regional and local factors in the restructuring of south-eastern Poland', in G. Grabher and D. Stark (eds) *Restructuring Networks in Post-Socialism: Legacies, Linkages, and Localities*, Oxford: Oxford University Press.

Herrigel, G. (1993) 'Power and the redefinition of industrial districts: the case of Baden-Württemberg', in G. Grabher (ed.) *The Embedded Firm: On the Socioeconomics of Industrial Networks*, London: Routledge.

Hodgson, G. M. (1993) *Economics and Evolution. Bringing Life Back into Economics*, Cambridge: Polity Press.

Ickes, B. W. and Ryterman, R. (1994) 'From enterprise to firm: notes for a theory of the enterprise in transition', in R. Campbell (ed.) *The Postcommunist Economic Transformation*, Boulder, CO: Westview.

Kemény, Sz. (1996) 'Competition, cooperation, and corruption: economic practice in a transforming market in Hungary', paper prepared for the Tenth International Conference of Europeanists, Chicago, March 1996.

Koch, T. and Thomas, M. (1996) 'The social and cultural embeddedness of entrepreneurs in East Germany', in G. Grabher and D. Stark (eds) *Restructuring Networks in Post-Socialism: Legacies, Linkages, and Localities*, Oxford: Oxford University Press.

Kornai, J. (1990) *The Road to a Free Economy*, New York: Norton.

—— (1992) 'The postsocialist transition and the state: reflections in the light of Hungarian fiscal problems', *American Economic Review* 82: 1–21.

Kuczi, T. and Makó, C. (1996) 'Towards industrial districts? Small-firm networking in Hungary', in G. Grabher and D. Stark (eds) *Restructuring Networks in Post-Socialism: Legacies, Linkages, and Localities*, Oxford: Oxford University Press.

Latour, B. (1988) *The Pasteurization of France*, Cambridge, MA: Harvard University Press.

Lewontin, R. C. (1982) *Human Diversity*, New York: Scientific American Books.

Luhmann, N. (1986) 'The autopoiesis of social systems', in F. Geyer and J. v. D. Zouwen (eds) *Sociocybernetic Paradoxes. Observation, Control and Evolution of Self-Steering Systems*, London: Sage.

McDermott, G. (1996) 'Renegotiating the ties that bind: the limits of privatization in the Czech Republic', in G. Grabher and D. Stark (eds) *Restructuring Networks in Post-Socialism: Legacies, Linkages, and Localities*, Oxford: Oxford University Press.

March, J. G. (1991) 'Exploration and exploitation in organizational learning', *Organization Science* 2: 71–87.

Mayr, E. (1984) 'Typological versus population thinking', in E. Sober (ed.) *Conceptual Issues in Evolutionary Biology*, Cambridge, MA: MIT Press.

Miner, A. S., Amburgey, T. L. and Stearns, T. M. (1990) 'Interorganizational linkages and population dynamics: buffering and transformational shields', *Administrative Science Quarterly* 35: 689–713.

Murrell, P. (1991) 'Can neoclassical economics underpin the reform of centrally planned economies?', *Journal of Economic Perspectives* 5: 59–78.

—— (1992) 'Evolution in economics and in the economic reform of the centrally planned economies', in C. Clague and G. R. Rausser (eds) *The Emergence of Market Economies in Eastern Europe*, Oxford: Basil Blackwell.

Petruno, T. (1994) 'Is NYSE being kept too cool under its collar?', *Los Angeles Times*, 26 January.

Powell, W. W. (1990) 'Neither market nor hierarchy: network forms of organization', in B. Staw and L. L. Cummings (eds) *Research in Organizational Behavior*, Greenwich, CT: JAI Press.

Poznanski, K. Z. (1993) 'Restructuring of property rights in Poland: a study in evolutionary politics', *East European Politics and Societies* 7: 395–421.

Pyke, F. and Sengenberger, W. (1992) *Industrial Districts and Local Economic Regeneration*, Geneva: International Institute for Labour Studies.

Róna-Tas, Á. (1994) 'The first shall be last? Entrepreneurship and communist cadres in the transition from socialism', *American Journal of Sociology* 100: 40–69.

Sabel, C. F. (1992) 'Studied trust: building new forms of co-operation in a volatile economy', in F. Pyke and W. Sengenberger (eds) *Industrial Districts and Local Economic Regeneration*, Geneva: International Labour Organization.

Schumpeter, J. A. [1912] (1961) *The Theory of Economic Development*, Cambridge, MA: Harvard University Press.

Scott, A. J. (1996) 'Regional motors of the global economy', *Futures* 28: 391–411.

Sedaitis, J. (1996) 'Network dynamics of new firm formation: developing Russian commodity markets', in G. Grabher and D. Stark (eds) *Restructuring Networks in Post-Socialism: Legacies, Linkages, and Localities*, Oxford: Oxford University Press.

Simmel, G. [1923] (1950) *The Sociology of Georg Simmel*, New York: Free Press.

Smith, A. (1995) 'Regulation theory, strategies of enterprise integration and the political economy of regional economic restructuring in Central and Eastern Europe: the case of Slovakia', *Regional Studies* 29: 761–72.

Smith, J. M. (1984) 'Optimization theory in evolution', in E. Sober (ed.) *Conceptual Issues in Evolutionary Biology*, Cambridge, MA: MIT Press.

Stark, D. (1990) 'Privatization in Hungary: from plan to market or from plan to clan', *East European Politics and Societies* 4: 351–92.

—— (1996) 'Recombinant property in East European capitalism', *American Journal of Sociology* 101: 993–1027.

Stark, D. and Bruszt, L. (1995) 'Network properties of assets and liabilities: inter-enterprise ownership networks in Hungary and the Czech Republic', Working Papers on Transitions from State Socialism, Einaudi Center for International Studies, Cornell University.

Szelenyi, I. and Szelenyi, S. (1995) 'Circulation or reproduction of elites during postcommunist transformation in Eastern Europe: Introduction', *Theory and Society* 24: 697–722.

Thévenot, L. (1985) 'Rules and implements: investment in forms', *Social Science Information* 23: 1–45.

Weick, K. E. (1976) 'Educational organizations as loosely coupled systems', *Administrative Science Quarterly* 21: 1–19.

—— (1977) 'Organization design: organizations as self-designing systems', *Organizational Dynamics* 6: 31–45.

Weizsäcker, C. and Weizsäcker, E. U. (1984) 'Fehlerfreundlichkeit', in K. Kornwachs (ed.) *Offenheit – Zeitlichkeit – Komplexität. Zur Theorie der offenen Systeme*, Frankfurt a.M.: Campus.

Williamson, O. E. (1985) *The Economic Institutions of Capitalism*, New York: Free Press.

—- (1993) 'Transaction cost economics and organization theory', *Industrial and Corporate Change* 2: 107–56.

4

DIFFERENTIAL DEVELOPMENT, INSTITUTIONS, MODES OF REGULATION AND COMPARATIVE TRANSITIONS TO CAPITALISM

Russia, the Commonwealth of Independent States and the former German Democratic Republic

Michael Dunford

Introduction: global and European spatial economic inequality

The year 1989 was an important turning-point in the trajectory of the European space economy in that it saw the abandonment of a communist project, one of whose goals was to close a deep-seated economic development divide between East and West. The aim of this chapter is to outline the evolution of this divide, to explain why communism failed in the end to close it in a durable manner, and to indicate and explain the impact of transition. The chapter examines two very different cases: that of Russia with its long-established communist order in which an attempt was made to create new capitalist institutions virtually overnight, and that of the former German Democratic Republic (GDR) whose transition involved the whole-sale importation of the well-developed institutions of the German social market economy.

To indicate just where the communist world stood in developmental terms, 1991 and 1989 data for Gross Domestic Product per capita measured at Purchasing Power Standards, and expressed as percentages of the figures for the United States (US) for a series of world regions, are plotted in Figure 4.1. The 1989 data indicate the existence of a clear hierarchy, with high scores for Western Europe, the US and a number of the 'new countries' that Europeans colonised in the nineteenth century (the United States, Canada,

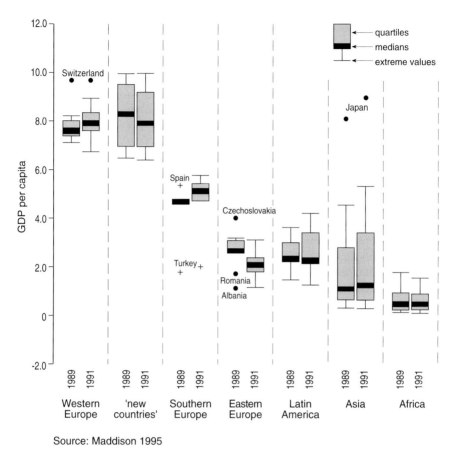

Source: Maddison 1995

Figure 4.1 GDP per capita by major regions, 1989 and 1991

Australia and New Zealand). Beneath these two groups lie the peripheral southern European Union (EU) member states (Greece, Spain and Portugal) and Ireland (the 'Cohesion Countries'). The seven East European countries included were well behind these three groups, but ahead of other world regions.

The 1991 data differ quite significantly, however, from the data for 1989. The relative position of most West European countries improved, as did those of three of the four EU Cohesion Countries. The most striking change, however, was the collapse of the East European economies. In 1989, this group (including the former GDR and Albania) lay ahead of Latin America and most Asian economies. In 1989–91, however, Bulgaria declined from 29 to 19 per cent of US per capita GDP, Czechoslovakia from 40 to 32, Hungary from 31 to 26, Poland from 26 to 22, Romania from 18 to 12, the USSR from 32 to 22, and Yugoslavia from 27 to 18. Transition, therefore, brought dramatic short-term decline, after a long period of slow relative decline under communism.

To help identify the causes of these disparities in economic development, GDP per capita can be divided into two elements: an element that depends on productivity differentials; and an element that depends on differentials in the employment rate (the percentage of the population in employment). More formally:

$$\frac{Gross\ Domestic\ Product}{Resident\ Population} = \frac{Gross\ Domestic\ Product}{Employed\ Population} \times \frac{Employed\ Population}{Resident\ Population}$$

Differences in productivity measure the average wealth created per person employed, which may vary because of productivity differences within a single sector or to differences in sectoral/functional specialisation. Differences in the employment rate, defined as the share of the population in employment, reflect variations in the capacity of an economic system to mobilise its human potential (see Dunford 1996).

Figure 4.2 plots the logs of the productivity and employment rates for most of the economies represented in Figure 4.1. The downward sloping lines represent lines of equal GDP per capita, with the figures increasing from bottom left to top right. Most of the West European economies are clustered in the top right-hand corner of the graph (though there are significant differences between them) due to the coexistence of relatively high productivity and employment rates. As far as the former communist countries are concerned, the employment rate is generally high, reflecting the high rates of participation in these countries. In this respect, these economies are close to the West European model. Their rates of productivity are, however, particularly low. It is these productivity differentials which explain their relatively low levels of GDP per capita.[1]

To understand the recent sharp deterioration in the relative position of the East European countries, it is important to place them in a long-term context, in particular as the distinctive trajectories of the European communist economies are in part a consequence of choices made in the pursuit of faster industrialisation and development. The context of these developments in the twentieth century was the very uneven character of nineteenth- and early twentieth-century industrial development.

As Maddison (1995) has shown, until 1820 most economies were predominantly agricultural, and rates of productivity growth were just sufficient to sustain the standards of living of growing populations. After 1820, however, rates of growth accelerated in the western half of Europe and the countries of the new worlds. Between 1820 and 1992, GDP per capita increased thirteenfold in Western Europe, and seventeenfold in the new countries (see Table 4.1). In the peripheral countries of Europe, development

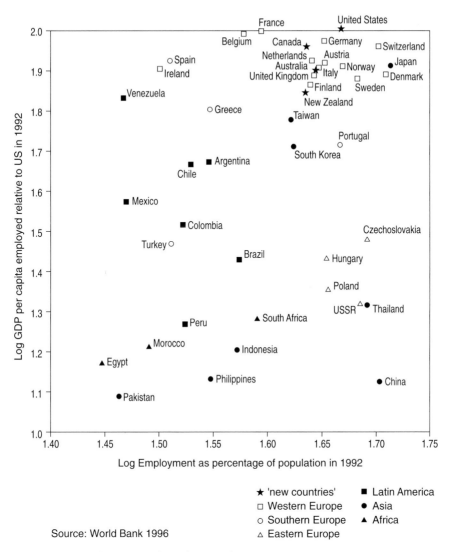

Figure 4.2 Employment and productivity by country, 1992

was slower, though in the period as a whole the per capita GDP of Southern Europe increased tenfold, leaving it in third position in 1992. Eastern Europe, which was fourth in 1820, retained its position until the economic collapse that accompanied the fall of communism, which saw it drop into fifth position. In Latin America, per capita GDP increased sevenfold. Asia, which was sixth in 1820, was sixth in 1992, while Africa (which was the weakest area economically in 1820) was relatively weaker in 1992.

As Figure 4.3 shows, however, growth was not regular. In the period up

Table 4.1 Economic performance of the major regions, 1820–1992

	Population (millions)			GDP per head (1990 $)			GDP (billion 1990 $)		
	1820	1992	Multiplier 1820–1992	1820	1992	Multiplier 1820–1992	1820	1992	Multiplier 1820–1992
Western Europe	103	303	3	1,292	17,387	13	133	5,255	40
'New countries'	11	305	27	1,205	20,850	17	14	6,359	464
Southern Europe	34	123	4	804	8,287	10	27	1,016	38
Eastern Europe	90	431	5	772	4,665	6	69	2,011	29
Latin America	20	462	23	679	4,820	7	14	2,225	161
Asia and Oceania	736	3,163	4	550	3,252	6	405	10,287	25
Africa	73	656	9	450	1,284	3	33	842	26
World average				651	5,145	8			
World total	1,068	5,441	5				695	27,995	40

Source: Maddison 1995: 18.

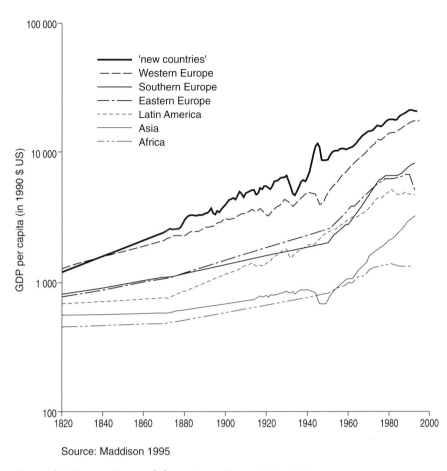

Source: Maddison 1995

Figure 4.3 Economic growth by major region, 1820–1992

to 1950, Western Europe lost ground relative to the new countries, but closed the gap in the *trente glorieuses* (the thirty-year 'Fordist' golden age that followed the Second World War) and in the subsequent phase of crisis (marked by a halving of rates of productivity and output growth). Southern Europe dropped behind Eastern Europe and Latin America until 1960, but then made up for its relatively slower growth. Asia, which had dropped behind Africa in the Second World War, subsequently overtook it and left it well behind as its growth accelerated after 1950. After the Second World War, therefore, growth was most marked in two areas that flanked the communist world (Western Europe and East Asia), that were of strategic significance and whose modernisation and development received large-scale support and assistance from the world's dominant economic power, the US.

What Figure 4.3 also shows is that disparities in GDP per head between the richest and poorest groups of countries have increased constantly. Within this pattern, however, there were several cases of convergence. After 1945, Western Europe and Southern Europe closed the gap on the United States, as did Japan and later Taiwan, South Korea, Thailand and other Asian countries. After 1929, in the case of the Soviet Union, and after the Second World War, in the case of the countries that fell under Soviet influence, convergence occurred. In the early 1970s, however, slower growth set in. At first it was offset by international borrowing, though the indebtedness that followed added a further dimension to the economic crisis of the European communist world in the 1980s.

In the next section I shall argue that the trajectories and the crises of the European communist economies and societies reflected the specific features of their political economies, and that the structures that emerged were profoundly different from those of other less developed European peripheries. A consequence is that the scenarios for their future development and the likely impact of market-driven reform programmes are also different from those of other groups of countries (whether in Asia or Southern Europe), not simply because of the different context in which transition is taking place, but also because of the existence of profound differences in the structures and modes of economic regulation and the evolutionary nature of economic and institutional change (see Grabher and Stark in this volume; Smith and Swain in this volume). In particular, I argue that the programme of transition is certainly, in the case of the former Soviet Union, a complete failure for three main reasons: first, because the programme involves a false understanding of the nature of communist economies; second, because it failed to anticipate the impact of structural adjustment on an economy that was not already a market economy; and, third, because it was associated with a failure to appreciate the nature of the institutions on which capitalism depends and of the mechanisms (of primitive accumulation) involved in the creation of capitalism. If the Russian case exhibits consequences of a misapprehension of the institutions on which capitalism depends, the former GDR is a case which shows that the rupture created by the overnight introduction of a fully developed set of social market institutions also has perverse and unexpected effects. These effects arise because of a misapprehension of the way effective institutions emerge and of their mode of integration with learned rules of behaviour in a communist order. Together, these two cases make it possible to highlight the difficulties, and to question the underpinnings, of neo-liberal strategies of transition.

The regulation and crisis of 'communist' pathways to development: the Soviet model

The trajectories of the communist economies and societies of Eastern and Central Europe are in many ways quite distinct from those of other less developed countries, in part because a critical factor that explained the transition to communism was a desire to de-link from the international capitalist system in order to permit industrialisation and development. As the Soviet Union was an agrarian economy with abundant mineral resources, the principle of comparative advantage implied that it should specialise in agriculture and natural resource exploitation, rather than industrial activities. To specialise in these activities was seen, however, as a recipe for under-development: modernisation and development depended, it was thought, on industrialisation, on the creation of the resource endowments to enable industrial expansion, and on the exploitation of the increasing returns associated with industrial activities. From the late 1920s, therefore, the Soviet Union chose a distinctive development path that involved the creation of a centralised command economy, a nationalisation of capital, the collec-tivisation of agriculture, and the establishment of a political order in which a Communist Party ruled, according to its official ideology, in the name and on behalf of the working class.

The fundamental aim of the command system was to accelerate indus-trialisation through a rapid expansion of the capital stock and a high degree of mobilisation of labour and raw materials, with little emphasis on the efficient use of resources. An implication of this objective was a high rate of investment (the devotion of a large share of resources to investment). Another implication was an increase in the share of the population in employment and a major transfer of labour from agriculture to industry. What resulted in the Soviet Union in the 1930s was dramatic industrial growth, while after the economic and demographic devastation of the Soviet Union in the Second World War (with the loss of some 28 million people), recourse was made to the Stalinist economic methods to reconstruct its economy and that of the economies that fell into the Soviet sphere of influ-ence in the post-war period. The consequence was rapid, but unbalanced, reconstruction and growth, with high rates of accumulation placing strong pressures on living standards. Only after the death of Stalin were the first of numerous attempts at reform made. Almost all of the reform efforts fell short of expectations, and growth, according to official Soviet figures, progressively slowed down.

The actual record of the Soviet and other communist economies is a subject of some dispute (Table 4.2; see also Table 4.4). Official indicators of economic performance suggest that there was a slow-down in growth in the 1970s and 1980s. Estimates corrected for repressed inflation suggest that the deterioration in economic performance and efficiency was much more

Table 4.2 Growth of the Soviet economy

	Average annual % rate of growth 1971–80	Average annual % rate of growth 1981–5
National income (official Soviet estimates)[a]	5.30	3.20
Gross National Product (CIA estimates)[b]	3.70	2.00
National Income (Khanin's estimates)[c]	2.00	0.59
Gross Social Product (Sapir's estimates)	3.50	0.65

Sources: Sapir 1994: 95; Nove 1989: 390–8.
Notes
a Official Soviet estimates were based on the turnover (rather than value added) of all enterprises and made no correction for illegal increases in industrial prices (which increase measured turnover) or reductions in product quality.
b CIA (Central Intelligence Agency) estimates put Soviet growth at an average of some 5 per cent per year in 1951–71.
c Khanin's recomputation made allowance for price increases not adequately reflected in official price indices which accordingly led to an overestimation of growth rates. Khanin's computations have been criticised and are probably too low (see Nove 1989: 393–4).

pronounced, with a slow-down in productivity growth (output per person employed) and in the efficiency of capital investment (output per unit of investment) due to delays in the completion of, or the failure to complete, capital projects. To get the same additional output, there was a need to invest more, which indicated the existence of problems in the use of resources and, in so far as attempts were still made to sustain growth, in the availability of resources. This crisis did not just affect activities at the margin (food processing and light industries) but also investments to develop new or maintain existing productive (transport and communications) and social infrastructures (health and housing), and investments in the core mechanical industries.

A number of factors have been advanced to explain the stagnation of the Soviet Union and other communist economies and the crisis to which it led. Traditionally, the stagnation of productivity and output growth, and the fall in the efficiency of investment, were explained in terms of the perceived structural inability of the communist command system to put in place a model of growth that was centred on an intensive use of resources (where productivity growth occurs as a result of innovation and learning within a particular economic sector), rather than on an extensive use of resources (where productivity increases stem from a transfer of resources from low productivity activities such as agriculture to higher productivity activities) in which the system had proved very effective (see Smith and Swain in this volume). Another explanation attributes the crisis to several sets of pressures imposed by the West, of which the most important were the imposition of an arms race that the less-developed communist economies could not sustain, restrictions on the import of advanced technologies, and the depiction of elite consumer life styles in the increasingly global media (with the message that embracing capitalism would put these life styles within the reach of all). While these factors played a role, there were also endogenous mechanisms of development and structural disequilibria which played an important role in the crisis and in the subsequent trajectories of transition.

As Sapir (1994) has argued, the command economies were not so much planned economies as economies characterised by shortages and by certain priorities. In communist economies, shortages were a system characteristic. First, there was excess demand for all kinds of goods in both consumer and producer markets. In the case of consumer goods, the consequence was the existence of queues, waiting lists, forced saving and a black economy which had negative effects on economic incentives. In the case of producer goods, the consequence was endemic uncertainty about the delivery of supplies in the right quantity, of the right quality and at the right time. Second, there was excess demand for labour: in communist systems the demand for labour was constrained not by the demand for the output workers produced and the real wage, as in market economies, but by the amount of fixed plant and equipment.

The existence of shortages was not simply a consequence of repressed inflation (which prevented prices from rising to market-clearing levels) but also of the rules, conventions, modes of decision-making and types of conduct to which the system gave rise. In particular, the existence of ambitious investment plans designed to maximise volumes of investment and of rapid rates of capital stock growth led to surges in demand for labour and materials that exceeded supplies at full employment. As a result, growth slowed down, projects were not completed and new ones were cancelled. The consequence was sharp investment cycles, whose roots lay in the fact that enterprises in construction and investment goods sectors were essentially over-financed/capable of effectively creating money through barter transactions causing excess demand for inputs and labour. As inflation was repressed, and as industrial materials were rationed administratively through informal wholesale markets, speculative hoarding and excess capacities were encouraged, further exacerbating shortages (Harrison 1984).

There was a second closely related dimension to this problem. As everything that could be produced was sold (as output was validated *ex ante* through the planning system and not *ex post* through sale on a market) the vulnerability of communist enterprises resided not in an inability to sell what they produced but in an inability to purchase the inputs they required. As a result there was a tendency for excess demand to appear upstream and for competitive behaviour to develop in relation to the acquisition of supplies.

The struggle for supplies revolved around the activities of the managers of enterprises and local administrators who sought to use their power to secure additional resources to make the attainment of plan targets easier and to increase their bargaining power in subsequent wholesale trading. The consequences were stock-building, hoarding and an exacerbation of supply shortages. These types of behaviour were another cause of the cyclical movements which characterised communist economies. Cycles were in other words endogenous movements induced by the behaviour of micro- and meso-economic actors and the functioning of the investment cycle (Sapir 1994, 1995: 437; Sapir, Badower and Crespeau 1994: 120–2).

To deal with this situation, Soviet-style economies developed systems of priorities which guaranteed a greater degree of security to certain enterprises. In the Soviet Union, the priority sectors were made up of 'three concentric circles' of enterprises that formed the military-industrial complex: an inner circle of industries controlled by the Ministry of Defence; a circle of reputedly civil industries closely related to defence efforts; and a third circle made up of their supply industries (Sapir 1994). These industries grew faster than the rest. In 1988 they employed some 20 million people, of whom some 4.7 million were involved directly in arms production. Beyond this military-industrial complex, upstream industries making basic goods were in a position of influence due to the dependence of the rest of the industrial system on their activities. Consumer-good industries in particular, and downstream industries in general, were weakly developed, with the result that there were few integrated *filières* of production outside the defence sectors.

The existence of shortages and of these priority principles had a number of other important economic and geographical consequences. First, supply uncertainty made vertical integration within immense factories more efficient than subcontract arrangements (see Pickles in this volume). Communist enterprises were consequently much larger than their western counterparts. (A second factor that determined the size of Soviet conglomerates was the view that competition and duplication were wasteful.) At the same time, a large share of their employees were involved in activities unrelated to the principal production of the plant, and a large share of the consumption of the workforce depended on channels of distribution within the enterprise. The Soviet Union, therefore, lacked a tissue of small and medium-sized firms and the flexibility it offers (see Grabher and Stark in this volume). For related reasons, services were underdeveloped, with most of the services that did exist being provided by enterprises for their worker collectives: funds equal to up to 8 per cent of the wage bill of enterprises were used to provide housing, child-care facilities, rest homes, clinics and other social services (Peaucelle 1995: 445).

Second, the system was characterised by permanent full employment. In communist countries, work was a constitutionally guaranteed right and obligation of every citizen, and shortages of labour were common. One consequence was the implementation of regional investment programmes designed to mop up reserves of labour throughout countries such as the Soviet Union, especially in the period up to 1977 when there were administrative controls on rural–urban migration. Another was the use, at first, of forced labour and, later, of high wages and appeals to politically conscious young people to mobilise the workforce required to exploit the resource-rich north. Another consequence was a tendency for enterprises to hoard labour, especially with the appearance of absolute labour shortages in the 1960s. Indeed, there was also a tendency for enterprises to be disinclined to innovate where innovation would alter their skill requirements and make them more vulnerable to labour supply shortages.

In the third place, the absence both of competitive labour market discipline and of effective alternative incentive systems had a negative impact on the incentive to work. As Harrison (1984: 125) pointed out:

> through [the] receipt of wage incomes provided for by the excessive financing of industry, households too . . . [were] overfinanced in relation to the consumer goods and services available. Their real entitlement . . . [exceeded] supply. The result . . . [was] the unintended accumulation of household cash balances, [a failure to meet the consumption entitlement its citizens acquired through labour,] loss of motivation to work for higher pay and encouragement of illegal exchange on retail markets.

Another element of this equation was the provision of a wide range of collective services for which users did not pay, or did not pay the full cost (health, education, public transport and culture). As consumers, individuals did not pay for services that were taken for granted and resented the lack of goods against which earned income could be exchanged. A result of the consequent lack of effort was the effective reappearance of disguised unemployment and a loss of dynamic efficiency, sometimes rectified through periodic political campaigns (see Harrison 1984: 126).[2]

The mode of functioning of the system accordingly created a number of deformations in the economic structures of communist societies: the priority given to defence-related industries; the dominance of large and relatively self-contained enterprises; the lack of small and medium-sized enterprises; and an underdevelopment of services. These deformations, which Sapir (1994) considers more important than the level of development, played a major role in later processes of transition, as did the large regional disparities (see Table 4.3) whose roots were, however, deeper.

The Russian Empire was characterised by the existence of several poles of development in the west. During the Second World War, and in the 1960s and 1970s, Soviet leaders made considerable efforts to develop Siberia.

Table 4.3 Regional disparities in the former Soviet Union

	Population (thousands)	GDP (millions of 1990 US dollars)	GDP per head (USSR = 100) (1990 US dollars)
Armenia	3,400	16,363	83
Azerbaijan	7,250	30,026	72
Belarus	10,300	65,619	110
Estonia	1,600	13,832	149
Georgia	5,500	29,520	93
Kazakhstan	16,850	98,513	101
Kyrgyzstan	4,500	13,158	50
Latvia	2,700	21,423	137
Lithuania	3,750	27,327	126
Moldova	4,400	20,242	79
Russia	148,800	990,360	115
Tajikistan	5,400	13,158	42
Turkmenistan	3,750	11,471	53
Ukraine	52,000	257,079	85
Uzbekistan	21,000	78,777	65
USSR	291,200	1,686,868	100

Source: Maddison 1995: 153.

Nevertheless, in the 1980s, there was a wide divide between a developed north-west and an underdeveloped south-east, represented most strikingly by the divide between the Baltic republics and Central Asia. In 1991, nearly 69 per cent of the population and almost 74 per cent of output were accounted for by Russia and the Ukraine, while GDP per head varied from almost 150 per cent of the Soviet average in Estonia to 42 per cent in Tajikistan (Maddison 1995: 153). At the point when it collapsed, the Soviet Union was characterised by large disparities between old industrialised areas in the north-west and Central Asia, to which one could add a part of Transcaucasia and, in particular, Azerbaijan.[3]

Sapir (1995) has argued that these difficulties (associated with shortages, with the system of priorities, with the asymmetries they created and with the modes of regulation of economic and social life) were intensified in the 1960s and early 1970s with the result that economic fluctuations started to affect the core mechanical industries. By the late 1970s, the growing disintegration and lack of articulation of institutions, rules, conventions and modes of microeconomic behaviour (which were due to the relative success of the system in converting a rural country into an industrial power) led to the appearance of a crisis of regulation (see Smith and Swain in this volume).

Perestroika was a radical attempt to reform and democratise the communist system and to accelerate economic growth in the face of this crisis. What resulted in 1986–9 was, on the one hand, an attempt to extend the traditional mode of development of the system through increased investment and growth and, on the other, an attempt to restructure it through political reform to weaken defenders of the old order, and economic reform to grant increased autonomy to enterprises, to give greater financial autonomy to republics, to develop co-operatives to permit the development of a private sector, and to reform the banking system, allowing the appearance of some 2,000 independent banks. The reforms, whether in agriculture, industry or in relation to the development of co-operatives, did not have the anticipated economic effects but did disorganise the former system. The most important consequence was an even deeper economic crisis reflected, first, in loss of financial control. In particular, public expenditure increased faster than revenue, increasing the public sector budget deficit, which was financed through the uncontrollable creation of money by independent banks. As a result, inflation soared on unregulated markets into which goods were channelled due to the immense price differentials that opened up between official and unregulated prices. In 1990, a recession saw GDP fall by 3.5 per cent. With greater shortages, speculation was rife and, increasingly, resort was made to local solutions as some republics ceased paying taxes and sought to secede, further exacerbating the budgetary crisis and fragmenting the national space economy. What followed was the break-up of the *perestroika* coalition and of the Soviet Union itself, and a switch to a neo-liberal programme of transition to capitalism.

Transitions to capitalism: from the Soviet Union to Russia and the Commonwealth of Independent States (CIS)

The years 1992–3 saw the introduction in Russia of economic shock therapy, modelled on the programme implemented in Poland in 1990, though in very different economic conditions. Amongst the immediate reasons for this choice[4] was the view that inflation and shortages were due to excess demand and price rigidity in the state trading system. The conclusion was that the restoration of macroeconomic equilibrium simply required the liberalisation of prices and reductions in the supply of money (which involved attempts to restrict the supply of credit and reduce the public sector deficit by reducing public expenditure and privatising public assets). At the same time, it was decided to make the rouble convertible and open up the country to imports.

This programme of economic liberalisation and stabilisation also amounted, however, to a more fundamental change in the sense that the aim was to replace the central planning system by the market, and a planned/administered allocation of resources by market competition in a context of strict controls over the supply of money. Alongside this goal of creating a market economy, a second fundamental aim was pursued: the replacement of collective property by private property through a programme of privatisation.

According to its advocates, privatisation would create clear property rights and lead to the emergence of a dynamic, profit-seeking entrepreneurial class which would increase the output of the goods consumers wanted and would innovate to reduce their costs to below those of their competitors. At the same time, the new market system was expected to discipline workers and create strong incentives for them to work harder for higher wages and move from inefficient public to efficient private enterprises. While a short-term deterioration in standards of living and the possibility of popular resistance were envisaged, the architects were confident that it would not take long for living standards to improve, that the former communist economies would converge on those of their western neighbours and that support for capitalist institutions would increase.

The intention and expectation were, in short, the replacement of a communist (central planning plus collective property) by a capitalist economic order (markets plus private property) which was considered intrinsically more efficient and more effective at raising standards of living. In addition, there was to be the replacement of authoritarian by representative democratic politics, with the new political institutions increasing their legitimacy and popularity as a result of the improvements in living standards that, it was suggested, the economic reforms would deliver.

The case of the former Soviet Union is a clear example of the naïvety of this vision of the nature of capitalist societies and of a transition from communist to capitalist societies. First, this vision is somewhat disingenuous, as it fails to acknowledge the strategic interests of the West in opening up markets and creating investment opportunities for western capital in former communist countries. The existence of this oversight was particularly remarkable in view of the fact that, as Gowan (1995: 3–4) has indicated, the debate was fundamentally concerned with the ways in which the West should reshape Eastern and Central Europe and with the creation of 'an international environment in which the domestic aspect of . . . [shock therapy] would become the only rational course for any government to pursue'. Second, the programme of transition to which it gave rise has so far proved, in the case of the Soviet Union, a complete failure, in part because it involves a false understanding of the nature of communist economies, and in part because it entails a failure to appreciate the nature of the institutions on which capitalism depends and of the mechanisms (of primitive accumulation) involved in the creation of capitalism.

The sequence of events that followed the implementation of shock therapy, and which can be summarised under six headings, differed profoundly from the stated expectations of its advocates.

1 The reforms first led to rapid changes in relative prices to which most

enterprises could not adapt (in part as communist enterprises relied on other channels of information and communication and, in particular, on administrative, political and interpersonal networks (see Smith and Swain in this volume; Begg and Pickles in this volume)), an end of subsidies and a disorganisation of inter-firm relations due to the suspension of state orders. In the absence of a developed credit system capable of tiding enterprises over a period of collapsing domestic markets, the result was widespread insolvency. A state of insolvency was, as Gowan (1996: 133–4) suggests, a stimulus to the sale/privatisation of some enterprises into the hands of western capital. What also resulted, however, was a dramatic increase in payment in kind and in particular of unpaid inter-firm debt (Peaucelle 1995: 449). This expansion of inter-enterprise arrears (which reached some 3,000 billion roubles in July 1992) created a serious risk of economic collapse and forced the government to take special action to net out arrears between enterprises and reverse some of its earlier measures. A further consequence was that inflation remained high, while shortages, especially in the wholesale sector, did not disappear. In effect, in Russia the inflationary dynamic is quite different in that increased prices lead to increased (private) money creation (in the shape of inter-firm debt) rather than in the opposite direction (Sapir, Badower and Crespeau 1994: 140–1). The World Bank (1996: 40) argues in relation to this period that there was insufficient financial discipline. As Burawoy (1996: 1110–11) shows, however, in a third phase of transition, when hard budget constraints were enforced, there was a de-monetisation of economic transactions and an increase in the role of Mafia-type organisations.

2 A second consequence of shock therapy (and also of the disintegration of COMECON (Council for Mutual Economic Assistance), on the one hand, and the dismantling of trade protection regimes which exposed domestic producers to subsidised western export drives, on the other) was a dramatic drop in industrial output and GDP. In Russia, some of the largest drops in industrial output were in the food sector and other consumer-good industries, due to the fall in real consumer incomes and demand, and increases in the cost and ruptures in the supply of essential inputs (as a result of the restructuring of supply chains and distance, transport and payment difficulties). Services, which were expected to offer a great deal of scope for the creation of new enterprises (see Bradshaw, Stenning and Sutherland in this volume), were also affected by the fall in income. Construction suffered from the reduction in orders from enterprises and municipalities in spite of the fact that additional housing stock is a precondition for greater mobility. There were also sharp regional differences in the impact of the reforms depending to a significant extent on the sectoral specialisation of regional economies: generally speaking, there was an improvement in the relative positions of areas in the north (because of their industrial specialisation) and

of areas in the east near the vast Chinese and East Asian markets (especially as the eastern frontiers were closed until the mid-1980s); conversely, areas to the south-west and areas dependent on consumer-good industries lost ground (see Bradshaw, Stenning and Sutherland in this volume).

As a measure of the magnitude of lost output, Table 4.4 records trends in GDP in the CIS and, for comparative purposes, in other transition economies. In 1995, Russian GDP stood at just 60 per cent of the 1989 level, and at 19–82 per cent in the CIS. These figures, however, assume unrealistically that growth would have ended in 1989. Column 4 of the table records 1995 GDP as a percentage of what it would have been if pre-transition growth rates had continued. On this indicator, the range for the CIS is 18–61 per cent. Of the European transition economies, Slovenia's position was strongest at 94 per cent. All these figures give, however, no real impression of the loss in output that stemmed from transition and which will have to be recouped. Columns 5 and 6 record the net present value of forgone GDP on the assumptions of no growth (column 5) and a continuation of 1981–9 growth rates (column 6) expressed as a percentage of 1989 GDP. The losses are extraordinary. Russia, for example, has lost between 92 and 136 per cent of its 1989 GDP. Armenia, Azerbaijan and Georgia have lost 216–47 per cent. The countries that suffered least on this indicator were Slovenia and Hungary. Most studies of the impact of transition content themselves with simple comparisons of current with 1989 growth rates. Comparisons of the present value of GDP under transition and non-transition strategies indicate, however, that net progress will occur only if exceptionally high rates of growth are sustained over very long periods of time. With, for example, a discount rate of 10 per cent and on the assumptions that, without transition, growth would have continued at a mere 0.5 per cent per year and that after 1995 post-transition growth will continue at 5 per cent per year, there will be no net gain until 2005 in Poland and Slovenia, 2009 in Hungary and 2013 in the Czech Republic. Almost all the CIS countries will see no net gains in cumulative output in the first quarter of the next century. The crash of the early 1990s represents in quantitative terms an extraordinary reversal and much more than a 'transitional recession'.[5]

3 The increase in prices, incomplete indexation of pensions and other welfare payments and decline in output led to spectacular falls in real income for large sections of the population, quickly wiped out the savings of most people, prompted the least well-off to sell off their possessions and resulted in dramatic increases in inequality. Average real incomes declined, with almost 73 per cent of the Russian population receiving an income of less than one-half of the average. Furthermore, 1.1 per cent of the population received 27.7 per cent of all income, and 35 per cent received less than the subsistence minimum (Sapir et al. 1994: 154), while the new rich are

thought to export legally $20 billion each year. The social impact of this restructuring went much further. The impoverishment of large sections of the population and increases in inequality were reflected in the reappearance of diseases once regarded as conquered (such as diphtheria, tuberculosis, polio and, in some areas, cholera). With sharp increases in rates of mortality, a decline in birth-rates and net emigration, the population declined by 2 million. Male life expectancy fell from 64 to 58 years in 1990–4 (World Bank 1996: 128). There was a remarkable decline in the status of professional, scientific and technical occupations. Access to wealth depended rather on contacts with western capital, connections with corrupt officials, speculation or involvement in criminal activities involving extortion, drug trafficking, prostitution, etc. There were also great regional differences in standards of living arising from two factors: first, employees in extractive industries, especially in oil and gas, where nominal wages were above average, reinforced their advantage through the strategic importance of these industries and the fact that official and illegal exports provided enterprises with flows of income to pay their employees (see Pickles in this volume); second, markets remained segmented, and so prices were not equalised across different cities and regions. As the results of elections and polls show, the expectation that transition would increase affluence and support for reform, or at the very least would create a sufficiently large and stable minority capable of politically guaranteeing further capitalist development, also proved ill-founded: instead there was a clear shift of opinion against privatisation and market reforms and against the neo-liberal political forces that had pursued them.

4 The measures designed to privatise the Russian economy had effects that their liberal advocates did not officially anticipate. These measures were of two kinds. First, in the case of medium and large-scale industrial enterprises, voucher privatisation was adopted. The first step was the distribution at the end of 1992 of vouchers worth 10,000 roubles to Russian citizens, and the second was the sale of about half of all industrial enterprises. The result was the sale of industries either to their employees and managers (see Peaucelle 1995: 445) or to investment funds created to manage the vouchers and to assume control of privatised firms. At the same time, industrial enterprises that remained in the hands of the state were made into shareholdings to open the way for later partial privatisation. The change in ownership, however, did not lead to the emergence of a group of Schumpeterian entrepreneurs. Described in a report by the head of the Privatisation Commission, Polevanov, as 'pauper-entrepreneurs' who 'cannot survive without state protection' (cited in Burbach, Núñez and Kagarlitsky 1997: 120), the new owners did not have sufficient capital to modernise the enterprises they acquired, and were disinclined to make long-term investments because of the weakness of demand and the disorganisation of economic life. Instead,

Table 4.4 Transition and GDP growth[a]

	Average annual growth 1971–80	Average annual growth 1981–9	1995 GDP as % of 1989	1995 GDP as % of 1995 non-transition GDP[b]	Net present value of forgone GDP with no growth as % of 1989 GDP[c]	Net present value of forgone GDP at 1981–9 growth rates as % of 1989 GDP[d]
Armenia	14.5	3.5	37.9	30.8	−184.0	−235.9
Azerbaijan	21.5	2.9	28.3	23.9	−173.1	−215.7
Belarus	6.6	5.0	54.1	40.4	−81.2	−157.1
Estonia	5.1	0.2	58.4	57.7	−144.6	−147.4
Georgia	6.8	1.2	19.0	17.7	−229.2	−246.4
Kazakstan	4.4	2.0	44.6	39.6	−117.1	−146.0
Kyrgyz Republic	4.4	4.0	47.3	37.4	−91.8	−151.6
Latvia	4.7	3.7	50.7	40.8	−132.2	−187.3
Lithuania	4.6	1.8	44.9	40.3	−158.2	−184.2
Moldova			43.6	43.6	−146.0	
Russia	6.5	3.0	60.0	50.2	−91.9	−136.0
Tajikistan	4.9	3.3	33.8	27.8	−152.5	−201.3
Turkmenistan	4.0	4.0	77.0	60.9	−44.3	−104.1
Ukraine			45.8	45.8	−117.3	
Uzbekistan	6.2	3.4	82.4	67.4	−32.3	−82.6
Albania		1.7	74.3	67.1	−125.8	−150.3
Bulgaria		4.9	74.4	55.9	−93.6	−167.9
Croatia			74.4	74.4	−82.6	
Czech Republic		1.8	85.1	76.4	−63.8	−89.8
Hungary	4.6	1.8	88.0	79.0	−47.1	−73.1

Table 4.4 continued

Macedonia			62.9	62.9	-87.0	-88.9
Poland	7.6	2.6	98.8	84.7	-50.9	-103.7
Romania		1.0	78.7	74.1	-89.4	-111.6
Slovak Republic		2.7	84.0	71.6	-72.1	
Slovenia			94.3	94.3	-37.4	
China	5.5	11.1	178.1	94.7	147.0	-38.8
Mongolia		5.7	88.4	63.4	-55.6	-143.1
Vietnam		4.4	154.6	119.4	105.2	39.0

Source: Elaborated from data in World Bank 1996: 173.

Notes

a The underlying data for 1989–94 are World Bank data. The data for 1994 are European Bank for Reconstruction and Development data. All the data are reported in World Bank (1996).

b Non-transition GDP in 1995 (computed on the assumption that 1981–9 rates of growth would have continued) as a percentage of actual 1995 GDP.

c Net present value of total forgone GDP over 1990–5 discounted at 10 per cent per year expressed as a percentage of 1989 GDP.

d Net present value of total forgone GDP over 1990–5 (computed on the assumption that 1981–9 rates of growth would have continued) discounted at 10 per cent per year expressed as a percentage of 1989 GDP.

the new owners resorted to a range of other strategies: redundancies were threatened to secure subsidies or preferential loans; attention was directed to short-term trading, arbitrage and speculation or to investments in the financial sector (cf. Smith 1994); assets were sold off, sometimes in export drives that risked destabilising international markets or that risked a dangerous proliferation of nuclear and weapons technologies, and often at the expense of long-term interests.

Second, in the case of small-scale catering, distribution, retailing and housing, a 1992 programme was introduced to privatise 100,000 small firms and 15,000 medium–large ones. These enterprises employed 30 million of the total workforce of some 76 million. In most cases, staff and managers retained control, though the former *nomenklatura* and criminal organisations also played an active role in privatisation.

Whether large industrial or small enterprises were privatised, the process was associated with systematic and widespread corruption and with a dramatic increase in the role of criminal organisations in economic life (creating what many are calling 'Mafia capitalism'). The structures of power and knowledge were not overturned. Instead, the new owners were closely linked to old party and state *nomenklatura* which gave political protection, information and economic privileges. As Stark (1996) has indicated in the case of Hungary, the reforms designed to privatise the Russian economy did not create new entrepreneurs but redistributed property amongst an elite, creating what he has called 'recombinant property' (see Smith 1998, for a parallel discussion of the Slovak experience). At the same time, the counter-vailing power of the party-state system, which had acted as a check on the monopolistic power of large conglomerates, withered away, reinforcing their monopoly power (Burawoy 1996).

5 As well as price liberalisation and privatisation, the strategy mapped out by Russia's ultra-liberal leaders and their western advisers involved the minimisation of government intervention, including the destruction of central planning and state welfare institutions and major cuts in public expenditure. The consequences were several-fold. First, there was a decline in the legitimacy and effectiveness of public action, which actually undermined the project of developing a market economy and capitalist state, and created a space for the rise of 'Mafia' organisations and agencies for private protection, extortion and the enforcement of contracts (see Burawoy 1996: 1110–11 and Smith 1997). Second, disabling the state to accelerate the creation of markets was one of the factors that explains the decline in output and an economic disaster that Burawoy (1996: 1105) calls 'economic involution'.[6] Third, the fiscal crisis of the state was exacerbated as the recession reduced fiscal revenues more than cuts in government programmes reduced expenditures. Finally, there was an erosion of fiscal transfers as administrative units suspended or sought otherwise to minimise the transfer

of taxes to the centre, and the centre curtailed its expenditures, undermining the financial solidarity necessary to limit regional differentiation and preserve national unity.[7]

6 Transition also reinforced a number of processes of economic and political differentiation and fragmentation: on the one hand, the spatial impact of economic restructuring was uneven; on the other, there was a reinforcement of local interests and demands for greater autonomy. In Siberia, for example, Sapir, Badower and Crespeau (1994) identify the emergence of a 'golden triangle of local interests' made up of powerful, large enterprises in strategic resource-related sectors, democratically elected municipalities or local authorities which are more independent of the centre than in the past, and local military commanders who must hire out their personnel and equipment to acquire food and fuel in a situation of reduced public expenditure. In these areas, this coalition of local elites and the local population has a clear interest in seeking ways of retaining the rents associated with the exploitation of local resources. In the eyes of Burbach, Núñez and Kagarlitsky (1997: 121–2), mechanisms of this kind reinforce what they call 'Kuwaitisation': a process in which trans-national capital and local elites establish a series of 'strong points' in wider territories that are doomed to serve as peripheries to the developed West (see also Sidaway and Power in this volume). These privileged poles have some prospect of reaching western standards of living, but to realise this goal must separate themselves from their less privileged neighbours and thwart all efforts to redistribute their wealth to less developed areas.[8]

Official expectations about the outcome of the neo-liberal sequence of liberalisation, opening up the economy to international competition, monetary stabilisation, privatisation and 'structural adjustment' reflect, first, a simplistic vision of the Soviet system, second, a naïve view of the necessary and sufficient conditions for the creation of a market society, and third, a failure to understand the impact of structural adjustment on an economy that was not already a market economy. As Sapir (1995: 439, my translation) has argued,

> The emphasis on relative prices quite simply ignores the existence of other channels of communication of information and especially the implicit nature of much information . . . [and] resulted in an incapacity to understand the effects of rapid liberalisation policies accompanied by an almost instantaneous opening up of their economies. The depressive effects of such measures were underestimated, and the emergence of phenomena of perverse restructuring, that is of specialisation on raw materials and semi-finished goods, wrong-footed the forecasts and the forecasters.

In the same way, the concentration of attention on state property as the main characteristic of Soviet-type economies led to a naïve approach to the process of privatisation. Emphasis was placed on the most rapid possible modes of privatisation in the hope that privatisation would lead to microeconomic restructuring. Reality did not hesitate to prove just how wrong that view was. The system of voucher privatisation, extensively used, did not lead to restructuring. It simply perpetuated a situation in which property rights, in the private as in the public sector, are soft.

First, there was a comprehensive liberalisation of commercial mechanisms in a situation in which there were no institutions capable of ensuring market discipline. One of the first problems that appeared was the absence of discipline over payment, because of the lack of an effective monetary system of inter-firm payments.

Second, in a system in which there were significant structural rigidities, where shortages, market segmentation and monopolistic structures endured, and where economic power resided in a capacity to control the use of resources, the dramatic changes in prices and the depression of demand led to intense conflicts over the distribution of income and strong cost-inflation on the one hand, and increased economic noise, confused relationships between efficiency, solvency and liquidity and increased transaction costs on the other. As a result, new long-term irreversible investment was discouraged and arbitrage, speculation and perverse restructuring were stimulated.

Finally, the ineffectiveness of monetary instruments, and the limited degree of flexibility of the productive sector, confine governments using International Monetary Fund approaches to structural adjustment to attempts to control public expenditure. As Sapir (1995: 439–40) indicates, the consequence is an obsessive and destructive preoccupation with the reduction of state spending. At the end of the day, a collapse in public investment will constrain private investment, a lack of public action will undermine notions of legality and give rise to rampant corruption, while reduced social spending will precipitate a social and territorial crisis which may well destroy the country.

What this experience makes clear is that there are two critical institutional prerequisites for a decentralised market economy. The first and most crucial is a well-organised system of payments and market institutions which centralise information about the individual demands and supplies of each group of goods and services and make them compatible (and not the Walrasian auctioneer of general equilibrium theory, whose closest real-world counterpart is paradoxically a benevolent Gosplan which centralises information about all excess demands – demand less supply of each good and service – and sets prices so as to set all excess demands to zero). Economic co-ordination depends, in other words, first on the existence of financial institutions which debit and credit the accounts of economic actors who engage in decentralised bilateral transactions and a central bank to clear net inter-bank transactions. Second, it depends on the existence of market institutions which enable the transactors to assess the quality of goods and services, the creditworthiness of purchasers, the reliability of the supplier characterised by their public/private character, differences in the degrees of collusion between producers and consumers, differences in the cost and in the asymmetry of information, etc. (see Boyer 1996). As Boyer (1996: 13) remarks, transition depends on an understanding of the 'necessary and sufficient conditions for a viable capitalist system' and of the 'complex process of implementation of capitalist institutions'.

The second institutional precondition is the development of an effective governmental and fiscal system. If the fiscal system is in tatters, it is impossible to control government deficits. And if the macroeconomic context is unstable, private adjustments are incapable of compensating for the lack of provision of essential collective goods, which include not just an efficient payments and monetary system but also a system of justice and accurate accounting. The development of an efficient state system is clearly a fundamental requirement for microeconomic restructuring and the creation of an effective economic order. There is, as Boyer (1996) argues, no invisible hand in the selection of basic capitalist institutions: market exchange is not a natural phenomenon but a social construction.

In the case of Russia, the situation is more complicated in that what has emerged is a capitalism that involves a combination of rentier, merchant, comprador and Mafia elements: a cul-de-sac in which the only way forward involves the development of a (nationalised) banking system to establish a system of payments and to play an active role in supervision of enterprises and the enforcement of market discipline; a renationalisation of major monopolistic conglomerates; the learning of new rules, norms and conventions which require majority consensus and not the deployment of executive authority aided and abetted by western advisers; a more managed restructuring of the productive system with the establishment of clear long-term economic intelligence, action plans, and public orders; and a major programme of public investment.

Transitions' to capitalism: transplanting viable capitalist institutions into the former GDR

In the last section, I argued that in the case of Russia the erosion of state intervention has created a situation in which the institutional conditions on which an effective capitalist economic order depends are absent, in part because the neo-liberal advocates of transition had little awareness of the institutions required for capitalism to function. In this sense, the situation in Russia stands in stark contrast to that of the former GDR, which was annexed and integrated overnight into the Federal Republic of Germany in October 1990, and into which all West German institutions were transplanted alongside a specialised privatisation institution (Treuhandanstalt), inherited from the 1990 Modrow government and endowed with a threefold task of privatising, restructuring and liquidating the property of the former East German state.[9]

German unification and the wholesale transplantation of West German institutions to the former GDR led, however, to an unprecedented recession only compensated by huge transfers from West Germany (and indirectly from the rest of the EU as a result of the forcing up of European interest rates). In spite of the coherence of the institutions established in the former GDR, transformation has proved extremely problematic, is far from being achieved, and will probably require decades.

The first major consequence of unification was a very large fall in net manufacturing output. In 1990, there was a 29.2 per cent drop in the current value of the output of the mining and manufacturing industries (see Table 4.5). The greatest slumps occurred in the consumer-goods sector, especially in textiles, food and light industries, because of the shift towards western consumer goods. Output also fell in the iron and other metal processing industries due, once again, to the collapse of the domestic market for consumer goods, though in these industries the difficulties of sectors which purchased semi-finished metal inputs (automobile, agricultural machine, furniture and textile industries) and the fall in exports to traditional East European markets also played a role. At first, industries such as pharmaceuticals, machine-making, automobiles and electrical engineering, dominated by large industrial combines (*combinate*), suffered less because major contracts for their output were upheld (see also EC 1996: 3–4).

The collapse of manufacturing and mining output and of the transport sector were reflected in a drop in GDP. In 1990, current price GDP fell by 19.6 per cent and real GDP by 14 per cent (Table 4.6). GDP declined by an even more dramatic 31.4 per cent in 1992. In all subsequent years until 1995, however, real GDP increased, thereby sharply differentiating the GDR from the other transition economies.

In contrast to the CIS, the slump in output led to a drastic decline in employment. In 1989, some 9.6 million people were employed in the GDR (see Table 4.5). This figure fell sharply to 8.7 million in 1990, and 6.1 million in 1993 when it stood at just 62 per cent of the 1989 average. By 1993, agriculture had shed 77 per cent of its 1989 workforce, mining and manufacturing had lost 50 per cent and trade and transport had shed some 17 per cent. Within these sectors, there were changes not just in their size but in their composition. Manufacturing firms disposed of facilities external to production such as health care, crèche, holiday, catering and library facilities. Within trade and transport, job reductions stemmed in part from a restructuring of the distribution system, including the railways and postal services.

Table 4.5 Output and employment in the former GDR: the immediate impact of unification

	Gross value added (in current prices)		
	1989	*1990* DM billions	*1990* % change over previous year
Agriculture and forestry	11.0	7.3	−33.6
Mining and manufacturing	152.6	108.0	−29.2
Construction	21.3	19.8	−7.2
Trade	16.7	13.7	−18.0
Transport	24.2	17.4	−28.3
Services	36.3	37.7	3.9
Public sector	20.1	21.9	9.1
Private non-profit organisations	3.5	4.0	13.8
Gross domestic product	285.6	229.7	−19.6
Income from nationals abroad	0.0	2.3	−
Gross national product	285.6	232.1	−18.8
Employment	*1989* ('000 persons)	*1990* ('000 persons)	*1990* % change over previous year
Agriculture and forestry	960	809	−15.7
Mining and manufacturing	3,655	3,194	−12.6
Construction	598	487	−18.6
Trade	732	664	−9.3
Transport	677	642	−5.2
Services	1,086	1,091	0.5
Public sector	1,746	1,678	−3.9
Private non-profit organisations	187	171	−8.6
Total (domestic concept)	9,641	8,736	−9.4

Source: DIW 4/1992.[a]
Note

a The DIW publishes regular economic data in its *Wochenbericht*, especially in biannual reports on the state of the world and German economies (see, for example, DIW 1995). In this chapter, data from this publication are referred to by the number and date of publication of the issue used.

Growth occurred in some services: consumer services expanded due in part to the importance of transfer incomes, the underdevelopment of consumer services, the low cost and ease of start-up, and the fact that such services enjoy a greater degree of protection simply because they are less exposed to competitive pressures from supra-regional firms. In banking and producer services there was substantial growth. Within the state apparatus and tertiary education there was a major restructuring of personnel, as individuals whose background and experience were considered unsuited to market societies

Table 4.6 Growth of income, prices and unit labour costs in East and West Germany, 1990–5 (percentage change over previous year)

	1990	1991	1992	1993	1994	1995
Real GDP (at 1991 prices)						
New German Länder	−14.0	−31.4	9.7	7.2	8.5	5.3
Old German Länder	4.7	3.7	1.6	−1.9	2.4	1.6
Prices of consumer goods						
New German Länder	–	19.1	16.5	8.4	3.7	2.1
Old German Länder	–	4.1	4.5	3.2	2.7	1.8
Income from employment						
New German Länder	4.1	9.3	20.1	10.3	8.5	7.0
Old German Länder	7.5	7.9	6.0	1.0	1.2	2.5
Employed labour force *(domestic)*						
New German Länder	−10.3	−19.0	−12.1	−2.8	1.1	1.1
Old German Länder	2.9	2.6	0.9	−1.5	−1.3	−0.6
Quantity of labour *(per calendar month)*[a]						
New German Länder	−15.9	−34.2	−0.9	−0.5	0.4	−0.9
Old German Länder	1.3	1.5	1.7	−3.1	−2.2	−1.7
Productivity						
New German Länder	2.2	4.1	10.7	7.8	8.8	6.3
Old German Länder	3.4	2.1	−0.2	1.3	4.6	3.3
Unit labour costs[b]						
New German Länder	17.8	39.4	10.3	3.0	−0.7	1.6
Old German Länder	2.8	4.3	4.5	2.8	−1.2	0.9

Source: DIW 1–2/1990, 1–2/1991, 1–2/1992, 1–2/1993, 1–2/1994, 7/1995, 15/1995, 47/1995,18/1996, 25–26/1996, 43–44/1996, 8/1997.

Notes

a The quantity of labour depends on the number of people employed and the average number of hours worked.

b To compute unit labour costs the income from employment per hour of work was divided by productivity measured by the output produced in an average hour of work.

were replaced by individuals with an acquaintance with western ideologies and a knowledge of western legal and financial administration.

As a consequence of the overall decline and recomposition of employment, more than 3 million people joined the ranks of the registered unemployed, the under-employed and the hidden-unemployed, made up of individuals who took early retirement, people on job training programmes and individuals who were made redundant or left their jobs 'voluntarily' and have ceased to look for work (see Tables 4.5 and 4.6, which show the decline in employment, the decline in the size of the active population and the growth in unemployment in the former GDR). Indeed, were it not for early retirement and other special labour-market programmes, as well as the opportunities to commute to jobs in West Germany (see Table 4.7),

Table 4.7 Labour market accounts for West and East Germany in thousands of persons

	1992	1993	1994	1995	1996
West Germany					
Active population	30,938	31,275	31,221	31,047	30,982
Nationals in employment	29,130	29,005	28,665	28,482	28,186
Self-employed	3,067	3,071	3,086	3,099	3,112
Employees	26,063	25,934	25,579	25,383	25,074
Short-time workers		767	275	128	206
Unemployment (registered)	1,808	2,270	2,556	2,565	2,796
Unemployment rate in %	5.8	7.3	8.2	8.3	9.0
Net commuters	322	329	331	330	
Domestic employment	29,452	29,334	28,996	28,812	
East Germany					
Active population	7,894	7,357	7,456	7,433	7,448
Nationals in employment	6,724	6,208	6,314	6,386	6,279
Self-employed	417	462	500	523	539
Employees	6,307	5,746	5,814	5,863	5,740
On job-creation schemes	388	260	280	312	278
Short-time workers	370	181	97	71	71
Unemployment (registered)	1,170	1,149	1,142	1,047	1,169
Unemployment rate in %	14.8	15.6	15.3	14.1	15.7
Early retirement	808	853	650	374	180
Further training	427	345	241	243	230
Civil servants in 'Wartestand'					
Net commuters	338	325	326	325	
Domestic employment	6,386	5,883	5,988	6,061	

Source: DIW 15/1995, 44/1996, 17/1997.

unemployment would have exceeded 30 per cent. At the same time, few of the unemployed will ever get jobs. In this sense, transition has effectively cast a whole generation of people onto an economic scrap-heap.

A number of factors explain this dramatic decline in output and employment. The first factor was monetary integration and the choice of a high exchange rate of 1 Deutsche Mark for 1 Ost Mark. On one hand, this high rate of exchange enabled the inhabitants of the former GDR to secure a very favourable conversion of their wealth and income, reinforcing their allegiance to the new Germany and the political forces that carried through unification. On the other hand, at this exchange rate very few East German goods were competitive within the new domestic market. Furthermore, very few East Germans preferred to purchase East German goods.

The second factor, to which reference has already been made, was the fundamental change in the mode of international integration of the former GDR. Most East German enterprises and conglomerates used to export to

COMECON countries. The switch to the Deutsche Mark and the collapse of COMECON closed off their former markets, destroyed the previous international division of labour between 'communist' countries and forced former GDR firms to compete on the much more sophisticated West German and European markets. As these enterprises had not been prepared to compete in these new markets, the negative impact on output and employment was deepened.

The negative impact of the exchange rate decision and the loss of former markets were reinforced by two further factors which forced up the prices of East German goods and services. The first of these factors was the creation of a single national market which led to rapid increases in rents and in the prices of goods and services in the new German Länder, and a convergence in prices in the two parts of the new Germany. As Table 4.6 shows, the inflation rate was 19.1 per cent in 1991 and 16.5 per cent in 1992, compared with 4.1 and 4.5 per cent in the West German Länder and 5.5 and 5.0 per cent in the EU (see also EC 1996: 6).

The second factor affecting price levels was the rapid extension of West German employment law to the former GDR, and the progressive increase in East German wages and salaries to West German levels. As Table 4.6 shows, wages and salaries increased by 9.3 per cent in 1991, and 20.1 per cent in 1992, compared with 7.9 and 6.0 per cent in the old Länder. As a consequence, the monthly gross income per full-time employed person reached some 60 per cent of the West German figure in the first quarter of 1992 and just under 70 per cent by the end of 1992, with further but slower increases projected for subsequent years (EC 1996: 6). There were three reasons for these wage increases. First, because of the increase in prices, wages had to increase to prevent a fall in living standards. Second, increasing East German real wages and salaries to West German levels was an implication of the unification treaty. Third, a convergence in living standards was required to prevent massive migration from east to west and to limit commuting to work in western Länder.

Had real output per head increased as rapidly as real wages, unit labour costs or the efficiency wage (equal to hourly wages divided by output per hour) would have remained the same. In spite, however, of substantial privatisation and restructuring, eastern output fell, and productivity stagnated in the first few years: constant price GDP per hour of work stood at ECU 9.77 in the first half of 1990, 9.52 in the second, 9.08 in the first half of 1991, 10.4 in the second and 9.78 in the first half of 1992 (see EC 1996: 7 and Table 4.6). With productivity stagnant and real wages increasing rapidly, East German competitiveness declined, with unit labour costs increasing by 17.8 per cent in 1990, 39.4 per cent in 1991 and 10.3 per cent in 1992 (Table 4.6). In these circumstances, either prices are forced up, making an area's economic activities less competitive, or the profit margin is squeezed, making investment less attractive, or both. In East

Germany, prices increased and the rate of profit on eastern investment declined. Indeed, with comparable rates of productivity in other East European economies and much lower wages, the relative attractiveness of the new Länder to foreign investors declined.

In essence, the industrial production of the former GDR collapsed because of the almost immediate extension of the established institutional architecture of the Federal Republic of Germany to the new East German Länder. What differentiated the GDR from other transition economies was, however, not simply the scale of the shock to which it was subjected, but also the fact that there was a massive flow of transfers from the rich west to the east which resulted from the full participation of the new federal states in VAT receipts, the transfers under the German Unity Fund and social security transfers. This massive flow of resources (which is reinforced by EU Structural Fund expenditures in the former GDR) is the basis of a system of critical life-support subsidies for many East German enterprises. In addition, it is the source of the growth in GDP and household incomes, and has made change in the former GDR politically palatable. As Table 4.8 shows, household incomes in the two Germanies have converged, though income inequality in the former GDR has increased, without yet reaching West German levels. More than two-fifths of household disposable income consists, however, of transfer income, while a considerable share of wage income is paid by enterprises in receipt of government subsidies (DIW 1994).

Clearly, the transfer into the former GDR of the well-established and coherent architecture of the German social market economy did not create the results that were at least officially expected. As Boyer (1996) has argued, in a country such as the former Federal Republic of Germany, the economic specialisation of firms and regions, institutions and modes of economic conduct had evolved jointly and slowly. In the case of the former GDR, the divorce between the new institutions, modes of behaviour and the specialisation of its economy led to a major crisis that remains unresolved in spite of the strenuous efforts of the West German authorities, very large-scale programmes of state expenditure and an intensive process of learning. The fact is that enterprises, managers, workers, trade unions, financial institutions and governmental authorities were unable to take advantage of a coherent set of institutions for several sets of reasons. First, the large east–west wage differences needed to offset low rates of productivity in the east were socially and politically infeasible. Second, unification severed former economic relationships. Third, the modes of economic behaviour required for the efficient functioning of an institutional order have to be learned. To learn how to organise production, vocational training, the provision of credit and innovation systems takes much more time than is involved in the transfer of political and legal structures. As Boyer (1996: 11) concludes,

Table 4.8 Public financial transfers to East Germany (DM billion)

	1991	1992	1993	1994	1995
Financial transfers to state and local government[a]	112.0	133.0	154.5	146.5	161.5
German Unity Fund	35.0	36.0	36.5	36.0	–
Net central government spending arising out of unification[b]	66.0	85.5	106.5	99.5	113.5
Redistribution of VAT receipts among federal states	11.0	11.5	11.5	11.0	–
New system of inter-state financial compensation	–	–	–	–	48.0
Transfers to the social insurance institutions	21.5	29.0	24.0	33.5	32.5
Transfers from West to East German unemployment insurance	21.5	24.5	15.0	19.5	17.5
Transfers from West to East German pension insurance	0.0	4.5	9.0	14.0	15.0
Total financial transfers	133.5	162.0	178.5	180.0	194.0
Memo item: Borrowing by the Treuhandanstalt	19.9	29.6	38.1	37.1	–
Total financial transfers as % of East German GDP	64.8	61.7	58.9	53.8	55.0
Total financial transfers as % of German GDP	4.6	5.3	5.7	5.4	5.6

Source: DIW 42/1995.

Notes

a Excluding administrative aid by state and local government; excluding reduced revenue in West Germany due to entitlement to tax allowances for investment in the new Länder; excluding spending by the Treuhandanstalt and the subsidisation of interest payments on ERP loans.

b In 1995 including interest payments on the debts incurred by the Treuhandanstalt to the end of 1994.

capitalism cannot be created by treaty or law. Its efficiency derives from a slow trial and error process which cannot be compressed into a few years. Similarly some behaviours considered normal in capitalist societies are not specifically for individuals socialised into a quite different order. The institutions of capitalism are not the outcome of a big bang but of a long, painful and contradictory process.

Conclusion

The transition of former communist economies into market economies amounts to a challenge that is unparalleled and that certainly cannot be compared with the recent transformation of the southern peripheries of the EU. Spain, Portugal and Greece were authoritarian-capitalist economies, in which most of the institutions required for a functioning democratic market society were in place. All of these economies had experienced rapid growth and greater integration into the global economic order in the period prior to the collapse of authoritarian rule. As Poulantzas (1976) argued, the rise of trans-national capital was one of the forces leading to the quest for closer international economic integration and associated political change. By contrast, in the case of communist Europe, transition was driven by economic failure rather than the recent confrontation of economic constraints, integration into the capitalist world economy was largely confined to the negative forms associated with deep indebtedness and the institutions and modes of economic conduct differed quite fundamentally from those of established market societies.

To highlight this contrast with other European peripheries, emphasis has been placed on the distinctive characteristics of the political economies of former communist countries. I have argued that the tension between these inherited structures and the functioning of market systems is reflected in the severity of the impacts of market-driven reform programmes. The scale and the durability of the collapse in output and income were never publicly anticipated, by either the advocates of shock therapy or those who embraced it because there was a triple failure: first, to anticipate the impact of structural adjustment on economies that were not already market economies, second, to identify the nature of the institutions on which capitalism depends and, third, to understand that modes of economic conduct taken for granted in capitalist societies have to be learned. To this list, which is not exhaustive, should be added the fact that markets often fail and that developed industrial economies are mixed rather than pure market economies (see Dunford 1997). The Russian case thus exhibits consequences of the absence/under-performance of crucial institutions on which capitalism depends, while the former GDR shows that the overnight introduction of a fully developed set of social market institutions will fail because there is a reciprocal inter-dependence between economic structures, institutions and learned rules of behaviour. Together, these two cases highlight some of the critical contradictions and question the underpinnings of neo-liberal strategies of transition.

What these cases also show, however, are the possible contrasts in the impact of sharp and rapid programmes of transition which stem from the specificity of the newly created economic orders. The effects of transition differed profoundly between the former GDR, which received large net transfers from West Germany, and the successor states of the former Soviet Union. In this respect, the German model is closer to what happened in the western half of Europe after the Second World War. At that time, the communist challenge led the US to aid and facilitate the reconstruction of

Europe (and Japan on the other flank of the communist world). In a context of immense disparities in the export potentials of different economic blocs, the US permitted a partial suspension of the rules of liberal market competition, offering large transfers of resources and technology and tolerating protectionism to permit the development of indigenous productive capabilities.

These considerations give rise to two further conclusions. The first is that the future will perhaps be much more differentiated than simple models of a transition suggest and that the economic orders that emerge in Russia and Germany may well involve new combinations of institutions and economic structures. What also makes such an outcome probable is the fact that the developed market economies have themselves hit certain limits to their development (which an opening up of new markets and sources of raw materials in former communist countries will not enable them to supersede). These limits, associated with the rise in unemployment and exclusion and the tightening of environmental constraints, will ultimately require institutional innovation. The second is that the relatively successful models of reconstruction put in place after the Second World War did not draw on the neo-liberal principles of structural adjustment: in situations of transition and reconstruction, there are perhaps more choices than the neo-liberal advocates of shock therapy are prepared to admit.

Notes

1 A more developed analysis would involve a division of GDP per capita into three elements: the employment rate, the number of hours worked per worker and hourly productivity. In the USSR, annual hours worked stood at 107 per cent of the US figure, so that the hourly productivity divide was greater than the gap in output per person employed. This figure was very close to that of all the Cohesion Countries except Spain (120 per cent). Elsewhere in the world, quite high figures were often recorded, particularly for South Korea (176 per cent of the US level) and Taiwan (157 per cent). Of West European countries, only Denmark, Finland and Switzerland exceeded the US figure. All the others were lower (in part because of the relative importance of part-time work).

2 As Harrison also points out, women accounted for half of the state-employed workforce, but did 2–3 three times as much housework as men because of the unequal division of domestic responsibilities (see also Meurs in this volume). A consequence of this unequal division of work, of the support for female partic-ipation and of the relative economic security women enjoyed was falling birth-rates and higher divorce rates (Harrison 1984: 126–8).

3 Conversely, Kazakhstan should be separated from the less developed areas of Central Asia as it was an area of European colonisation whose advanced mining and industrial north (with its space-complex and firing ranges for nuclear and anti-aircraft missile trials) makes it resemble Siberia, rather than the areas further south.

4 President Havel of the Czech Republic is often cited as saying that 'it is impossible to cross a chasm in two leaps' (cited in World Bank 1996: 9). A much less flippant

response is to say that the safest and surest way is to build a bridge. To do so, however, takes time, and there is no doubt that ardent neo-liberals thought that the window of opportunity was perhaps a brief one, and that what mattered most was to lock the transition economies into a wider capitalist economic order and set in motion irreversible change, no matter what the cost. Perhaps also, proponents of rapid transition feared that there might be a viable 'third way' to which a gradualist path might lead and which, it should be remembered, was a central objective of the democratic opposition in the former GDR.

5 It is important to note, however, that in Russia and the CIS employment fell much less than output (in contrast to Central Europe where employment in the state sector fell sharply, and unemployment and non-employment increased (except in the Czech Republic)). In these countries, enterprises were reluctant to impose mass layoffs. Instead, workers remained formally attached to their enterprises, received low or zero wages and continued to receive various enterprise social benefits, depending for income on the informal sector (see World Bank 1996: 73–4).

6 For Burawoy (1996: 1114–15) the fundamental roles of government are those of establishing an array of political, economic and property relations, and the nature of budget constraints. In Russia, there *was* a change in property relations. At first, budget constraints remained soft, in part to limit centrifugal tendencies. The subsequent hardening of budget constraints encouraged a restoration of barter, regional autarchy, a strengthening of the Mafia and further involution. His conclusion is that there is no market road to a market economy, and that a Chinese road with hard budgets, collective property and a strong state may represent a more effective way of transforming communism.

7 Burawoy (1996: 1107–8) argues for centralisation of a contractually agreed share of local revenue, limited central redistribution to create hard local budget constraints on local authorities, and local retention of the surpluses earned as a result of the successful fostering of local economic development.

8 Burbach, Núñez and Kagarlitsky (1997: 122) argue that it is this process that explains the disintegration of all the East European federations whether it be the Baltic Republics in the case of the former USSR, Slovenia in the case of the former Yugoslavia, the Czech Republic in the case of the former Czechoslovakia and perhaps, in the future, resource-rich areas such as Yakutia in Russia.

9 By the end of 1990, the Treuhand had privatised some 400 enterprises. One year later this figure stood at 3,500, and at the end of 1992 it had reached 11,403. Of this total, 2,000 were management buy-outs. By the end of 1993, the Treuhand had accumulated debts of 300 billion Marks; the receipts from privatisation amounted to just 40 billion Marks (Le Gloannec 1995: 333).

Bibliography

Boyer, R. (1996) 'The seven paradoxes of capitalism. . . . Or is a theory of modern economies still possible?', seminar given at the University of Wisconsin-Madison, 18–19 November.

Burawoy, M. (1996) 'The state and economic involution: Russia through a China lens', *World Development*, 24, 6: 1105–17.

Burbach, R., Núñez, O. and Kagarlitsky, B. (1997) *Globalization and Its Discontents. The Rise of Postmodern Socialisms*, London and Chicago: Pluto.

DIW (Deutsches Institut für Wirtschaftsforschung Berlin) (1994) 'Gesamtwirtschaftliche und unternehmerische Anpassungsfortschritte in Ostdeutschland, Elfter Bericht', *Wochenbericht*, 31/94.

DIW (1995) 'Die Lage der Weltwirtschaft und der deutschen Wirtschaft im Herbst 1995', *Wochenbericht*, 42/95.

Dunford, M. (1996) 'Disparities in employment, productivity and output in the EU: the roles of labour market governance and welfare regimes', *Regional Studies*, 30, 4: 339–57.

Dunford, M. (1997) 'Divergence, instability and exclusion: regional dynamics in Great Britain', in R. Lee and J. Wills (eds) *Geographies of Economies*, London: Arnold.

EC (European Commission) (1996) *The Spatial Consequences of the Integration of the New German Länder into the Community*, Luxembourg: Office for Official Publications of the European Communities.

Gowan, P. (1995) 'Neo-liberal theory and practice for Eastern Europe', *New Left Review*, 213: 3–60.

Gowan, P. (1996) 'Eastern Europe, Western power and neo-liberalism', *New Left Review*, 216: 129–40.

Harrison, M. (1984) 'Lessons of Soviet planning for full employment', in K. Cowling, A. Ford, M. Harrison, K. Knight, M. Miller, W. Buiter, A. Roe, G. Renshaw, E. Ahmad, B. Sadler and P. Stoneman, *Out of Work. Perspectives of Mass Unemployment*, Coventry: University of Warwick, Department of Economics.

Le Gloannec, A.-M. (ed.) (1995) *L'état de l'Allemagne*, Paris: La Découverte.

Maddison, A. (1995) *Monitoring the World Economy 1820–1992*, Paris: OECD.

Nove, A. (1989) *An Economic History of the USSR*, second edition, London: Penguin.

Peaucelle, I. (1995) 'Firme ou artel? Vers un rapport salarial original en Russie', in R. Boyer and Y. Saillard (eds) (1995) *Théorie de la régulation. L'état des savoirs*, Paris: La Découverte.

Poulantzas, N. (1976) *The Crisis of the Dictatorships*, London: New Left Books.

Sapir, J. (1994) 'Les grandes puiassances économiques depuis 1945. Aspects de l'évolution économique de l'ex-URSS depuis 1945', *Les Cahiers Français*, 265: 93–100.

Sapir, J. (1995) 'Crise et transition en URSS et en Russie', in R. Boyer and Y. Saillard (eds) (1995) *Théorie de la régulation. L'état des savoirs*, Paris: La Découverte.

Sapir, J., Badower, A. and Crespeau, M. (1994) *L'expérience soviétique et sa remise en cause*, Paris: Bréal.

Smith, A. (1994) 'Uneven development and the restructuring of the armaments industry in Slovakia', *Transactions of the Institute of British Geographers*, 19: 404–24.

Smith, A. (1997) 'Breaking the old and constructing the new? Geographies of uneven development in Central and Eastern Europe', in R. Lee and J. Wills (eds) *Geographies of Economies*, London: Arnold.

Smith, A. (1998) *Reconstructing the Regional Economy: Industrial Restructuring and Regional Development in Slovakia*, Cheltenham: Edward Elgar.

Stark, D. (1996) 'Recombinant property in East European capitalism', *American Journal of Sociology*, 101: 993–1027.

World Bank (1996) *From Plan to Market: World Development Report 1996*, Oxford: Oxford University Press.

Part II

INDUSTRIAL RESTRUCTURING, UNEVEN DEVELOPMENT AND NETWORKS IN TRANSITION

5

INSTITUTIONS, SOCIAL NETWORKS AND ETHNICITY IN THE CULTURES OF TRANSITION

Industrial change, mass unemployment and regional transformation in Bulgaria

Robert Begg and John Pickles

Introduction

Since November 1989, Bulgaria has experienced a particularly difficult period of economic transformation, from a command economy sheltered by CMEA towards a fledgling market economy subject to the discipline of world trade. Industrial output has collapsed, falling by 66 per cent between 1990 and 1994 (Weiner Institut für Internationale Wirtschaftvergleiche 1995). Mass unemployment has become a fact of life as national unemployment has risen to around 20 per cent, with regional pockets much higher than that (NSI 1995: 94).[1] This sharp rise in unemployment has clear regional dimensions, and regional differentiation has deepened substantially, as it has in all reforming countries in CEE in the 1990s (Barta 1992; Murphy 1992; Pavlínek 1992; Zaniewski 1992).

In this chapter, we trace the process by which regional differentiation has occurred, and evaluate theories of regional growth and decline at two levels. First, authors (like Buckwalter 1995) have attempted to test neo-classical models of regional convergence against the Myrdalian concepts of circular and cumulative causation and regional divergence. We examine this account of regional dynamics in post-communist Bulgaria and find it wanting.

Second, we review hypotheses that describe regional patterns of unemployment as a by-product of the industrial restructuring of socialist systems of production. In these accounts, production decline, plant closure, industrial labour shedding, and rapid increases in the level of unemployment are explained as a result of the 'natural' adjustments of inefficient state socialist industries to the privatising of ownership, restructuring the organisation of

115

enterprises, and the rationalisation of production. In examining such theories we question these notions of economic restructuring and decline and show how, in the case of Bulgaria at least, the restructuring of employment and production has gone on – at least until the crisis year of 1996 – *without* substantial changes in ownership, organisational change, or rationalisation. Instead, we argue that the emergence of pockets of regional unemployment in Bulgaria is the result of historically and culturally specific processes for which an understanding of command economy structures is at least as important as a reliance on models of investment, international markets, or industrial rationalisation. That is, we argue (with Stark 1997) that recombinant institutional forms are emerging that build *on* and work *with* the ruins of communist institutions. Finally, we attempt to show how the restructuring of the spatial divisions of labour is itself embedded in a broader cultural politics of ethnicity, in which the impacts of branch-plant closure and mass unemployment have fallen most heavily on peripheral regions and ethnic minorities.

Industrial restructuring and regional convergence or divergence?

Much early work on CEE transitions was rooted in neo-liberal perspectives (for a review of these, particularly in the context of policies of shock therapy, see Gowan 1995). Among other things these assumed that management and workers in enterprises had certain types of interests and that these operated within a specific set of institutional arrangements, most of which were presumed to be – in the last instance – those of a liberal market economy distorted by the persistence of bureaucratic control exercised over the production process. Once this control (and its distorting effects) were removed, once the state was removed as a direct player at the point of production, once the discipline of markets removed the false pricing of a protected economy, and once incentives were established for individuals through privatisation, investment opportunities and the discipline of active labour and capital markets, rational economic behaviour of actors would be asserted, incentive structures would be 'normalised', and the revitalisation of regional economies would occur (Sachs 1994). Because CEE governments have neither the financial nor the political capital for strong interventions in regional or industrial policy, it is to the patterns of foreign direct investment and their effects on regional development that attention has turned. Changing patterns of investment capital have thus become an important marker of regional change. Nowhere has this been more true than in the attempts by regional scientists to account for regional change (e.g., Buckwalter 1995; Fassman 1992; Murphy 1992).

Traditional economic theory offers two mutually exclusive hypotheses regarding the shifting fate of regions during the transition. Neo-classical theories of regional convergence assert that in a national economy where the free movement of labour and capital is possible, investment decisions will lead to regional equilibrium (Williamson 1965; Richardson 1980). In a situation of relative disadvantage, lagging regions which have lower wages and hence a higher potential return on capital may attract new investment which in turn will reduce unemployment and increase wages. Through time this process will produce regional equilibrium. By contrast, Myrdal (1958; Kaldor 1975) argued that the movement of labour and capital is constrained by social and political forces. Poverty and regional inequality – far from diminishing in time – will persist because lagging regions cannot generate sufficient investment capital or provide the infrastructure necessary to attract investment capital. Here regional and social divergence and poverty in lagging regions are persistent, although not inevitable. Redressing regional inequality cannot occur without the intervention of non-market forces, in particular the state. Not surprisingly, then, policy analysts and scholars have been interested in the extent to which foreign investment is occurring in old industrial cores, new industrial districts, or lagging regions. That is, they are interested in whether theories of regional convergence arising from differences in labour market and input costs might be sustained, or whether foreign direct investment exacerbates old regional inequalities in transition economies.

Buckwalter (1995) has attempted to use patterns of direct foreign investment in Bulgaria to assess the extent to which the transition from a command economy to market economy has resulted in regional divergence. In his paper Buckwalter (1995: 291) proposes that neo-classical arguments suggest that transition would bring no 'drastic geographical concentration of prosperity, [that] Sofia's dominance of the national economy would not increase, and [that] marginal districts would suffer no disadvantage'. On the other hand, Myrdalian notions of cumulative causation would argue that 'drastic spatial adjustment would accompany the transition to a market economy'. Using foreign direct investment data, Buckwalter tests these competing economic theories of regional change and – although some divergence is apparent – finds that the available data allow no firm conclusions to be drawn.

Investment, institutions and capital in Bulgarian industry

Although Buckwalter does suggest that his analysis might overreach the evidence presented by the available data, we would also suggest that his analysis *misinterprets* the regional structure of the Bulgarian economy. Specifically, regional economic theory at this scale fails to understand the dynamics of industry and employment in creating the emerging regional disparities in the first place. Such theories emphasise the effect of investment on industrial production. But the economic situation in CEE since 1989 has been one of precipitous economic decline. In Bulgaria, disinvestment, rather than investment, has been the rule. Little industrial investment has occurred. In 1994, 91 per cent of all industrial output still came from the public sector (NSI 1995: 225), and the private investment in industry that has occurred has largely involved the acquisition by local agents of existing enterprises (firms) or plants (factories). Modernisation and greenfield development in industry are rare. Only 7.2 per cent of all transactions for privatisation of industry have been by foreign companies (Centre for the Study of Democracy 1995: 18). Total FDI in Bulgaria up to 1995 was only $470 million, of which $110 million was in hotels and transport (Privatisation Agency 1995: 15–16).

To assess the role of new investments in transforming the regional structure of the Bulgarian economy we would need to understand much more about the different forms of 'capital' that actually constitute 'investment'. However, a comprehensive discussion of capital in the CEE is not known to us. A fuller discussion might include such notions as: capital as access (e.g., the distribution of grain export permits in unannounced meetings); capital as position and contacts (e.g., the opportunities created by being director of a large trading company like Vinprom with US, English and German contacts, or a strategically important large enterprise such as Neftochim); and the conversion of political to economic capital, both legally and illegally. In Bulgaria, capital as money would be of multiple hues. Red capital, in the form of *nomenklatura* power and cash at first, and in the form of emerging holding companies (e.g., commercial investment and holding companies, such as Multi-group run by apparatchiks, including ex-security forces); black capital, which dates with red capital to pre-transition days (e.g. Gypsy black-marketeers, Turkish smugglers, urban commercial security agencies); and blue capital, the new democratically corrupt (e.g., permit racketeering at vacation resorts such as Sunny Beach, in which large amounts of foreign currency are spent), are probably all more important to the success of the industrial sector than true foreign investment at this time. In such an understanding of capital, investment decisions are shaped by established networks, institutional and regulatory structures, and actors operating under conditions of extreme uncertainty, and chance. And it is precisely these forms of capital – often eschewing the more obvious, formal routes of investment – that build on opportunities accessible through (and because of) established social networks and existing institutional arrangements that allow them to achieve a regional character.

In the neo-classical model, wage differentials created by unemployment and return to capital created by demand for investment cause regional convergence by shaping investment decisions. But neither labour nor capital markets as we would recognise them existed in Bulgaria between 1989 and 1996. Registered unemployment at the national level was 20 per cent in 1994. Even the capital, Sofia (the industrial heart of the country), had 5.7

per cent registered unemployment in March of 1994 (Labour Office 1994). Wage pressure from core to peripheral regions was not a limiting factor in investment decisions. In fact, labour and wage decisions (even as late as 1997) operate largely through a system of government controls and tripartite agreements. That is, labour markets are not yet developed.[2]

While this may lead us to focus on models of circular and cumulative causation, it would presume that investments in fixed capital reinforce existing regional imbalances. But if there is little formal investment, it is difficult to argue that regional divergence is occurring as a result of the actions of 'economic decision-makers' in any traditional sense of the phrase. Moreover, export-driven growth plays a prominent role in cumulative causation models (Thirwall 1980). To the extent that there are differences in domestic and foreign demand for regional products, regional economic divergence might occur. This would in turn require the existence of region-ally organised industrial agglomerations in particular sectors (see Smith 1998). Such agglomerations are one critical feature of a political economy of regional growth or decline (Murgatroyd and Urry 1983; Massey 1984; Scott 1988) and form the basis of parts of the more traditional descriptions of industrial restructuring (Chinitz 1961). Regional agglomerations are also critical in understanding the fate of CEE regional economies (Smith 1994, 1995, 1998). But, as we will describe below, such agglomerations often are not influenced directly by market forces in post-communist Bulgaria.

Sectoral regional agglomerations in industry that do occur in Bulgaria are important in two respects. First, the divergence or convergence of Bulgarian regions during the post-communist period depends on the location of troubled sectors. But whether agglomerations of failing sectors are found in core or peripheral regions is a result of prior politically determined allocation decisions. Second, if key CMEA industrial sectors are located in core regions, the failure of those sectors in international capitalist markets could well cause core decline and regional convergence. If they are located in peripheral regions, their decline would cause divergence. In turn, periph-eral regions may have marginal plants which fail in a market economy and this would cause divergence. But these changes say little about regional dynamics or the economic vitality of core or peripheral regions. The spatial impact of adjustment to loss of traditional markets and competition in new markets is a phenomenon determined by prior allocations of production units: economically deterministic models of convergence and divergence thus seem to offer little in the way of explanatory power.

In part, this lack of explanatory power can be explained by the specific structure of Bulgarian industry. The location of fixed capital reinforces the status quo. Limited capital investment, with persistent and ubiquitous high unemployment and poorly developed capital and labour markets, is the norm. Although it might be reasonable to argue that such conditions ought to reinforce existing inequalities at the scale of large regions, it would be too soon and too ambitious to argue that they would deepen them. Convergence or divergence processes with respect to industrial development may be visible in ten more years, but not now.

Part of this difficulty in explaining the emergence of regional unemployment pockets and regional industrial change in Bulgaria is that they are further confounded in two ways. First, they are confounded by the geographic scale of analysis at which data are generally available, the *oblast* (or regional) level. Second, they are confounded by the way in which large industrial enterprises dominate the Bulgarian economy. In Bulgaria there are at least two types of quite distinct regional change occurring in two kinds of economic space. We believe these economic spaces must be considered separately and in relationship to each other if we are ever to understand long-term regional economic change or the causes for patterns of unemployment. Moreover, two distinct (though related) politics of transition are at work in each. But before turning to this second issue, we will first consider the question of scale.

Scales of analysis and patterns of regional change

The regional convergence or divergence of industrial production in Bulgaria is an empirical issue. But to date published analyses of the issue have been ill-designed to answer the question because they have focused only on data available at the geographic scale of the *oblast* (regional state).[3] Neither convergence nor divergence with respect to the nine *oblasti* of Bulgaria can be demonstrated from this level of industrial output data. Certain other criteria are needed. First, regional economic convergence would be observable only to the extent that broad regional differences existed in 1989. Second, divergence or convergence would actually have to occur, and to demonstrate this we would need to show that convergence or divergence since 1990 was the consequence of market relations and did not mimic the pattern of convergence or divergence between 1985 and 1990.

There were regional differences in industrial output in 1990. How different they appear depends on the scale of analysis. Distinct regional differences in culture and economy typify Bulgaria, but the industrial economy is generally broadly scattered across the territory of the country, in part as a result of natural resource availability, but more importantly resulting from planning allocations to regions to 'build socialism' (Table 5.1). Contrary to Murphy's (1992) findings that industrial production in CEE states was concentrated in western regions and declined to the east, no such patterns exists for Bulgaria. In the post-Second World War push to industrialise, major producing centres were distributed to the centre and east of the country for social and strategic reasons, and employment-creating industries were allocated to border areas for social and strategic reasons. Sofia is the

Table 5.1 Regional distribution of industrial output by *oblast*, 1990 and 1994

Oblast	Output (current lev millions)		Share (%)		% change
	1990	1994	1990	1994	
Montana	3,354	23,067	7.3	5.2	−27%
Sofia City	6,110	59,704	13.2	13.6	+3%
Sofia Region	5,104	47,215	11.0	10.7	−3%
Lovetch	6,727	57,743	14.6	13.1	−10%
Plovdiv	6,583	61,018	14.2	13.9	−3%
Rousse	3,729	29,908	8.1	6.8	−16%
Haskovo	5,197	41,830	11.2	9.5	−15%
Varna	4,651	41,274	10.1	9.4	−7%
Bourgas	4,751	77,882	10.3	17.7	+72%
Total	46,206	439,641	100	100	
Coefficient of variation	0.22	0.33	.22	.33	

Source: National Statistical Institute 1991: 431; 1995: 393.

largest centre of population and industry, Varna and Bourgas are industrial port cities on the Black Sea, Plovdiv and Pleven (in the Lovetch *oblast*) are industrial centres of metallurgy, chemicals and agricultural processing, and Veliko Turnovo (Rousse *oblast*) is Bulgaria's sixth largest industrial agglomeration (Ganev 1989). Haskovo, Montana and Rousse are peripheral and lagging regions, but, despite regional levels of dominance and marginality, there are industrial cities and lagging regional economies within all nine *oblasti*.

Table 5.2 shows that the per capita output of Montana in 1990 (0.51) was marginally higher than that of Sofia (0.50) and Haskovo (0.49). Indeed, the per capita distribution of output in 1990, except for Lovetch, is remarkably uniform. Between 1990 and 1994, Montana, Rousse and Haskovo also showed the greatest loss of share, which might suggest that regional divergence was occurring (Table 5.1).[4] But in fact little divergence is apparent. If we exclude Bourgas from the data (where the massive Neftochim *combinat* distorts figures for the whole regional economy), no *obstina* changed its absolute share of production by more than 2 per cent. Moreover, the large percentage drops in share by Montana, Rousse and Haskovo (−27 per cent, −16 per cent, −15 per cent) correspond to small absolute changes (−2 per cent, −0.4 per cent, −0.3 per cent). Lovetch, an industrial powerhouse, lost an absolute share of 1.5 per cent – a figure higher than either Montana or Rousse. Most telling of all, 95 per cent of the variance in 1994 output (or share) is explained by the distribution of output in 1990.[5] Spatial *stability* not *divergence* seems to have been the norm. The emergence of pockets of

Table 5.2 Per capita output by *oblast*, 1990 and 1994

Oblast	Per capita output (current lev millions)	
	(/1000) 1990	(/100) 1994
Montana	0.51	0.37
Sofia City	0.50	0.50
Sofia Region	0.50	0.48
Lovetch	0.64	0.58
Plovdiv	0.51	0.50
Rousse	0.44	0.39
Haskovo	0.49	0.46
Varna	0.47	0.45
Bourgas	0.54	0.92
Coefficient of variation	0.16	0.19
Average absolute deviation	0.003	0.02

Source: National Statistical Institute 1991: 431; 1995: 393.

regional unemployment cannot, therefore, be explained in terms of changing patterns of industrial output or the working of broad market forces. Too little time has passed for market structures to have had any significant impact on regional change, and little change in the distribution of industrial output has occurred at a broad regional level.

Moreover, one region – Bourgas – has a highly distorting effect on these figures. The variance in industrial output for 1994 explained by the distribution of output in 1990 is 95 per cent without Bourgas (stability) but only 72 per cent (divergence?) if Bourgas is included. The coefficient of variation for per capita production climbs from 0.10 to 0.29 if the Bourgas *obstina* region is included (divergence?), but changes only from 0.11 to 0.115 with Bourgas omitted (stability).

Issues of scale are thus critical in interpreting the transformations in CEE (see also Wyzan 1993 on Bulgaria and Fan 1995 on China). Later we consider lower-level (*obstina*) data on unemployment, but for the moment it is worth while pushing harder against the two contradictory patterns of regional change found in Table 5.1. In fact, statistically these are a function of the increase in Bourgas's share of industrial output. But this gain in share by Bourgas is due almost entirely to its role in the chemical sector. The industrial share of the regional output of the chemical sector in Bourgas climbed from 36.6 per cent in 1990 to 71.4 per cent in 1994. Bourgas's chemical industry climbed from 29.2 per cent of the nation's sectoral share in 1990 to 48.5 per cent in 1994. Most disturbingly, for purposes of trying to understand broad trends in regional patterns of industrial and unemployment change, Bourgas's chemical sector climbed from 4.0 per cent of total national industrial output across all sectors in 1990 to 13.5 per cent in 1994. This remarkable increase in share of Bourgas's chemical sector is due to the increasing importance in a declining economy of one plant, the Neftochim petrochemical complex (see Pickles in this volume). In 1993, Neftochim employed 10,000 people (ILO 1994: 79), almost 4,000 more employees than the labour force of Kameno (the *obstina* or county in which Neftochim is located) and 10 per cent of the 115,258-person labour force of the nearby city of Bourgas.

Spatial divisions of industry I. Enterprise space

The geographic space in which much analysis of regional change has been carried out is the *oblast* territorial unit. But, as central planning recognised full well, while territorial organisation and administration were integrally related to the sectoral organisation and administration of industry, the two were not synonymous and sectoral planning exercised much greater power over industry than territorial planning. The economic space of production under state socialism, and the primary legacy in the period of transition, was therefore *enterprise space*. We use the phrase 'enterprise space' to refer to the structure of state socialist production organised around large integrated combines distributed regionally according to political as well as 'rationally based resource allocation constraints', and comprising production units whose volumes are so large that they often dominated (and continue to dominate) *oblast*-level data and sectoral data. It is this dominance, we would argue, that has led to the misleading analyses of regional change discussed above. Because of the scale and strategic importance of these enterprises, and the size of the 'prize' when these cash cows are finally privatised, industries within enterprise space have yet to be allowed to experience the effects of economic liberalisation, although market practices (such as the emergence of secondary party processing and second party importing of petrol) have certainly had an impact on producers such as Neftochim (see Pickles in this volume). Instead, industries have played a much more political role in the broader transformation of the country.

The extensive phase of Bulgaria's economic development occurred from roughly 1950 up to the early 1980s and was characterised by Stalinesque gigantism in both agriculture and industry. Between 1950 and 1974 over one million small farming plots had been aggregated into 130 agro-industrial complexes (Begg and Meurs 1997). Industry exhibits a similar pattern. By 1989:

> almost three-quarters of all workers were employed in enterprises with more than 500 workers, with one-fifth employed in enterprises with more than 3,000 workers. Indeed, the enterprises that had the most growth in the 1980s were those with more than 10,000

workers. Large-scale enterprises were so pivotal and powerful in their negotiations with government agencies, over such matters as prices for their products, subsidies, wages, that in spite of worsening economic conditions they had sufficient resources for hiring extra workers and retaining them.

(International Labour Organisation 1994: 22)

Table 5.3 presents the size distribution of industry for Bulgaria in 1988. In 1988 twenty-nine plants, 1.3 per cent of all establishments, produced 18.2 per cent of the total national output. These are the firms we include in 'enterprise space'.

A few huge enterprises dominate their sectors and, statistically, their regions. The Kremikovtsi iron and steel plant just outside Sofia employs 16,000 of the nation's 50,000 metallurgical workers. Iron metallurgy as a sector in Sofia *oblast* increased from 2.7 per cent of national industrial output to 5.7 per cent in 1994. Kremekovtsi and Neftochim together accounted for almost one-fifth of 1994's national industrial output. Other similarly large *combinats* include the Arsenal military works in Kazanlak (11,000 employees), the VMZ military works in Sopot (at its peak 20,000 employees), and Balkancar in Plovdiv.

Table 5.4 shows the concentration of production in two ways. The number of establishments by sector is indicated in the first column. Two indicators of regional concentration follow: *oblast*-level analyses of the coefficient of variation and the proportion of plants which would have to be redistributed to have an equitable geographic distribution.

The first six sectors are concentrated both geographically and in terms of production, and these correspond to what we have termed 'enterprise space'. The chemical industry is geographically dispersed, but is dominated by Neftochim in production terms. The leather and shoe, and wood products sectors are not easily classified. Sectors such as textiles, electrical components, agricultural processing, and machinery and metals are geographically dispersed and made up, in part, by a large number of smaller plants or

Table 5.3 Size distribution of Bulgarian manufacturing establishments, 1988

	No. of establishments	% share of total production	
less than 500	1,608	28.9	} Branch-plant space
500–1,000	336	22.5	
1,001–3,000	209	30.2	
3,001–5,000	20	6.6	} Enterprise space
5,001–10,000	7	8.2	
10,000 or more	2	3.6	

Source: Jones and Meurs 1991.

Table 5.4 Geographic diversity by *oblast* by sector, 1990

Sector	Number of plants	Oblast *distribution of plants*		
		Coefficient of variation	% redistribution	
Oil and gas	1	na	na[6]	
Coal	8	1.63	67	⎫
Fe metallurgy	13	1.35	47	⎪
Non-Fe metallurgy	28	0.86	34	⎬ Enterprise space
Glass and china	20	0.67	27	⎪
Paper	20	0.92	37	⎭
Chemical	120	0.29	12	
Printing	37	0.46	13	
Leather and shoe	50	0.53	22	
Apparel	98	0.42	16	
Building materials	145	0.34	14	⎫
Textiles	150	0.38	16	⎪
Electrical	246	0.42	17	⎬ Branch-plant space
Wood products	280	0.54	22	⎪
Food and beverage	356	0.23	8	⎪
Machine building	588	0.30	9	⎭

Source: National Statistical Institute 1991: 430.

workshops, what we are calling the 'branch-plant space economy'. The decentralisation of decision-making regarding layoffs to the enterprise and regional level has implications particularly for the firms in these sectors.

In deciphering the geography of CEE transition, Pickles (1995: 116, and in this book, following Clarke *et al.* and Burawoy) calls for an understanding of 'production politics', arguing that the transition is not so much one of 'breaks', but of a 'transformation of social relations, forms, and practices that carries important traces of the legacy of the previous 40 years' (Pickles 1995: 117). In enterprise space this transformation and legacy manifest themselves in a continuation of command economy relations for the largest enterprises. In terms of the future of the Bulgarian economy, these enterprises exercise an unusual level of economic and political power regionally and nationally, and – as a consequence – must be considered separately from smaller enterprises.

The Neftochim board of directors in 1993 was still staffed by deputy ministers from the Ministries of Industry, Finance, and the Council of Ministers. Kremikovtsi and Neftochim continue to operate with large debts. Neftochim in 1994 still owed the World Bank $109 million; Kremikovtsi owed 3.5 billion 1993 lev to the state; in 1993 VMZ in Sopot kept 3,000 workers on payroll who were not working (ILO 1994: 79–80).[7] Up to now such enterprises have been essentially protected from global market forces.

While many now have hard budget constraints and declining budgets, they continue to be run with the labour 'redundancy' of the command economy, protecting wage bills and worker loyalty, operating as political as well as production units (see Clarke *et al.* 1993 for similar claims about enterprises in Russia). They continue to be run, in some cases, by the very civil servants who ran them during the communist era, or – as in the case of Neftochim – by successive boards dominated by the political appointees of the latest governing party. To speak of them as corporations may be a useful metaphor, but to consider them private or in any way subject to normal market disciplines is naïve. In fact, the tighter the budget constraints the deeper the level of political negotiating with the central state and bartering with suppliers that is needed to survive.

These giant industrial units are removed from the economy in other ways. Neftochim and Kremikovtsi have few backward and limited forward linkages. Part of the gigantism which created these prodigies was the horizontal and vertical assimilation of production functions into giant *combinats*. In some sectors the *combinats* were spatially distributed to improve spatial equity, but many of these large enterprises were fully integrated, with multiple functions collected in one place. One consequence was the dominating presence of single *combinats* in the regional economy, but without forward and backward linkages into that economy. For example, Neftochim produces everything from octane to rubber tyres within a sprawling 1,300 hectare complex, but there were few local suppliers or subsidiaries until after 1991 (and even then many of these linkages were with small 'private' companies spun off from the large plant).

Moreover, despite the dominance and power of these industries, even their complete failure would have more regionally circumscribed impacts than *oblast*-level data would indicate. Neftochim, for example – despite its nearly 10,000-person labour force – draws the bulk of its workers from neighbouring towns and villages, with only small numbers commuting from outskirts of the *obstina* or neighbouring *obstini* in the region. This unusually concentrated labour force, plus the fully integrated nature of the *combinat* and the few direct forward and backward linkages into the local or regional economy, mean that Kameno and Bourgas *obstini* would absorb most of the impact of the closing of Neftochim. Based on 1991 employment figures, the hypothetical closure of the plant would have meant that unemployment in Kameno would have soared to 40 per cent or higher by 1994, while for Bourgas it would have increased to 5 per cent. Indirect effects would hit every service sector in those *obstini* and long-term unemployment might be chronic, but Aitos or Stara Zagora (*obstini* in the same *oblast*) would be largely unaffected.

In Bulgaria's enterprise space, linkages and power flow directly back to Sofia, where decisions are still made with respect to such plants. The enterprises rise imperiously in their regions, but have little direct impact on other sectors of industry whose organisational structures, political ties and geographies are quite different.

Textiles, apparel, machine building, electronics, agriculture and agricultural processing have an entirely different organisational structure, and consequently have played a very different role in the political economy of Bulgaria's post-communist transformation. It is to these that we now turn. In many ways it is on the branch structure of these sectors that the fate of the broader regional economies of Bulgaria will eventually revolve, and it is upon the reorganisation of these more broadly distributed industries that changing patterns of unemployment in Bulgaria do depend. To assess the changing patterns of unemployment we cannot use *oblast*-level data, but we must turn instead to the less abundant *obstini* data sets.

Mapping unemployment

One of the abiding characteristics of the slow transformation from command economy to free market in CEE has been high and persistent unemployment. World Bank (1991) projections of labour shedding in Bulgaria, even assuming a 'worst case scenario' (large labour redundancies and slow adjustment) have consistently underestimated actual experience (Table 5.5).

Not only is actual unemployment running about 7 per cent higher than the World Bank's worst case scenario, but there is every reason to believe that the estimates of actual unemployment are themselves low. The difference between the number of registered unemployed, as reported by the unemployment regional offices and the National Statistical Institute (680,200), and unemployment as measured in the 1992 Census of Bulgaria (783,500) suggests an undercount of about 15 per cent (ILO 1994: 35).

Demographically, the incidence of unemployment is unevenly distributed. For young men aged 15–24 the rate in March of 1996 was 37 per cent (NSI 1996). For school leavers in rural areas the rate was 57 per cent. The regional distribution of unemployment is also uneven.

Table 5.5 World Bank estimates and actual unemployment in Bulgaria (percentages), 1991–4

	%			
	1991	1992	1993	1994
World Bank estimates	1.7	7.8	10.9	
Actual unemployment	11.1	15.2	16.4	20.0

Sources: Weiner Institut für Internationale Wirtschaftvergleiche 1995; volume I: 123; *SRB* 1995: 50, 52.

In trying to explain broad regional patterns of unemployment in CEE, Fassman (1992: 55) identified three types of high unemployment region: (1) underdeveloped rural regions in peripheral locations; (2) old industrial cores; and (3) single-industry regions with local catastrophes. In his analysis of Bulgaria he argued, again based on *oblast*-level unemployment data, that the Plovdiv *oblast* is an example of an old industrial core region because in 1991 it was the only Bulgarian region with unemployment greater than 10 per cent. Fassman (1992: 55) argued that regionally varied unemployment in the CEE is typical in the early phases of transition, and that this mirrors western industrial restructuring in that high regional unemployment 'represents particular problem situations' that are masked by national level unemployment statistics (Figure 5.1). We would go further and argue that Fassman's own regional (*oblast*)-level unemployment patterns in fact mask the geographical structure of production within *oblasts* and hence the actual causes of district (*obstina*)-level variations in unemployment.

Figure 5.1 (after Fassman 1992: 55), showing the distribution of unemployment by *oblast* in 1991, would seem to justify Fassman's assertion. Fassman, however, makes two common errors. First, because he ignores the problem of scale, he misinterprets actual distribution of unemployment

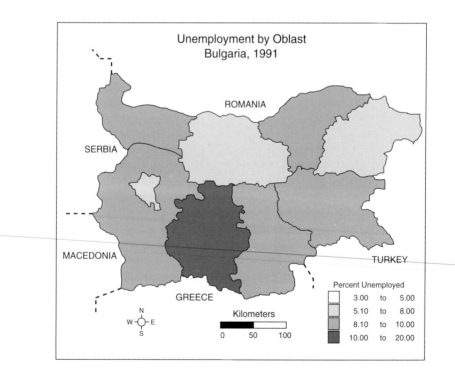

Figure 5.1 Oblast-level unemployment, 1991 (after Fassman 1992: 55)

in the Southern Rhodopes (the southern mountain region of Bulgaria bordering Greece). Second, because he works from generalisation of both scale and regional type, he misunderstands the causes of unemployment in the Southern Rhodopes.

Figure 5.2 shows the distribution of unemployment at the *obstina* level for April 1994. Although the unemployment for the Plovdiv *oblast* remained high (over 17 per cent), the high unemployment in the industrial core of Plovdiv now disappears. Plovdiv *obstina*, as opposed to Plovdiv *oblast*, shows unemployment of 6.1 per cent, only slightly higher than Sofia's 5.9 per cent. The source of unemployment in the Plovdiv *oblast* is not old industry in decline; in fact, by 1991 (the year Fassman based his arguments on) there had been no decline in core enterprise employment. Unemployment was entirely a function of plant closures and lay-offs from the largely Pomak *okrugi* of Smolyan and Parzardjik (Simsir 1988; NSI Smolyan 1995). Far from arising as a consequence of the decline of an old industrial core, the Plovdiv *oblast* shows high unemployment because seven rural and ethnic *obstini* show unemployment of greater than 30 per cent.

The overall pattern of unemployment in Figure 5.2 is clearly regional. However, it is not what we would expect given the patterns of relative

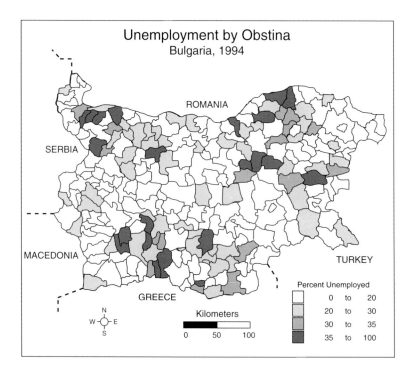

Figure 5.2 Obstina-level unemployment, 1994

129

industrial decline presented in Figure 5.1. Although the Montana and Rousse *oblasti* both show large declines in their share in the industry and have pockets of high unemployment, the Plovdiv and Varna *oblasti* (which fared relatively well between 1990 and 1994) also have *obstini* with high unemployment. If broad patterns of regional industrial decline do not explain the pattern of unemployment, what does?

We believe that the answer lies in the organisation of the branch-plant space economy under state socialism and during the transition. That explanation is grounded in three things: (1) the nature of industrial structure and practice under command economy institutions; (2) the recent history of economic reforms under communism; and (3) the differential responses by plant and regional managers to the specific process of restructuring after 1990. This pattern of unemployment in the transition years can, therefore, be understood only if command economy institutions and their effects are unpacked. This is not to say that market forces do not have an impact on the Bulgarian economy. They clearly do. But the transmission of largely external market impulses through the space economy of Bulgaria runs along the well-established conduits of the institutional arrangements of its command structure rather than along the ephemeral traces of emerging internal markets and their fledgling institutions.[8]

Spatial divisions of industry II: Branch-plant space

As we have seen, neo-classical regional convergence arguments have provided premature and inappropriate theoretical explanations for the changes occurring in the geographies of industrial production and employment in Bulgaria. We have argued that this results from a fetish for *oblast*-level data: the most accessible level of data available. *Oblast*-level data are too generalised to provide meaningful information about changes in some of these sectors of industry. For example, the branch-plant structure of textiles, apparel, electronics and agricultural processing – comprising core enterprises and networks of branch plants, workshops and/or processing co-operatives – has a distinctive 'regional' character. In the Bulgarian context this implies both *rural* and *ethnic* regions. In these sectors, patterns of regional convergence, rust belt decline, west–east increase in unemployment – all patterns based on *oblast*-level data – fail. Instead, the fortunes of the branch-plant economy depend largely on residual command economy institutions, social practices and ethnic differences. Identifying and understanding the role branch-plant economies have played in the rapid processes of regional change that are occurring, and the role of institutions, practices and cultural politics in these transformations will require a much finer grained analysis. For this we need a more detailed understanding of the organisational structure of industry and much finer grained plant and *obstina*-level data.[9]

Institutional legacies and changes in the regional structure of industries

Based initially on five-year plans, the goals and production quotas for enterprises were set by the Council of Ministers. Much of the quota setting occurred under the directive of the Council by bureaucrats within the Ministry of Industry. Ministries, in turn, negotiated quotas directly with some of the larger plants. Although the five-year plans which drove production quotas carried the force of law, they were strongly supported by an incentive and reward structure that combined material and moral incentives and established a tight relationship between performance and employee wages. Because overachieving on production quotas led to what McIntyre (1988b: 91) called a 'ratchet-like relationship' between one year's quota and another, plan fulfilment became the managers' primary goal. This in turn led to the types of behaviours which created the crisis of command economies: hoarding of labour and materials, storming, under-production, poor quality of output, and intense bargaining among enterprises and units within enterprises, between managers and workers, and enterprises and the central Ministry. Although the state supply network officially managed the flow of inputs among firms, personal networks, barter and corruption permeated the distribution process as enterprise managers negotiated the arcane rules of the command economy (Clarke *et al.* 1993; McIntyre 1988b; Pickles 1995).

These arrangements persist in the transition economy. Large firms which have long enjoyed a special status in the planning process continue to be protected by the central state because of their size, strategic importance, hard currency potential and confusion among politicians and bureaucrats over who has control. Although the vast majority of smaller plants enjoy no such shelter, they continue to be managed within a network of command economy relations. Distribution and supply networks for intermediate goods still depend on linkages. Subsidiary plants depend on parent plants for inputs and orders. Monopsony buying or marketing firms persist. Personal networks, barter and corruption have deepened in the face of supply shortages and increasing prices for inputs (see Smith and Swain in this volume on the recasting and endurance of pre-existing networks).

The early successes of the command economy in the initial stages of 'extensive' growth also resulted in the building of barriers to the production of high-quality consumer and technical goods, and rationalisation and intensification of the economy. Recognised as early as the 1960s in the Soviet Union, these problems led during the 1980s to a general pattern of putative reform throughout CEE. Reforms were of two types. In Hungary planners latched onto 'perfect computation' strategies where optimisation techniques were supposed to replace the hand of the market. More generally, command economies looked to the simplification of the quota system and the decentralisation of authority. Under such reforms, quotas were set in the simplest output terms and central planners released control of many decisions to enterprise managers. In Bulgaria these reforms led to the following: (i) decentralisation with much looser central supervision over enterprises and targets were limited to general production guidelines; (ii) enterprises were encouraged to think in terms of profits and foreign exchange earnings; (iii) industrial democracy within the workplace was encouraged (Walliman and Stojanov 1989); (iv) the Mineral Bank was established as a source of competitive investment funds for large and small firms (McIntyre 1988a; Jones and Meurs 1991); and (v) enterprises were encouraged to rely to a greater extent on internal financing of investments (Bristow 1996: 15).

Three elements of the 'New Economic Mechanism' reforms are particularly important in this regard (Wyzan 1988). First, the decentralisation of Ministry power to large industrial units meant that:

> The powers handed down by central authorities came to rest at a production level now made up of larger units, called production associations or *combinat*s. The association brought a number of previously independent enterprises under the direction of a single management, somewhat like branch-plants of a large modern Western corporation.
>
> (McIntyre 1988b: 93; see also Smith 1998 on the Slovak case and McDermott 1997 on Czechoslovakia)

This solution to problems of production (devolution of command from the top and concentration of power at the bottom) was to be repeated during the early years of transition. Second, during the last days of Zhivkov's reforms, Decree 56 was passed creating 'firms' as juridical persons. A variety of firm types were created which under Decree 56 'are established, recognised, and dissolved by the local district courts' (Wyzan 1988: 86; see also Freidberg and Zaimov 1997). Over 12,000 firms were established in 1989 alone under Decree 56, many of which 'had been set up by bureaucrats in the ministries and managers of state enterprises illegally to convert public property for themselves – the so-called "*nomenklatura* privatisations"' (Jackson 1991: 207). Third, one of the elements of the labour reforms of 1987 was the prescription of a wage bill to an enterprise rather than an hourly wage. Under such a system enterprise managers were allowed to choose between higher wages or full employment. As we will see later, the fate of 'branch plants' and 'workshops' depended on these interlinked reforms.

Industrial privatisation has not yet taken place in Bulgaria. The first sale of a state enterprise did not occur until February of 1993 when a maize-processing plant in Razgrad was sold to a Belgian firm for $45 million (Bristow 1996: 205). In 1995, 88 per cent of all industrial production was public sector (NSI 1996: 89). Between 1990 and 1994, a set of reforms was implemented, but – far from being a major break with the past – these were modelled on communist reforms of the 1980s. First, rather than face real privatisation of industry, reform was proposed under the banner of de-monopolisation (Bristow 1996: 85). Bulgartabak, for example, the huge state purchasing and producing monopoly in tobacco, was broken into twenty-two separate firms. There was little real competition involved as a consequence of this form of de-monopolisation, which tended instead to create smaller regional monopolies which in turn co-ordinated their activities informally through national associations (although in some cases, as with the tobacco monopolies, private contracts were set up with foreign firms). Even though real privatisation did not occur, the decentralisation of decision-making regarding employment and production did give more power to the enterprise at the regional level. This power was enhanced in 1991 when a total wage-bill mechanism was introduced by tripartite agreement among government, employers and trade unions. Average wages were no longer to be set by fiat. Instead enterprise managers could elect to allocate the wage bill according to the needs of the enterprise and could, for the first time, remain within its cap by laying off workers.

> [B]y late 1991 [a structure of tripartite wage commissions existed] in 11 ministries and 50 local districts. A major function at these levels was to negotiate how the national parameters were to be applied at lower levels – for instance, with regard to the trade-offs between the level of employment and the average wage for those in employment.
>
> (Bristow 1996: 45)

With the decentralisation of power came a retreat from production. Within the command economy structure plants had roles within a hierarchy. In apparel, for instance, some of the central plants organised production requiring basic labour in regionally dispersed workshops. Such workshops required little more investment than sewing machines. Infrastructural support, transport rebates and wage supplements were all provided under the mountain regions, border and/or depressed regions policies (Ministry of Construction and Territorial Development 1996). After 1991, faced with fixed wage bills, regional enterprises seem to have elected to close down the smaller, more marginal workshops and retain production and wages at central locations. In some cases (particularly textiles and machinery), not only would jobs atrophy at the most marginal tentacles of the enterprise, but equipment too would be recentralised to regional centres of production, preventing managers and workers from reorganising and keeping open the plant.

Networks, regional industrial change and the cultural politics of ethnicity

In discussing the broader concepts of networks as resources, Czako and Sik (1995: 227) argue that 'the half-century of communism actually strengthened pre-communist network oriented cultures'. In this view, command economies advantaged short-term profit seekers, particularly those who operated in the second economy or in the recesses of the formal economy, and rewarded social practices that drew heavily on strong social networks: 'socialization, self-organization, cartel building and maintenance ability, and quasi-corporate networking practices' (Czako and Sik 1995: 235).

Such networks are key to the power and practice of mid-level managers in post-communist Bulgaria. Czako and Sik (1995: 235) argue that 'just because the political regime changed in the short run, the institutional structures and operations of the "lower economy" hardly change'. Indeed, we would argue that this is precisely the current character of the formal sector in industry in Bulgaria today. In explaining the failure of command economy reforms in Hungary and Poland during the 1980s, Sachs (1994: 29) argues that reformers simply 'failed to realize that the "insiders" of the enterprises, the management and workers, could seriously distort enterprise behavior to their advantage'. He warns that 'if privatization proceeds too slowly, there is a risk that managers and workers . . . may come to view the enterprise simply as their own' (Sachs 1994: 82). Given the power to determine employment and production levels in branch plants and workshops, regional managers in Bulgaria have sought to preserve the employment base of local '*bliski*' first and extended 'enterprise *bliski*' second.

The notion of network is particularly strong in Bulgaria. In a survey of networks in CEE, Czako and Sik (1995) found Bulgarians to be more likely than Czechs, Poles or Hungarians to be engaged in reciprocal relations in their production and acquisition of food, construction materials and labour, transportation, or child care. They were almost twice as likely to use connections without payment to accomplish civil goals. They were more likely than Czechs, Poles or Russians to perceive the new capitalist elite as derived from the *nomenklatura* and likely to take advantage of others. Of particular importance here is the importance Czako and Sik found that Bulgarians placed on ethnic ties. In this section we build on these ideas of networks and ethnic affiliation and argue that the persistence of command economy structures, the importance of networks in Bulgarian social and economic relations, the process of industrial restructuring and employment decisions, and the particular regional character of ethnicity and branch plants combine to explain large parts of the differential patterns of unemployment found across Bulgarian *obstini*.

Figure 5.3 shows two indicators of ethnicity. First, it gives the percentage of the vote won in the 1990 parliamentary election by the Movement for

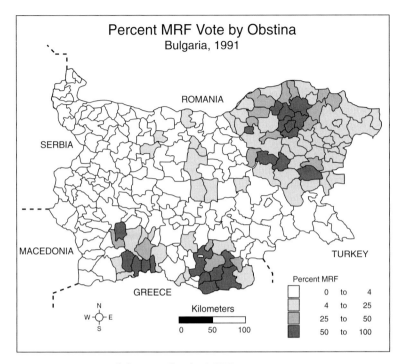

Figure 5.3 Percentage of the vote for the MRF, 1991

Rights and Freedoms. Although the Constitution makes ethnic political parties illegal, the MRF is led by a Muslim and considered by most to be a party of the ethnic Turks. In the absence of reliable census data on ethnicity, the vote is a good surrogate for the concentration of ethnic Turks in Bulgaria.

We cannot infer a direct causal relationship between unemployment and ethnic status from these maps alone, but the coincidence of unemployment and minority status is remarkably high. Table 5.6 shows the increasing concentration of the Turkish vote with increasing unemployment. The table takes the *obstina* as the unit of analysis and asks the question: to what extent does the proportion of MRF votes for an *obstina* go up as unemployment rises? The MRF vote clearly increases among those *obstini* with higher unemployment. The relationship is not linear, but the percentage of an *obstina* voting for the MRF is, on average, twice as high for the upper quartile of unemployment by *obstina*.

Table 5.7 displays these data for the ten *obstini* with the highest unemployment, where greater than 34 per cent of the labour force was registered as unemployed in 1994. Eight of the top ten *obstini* by unemployment have a high concentration of either Pomak or ethnic Turks, that is, of Muslims. For the twenty-three *obstini* with the highest unemployment (the

135

Table 5.6 Concentration of the Turkish vote with rising unemployment

Cut point (%) (N)	Cumulative frequency (unemployment, %)	Mean percentage for group	
		Unemployment	MRF vote
0 (256)	0	19.41	12.25
17.8 (126)	50	27.41	16.77
23.42 (62)	75	33.87	23.16
34.58 (23)	90	42.33	23.42

Sources: Labour Office 1994; Tzentralna Izbilratena Komicin 1991.

Table 5.7 Registered unemployment in the *obstini* with the ten highest rates, March 1994

Percent unemployment	Percent MRF vote	Pomak[10] concentration	Name
1 56	37		Antonovo
2 50	0		Medkove
3 49	2	Yes	Septemvri
4 48	55		Borino
5 44	58		Omurtag
6 44	45		Vetevo
7 44	49		Glavnitza
8 43	7	Yes	Rakitovo
9 43	4		Parmovay
10 39	14	Yes	Devin

Sources: Labour Office 1994; Tzentralna Izbilratena Komicin 1991.

upper decile in terms of unemployment) areas with concentrated Muslim populations make up 23.4 per cent.

Table 5.8 examines regional labour market segregation in June of 1993 (ILO 1994: 34). At the level of the *oblast*, unemployment ranged from under 10 per cent in the Sofia district to around 20 per cent in Plovdiv, Haskovo, Montana and Rousse. At the level of municipality (*obstina*) and regional labour office the discrepancies are even greater. The ILO reports that at this time, five regional labour offices had unemployment rates of over 40 per cent (ILO 1994). The spatial distribution of the regions of highest unemployment has little immediately apparent pattern. No urban–rural dichotomy is apparent. Devin and Kroumovgrad are peripheral, rural regions in the Rhodopes. Vetovo (Rousse) and Omurtag (Turgoviste) are both peri-urban *obstini* with development potential. Ljulin lies within the Sofia urban core which, in June of 1993, had an unemployment rate of only 8.8 per cent (ILO 1994). Dissimilar in many ways, these *obstini* yet share a common social characteristic. Table 5.8 lists these five regions and characterises them socially.

Table 5.8 Social characteristics of the regions with the highest unemployment
(> 40%), June 1993

Region	Salient social characteristic
Ljulin, Sofia	Main Gypsy area of Sofia
Devin, Plovdiv	Pomak region
Vetovo, Rousse	Rumelian Turkish region (45.39% MRF vote)
Omurtag, Rousse	Rumelian Turkish region (58.48% MRF vote)
Kroumovgrad, Haskovo	Turkish region (70% MRF vote)

Sources: Labour Office 1994; Simsir 1988.

Although these regions vary greatly in standard descriptors of vulnerability
to unemployment, they share a common demographic attribute: within these
regions exists a high concentration of some ethnic minority. In this context,
'the Muslim' exists in a separate Bulgaria – what Chatelot (*Le Monde*, 2 May
1997) has recently described as a new phase of indifference – and has corre-
spondingly suffered to a greater extent from regional industrial involution
(see Burawoy 1996). In rejecting a simplistic 'market forces' description of the
geographic distribution of labour shedding, we will not commit the equally
grievous folly of substituting a simplistic explanation like 'ethnic prejudice'.
This is not an explanation of the geographic pattern of labour shedding in
post-communist Bulgaria, but it does raise questions about the process.
That such prejudice might exist in the heart of the Balkans is a reasonable
proposition. The constitution of ethnic identity in Bulgaria is, however,
complex. In particular, the Rhodope region has a complex history of ethnic
tensions pre-dating and running through the communist period and an
important legacy from command economy institutions and practices (such as
a long history of resistance to agricultural collectivisation) which influence
the specific processes of economic reform and industrial restructuring that
have produced the current regional pattern of mass unemployment. It is in
the interaction of these patterns and processes – not in any one variable – that
an explanation of regional mass unemployment must be sought. But it is
precisely to the cultural politics of ethnicity that we now turn in order to
ask: (i) what have been the specific impacts of transformation on the Muslim
communities; and (ii) to what extent have the cultural resources of local
people and their own networks of association in Muslim enclaves and
peripheral regions been mobilised to deal with regional industrial collapse
and mass unemployment?

Impacts of the transformation

One explanation for this pattern of differentially high unemployment in
ethnic enclaves rests in the recent history of industrial development in

Bulgaria and in the pattern of decentralisation of authority and concentration of production at the sub-regional level that we describe above. Geographic equity has long been a goal of command economies (Berentsen 1979; Pleskovic and Dolenc 1982; Lakshaman and Hua 1987). Although imperfect in implementation, explicit policies for lagging regions existed in Bulgaria during the late 1970s and 1980s.

Bulgaria's post-1944 development was one of rapid industrialisation and urbanisation. By 1977, the Zhivkov regime was concerned enough with maintaining a stable agricultural base and rural population to implement key regions and key settlement policies. The key regions policy centred on the north-west (Montana), south-west and border regions with Greece and Turkey. These latter two regions contain a high number of Muslims. The Key Regions Settlement policy involved a programme of public buildings, services and the development of branch plants and workshops. The correspondence of investment to areas populated with Gypsies and Muslims was strong enough for these investments to be called 'the social industry policy'.

The branch plants so created worked to order from urban plants as suppliers of intermediate goods or jobbers. Workshops were frequently associated with apparel or textiles and employed the wives of men employed in mining or agriculture. The viability of these firms varied greatly in the 1990s, but their fate was uniform. With decisions on production and employment resting with bureaucrats and enterprise managers at the regional level, dissolution of Muslim workshops and factories occurred early – and almost totally.

In Ruen, a heavily Turkish region in the Bourgas *oblast*, textile workshops were formerly distributed throughout the villages of the *obstina*. By 1994, employment in workshops at the *obstina* level had dropped from nearly 1,000 to less than 200, leaving some villages with 90–95 per cent unemployment (see Pickles in this volume). In Dospat, a Pomak region, an apparel factory with standing contracts for Italian firms was closed down and the equipment shipped to Plovdiv. For the southern Rhodopes and Rousse, the early dissolution of tobacco co-operatives and the break-up of Bulgartabak, devastated the predominantly Turkish tobacco-growing population and ultimately contributed to the fall of the Berov government (see Meurs and Begg, this volume).

Networks of association and local cultural resources

Typically branch plants in peripheral regions like the Rhodopes have been seen by government officials and western analysts as prime candidates for closure: they are seen to have marginal equipment, low levels of efficiency, and high costs of production (fieldnotes 1995, 1996). To see these plant closures as a result of the working of a market calculus of efficiency and distance costs, however, misses the very nature of Bulgarian industry and Bulgarian society. Although some branch plants certainly fit this description, others have or had relatively new equipment and systems of production and quality control sufficient to satisfy European markets. In cases where equipment was not removed or sold off, some workshops, shoe factories and machine shops in the Haskovo *okrug* have established independent contracts either domestically or with Greek and Italian firms. In cases where the border regions policy had resulted in the location of high-tech electronics equipment in remote regions (such as Malko Turnovo), factories became the centre of independent entrepreneurial activities, including banking and cross-border trading.

Two important conclusions can be drawn here. First, plant closure in peripheral regions had as much (if not more) to do with core enterprise politics in the struggle over the conditions of privatisation and the protection of assets in a declining economy than it did with rationalisation of unproductive, inefficient and/or costly production systems left as a legacy of the branch-plant economy created under communism. But second, in cases where core enterprises failed to close branch plants and workshops, where managers and workers managed to sustain operations and/or where local state officials supported local producers against the diktats of the core enterprise, some successful enterprises have emerged.

It is also in these socio-ethnic relations and the regionally specific experiences of Muslim *bliski* (close personal and social ties embedded in broader networks of association (Asen Valiski 1996: personal communication)) that we find a new entrepreneurialism and a revitalised cultural politics struggling to protect threatened enterprises and sustain local communities. Efforts to create any kind of defence against the economic crisis brought about by the collapse of the branch-plant economy have forced residents of peripheral regions to draw on whatever cultural resources are at their disposal and deepen them wherever possible. This involves strengthening economic ties with Turkey through the complex family linkages forged under successive expulsions and return migrations. It includes an extension of household and community dependence on remittances from migrant workers. It includes protecting and enhancing networks of social power (be they organised through the family, religion, politics, or trade) to consolidate and protect whatever economic gains can be made. In part, this involves a turning to new economic activities and forms, of which barter (especially of potatoes for other foodstuffs) and the formalising of informal activities (such as the recent extension of the traditional practices of mushroom-gathering into the export economy for the European Union) play an increasingly larger role. But it is also in this context that managers and workers in workshops and former branch plants are struggling to find new industrial linkages and logics to sustain themselves in the current transformation.

Conclusion

The deposing of Todor Zhivkov in November 1989 in a closed meeting of the politburo has often been characterised as a clear and abrupt demarcation between an old rigid communist regime and a new form of capitalist regulation; between a legacy of despotism and an era of democracy. Certainly the power of such symbols, the toppling of dictators or the falling of walls, were essential mobilising sites for political transformation (see Kubik 1994). But they can also obfuscate analysis of an economy in transition. We argue that the current state of the Bulgarian economy is best understood as a slow and incomplete transformation of a system that was already in crisis and already undergoing change. This is not to argue that actors are not responding to crisis in economically rational ways; indeed, daily life in Bulgaria constantly requires responses by individuals and groups to complex and uncertain changing circumstances. But in theorising *this* transformation we would be well served to think of economic change in terms of networks of association developing wars of position in a strategic geopolitics in which locality and heritage provide common resources on the basis of which people respond to uncertain and difficult circumstances. And it is these complex networks and legacies that produce the patterns of regional change we see emerging. More specifically, the transformation of the regional economies of the southern Balkans is an economic transformation rooted in the social and institutional legacies of central planning, and as such our analyses and policy prescriptions must be couched in terms of metaphors and ways of thinking that take their meaning from the institutional practices of command economies.

From a neo-classical perspective, it does not matter who the new private owners are as state property is sold off. In building market regulation and market institutions, it does not matter who passes the laws, so long as they are passed. For us, however, such things are crucial. Indeed, it is difficult to understand why the cadre of *nomenklatura* who *were* the state during the period of the command economy should (or even can) relinquish either their networks or their practices without force or compensation. Nowhere in CEE has the ballot or the IMF exerted sufficient pressure to dislodge completely either the old networks of co-operation or some of the systems of discipline and disincentive inherent within the institutional structure of the command economy. In Bulgaria those institutions still dominate in industry.

We have argued that industrial restructuring and labour shedding occur at two levels of the economy. Enterprise space is a metaphor for a distributed group of large factories whose importance to the economy and embeddedness within the control structures of a command economy make them unique. They are sheltered from market forces, produce a significant proportion of national output, and continue to employ large numbers of people. They are tied to the metals, chemical, weapons sectors. Branch-plant space is a metaphor for a broadly scattered group of enterprises, branch plants and workshops. These typify the organisational structure of the apparel, textile, electronics, machine and agricultural processing industries. These plants are more vulnerable to the indirect hand of the market. The market is felt, however, through a hierarchical bureaucracy that persists as a remnant of the command economy, and which has used the rhetoric of cost, ageing equipment and inefficient locations to legitimise their garrisoning of resources and political power in the core enterprises at central locations. Certain regions have proved more vulnerable in a situation in which decisions regarding labour shedding at the regional level have been decentralised to enterprise managers, as a result of which power, employment, control over the wage bill, and production have been further aggregated over time in those locations that served as the heart of extended production complexes and branch-plant economies. These locations correspond to the major urban centres in each of the regions of the country.

For a mode of regulation to be stabilised

> there must exist a materialisation of the regime of accumulation taking the form of norms, habits, laws, regulation networks and so on that ensure the unity of the process . . . this body of interiorised rules and social processes is called the mode of regulation.
> (Lipietz 1987, 1991: 23)

For forty years command economies developed an intricate system of rewards and sanctions which produced dramatic extensive industrialisation. The mode of regulation which governed that growth was a plan-driven system of central controls embedded in sets of social and political relations that linked apparatchiks, workers, plants and regions. It was a complex system of overlapping national and regional networks. It was a set of institutions and habits that are slow to change and almost impossible to erase overnight (despite the wholesale destruction of industry in some regions). Sachs (1994: 28) described the Polish and Soviet reforms of the 1980s in ways that apply equally well to the Bulgarian reforms of the same period: 'The main step was enterprise reform, promoted in both countries in the late 1980s, designed to give greater freedom to the state enterprises to set wages, inputs, and outputs.' These reforms were partial and unsuccessful. For Sachs (1994: 29) the failure of the 1980s reforms is a failure on the part of party leaders

> to realise the insiders of enterprise, the management, the workers could seriously distort enterprise behaviour to their own advantage and . . . increase their income at the expense of the state by absorbing whatever income flow and whatever assets they could from state enterprises.

The reforms of the 1980s and the transformations of the 1990s failed to recognise the potential of the established hierarchy, as well as new actors intent on appropriating the new spaces of power, to influence these processes of change. Liberalisation has not only permitted a conversion of political to economic capital and formed the basis for emerging class distinctions (what Szelenyi (1988) has referred to as 're-embourgeoisement'), but has encouraged outright competition between core industries and workers (i.e., those at the heart of distributed production complexes and branch industries) and those in peripheral regions. The legacy of state planning and its Stalinist/Zhivkovian assimilation policies now compounds regional institutional competition for scarce and declining resources with the geography of ethnicity.

The nature of the current transformation depends on how legislatures, political parties, bureaucrats, enterprise directors, plant managers and other interest groups respond to the complex challenge of constructing a new mode of regulation, transforming the institutional linkages and social and spatial practices of state socialism as they do. Only at the level of the enterprise structure and the local region can we discern this complex set of relations. In this interpretation, regional mass unemployment does not stem from the decline of old industrial cores (although they may also be declining), or from the rationalisation of inefficient and costly distributed branch-plant economies (although branch plants have been closed down). Instead, it is in the social, geographical and institutional arrangements of state socialism, and the positional wars over scarce resources in the period of transformation, that we must look for explanations of mass unemployment.

Acknowledgements

The research on which this chapter is based was funded by National Science foundation grants nos. INT-9021910 and SBR 9515244. We would like to thank Mieke Meurs, Dimitrina Mikhova, Christo Ganev and Stefan Velev for their help with parts of this research, and Adrian Smith for helpful comments on the chapter.

Notes

1 Unemployment statistics are notoriously unreliable, and under-report actual unemployment, especially because they do not record the discouraged unemployed and those who have assumed that, having lost a job, they are practically 'retired'.
2 The regulation of wages by ministries during the late 1980s and early transition period is central to our understanding of the spatial distribution of unemployment in Bulgaria. Briefly, with exceptions, the wage bill is determined by ministry and the plant managers allowed discretion over wage–job trade-offs.

What is important to our explanation is: (1) the continuity of wage bill discipline between regimes; (2) the devolution of discretion to the firm and factory level; and (3) the relationship of the institutional and spatial structure of branch plants. The role of wages in such a discussion is substantially different from the role it plays in neo-classical models of regional convergence.

3 Bulgaria's principal administrative division is into geographic units consisting of nine *oblasti* (singular *oblast*), equivalent to provincial governments or regional administrations, and 277 *obstini* (singular *obstina*) equivalent to counties. The land area of Bulgaria is 110,631 square kilometers, about the size of the state of Pennsylvania. See Koulov (1993) for a more detailed description of Bulgarian administrative divisions past and present.

4 The coefficient of variation for 1990 increases by 50 per cent, from .22 to .33 in 1994. For per capita production the coefficient of variation increases from .10 to .29. The most immediate conclusion to be drawn from these data is that Bourgas's 7.4 per cent increase in absolute share has been at the expense of the rest of the country (*SRB* 1991: 414, 431; 1995: 412, 424).

5 With Bourgas included $R^2 = 4.4$, but without Bourgas $R^2 = .95$.

6 Since these data indicate the percentage of plants that would have to be relocated to create a more even distribution, it is meaningless to calculate figures for sectors with one plant.

7 In September of 1996 Kremikovtsi had 16,000 people in a 1962 Soviet steel plant. Its continuing debt had led to the failure of one of its main bankers, Economic Bank. At the behest of the World Bank and IMF, Kremikovtsi will become one of seventy-one state firms without a credit line.

8 Even the banks operated until recently as they have since the 1980s (first as state banks and later as private banks) as virtually unrestricted loan agencies for debt-ridden enterprises.

9 Both have been extremely difficult to obtain. *Obstina*-level unemployment data were only available in 1995, and then only through 'special' channels. Plant-level data on employment and lay-offs have never been made publicly available in any systematic fashion.

10 For current purposes, regions are considered Pomak if cited as such in Amnesty International (1986) and Simsir (1988). The Pomak are Muslims of Bulgarian heritage who did not generally vote for the MRF.

References

Amnesty International (1986) *Bulgaria: Imprisonment of Ethnic Turks*, London: Amnesty International.

Barta, G. (1992) 'The changing role of industry in regional development and regional policy in Hungary', *Tijdschrift voor Economische en Sociale Geografie* 83, 5: 372–9.

Begg, R. and Meurs, M. (1997) 'Writing a new song: state policy and path dependence in Bulgarian agriculture', in I. Szelenyi (ed.) *Comparative Perspectives on Decollectivization: Russia, Bulgaria, Hungary, China, and Cuba*, Boulder: Westview Press.

Berentsen, W. H. (1979) 'Regional planning in the German Democratic Republic: its evolution and goals', *International Regional Science Review* 4, 2: 137–54.

Bristow, J. A. (1996) *The Bulgarian Economy in Transition*, Cheltenham: Edward Elgar.

Buckwalter, D. W. (1995) 'Spatial inequality, foreign investment, and economic transition in Bulgaria', *Professional Geographer* 47, 3: 288–98.

Burawoy, Michael (1985) *The Politics of Production*, London: Verso.

Burawoy, Michael (1996) 'The state and economic involution: Russia through China lenses', *World Development* 24, 6: 1105–17.

Centre for the Study of Democracy (1995) *Evaluation of Privatisation Results for 1994*, Sofia: Centre for the Study of Democracy.

Chinitz, B. (1961) 'Contrasts in agglomeration: New York and Pittsburgh', *American Economic Review* 51: 279–89.

Clarke, S., Fairbrother, P., Burawoy, M. and Krotov, P. (1993) *What About the Workers? Workers and the Transition to Capitalism in Russia*, London: Verso.

Fan, C. C. (1995) 'Of belts and ladders: state policy and uneven regional development in post-Mao China', *Annals of the Association of American Geographers* 85, 3: 421–49.

Fassman, H. (1992) 'Phaenomene der Transformation – Oekonomische Restrukturierung und Arbeitslosigkeit in Ost-Mitteleuropa', *Petermanns Geographische Mitteilunen* 136: 49–59.

Friedberg, James and Zaimov, Branimir (1997) 'Politics, environment and the rule of law in Bulgaria', in K. Paskaleva, P. Shapira, J. Pickles and B. Kulov (eds) *Bulgaria in Transition: Environmental Consequences of Political and Economic Transformation*, Ashgate: Aldershot.

Ganev, Christo (1989) 'The urban process and the appearance of agglomerations in Bulgaria', *Socio-economic Planning Sciences* 23, 1: 17–22.

Glenny, M. (1993) *The Rebirth of History*, London: Penguin Books.

Gowan, P. (1995) 'Neo-liberal theory and practice for Eastern Europe', *New Left Review* 213 (September/October):

International Labour Organisation (ILO) (1994) *The Bulgarian Challenge: Reforming Labor Market and Social Policy*, Budapest: International Labour Organisation.

Jackson, M. R. (1991) 'The rise and decay of the socialist economy in Bulgaria', *Journal of Economic Perspectives* 5, 4: 203–9.

Jones, D. and Meurs, M. (1991) 'On entry of new firms in socialist economies: evidence from Bulgaria', *Soviet Studies* 43, 2: 311–27.

Kaldor, N. (1975) 'What is wrong with economic theory', *Quarterly Journal of Economics* 89, 3: 347–57.

Koulov, B. (1993) 'Tendencies in the administrative territorial development of Bulgaria', *Tijdschrift voor Economische en Sociale Geografie* 83, 5: 390–401.

Kubik, Jan (1994) *The Power of Symbols against the Symbols of Power: The Rise of Solidarity and the Fall of State Socialism in Poland*, University Park, PA: Pennsylvania State University Press.

Labour Office (1994) 'Unemployment by obstina', Sofia: Bulgarian Labour Office, mimeo.

Lakshaman, T. R. and Hua, C. (1987) 'Regional disparities in China', *International Regional Science Review* 11, 1: 97–104.

Lipietz, A. (1987) *Mirages and Miracles*, London: Verso.

Lipietz, A. (1991) 'New tendencies in the international division of labor: regimes of accumulation and modes of regulation', in M. Storper and A. J. Scott (eds) *Pathways to Industrialization and Regional Development*, New York: Routledge.

McDermott, G. A. (1997) 'Renegotiating the ties that bind: the limits of privatisation in the Czech Republic', in G. Grabher and D. Stark (eds) *Restructuring Networks in Post-Socialism: Legacies, Linkages, and Localities*, Oxford: Oxford University Press.

McIntyre, R. (1988a) 'The small enterprise and agricultural initiatives in Bulgaria: institutional incentives without reform', *Soviet Studies* 40, 4: 602–15.

McIntyre, R. (1988b) *Bulgaria: Politics, Economics, and Society*, London: Pinter.

Massey, D. (1984) *Spatial Divisions of Labour*, London: Macmillan.

Ministry of Construction and Territorial Development (1996) *Regional Policy in the Republic of Bulgaria*, Sofia: Ministry of Construction and Territorial Development.

Murgatroyd, L. and Urry, J. (1983) 'The restructuring of a local economy: the case of Lancaster', in J. Anderson, S. Duncan and R. Hudson (eds) *Redundant Spaces in Cities and Regions? Studies in Industrial Decline and Social Change*, London: Academic Press: 67–78.

Murphy, A. B. (1992) 'Western investment in East-Central Europe: emerging patterns and implications for state stability', *Professional Geographer* 44, 3: 249–59.

Myrdal, G. (1958) *Economic Theory and Underdeveloped Regions*, Bombay: Vora and Company.

National Statistical Institute (NSI) (1991–6) *Statistical Yearbook*, Sofia: Republic of Bulgaria.

National Statistical Institute (NSI) (1995) *Statistcheski cbornik Smolyan '94*, Smolyan.

Pavlínek, P. (1992) 'Regional transformation in Czechoslovakia: towards a market economy', *Tijdschrift voor Economische en Sociale Geografie* 83, 5: 361–71.

Pickles, J. (1995) 'Restructuring state enterprises: industrial geography and Eastern European transitions', *Geographische Zeitschrift* 83, 2: 114–31.

Pleskovic, B. and Dolenc, M. (1982) 'Regional development in a socialist, developing and multinational country: the case of Yugoslavia', *International Regional Science Review* 7, 1: 1–24.

Privatisation Agency (1995) *Privatisation in Bulgaria*, Sofia: Privatisation Agency.

Richardson, H. W. (1980) 'Polarization reversal in developing countries', *Papers of the Regional Science Association* 45: 67–85.

Sachs, J. (1994) *Poland's Jump to the Market Economy*, Cambridge, MA: MIT Press.

Scott, A. J. (1988) *New Industrial Spaces*, London: Pion.

Simsir, B. N. (1988) *The Turks of Bulgaria*, London: K. Rustum and Brother.

Smith, A. (1988) *Constructing Capitalism: Industrial Restructuring and Regional Development in Slovakia*, Cheltenham: Edward Elgar.

Smith, A. (1994) 'Uneven development and the restructuring of the armaments industry in Slovakia', *Transactions of the Institute of British Geographers* 19: 404–24.

Smith, A. (1995) 'Regulation theory, strategies of enterprise integration and the political economy of regional economic restructuring in Central and Eastern Europe: the case of Slovakia', *Regional Studies* 29, 8: 761–72.

Statistical Reference Book (*SRB*) (1991 and 1995) Sofia.

Stark, D. (1997) 'Recombinant property in East European capitalism', in G. Grabher and D. Stark (eds) *Restructuring Networks in Post-Socialism: Legacies, Linkages, and Localities,* Oxford: Oxford University Press: 35–69.

Szelenyi, I. (1988) *Socialist Entrepreneurs,* Madison: University of Wisconsin Press.

Tzentralna Izbrilratena Komicin (1991) *Buletin za resultate ot izborite za napoddin predstaviteli provedeni na 13 oktombri 1991 godini,* Sofia: Tzentralna Izbrilratena Komicin.

Walliman, I. and Stojanov, C. (1989) 'Social and economic reform in Bulgaria: economic democracy and the problems of change in industrial relations', *Economic and Industrial Democracy* 10, 3: 361–78.

Weiner Institut für Internationale Wirtschaftsvergleiche (1995) *Countries in Transition 1995,* Vienna: Weiner Institut für Internationale Wirtschaftsvergleiche.

Williamson, J. G. (1965) 'Regional inequality and the process of national development', *Economic Development and Cultural Change* 13, 4: 3–45.

World Bank (1991) *Bulgaria: Crisis and Transition to a Market Economy,* Washington: World Bank.

Wyzan, Michael (1993) 'Economic transformation and regional inequality in Bulgaria: in search of a meaningful unit of analysis', paper presented at the American Association for the Advancement of Slavic Studies, Honolulu: 19–22 November.

Zaniewski, Kazimierz (1992) 'Regionl inequalities in social well-being in Central and Eastern Europe', *Tijdschrift voor Economische en Sociale Geografie* 83, 5: 342–52.

6

ECONOMIC RESTRUCTURING AND REGIONAL CHANGE IN RUSSIA

Michael Bradshaw, Alison Stenning and Douglas Sutherland

Introduction

The processes of decentralisation and regionalisation, which accelerated the collapse of the Soviet Union, are now dominant forces in Russia's political economy. Regions, as represented by their executive and legislative bodies, are themselves increasingly seen as important political and economic actors. Thus, one of the key tasks for post-Soviet geography and economics is to try to explain regional patterns of economic change across the Russian Federation. To that end, this chapter offers an exploration of the relationship between economic restructuring and regional change in Russia at both the national and regional scales. The chapter is divided into two major sections: the first section examines the interrelationship between economic and regional change at the macro-scale, across the majority of Russia's eighty-nine federal units; the second section examines the politics of economic change in one region: Novosibirsk in West Siberia. The different scales of analysis adopt different methodologies, and reflect different approaches towards theorising transition. At the macro-scale, statistical analysis is used to examine the relationship between a variety of economic indicators of decline and renewal, the aim being to relate macro-structural changes in the Russian economy to processes of regional economic change. However, we are well aware of the limitations of this approach and the place-specific analysis presented in the second part of the chapter serves to exemplify the need for local-level study, with its focus on local politics and culture as constituting elements in the process of regional economic change. Thus, this parallel approach to understanding regional change highlights the dangers of over-generalisation at the macro-scale and of focusing upon purely economic aspects of transformation; however, this study also warns against the over-extension of the findings of case study research. Ideally the two should go hand in hand, enabling an understanding of economic life as a set of social and cultural practices.

Transitional recession and economic restructuring

All the countries undergoing economic transformation in Central and Eastern Europe and the former Soviet Union have suffered, and in many cases continue to suffer, a decline in output regardless of their initial macroeconomic starting positions and experiences of introducing reforms (World Bank 1996a: 26–9). As a result a 'J-curve' hypothesis has been advanced to explain the initial contraction of output, often described as a 'transitional recession' followed by recovery as market mechanisms begin to be the dominant method of economic co-ordination (Brada and King 1992). Possible causes for the initial contraction in economic activity include shifts in demand, the disruption of supply linkages and the collapse of the volume of trade as a result of the dismantling of the CMEA (Council for Mutual Economic Assistance) and the Soviet Union. As Figure 6.1 shows, the reality of post-Soviet economic recovery has been somewhat different: for the most part the economies of Central Europe seem to have bottomed out and are now on the road to recovery, while the Commonwealth of Independent States (CIS) members seem to be experiencing more of an 'L-curve' – dramatic decline followed by a prolonged economic slump (Lavigne 1995: 152). The different experiences of Central and Eastern Europe and the CIS suggest that evidence of economic recovery is harder to discern in the CIS.[1] Equally, there are substantial variations within these regional groupings (see Dunford in this volume).

A major contributory factor to the output decline, which was underestimated at the onset of economic transformation, was the degree of

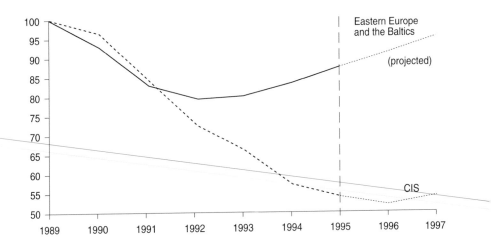

Source: EBRD Transition Report 1996: 111

Figure 6.1 Real GDP for CIS and Eastern Europe, 1989–97

institutional reform necessary. Key issues such as the creation of transparent and stable property rights (which remains a problem in Russia), effective corporate governance, financial sector development, contract enforcement and the development of a regulatory environment in which policy reform and the transmission of even basic economic information could be achieved were not resolved overnight, nor could they be. There is no quick-fix transition from plan to market (Stark 1995). To a large extent, the ability of individual countries, and even regions in a Russian context, to resolve these issues favourably has determined the length of the initial decline and the speed of any subsequent recovery, and as such helps to account for some of the variation in 'transition performance' across countries.[2] However, it would be wrong to think that the starting point for transition was an institutional void. Many 'Soviet' institutions have in fact remade themselves to fill new roles in the emergent market economy, and others have continued to function much as before (see Smith and Swain in this volume). Thus, processes of institutional adaptation and adoption are at work (Grabher and Stark in this volume) and, as the Novosibirsk case reveals, the nature of local institutional thickness may be an important contributor to a region's economic performance (see Amin and Thrift 1994).

Macroeconomic trends

Despite the poor performance of the post-Soviet republics relative to the economies of Central Europe, by the end of 1996 the Russian government had achieved some notable macroeconomic successes. Inflation was reduced from 2,509 per cent in 1992 to a provisionally estimated 22 per cent in 1996. The imposition of a rouble corridor resulted in greater exchange rate stability from the end of 1995. Furthermore, Russia's extreme wealth of raw materials, coupled with low energy and labour costs for primary processing, meant that Russia was less adversely affected by the disruption of trade links with former CMEA partners and republics of the Soviet Union by being able to transfer much of its external trade to Western Europe and North America. Between 1992 and 1995, for example, OECD trade with Russia increased more than 50 per cent, in dollar terms, whereas trade with the traditional partners either fell or remained sluggish. This helped to sustain the level of resource production and primary processing, but also served to increase the level of import competition for the inefficient manufacturing and consumer goods sectors of the Russian economy. Consequently, a positive foreign trade balance has risen annually from just over $10 billion in 1992 to an estimated $40 billion in 1996, although this is tempered by large-scale 'capital hover' – flight capital in offshore bank accounts waiting for conditions to justify investment in the Russian economy, amounting to some $10 billion annually. However, the appearance of countries such as Switzerland and Cyprus in the top ten foreign investors in Russia suggests that Russian capital deposited in offshore bank accounts is now starting to return in order to realise new investment opportunities in Russia. Despite the dubious origins of this flight capital, this is further evidence of increased stability in the Russian economy.

While there are signs of economic stabilisation, recovery remains elusive. In 1996 real GDP declined by a further 6 per cent. Industrial production fell by a further 5 per cent in 1996, which masks a gradual shift in the composition of output from large-scale enterprises to smaller-scale activity. The relatively unreformed agricultural sector continued to suffer real declines in output. Furthermore, there has been a severe collapse of investment activity to 45 per cent of the 1992 level by the end of 1996. However, while the central state, in the form of the federal government, has cut its share of investment, the local state, in the form of republic/*oblast* administrations and city governments, has become more involved in financing local economic development. What is more problematic is the reluctance of Russia's new private financial sector to fund industrial restructuring. At present the shortage of capital investment is undoubtedly hindering the growth of new forms of economic activity.

As a result of persistent production decline, coupled with a reduction in state subsidies, and a shortage of funds to finance new growth, there is now a steady growth in the rate of unemployment, based on the ILO methodology (see Goskomstat 1996: 60–3), from 4.8 per cent in 1992 to 9.2 per cent in 1996. Regional variation in Russian unemployment is, however, vast – ranging from 5.8 per cent in Moscow at the end of 1995 to around 24 per cent in Karachaevo-Cherkess and North Ossetia in the North Caucasus. Given that official figures do not account for the widespread phenomenon of 'hidden unemployment', it is clear that economic hardship is now widespread in Russia (McAuley 1995).

Key characteristics of economic restructuring in Russia

Despite its somewhat sluggish pace, there has already been substantial structural change in the Russian economy (see OECD 1995: 3). The Soviet-type economy can be characterised as being hyper-industrialised with, in comparison to western industrial economies, a greater emphasis on heavy industry and military-industrial sectors and a relative neglect of light industry (particularly consumer goods production) and the service sector. The collapse of the Soviet system is therefore likely to bring about a change in the macroeconomic structure of the Russian economy (Sutherland and Hanson 1996: 368–74). The composition of output is increasingly shifting from large-scale industrial output, the backbone of the Soviet economy, to smaller-scale manufacturing activity and new services. At the same time, the primacy of the military-industrial complex is being replaced by the primacy of the consumer and the international market-place. If one takes into account the shift in relative prices, from those reflecting planners' preferences to those reflecting the 'true' cost of production, then the typically reported sectoral declines take on a somewhat different hue (Gavrilenkov and Koen 1994). Catastrophic absolute declines in light industry are translated into a constant share of industrial output, while food processing is shown to have grown substantially rather than declined. In the case of light industry, this is because its output was overpriced by Soviet planners. Today, light industry prices are only 30 per cent of the 1991 prices in real terms (*Russian Economic Trends* 1996: 55). This serves to demonstrate the problems involved in trying to discern what is actually going on in the Russian economy. In agriculture similar problems cloud the picture over whether an output decline is the result of another process. For example, the historically high per capita consumption of meat is declining as consumer purchasing power declines, reducing the need for both livestock and grain for animal feed, while at the same time imports of meat are growing as consumers favour higher-quality imported products (Sedik, Foster and Liefert 1996). Whether such compositional changes in agriculture are desirable is a moot point, though output declines of these products can at least be understood as rational at the scale of the national economy and the individual consumer.

Dynamics of regional collapse and decline

This section seeks to examine the regional dimensions of economic decline, identifying those types of regions which have suffered the most during 'transitional recession'.

Production decline

For the first three years after the onset of economic transformation, industrial production fell at a faster rate than overall GDP decline. In 1995 and 1996 the picture was reversed, with industry showing more modest declines. Industrial decline in the first eleven months of 1996 was −5.0 per cent, whereas real GDP decline was −6.3 per cent. The overall declines in output experienced by the various industrial sectors resulted from the cessation of 'pure socialist output' (goods that would not be produced in a market economy) along with the loss of market share to foreign products as producers faced for the first time the consequences of foreign trade liberalisation.

Between 1992 and 1994, patterns of industrial decline varied significantly at the regional level, from −3.5 per cent in the Republic of Sakha (Yakutia) in the Russian Far East to −63.2 per cent in the Republic of Dagestan in the North Caucasus (Figure 6.2). In general, regions with strong raw material bases or an interventionist local administration performed better than the Russian average, while regions more dependent on agriculture and weaker industrial sectors suffered the greatest output declines (Table 6.1). It is also

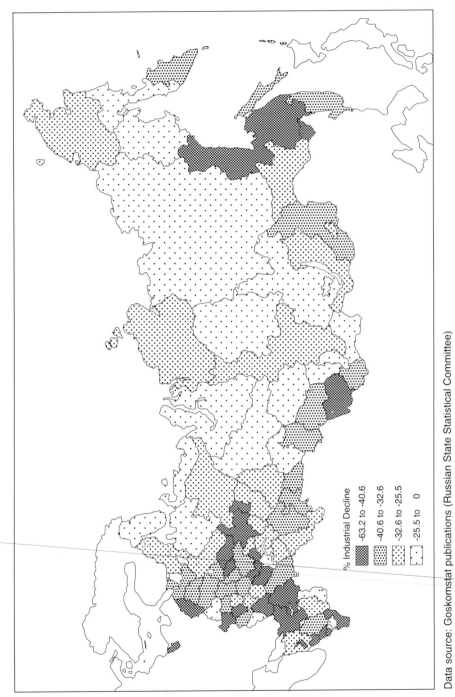

% Industrial Decline
-63.2 to -40.6
-40.6 to -32.6
-32.6 to -25.5
-25.5 to 0

Data source: Goskomstat publications (Russian State Statistical Committee)

Figure 6.2 Decline in the physical volume of industrial output between 1992 and 1994

Table 6.1 Ranking of regions by key variables

Industrial output 1994/92 (RF[a] = −32.1)		Industrial employment 1994/92 (RF = −12.9)		Export growth relative to Russian average 1992/94		Employment growth 1992/94 (RF = −5.0)		New service growth (RF = 7.4)		Unemployment % 1992 (RF = 4.7)	
−3.5	Sakha	8.3	Yamal-Nenets	200.2	Altay Republic	16.5	Adygeya	54.2	Moscow	3.4	Samara
−3.9	Nenets	2.4	Sakha	18.6	Kalmykia	8.6	Moscow	23.1	Adygeya	3.2	Tambov
−8.0	Koriak	2.2	Khanty-Mansy	17.2	Kabardino-Balkar	3.3	Tyumen	21.6	Perm	3.2	Tatarstan
−11.0	Evenk	0	Nenets	13.2	Kurgan	1.9	Chita	19.5	St Petersburg	3.2	Khakasia
−11.2	Ulyanovsk	0	Karachevo-Cherkess	10.8	Krasnoyarsk	0.6	Altay Republic	18.0	Sverdlovsk	3.1	Belgorod
−13.6	Yamal-Nenets	0	Taimyr	8.4	North Ossetia	−1.4	Nizhniy Novgorod	13.3	Moscow *oblast*	2.9	Vologda
−16.7	Khakasia	0	Evenk	3.6	Khakasia	−1.4	St Petersburg	13.1	Samara	2.8	Orel
−17.0	Khanty-Mansy	0	Aginsk Buryat	3.2	Vologda	−1.6	Volgograd	10.9	Volgograd	2.4	Kursk
−17.7	Belgorod	−3.1	Tyumen	2.9	Lipetsk	−1.8	Moscow *oblast*	10.9	Kaliningrad	2.3	North Ossetia
−17.8	Tomsk	−4.3	Astrakhan	2.5	Kemerovo	−2.0	Krasnodar	10.4	Novosibirsk	0.7	Sakhalin
−47.0	Bryansk	−20	Koriak	0.4	Krasnodar	−13.1	Karachevo-Cherkess	−7.1	North Ossetia	27.8	Chukchi
48.5	Rostov	−20.6	Orel	0.4	Chita	−13.1	Rostov	−7.6	Altay Republic	12.5	Dagestan
−48.9	Kaliningrad	20.7	Khabarovsk	0.4	Sakha	−13.2	Murmansk	−8.7	Penza	9.4	Kabardino-Balkar
−49.3	Mordova	−21.4	Tuva	0.3	Magadan	−13.6	Pskov	−8.8	Kalmykia	7.6	Adygeya
49.5	Altay Kray	−23.7	North Ossetia	0.3	Karachevo-Cherkess	−13.9	Udmurt	−9.2	Buryatia	7.5	Tuva
50.4	Kabardino-Balkar	−25	Pskov	0.3	Mordova	−14.0	Kurgan	−10.4	Tuva	7.3	Kalmykia
−52.1	Khabarovsk	−25.9	Dagestan	0.3	Tomsk	−14.3	Tambov	−11.1	Sakha	7.0	St Petersburg
−52.6	Jewish	−35.8	Kamchatka	0.2	Volgograd	−15.9	Kabardino-Balkar	−11.4	Kurgan	6.9	Tomsk
−52.9	Adygeya	−38.9	Chukchi	0.2	Dagestan	−23.1	Chukchi	−15.2	Dagestan	6.9	Kamchatka
−63.2	Dagestan	−39.3	Magadan	0.1	Adygeya	−24.9	Magadan	−18.7	Magadan	6.8	Leningrad

Source: Goskomstat publications, Russian State Statistical Committee.

Note

a RF = Russian Federation.

worth noting that interventionist local administrations have taken a variety of approaches, the most oft-quoted comparison being that of Ulyanovsk, which tried to resist marketisation by protecting the local economy and subsidising prices, and Nizhniy Novgorod, which embraced market reforms and sought to attract foreign investment (see Layard and Parker 1996: 243). As the second part of this chapter reveals, even conservative local administrations are embracing, or being forced to adopt, market solutions in an attempt to stimulate economic regeneration. Thus, at the local scale, economic performance is shaped by a region's resource endowment, its economic-geographic position in the new market economy, its economic structure, the calibre and vision of the local leadership and the nature of their relationship with the federal government in Moscow.

De-industrialisation

The downsizing of industrial employment is a continuing feature of the Russian economy. Between 1992 and 1994 industrial employment fell by 12.9 per cent, masking regional variations of growth in raw-material-rich regions and declines in the Russian periphery. At the national level industrial output decline is important in determining declines in employment growth. Statistical tests which examine whether movements in one data series precede those in another data series have revealed that changes in industrial output precede changes in industrial employment. Those tests also reveal that the lag is very short, being most significant after one month and almost completely eroded after four months.

Examination of the effect across the regions reveals that, although this effect is statistically significant, its degree of magnitude is limited – a 1 per cent decline in industrial output, for example, would result in a 0.18 per cent decline in industrial employment. A more influential factor is the initial unemployment rate (1992 in this case) – the higher the initial rate of unemployment, the larger the subsequent decline in industrial employment, indicating that regions in a weak position at the beginning of economic transformation are bearing a large share of de-industrialisation. Additional tests for regional dependence on weak resource-based industry, harsh environmental conditions (northern regions) and 'rust belt' regions also indicate that these have negative influences on industrial employment (Table 6.2 and Figure 6.3).

Unemployment

Analysis of the growth of unemployment is complicated by the existence of considerable under-employment, highlighted by significant levels of administrative leave and short-time work (Blasi, Kroumova and Kruse 1997).

Table 6.2 Factors affecting regional change

Growth opportunities	
Variable	*Effect on employment growth*
New service growth	Strong positive effect
Export growth	Weak positive effect
North West & Central economic regions	Moderate negative effect
Variable	*Effect on labour force growth*
New service growth	Strong positive effect
Export growth	Weak positive effect
Volga economic region	Moderate positive effect

Decline	
Variable	*Effect on industrial employment growth*
Industrial output decline	Weak negative effect
1992 unemployment rate	Moderate negative effect
Far East & North economic regions	Strong negative effect
North West & Central economic regions	Moderate negative effect
Variable	*Effect on unemployment growth*
Industrial output decline	Weak negative effect
1992 unemployment rate	Strong positive effect

Note:
See Appendix for further details on the statistical analysis.

As a proxy for the level of unemployment (given the problems with the official data) we have taken the residual between the labour force (the economically active population) and the number employed and noted change over time, with the aim that this will reflect the 'true' change (in terms of magnitude) in a regional economy. As we have seen above, declines in industrial output are an important variable in determining unemployment growth, with a larger effect than just on the industrial labour force. This probably reflects a situation where the collapse of linkages related to the traditionally important industrial sector in the region, which in many instances dominated the local economy, adversely affects other branches of the local economy. Thus, the effects on industrial employment and total employment in the region move in opposite ways. One possible explanation for this is that regions performing poorly at the beginning of economic transformation had to restructure by sectoral shifts in the regional employment pattern or through out-migration of the population in search of work opportunities elsewhere. The most obvious example is the Russian North, where the reduction of state subsidies has led to the wholesale closure of entire sectors of the economy and substantial out-migration.

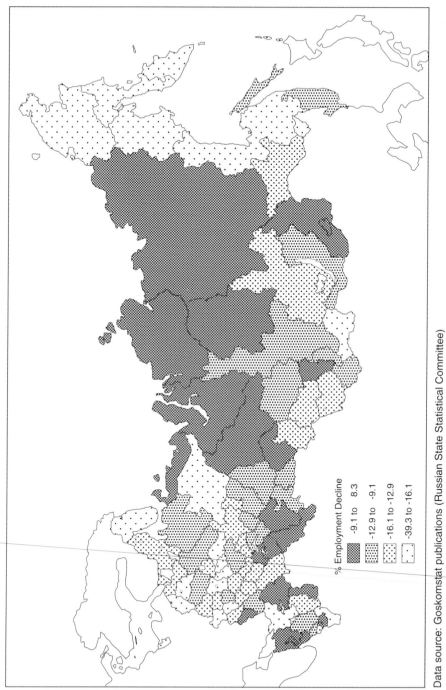

Data source: Goskomstat publications (Russian State Statistical Committee)

Figure 6.3 Decline in industrial employment between 1991 and 1994

The relationship between unemployment and production decline is complex. It may be the case that regions dominated by enterprises that are experiencing substantial production decline, but are retaining their workforce on the books, are simply not responding 'rationally' to market signals. These enterprises are able to postpone labour shedding thanks to subsidies from the federal and/or local government and because the workforce is willing to put up with this hidden unemployment. However, it is increasingly clear that the state, at both the national and local levels, can no longer afford to subsidise loss-making enterprises and that the workforce is now seeking alternative employment opportunities or protesting about unpaid wages (World Bank 1996b). Thus, unless new forms of economic activity can provide new job opportunities, there is substantial 'pent-up' unemployment still hidden in the industrial economy. According to the OECD (1996:17), 1.7 million people were on short-time work in 1992, rising to 4.8 million in 1994. According to one assessment of the Russian economy, in early 1997 some 10.6 million people, or 14.2 per cent of the labour force are on administrative leave and/or short-term work (Bush 1997).

Dynamics of economic and regional growth

One source of new economic opportunity in post-Soviet Russia is hypothesised to arise from sectors of the economy that were underrepresented in the centrally administered economy. Analysing the structure of the Soviet economy in the late 1980s, Schroeder (1987: 240) identified what she called the 'service gap'. By international standards, and relative to domestic demand, the service sector was underdeveloped. Consumer, producer and financial services were particularly underdeveloped and the growth of such services is expected to present many opportunities for employment creation and the restructuring of economic activity. Thus, a key component of the restructuring of the Russian economy is a process of 'tertiarisation'. This does not mean that there will not be a place for a rejuvenated industrial economy in Russia, for it is widely recognised that a healthy service sector requires a buoyant manufacturing economy; rather that the processes of marketisation provide additional opportunities for the development of new service activities. Thus, the expansion of the services sector may be seen as a key growth dynamic in the Russian economy. The development of small and medium-sized enterprises is also expected to play a positive role by capturing intra-sectoral shifts in economic activity. Small and medium-sized enterprises are frequently 'spun off' from failing state enterprises, but are also the main means by which new entrepreneurs enter the economy. Consequently, the health of the SME sector in a region may be a good indicator of the nature of the local economic environment as well as attitudes towards the promotion of private enterprise. The final issue analysed here is the new opportunities offered by foreign direct investment and foreign trade. As relative prices change and the diktat of the planners no longer governs foreign economic activity, the geography of this sphere is undergoing change (Bradshaw 1991, 1995). The opening up of the Russian market is creating new opportunities for foreign business to set up local production, an option outlawed during the Soviet period. Given the importance of raw materials in the Russian economy, the major trading regions are not expected to change, though new opportunities will be presented to other regions. At the same time, increased foreign trade activity may benefit the major gateway regions of St Petersburg in the north-west, Krasnodar in the south and Vladivostok in the east. However, at present Moscow serves as the main entrepôt for the Russian economy, in terms of both commodity and capital flows.

Service sector development

As many of the services associated with a market economy were either particularly underdeveloped or virtually absent in the Soviet Union, the development of services is expected to confer new employment opportunities and growth potential for the economy.[3] It is also possible that many of the trading activities that have emerged in post-Soviet Russia have simply surfaced from the black economy during the Soviet period. In other words, service sector activity was systematically underreported during the Soviet period. As Table 6.1 reveals, many of the regions with the largest growth are those which are now emerging as important regional service centres – Samara in the Volga region, Sverdlovsk (Yekaterinburg) in the Urals and Novosibirsk in Siberia. The regions that have performed badly in this respect tend to be the poor peripheral regions of the North Caucasus and Siberia. New service growth is strongly associated with total employment growth (Table 6.2), suggesting that those regions which show high levels of service sector growth are also showing a growth in the total number employed.

Small and medium-sized enterprises[4]

In the Soviet Union there was a relative absence of firms and employment in the small and medium-sized enterprise (SME) sector. However, while rapid growth in the number of SMEs was initially experienced, this fast growth soon petered out to such a degree that in 1995 the number of SMEs and the proportion of the labour force employed in this sector stood at only 11 per cent (compared with a Višegrad average of greater than 23 per cent and a European Union average of 57 per cent of the labour force).

There are diverse regional experiences in the development of the SME sector. The metropolitan areas of Moscow and St Petersburg have relatively high shares, with 22 and 17 per cent of the local labour force employed in this sector respectively, while a region generally accepted as a 'reform leader', Nizhniy Novgorod, has only 7 per cent (see Figure 6.4).

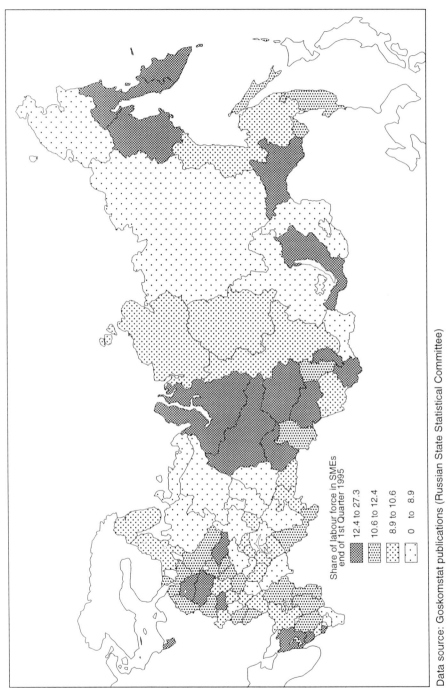

Data source: Goskomstat publications (Russian State Statistical Committee)

Figure 6.4 Share of labour force employed in SMEs at the end of first quarter 1995

Several features seem to be influencing the development of SME employment, such as the level of productivity in a region and the education level of its population. Location in Siberia also has a positive influence. This is expected to be due to increasing transportation costs making it necessary to produce many items locally that were formerly imported into the region. Negative influences on employment creation include high real wages and levels of unemployment, and increasing levels of state-directed investment crowding out the private sector. Growing levels of crime also inhibit the development of employment in the SME sector. In the development of state-sector SME employment all these relationships tend to be reversed, implying that in regions where the 'state' plays an active role there are lower levels of productivity per worker employed in SMEs.

Foreign economic activity

The increasing internationalisation of the economy is a key feature of post-Soviet Russia. During the Soviet period the state monopoly over foreign trade served to isolate domestic enterprises and regional economies from international competitive forces. During the early years of *perestroika* reforms were introduced to allow the creation of joint ventures between Soviet enterprises and foreign companies. Since then the Russian government has allowed foreign companies to create 100 per cent foreign-owned subsidiaries and foreign investors have been allowed to purchase shares in Russian enterprises. At the same time as foreign capital has been encouraged into the Russian economy, control over foreign trade activity has been liberalised. Whilst previously decisions concerning foreign economic activity were made in Moscow by the Ministry of Foreign Trade and its foreign trade organisations, now individual enterprises can enter directly into import/export activity and local government has become involved in the management and promotion of foreign trade.

Given the relatively narrow commodity structure of Russian exports (in 1995 mineral products, including energy, accounted for 40.2 per cent of non-CIS exports, and metals and precious stones a further 29.9 per cent), it not surprising that the major oil- and gas-producing regions, principally Tyumen *oblast*, which accounted for 36.8 per cent of Russia's exports in 1995, and mining regions such as the Republic of Sakha (Yakutia), which is the sole producer of diamonds in Russia and accounted for almost 10 per cent of Russia's exports, dominate the geography of export production. Previously, the revenues from such activity were hoarded by Moscow; now an increasing share remains in the region, as demonstrated by the growing correlation between exporting regions and the geography of imports and hard currency reserves. Those same resource-producing regions have also attracted a large share of the foreign investment (see Bradshaw 1991, 1995, 1997). However, the increasing attraction of Russia as a market to sell goods has also resulted in consumer-oriented inward investment in European Russia.

The measure used in our analysis examines the extent to which exports in any given region have grown relative to the Russian average. Given the relative stability of Russia's export structure and the role of the resource regions, they do not figure in Table 6.1; instead we see those regions which have sought to increase their export activity. While some of the regions listed in Table 6.1 are there simply because their exports have grown from small bases, this analysis does pick out those regions which have boosted export activity on the basis of ferrous and non-ferrous metals (aluminium) production and processing. The collapse of domestic demand for iron and steel and aluminium, together with low production costs, have provided a major export opportunity for regions such as Vologda and Kursk. None of the regions listed in Table 6.1 would suggest that Russia's manufacturing industry is in a position to expand export activity, the one exception being elements of the military-industrial complex.

Thus, de-industrialisation in Russia seems to be spawning both a retreat into the resource economy and the expansion of the service sector, each with a different geography; while the geography of decline seems closely aligned to heavy industry and manufacturing. In the second section of the chapter attention is turned to Novosibirsk, a region lacking in resource wealth and beset by the problems associated with the geography of decline. Novosibirsk does not figure as a key region in the analysis presented above, except to say that it emerges as a possible service centre in West Siberia. As we shall see, from the viewpoint of the region, the future for Novosibirsk is seen to lie, in part, in the development of its service economy as it seeks to establish itself as the financial capital of Siberia.

Restructuring and regional change: the case of Novosibirsk

It would seem that the patterns of economic growth and decline in Novosibirsk bear out the regional-level analysis carried out in the first section of this chapter. However, such analysis is unable to assess the policy responses of local elites, and it is with this issue in particular that this second section deals.

Constructing a Soviet city

The city of Novosibirsk (Novonikolaevsk until 1925) was founded in 1893 with the crossing of one of Russia's major north–south river routes by the Trans-Siberian railway and its pre-revolutionary growth was built largely on trade and transport. By 1917 the city had experienced little industrial growth. Thus, Novosibirsk's development has been characterised by a number of the key features of the Soviet industrialisation project. In the first five-year plans, the city's development was inextricably linked to the development of the Ural–Kuznetsk combine, a project assigned ultimate importance by both Lenin and Stalin for the strategic and economic goals of the new Soviet state, with the explicit aim of exploiting the resources of the east for the benefit of the country as a whole. Novosibirsk itself was to become a centre for machine building, that sector which Lenin identified as the very heart of the socialist industrialisation project (Protopopov 1948:17). Novosibirsk's heavy industrial base was further strengthened by the evacuation of industry from European Russia during the Second World War. From 1941 to 1942 the value of industrial production in the city increased fourfold. To a significant extent, then, Novosibirsk's historical development was structured by national strategic interests at the expense of local needs. Recent literature, however, highlighting the existence of 'a multiplicity of social relations that did not conform to officially prescribed hierarchical patterns' (Stark 1992: 300), undoubtedly brings into question the functional nature of the Soviet Union's national and strategic industrialisation policies. Nevertheless, initial investigations in Novosibirsk suggest that such social relations were marginal in their impact on development agendas, at least until the late Soviet era, scarcely increasing local control over the economy.

Despite this, on the eve of reform Novosibirsk was a major industrial centre of regional and national significance, with a population of 1.44 million. Approximately half of all employment remained in industry, with 54 per cent of industrial workers employed in defence-oriented enterprises. The city also possessed a very important scientific centre, employing well over 5,000 scientists and competing at a world scale in certain fields, and functioned as a major trade and transport node, adding air and road routes to already existing river and rail networks. However, like many other Russian regions, Novosibirsk has experienced massive economic decline in the last few years. Production continues to decline in the majority of the city's large industrial enterprises, creating unemployment levels of close to 14 per cent (*Novosibirskie Novosti*, 24 Februay 1996). Employment has fallen particularly heavily in military enterprises and a real threat of bankruptcy promises even greater job losses in the near future. Whilst this apparent 'transitional recession' could have been expected, given the city's inherited economic structure and the nature of post-socialist reforms, what is perhaps more unexpected is the nature of local economic strategies and initiatives proposed by local politicians, administrators, industrialists and entrepreneurs.

Building a post-Soviet future

The decentralisation of economic management which has accompanied the processes of marketisation has provided local actors with opportunities to deliberate upon, articulate and implement a whole range of ideas for post-Soviet development. In Novosibirsk, the focus has been upon four key strategies:[5] internationalisation; the development of science and technology; the creation of a major financial services centre; and the promotion of Novosibirsk as a political and administrative counterweight to Moscow.

Internationalisation

After years of exclusion from world markets, the regional administration has made an explicit commitment to increase Novosibirsk's participation in the international division of labour, not for its own sake but as 'one of the key directions for the solution of the tasks of social and economic development confronting the region' (Novosibirsk *oblast'* 1995: 4). Through the development of Novosibirsk's international airport and associated container port, increased exports (in prioritised knowledge- and labour-intensive industries such as radio-electronics and textiles) and an attempt to capitalise on the city's position as a gateway between Europe and Asia, it is hoped that Novosibirsk will be able to participate in 'the most contemporary tendencies in the development of world capital flows' (Novosibirsk *oblast'* 1995: 4). It is argued that such participation will contribute to the technological development of Siberian industry and the widening of the city's income and employment base. However, it is interesting to note that much of the rhetoric surrounding increased international economic activity borrows heavily from discussions of the import-substitution and subsequent export-orientation of East Asian states. Moreover, the expansion of exports is very much oriented towards Asian markets, building on a long-term association with China and Mongolia. Internationalisation of the local economy appears to represent, therefore, an attempt to argue for inclusion in 'Asia' rather than (or as well as) 'Europe'.

These plans for the internationalisation of the local economy experience virtually no opposition within Novosibirsk. From all perspectives (labour, administration, business, etc.), entering the world economy and becoming a full global player is seen not only as inevitable but also beneficial. Although many actors do recognise the problems with implementing such plans (not least, the impact of international competition on a largely uncompetitive economy and the need for major capital outlays and tax concessions in an already cash-starved region), attention seems to be focused on adjusting to a globalised economy, rather than devising radical alternatives. The manager of one light industrial firm argued, for example, that long-term goals should be explained to 'the people', making the point that if the city saves and invests in infrastructure for one year then 'things will be different' – but this is a difficult argument to sustain when public-sector wages have not been paid for six months.

Science and technology

With the loss of the Cold War rationale for an extraordinarily bloated military-industrial complex (MIC), central state funding for fundamental and applied science has effectively collapsed. In Novosibirsk, state funding for the MIC has fallen by up to 90 per cent, as research contracts are terminated and state expenditures are concentrated on more immediate needs such as unpaid wages and a hugely expensive military effort in Chechnya (see *OMRI Daily Digest* 1996; Stone 1996). Under these conditions, there are justifiable fears that the concentration of research and highly skilled workers in Novosibirsk will be lost. It is barely possible to survive in science today without foreign grants and many Novosibirsk scientists are being tempted to emigrate. It is argued that, with investment, Novosibirsk's science could compete on a world scale once again, but without it Novosibirsk and Russia might lose something in which they have invested a great deal, and which could contribute much to the city's future.

In answer to this, legislation has been passed which supports the creation in Novosibirsk of a technopark aimed at encouraging the commercialisation of innovation and the development of market-oriented research. The technopark also represents an attempt to renew the links between science, industry, education and regional development upon which the Novosibirsk science centre was originally founded (see Ibragimova and Pritvits 1989; Ibragimova 1978). Moreover, the technopark is upheld as a possible alternative to dependence on foreign funding and support for development plans (*Novaya Sibir'* 1996). The 'American solution' of grants and contracts for the survival of Siberian science, for example, is rejected as a temporary, unstable and selective solution, which must be replaced not by a reformed system of central state funding, but by a more local, more permanent, more long-term and more inclusive solution, which not only builds on local expertise and existing R&D institutions, but also aims to secure Novosibirsk a knowledge-intensive market niche into the twenty-first century.

Finance and financial services

The accumulation of funds for investment is clearly a critical factor in local economic development and it is widely claimed that Novosibirsk can and should develop as a major regional financial centre which will attract and manage funds for investment, innovation and development, not only for Novosibirsk, but also for the whole of the Russian east. The liberalisation of banking in the Russian Federation has led to a highly uneven development among commercial banks – two-fifths of all banks are located in Moscow, along with 80 per cent of financial resources. However, a number of secondary, regional centres have emerged and it this status to which Novosibirsk aspires. Over 50 independent banks, approximately 100 investment companies and 9 investment management funds are already located in the city and Novosibirsk is also home to 2 major stock and currency exchanges, including the first, largest and most important east of the Urals. Both exchanges are capable of trading shares from throughout Siberia and the Russian Far East, acting as intermediary with the outside world, using electronic trading technologies which link Novosibirsk to other regional and international exchanges.

Nevertheless, the concentration of financial and economic activity in Moscow means that Novosibirsk's exchanges in fact deal in very little of non-local importance and primarily with local government bonds. The real benefits of financial service development so far remain negligible, although there is still potential for finance-related growth (consider the importance attached to financial services in the West), so local authorities are active in encouraging financial development and directing that activity towards the perceived interests of the local population, most particularly through an advisory committee established under the regional governor. Moreover, the apparent importance of financial-sector growth in Novosibirsk highlights the influence of discourse in economic development strategies. A highly developed financial sector strengthens Novosibirsk's position as an 'important city' in Russia's increasing integration into world markets and international financial institutions. It suggests that Novosibirsk is competing successfully with Moscow and St Petersburg and is not completely excluded from world city networks. The expansion of financial services is presented as critical for Novosibirsk's finding and maintaining paths for growth in the next millennium. Playing the games of the deregulated markets of neo-liberalism suggests that Novosibirsk is not simply a bystander, consigned to the dustbins of history by its antiquated and obsolescent heavy industry, but a city which is part of global networks.

Political/administrative centre

The development of financial services is closely linked to the widely promoted idea that Novosibirsk is the capital of Siberia. In fact, Siberia has no capital, but Novosibirsk, as the largest city in Siberia and the Far East, assumes this status. Novosibirsk's attempts to promote its position as a gateway for international and financial activity are just a small part of the city's bid for wider political status. This endeavour can be seen to have a number of bases. Through the development of administrative functions, it is hoped that there will be an increase in the role of service sector. However, of critical importance, perhaps, is a blatant attempt to gain influence at the federal and international level. The quality of political connections continues to play a major role in economic development. On more than one occasion the regional governor, Vitalii Mukha, has used threats of Siberian regionalism, through the Siberian Agreement,[6] against Yeltsin in Moscow to improve

Novosibirsk's position and reinforce perceptions of Novosibirsk as a regional force to be reckoned with.

The city's elites are making an explicit attempt to establish Novosibirsk as a gateway for international market access and for the financing of those markets, as a node in the rapidly expanding global flows of finance, information and goods, combining this with an effort to assure Novosibirsk the political status to defend its interests and implement its strategies. The implications of this are that policies for economic regeneration are increasingly being articulated through competitive localisms as cities and regions engage in heightened rivalry in an attempt to 'hold down the global' (Amin and Thrift 1994).

Economic restructuring and social change

In response to the patterns of economic change outlined in the first half of this chapter, politicians, administrators and industrialists in Novosibirsk can be seen to be proposing a whole gambit of ideas and strategies for paths 'out of the crisis'. Whilst initiatives such as the promotion of high technology development and the expansion of the city's international airport build upon Novosibirsk's historical and geographical experience, other strategies (for example, the creation of a major currency and stock trading centre and the establishment of a free economic zone) relate more closely to dominant discourses of economic development borrowed from (or imposed by) the West with little understanding of the Russian or Siberian context.

Moreover, despite the enunciation of ideas for the promotion of alternative growth sectors, in practical terms most of the regional and city administrations' attentions remain focused on plans for the sustenance and survival of large-scale, heavy industry (particularly through the conversion of former defence industries), the payment of wages in the public sector and attempts to ensure heating and water for the region's population in the winter months. The vast majority of local and regional budgetary income is spent on providing the population with basic services – housing, heating, transport, pensions, etc. This is reflected in the fact that the regional administration currently has two concrete policy programmes – one for budget fulfilment and one for social policy – and it is through these two programmes that the economic development goals of the *oblast* are accomplished.

Given the severity of the 'transitional recession' and the radical nature of the anticipated transformations, it cannot be expected that economic and political actors will encourage economic restructuring in ways that proponents of capitalist transition might see as rational. In addition to the challenges of domestic and international economic change faced in the West (such as altered competition, increased internationalisation and the emergence of new sectors for growth), the Russian Federation (and other post-socialist states) are experiencing fundamental changes in the nature of, and rationale behind, their economic systems. The processes of marketisation represent in essence the construction, development and expansion of capitalist social relations, and, as such, must be seen as the rearticulation of social, political and cultural practice. Furthermore, those processes must be seen to be taking place within the frameworks of inherited institutional contexts, not only in the form of concrete organisations but also as 'embedded routines' (Stark 1992; Smith and Swain in this volume). The development of capitalist social relations in the Russian Federation involves, therefore, the adoption and adaptation of whole new sets of rules and behaviours, many of which are wholly unfamiliar to local leaders and other political and economic actors. Whilst there is undoubtedly some translation of 'market rationality' and the discourses and structures of the world economy, we must recognise and accept the persistence and development of Soviet-type and Soviet-era behaviour and social relations (for more on the legacies of socialism and the diversity of transformation processes see Grabher and Stark in this volume and Begg and Pickles in this volume).

Conclusions

Culture, politics and society are often written off as explanatory tools which are only employed in the final analysis to account for change which we cannot understand in purely economic terms. However, such a position betrays a fundamental misunderstanding of the nature of 'the economic', and hence of economic change. As Block (1990) states, it is impossible to understand economic practices without *simultaneously* attempting to understand the cultural and social practices which underpin them. He notes that 'the "economistic fallacy" imagines that capitalist societies do not have cultures in the way that primitive or premodern societies do' (Block 1990: 27). Such an imagination reinforces the orthodox view that the imposition of capitalist structures and institutions (such as private property, the competitive imperative and a monetised accounting system) will force the universal development of free market rationality, and if it does not, it is simply necessary to wait, tinker with the system and delete any irrationalities. That those so-called irrationalities are embedded within local economic practice is not a view which is frequently countenanced, and whilst advisers and commentators within the European Bank for Reconstruction and Development and the World Bank, for example, have begun to accept the importance of historical and institutional legacies in the processes of post-socialist transformations, those legacies are still seen as marginal and secondary (see EBRD 1996 and World Bank 1996a). This chapter has aimed to explore the relationship between structural economic change and the nature of regional transformations and began by analysing the economy in tight economic terms through a macro-scale review of regional change. The second section of the chapter has, however, demonstrated that there is no simple link between structural economic change and the choice of futures at the local level. Each region faces a choice of futures, constituted as much through social, political and cultural practice as through the purely 'economic'. Thus, any attempt to understand the path and process of economic transformation and differentiation in the Russian regions must take account of this.

Appendix

All the regressions reported here are standard least squared analyses. However, cross-sectional analysis is particularly susceptible to problems of heteroskedasticity, whereby the disturbance variance in the regression is not constant across observations (usually larger or richer economic units display greater variation). In order to correct for this the data have been expressed in logarithmic form, and an additional procedure (White's heteroskedastic consistent standard errors and variance) was used.

A second problem of spatial autocorrelation was corrected for by the addition of appropriate dummy variables for the larger macro-regions. The dummy variable was selected by using macroeconomic regions where conditions such as harsh environmental conditions, or groupings of 'rust belt' regions, were expected to have similar impacts on the regressions.

Acknowledgements

Michael Bradshaw and Douglas Sutherland wish to acknowledge support from the Economic and Social Research Council (ESRC), research grant number R000236389. Alison Stenning wishes to acknowledge support from the ESRC in the form of a Research Studentship which funded her fieldwork in Novosibirsk.

Notes

1 Other factors complicating the analysis of transformation and restructuring are measurement problems (Berg 1993; Lequiller and Zeischang 1994). Central planning in the Soviet Union conferred an economic structure with relatively few small economic units, in comparison with western economies, that was resistant to rapid change. As a result the focus was placed on large enterprises linked to particular industrial ministries (see Begg and Pickles in this volume). The statistical agencies have, therefore, relatively limited experience of estimating the share of small-scale output. Having concentrated on censuses of large-scale enterprises in the past, they now lack the technical ability to capture current changes in the economy adequately. Thus, official statistics are a far more effective measure of the decline of the former state sector than they are the newly

emergent private sector. The bottom line is that we do not have an accurate measure of the level of economic activity in Russia at present.

2 The various 'liberalisation indices' used in the World Bank (1996a) and EBRD (1996), among others, are attempts to capture such developments.

3 In our measure of new services we have included employment in trade and catering, general commercial activity and real-estate operations, the financial sector, and transport and communications. Transport and communications already had large shares of employment prior to the collapse of the centrally administered economy and if one were examining levels of new service employment this might introduce a significant bias in favour of the raw material-extracting regions in the north and east of the country where large transport sectors existed. This sector was included as considerable opportunities for new forms of transportation exist (North 1997) and because the communications infrastructure is particularly weak.

4 This section is based on Sutherland (1997).

5 The information and arguments put forward in this section are based on interviews carried out in Novosibirsk between March and August 1996, and on a review of local media sources.

6 Mukha was the first chair of the Siberian Agreement in 1991, and has recently been re-elected to the post.

Bibliography

Amin, A. and Thrift, N. (eds) (1994) *Globalization, Institutions and Regional Development in Europe*, Oxford: Oxford University Press.

Berg, A. (1993) 'Measurement and mismeasurement of economic activity during transition to the market', in M.I. Blejer *et al.* (eds) *Eastern Europe in Transition: From Recession to Growth? Proceedings of a Conference on the Macroeconomic Aspects of Adjustment, Cosponsored by the International Monetary Fund and the World Bank*, World Bank Discussion Paper No. 196, Washington, DC: World Bank.

Blasi, J., Kroumova, M. and Kruse, D. (1997) *Kremlin Capitalism: Privatizing the Russian Economy*, Ithaca, NY: Cornell University Press.

Block, F. (1990) *Postindustrial Possibilities: A Critique of Economic Discourse*, Berkeley: University of California Press.

Brada, J.C. and King, A.E. (1992) 'Is there a J-curve for the economic transition from socialism to capitalism?', *Economics of Planning*, 25, 1: 37–53.

Bradshaw, M.J. (1991) 'Foreign trade and Soviet regional development', in M.J. Bradshaw (ed.) *The Soviet Union: A New Regional Geography?*, Belhaven Press: London.

—— (1995) *Regional Patterns of Foreign Investment in Russia*, London: Royal Institute of International Affairs.

—— (1997) 'The geography of foreign investment in Russia, 1993–1995', *Tidschrift voor Economische en Sociale Geografie*, 88, 1: 77–84.

Bush, K. (1997) 'The Russian economy in February 1997', paper presented at the Annual BASESS Conference, Fitzwilliam College, Cambridge, 12–14 April.

European Bank for Reconstruction and Development (EBRD) (1996) *Transition Report 1996: Infrastructure and Savings*, London: EBRD.

Gavrilenkov, E. and Koen, V. (1994) 'How large was the output collapse in Russia? Alternative estimates and welfare implications', *International Monetary Fund Working Paper*, WP/94/154.

Goskomstat (1996) *Metodoligicheskie polozhniya po statistike* (Methodological basis for statistics), Moscow: Goskomstat.

Ibragimova, Z. (1978) *Ne slavy radi, a pol'zu dlya . . . O budnyakh sibirskoi nauki* (Not for the sake of glory, but in the interests of . . . On the routine work of Siberian science), Novosibirsk: Zapadno-sibirskoe Knizhnoe Izdatel'stvo.

Ibragimova, Z. and Pritvits, N. (1989) *Treugol'nik Lavrent'eva* (Lavrent'ev's triangle), Moscow: Sovetskaya Rossiya.

Lavigne, M. (1995) *The Economics of Transition*, Basingstoke: Macmillan.

Layard, R. and Parker, J. (1996) *The Coming Russian Boom: A Guide to New Markets and Politics*, London: Free Press.

Lequiller, F.I. and Zeischang, K.D. (1994) 'Drift in producer price indices for the former Soviet Union countries', *IMF Staff Papers*, 41, 3: 526–32.

McAuley, A. (1995) 'Inequality and poverty', in D. Lane (ed.) *Russia in Transition: Politics, Privatisation and Inequality*, Harlow: Longman.

North, R.N. (1997) 'Transport in a new reality', in M.J. Bradshaw (ed.) *Geography and Transition in the Post-Soviet Republics*, Chichester: Wiley.

Novaya Sibir' (1996) 'Sekret russkoi zhivuchesti: amerikanskaya razgadka' (The secret of Russian vitality: the American solution), 21 March: 8.

Novosibirsk *oblast'* (1995) *Kontseptsiya razvitiya vneshekonomicheskoi i mezhregional'nikh svyazei Novosibirskoi oblast'* (Conceptions for the development of foreign economic and interregional links of Novosibirsk *oblast*), Novosibirsk.

Novosibirskie Novosti (1996) 'Bezrabotitsa kak sostavlyayushaya rynochnikh otnoshenii' (Unemployment as a component in market relations), 24 February: 2.

OMRI Daily Digest (1996) 'Academicians, teachers protest', no. 118, 11 October.

Organisation for Economic Co-operation and Development (OECD) (1995) *The Russian Federation 1995*, Paris: OECD.

—— (1996) *Labour Restructuring in Russian Enterprises: A Case Study*, Paris: OECD.

Protopopov, N.N. (1948) *Novosibirsk*, no publication details.

Russian Economic Trends (1996) 5, 2: 55.

Schroeder, G. (1987) 'USSR: towards the service economy at a snail's pace', in Joint Economic Committee, Congress of the United States, *Gorbachev's Economic Plans* vol. 2, Washington, DC: US Government Printing Office.

Sedik, D., Foster, C. and Liefert, W. (1996) 'Economic reforms in agriculture in the Russian Federation, 1992–95', *Communist Economies and Economic Transformation*, 8, 2: 133–48.

Stark, D. (1992) 'The great transformation? Social change in Eastern Europe', *Contemporary Sociology*, 21, 3: 299–304.

—— (1995) 'Not by design: the myth of designer capitalism in Eastern Europe', in J. Hausner, B. Jessop and K. Nielsen (eds) *Strategic Choice and Path-Dependency in Post-Socialism*, Cheltenham: Edward Elgar.

Stone, R. (1996) 'Vozrozhdeniye nauchnogo gorodka' (Rebirth of a science city), *Nauka v Sibiri*, May: 9.

Sutherland, D. (1997) 'Small and medium sized enterprises in the Russian regions: employment constraints and incentives', *Russian Regional Research Group*

Working Paper, No. 7, School of Geography and Centre for Russian and East European Studies, University of Birmingham.

Sutherland, D. and Hanson, P. (1996) 'Structural change in the economies of Russia's regions', *Europe–Asia Studies*, 48, 3: 367–92.

World Bank (1996a) *From Plan to Market: World Development Report 1996*, New York: Oxford University Press.

—— (1996b) *Fiscal Management in Russia*, Washington, DC: World Bank.

7

RESTRUCTURING STATE ENTERPRISES

Industrial geography and Eastern European transitions

John Pickles

The twentieth century began in 1917 and ended in 1989.
(Altvater 1993: 2)

Commentaries which focus on the political conditions of the transition to a market economy underestimate the capacity of the [Soviet-style] economy to reproduce itself and resist transformation.
(Burawoy and Krotov 1993: 58)

[In Bulgaria] the state enterprise sector . . . shed excess labor faster than in perhaps any other reforming country.
(Borensztein, Demekas and Ostry 1993: 9, quoted in Wyzan 1993: 5)

Introduction

In the twentieth century the countries of Central and Eastern Europe (CEE) experienced two distinct modernising transformations of their space economies: a pre-1948 capitalist modernisation (exacerbated in the early 1940s in countries such as Czechoslovakia by heightened investment by and demand from the German war economy)[1] and a statist, centrally planned modernisation underpinned by the productivist logic resulting from economic integration into CMEA, with its associated regionally organised system of production and divisions of labour. Since the political changes of 1989, and with the liberalisation of economic policies that has followed in their wake, the same countries are now experiencing a third modernising of their political economies, including major transformations in their industrial geographies.

The characteristics and form of capitalist relations and geographies this third round of modernisation is engendering are far from clear, and pose important questions about the ways in which we conceptualise the transition from centrally planned to market-oriented economies, or from state socialism to reform capitalism in the late twentieth century. In particular, this renewed round of modernisation calls for a closer engagement between transition theory (derived from the transformations occurring in the countries of CEE and the former Soviet Union) and industrial and regional restructuring theory (derived from the transformations of western capitalism in the countries of Western Europe and North America). That is, we need to ask what forms this geography of transition and modernisation are taking.[2]

Such an engagement between transition and restructuring theories is only now beginning. As Burawoy and Rutland have pointed out, 'transitology'[3] has to a large extent remained embedded within the same ideological frameworks that marked Sovietology during the Cold War and continue on in celebratory and hegemonic fashion presupposing or lauding the liberatory potential of liberal market arrangements and the democratising effects of pluralist parliamentarianism.[4]

In the countries of CEE there are, of course, important differences in history, geography, society and economy that militate against general statements that apply equally to all parts of the region. Indeed, these issues of local specificity provide geographers and others with some of their most important contemporary theoretical and practical challenges to understand better the role of locality in the constitution of particular places undergoing modernisation (see Pavlínek, Pickles and Staddon 1994). Yet until recently the social and spatial organisation of industrial production, the labour process model that typified the 'Sovietisation' of the economies and societies of CEE, and the industrial decline and reorganisation that now mark their transition to a liberal political economy have largely been ignored by geographers and regional scientists. Instead, geographical analysis has tended to focus more on national-level accounts of sectoral growth and decline, employment and unemployment patterns, and international trade and foreign direct investment issues. Missing from these geographical accounts have been what Michael Burawoy calls production politics, and a corresponding focus on the level of the social relations within and between enterprises and the relations between enterprises and the communities within which they are embedded.

Thus, to date – and in contrast to industrial geographies of the capitalist world[5] – little systematic published work has yet emerged from transition studies or industrial geography addressing the restructuring of regional production systems, industrial geographies and labour geographies of CEE in the 1980s and 1990s, and little in the way of systematic analysis of the regional political economy of transition is available.[6] Where any engagement with western industrial restructuring and industrial geography literature has emerged, it tends to have become captured by the aesthetic appeal and development possibilities of post-Fordist and flexible specialisation models now extant in a wide array of western literature. In this literature, industrial and regional policy is represented in terms of: (i) an initial phase of Stalinist mass-production systems, which, because of the political barriers of the command society, were unable to modernise in the 1970s in line with western economies; and (ii) a current phase of restructuring involving the (much hoped for) penetration of post-industrial economic practices into this inefficient and socially bankrupt landscape.[7] In this interpretation of a newly emerging 'post-industrial' economy, the metropolitan cores are seen to provide the pivots of dynamic economic axes while the older industrial regions experience de-industrialisation and emerge as the new rust-belts of CEE. The newly emerging economy is (or will be), in this model, one of industrial renewal and modernisation through the development of greenfield sites, industrial parks and sunbelts fostered in particular by foreign direct investment.[8] This model of post-Stalinist restructuring (modelled on post-Fordist literatures) represents a modernising both of industrial landscapes and of regional development theory, and has proved attractive to policy-makers throughout the region. Industrial parks, tax free zones, free ports and new greenfield sites now receive a great deal of government encouragement and public investment to attract foreign manufacturers. In regions such as western Hungary theorists already see a Third Italy flexibly specialised region of horizontally integrated small firms emerging, linked directly to a similarly emerging economic region across the border in Austria. This scripting of a post-Fordist industrial landscape does, however, pose serious problems at the level of empirics.

This chapter questions these applications of post-Fordist and flexible specialisation models of industrial change to these transition economies, and the projections for future industrial geographies that are being drawn from liberal economic and political models of transition: collapse of inefficient and 'politically' planned industrial sectors, growth in those sectors and regions that exhibit comparative advantages, and a shift in those disadvantaged regions from older, 'social' industries into 'new' industries geared to tourism, services, information processing and communications. Specifically, the chapter argues for the importance of a detailed understanding of the production politics of state enterprises under central planning in any analysis of transitional restructuring of their industrial geographies.[9] Thus, this chapter has two primary echoes: first, echoing Altvater above, that the nature of political life and form of economic development that typified the period from the Russian Revolution to the European revolutions of 1989 have now come to an end (in both the East and the West) and modernity itself is undergoing transformation across the globe.[10] Second, echoing Burawoy, the transition that is occurring in this post-socialist era is not one of epochal 'breaks', but a *transformation* of social relations, forms and practices that carries important traces of the legacy of the previous forty years. In this sense, what is emerging cannot yet be thought of in terms of Western European models, but must be thought of – at least as a possibility – as new or hybrid forms articulating elements of both eras and regimes in regionally and sectorally specific ways.[11]

The central goal of this chapter, then, is to highlight the actually occurring transition. Specifically, I wish to question three key assumptions in much contemporary transition theory, each of which has important consequences for the ways in which economic (and specifically industrial) change is understood: (i) assumptions about the nature of power in a command economy; (ii) assumptions about the nature of political transition from a command society; and (iii) assumptions about the political economy of industrial restructuring. To unmask these assumptions I shall ask: How was power deployed and control exercised in a command economy? What were the ways in which the central state controlled state enterprises, state enterprises controlled communities, and state apparatuses controlled populations? In what ways were these systems of control mediated? And how did large state enterprises such as Neftochim Bourgas mediate the power of the state, and local communities mediate the power of state enterprises and the central state (if they were able to do so at all)? In other words, by clarifying the conditions within which state industries functioned, and by focusing on the legacies of central planning in the current conjuncture, I seek to ask, which transition is now occurring?

Control systems and the command economy

Current theories of transition which address the restructuring of economic relations from a bureaucratic command society to a market-oriented economy usually emphasise a particular model of the bureaucratic command economy (Figure 7.1). In this model the flow of power in a bureaucratic administrative command system is hierarchical and relations of dominance and control are unidirectional – flowing from the party apparatus (and particularly the central committee of the party) through the apparatuses of the state (particularly the ministries) to the enterprises and lower organs of state government (at the regional and local levels). In this model enterprises and communities are controlled by the party and the state apparatuses through quota requirements, managerial appointments, reporting regulations and strict state sanctions. As the sites of value creation, state enterprises, particularly the larger ones, achieve a certain autonomy from local and regional communities and, under the protection of the central state apparatuses (particularly the party and the ministry), are able to exercise certain claims on the communities (for example, for the provision and housing of the labour force) and can operate relatively free from local community oversight or sanction (particularly relating to matters such as land acquisition, environmental pollution, etc.).

175

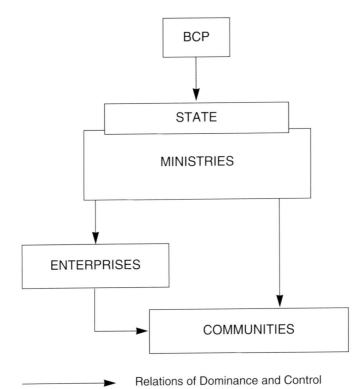

Figure 7.1 The flow of power in a bureaucratic administrative command system

In this model, the central state controls and dominates the enterprises (or *combinats*) and the state and the enterprises exercise relations of dominance and control over the communities. Local communities, particularly the smaller communities and those near to large state enterprises or other operations (such as military bases), are paralysed and dominated in this hierarchy of power. In this sense, the communities are captured (see Pickles 1992a). This model of capture of local communities (and workers) within a productivist system is one in which meeting quotas set by the central Ministry of Industry is placed above all other considerations, particularly considerations relating to worker and community health and safety. Large state enterprises provide the exemplars of this kind of productivist model of power and control over workers, localities and populations, and – because of the political nature of production decisions – these enterprises come to typify the inefficient and irrational incentive structures of state industries (see Kornai 1992).

Production politics and the case of Bulgaria

In part, this model of control and domination captures well one aspect of the industrial geography and sociology of Bulgaria. Between 1948 and 1989, Bulgaria in effect became a newly industrialised economy through the determined efforts of and investments by the bureaucratic command state. In 1946, industry accounted for about 30 per cent of GDP, but in 1989 it accounted for about 60 per cent, double the share of GDP that is typical for countries with the same per capita income. Industrial growth was concentrated in heavy industry, particularly in metallurgy, chemicals, textiles, wood and paper, and engineering, but also with a strong electronics (especially computer) industry.

The story of Neftochim illustrates well the role of state enterprises in the current period of transition. Neftochim is a fully integrated petrochemical, chemical and polymer *combinat* of immense size.[12] Neftochim Bourgas was conceived and built as a large, fully integrated state enterprise with a capacity far exceeding original estimates of domestic need. The installation was built by the state and 70 per cent of its capital equipment was of Soviet or Eastern European origin, designed for the largest production plants, with the aim of re-exporting processed and subsidised Soviet oil.[13] The plant was operated by a management council responsible directly to the Ministry of Industry, from whom its budget was derived. In the past it produced or distributed all of Bulgaria's petrol, and chemical and polymer products, and was also a major earner of foreign currency through the re-export of processed oil. In all these senses Neftochim was (and remains) a strategic national industry, and thus fell even more closely under the jurisdiction of the central ministry in Sofia.

Moreover, the plant was also a major foreign currency earner and for many groups associated with it (from workers to managers to environmental inspectorates) the *combinat* was a 'Swiss cow' which was systematically milked of surplus capital through high wages, benefits, capital transfers and large fines. The result is an ageing and inefficient plant, hazardous work conditions and polluted communities. Conditions are so poor, and recognised to be so, that labour contracts limit the time workers may work in particularly hazardous plants, retirements are set at 47 for women and 52 for men, and wages are pegged to grades of workplace toxicity and hazard to reward financially those workers whose health is more severely compromised.[14]

The enterprise thus reflects the character of Bulgarian industrial development between the 1960s and the 1980s: large single plants built in the rural periphery of large urban centres, operating under special protection from the central ministry, and largely immune to the regulations of the Ministry of Environment and local government officials. Links to the Communist Party were strong, through both the state and individuals associated with the plant, and the powerful national and local interests associated with it protected its budget, supplies, markets and practices.

The model of hierarchy and control (above) partly captures the nature of decision-making within the bureaucratic command state and the ways in which state enterprises operated. Moreover, this hierarchical model provides the central metaphor for transitional theories of the current period of restructuring: here the period of transition is marked by a series of political and economic reforms that will open up the space for the entrepreneurial activity and market-driven decisions that will overcome the rigidities and inefficiencies of command planning. In this model of transition to a liberal polity and economy, de-communisation of the state and the opening up of multi-party democratic electoral politics will provide the general conditions for the decentralisation and de-monopolisation of the bureaucratic command administrative system and the liberalisation of the economy. Such an autonomous representative government and independent civil service will permit the emergence of, and enforcement of, a rational framework within which civil society can be regulated by government agencies: pollution can be monitored and punishment exacted by state apparatuses at the regional level, like the regional environmental inspectorates.

In this model, privatisation and marketisation will permit the emergence of horizontally and vertically linked private and state firms in which the incentives of competition for materials, labour, technology and markets guide decision-making at the plant level. And democratisation and decentralisation of power to communities will permit the regulations of the state to be administered and local regulations and controls to be propagated and acted upon. Communities will thus become empowered to exercise their own powers over recalcitrant polluters or to enter into negotiations with local enterprises to calculate appropriate trade-offs between environmental impacts and economic vitality.

While such liberal models of restructuring are widely reproduced, they seem to me to be models largely embedded in a normative analytics of ideal transitions. Little evidence yet supports such models of transition. Instead, to answer the question 'What is the nature of the actually occurring transition?' we need to know much more about three related aspects of enterprise behaviour under central planning and in the transition. First, we need to know more about the ways in which the bureaucratic command economy operated at the enterprise level and how enterprises functioned in terms of direct coercion and brute force through a hierarchy of command and control. Second, we need to know how relations between the enterprise and the local, regional and central levels of the bureaucratic command administrative system were organised beyond the exercise of force. Specifically, we need to understand how this system of hierarchical command and control was established and maintained by complex relations of negotiation among enterprises and between enterprises and various levels of the state administrative apparatus. Third, we also need to know how this legacy of social relations within which the enterprise was embedded continues to function now, or whether and how it is in the process of being transformed.

Enterprise behaviour in a shortage economy

The centrally planned state enterprise operated under soft budget constraints: the budget for operations was always the subject for bargaining and contestation between the enterprise managers and directors and the various branches of the bureaucracy. Soft budgets included soft subsidies, taxation, credit, administrative pricing, and materials and labour allocation, and negotiations involved both the ministry directly responsible for the enterprise, as well as the fiscal authorities, banking sector and pricing organisations. As Janos Kornai (1992: 142, n. 19) points out:

> This manoeuvring involves a lot of legwork, searching for connections and supporters . . . [and] produces . . . 'rent-seeking behaviour.' Although this is costly and tiresome, it can be lucrative. It may be worth spending more time in the 'corridors of power' than on the factory floor or in the office on sales negotiations.

One consequence of this politics of subsidies was a struggle between the enterprise and the central government over the appropriation of surplus from the enterprise and the allocation of resources to it. The central authorities generally sought to maximise the surplus extracted from major industries to support the wider functions of government, while the enterprises struggled to minimise the resources taken from them and maximise their annual budget allocations (Figure 7.2).

Since labour and supplies were both difficult to obtain, managers of state enterprises also entered into arrangements with workers and suppliers to guarantee access to both. Kornai (1992: 300–1) describes these labour practices as 'the hoarding tendency and the existence of unemployment inside the factory gates'. With uncertainties in supply, management needed to maintain access to a sufficient pool of workers to guarantee that quotas were met by end of the month speed-up of production, even at the expense of a surplus, under-employed workforce for the rest of the time.

Thus, state enterprises under central planning operated within at least three bargaining arrangements: vertical bargaining with the governing institutions over surplus extracted and subsidies provided; complex horizontal linkages with supply firms to guarantee that periodic quotas could be met; and internally negotiated 'flexible' work regimes with worker collectives that ceded some shopfloor control to workers and provided a wide range of incentives such as cheap food, health care and recreational facilities (social reproduction in the factory) in return for efforts to meet production quotas (Figure 7.3).

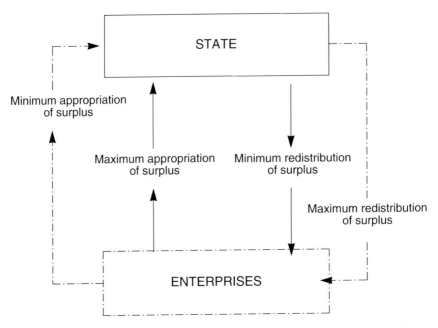

Figure 7.2 The structure of coercive bargaining between the state and enterprises

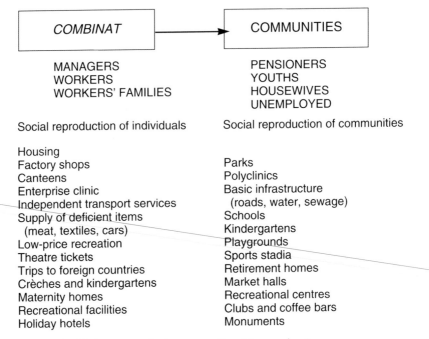

Figure 7.3 Embedded asymmetrical systems of social reproduction

Consequences of transition: regional and industrial restructuring

The post-socialist industrial transition in CEE has been largely about the decline of the multi-plant, multi-region large state industries. Operating as the norm of centralised industrial production and financial management, the large state industries developed from the 1960s to the 1980s have been unable to sustain their CMEA-based organisational production structure in the new era of fragmentation. Large parts of the productive machinery of the countries of CEE have collapsed, employment in manufacturing has declined precipitously. The multi-plant structure has disintegrated and the branch-plant economy (set up as social industries to support regional employment needs in peripheral regions) has largely collapsed, increasingly being replaced by a structure of centralised, protected, powerful, or relatively resilient core plants.

This desertion of the branch-plant economies has been exacerbated by the political and economic crises of the state. Between 1989 and 1995 Bulgaria had seven different governments and people speak often of 'reform fatigue'. Such reform fatigue was also accompanied by deepening economic and political crises which led in the November 1994 national elections to the landslide victory of the Bulgarian Socialist Party. The debt burden incurred by the reform governments since 1991 and the social and economic costs of repaying these debts incurred during the 1980s were exceedingly high, and severely restricted any plans for restructuring of older state enterprises. In 1993, $9 billion dollars was owed to the London Club alone and the total foreign debt amounted to $16,500 per capita (at a time when the average annual wage was about $1,440). To deal with this debt burden the government in 1994 vouched to melt the hard currency reserve to the dangerously low level of $200 million; to extract additional revenues from salaries, pensions and social benefits to the tune of 30–40 stotinki per lev per annum; to cut expenditure on the army and the police; to make cuts in government expenditure on education, health, science and culture; and to maintain a freeze on all new investments in production and infrastructure. Casualties of the fiscal crisis included the large state enterprises, whose budgets and subsidies were curtailed.[15]

Second, national production collapsed. The post-war Bulgarian economy was tied particularly closely to the regionally integrated markets of CMEA (in the 1980s, 80 per cent of Bulgaria's trade was with CMEA countries). With the collapse of these markets industrial and agricultural production declined precipitously. In 1990 alone industrial output fell by 40 per cent, and between 1989 and 1993 total economic output fell by over 50 per cent. Industrial capacity utilisation dropped from 80–5 per cent to 55–68 per cent depending on sector (Table 7.1). Since 1989–90 Bulgaria has shed labour faster than any other reforming country. Moreover, the effects of the decline

Table 7.1 Average utilisation capacities

Over 90%:	Printing
80–90%:	Glass, china, earthenware
70–80%:	Leather, furriery, footwear Clothing Ferrous metallurgy
60–70%:	Non-ferrous metallurgy Coal Logging and wood products Textiles and knitwear
50–60%	Pulp and paper Building materials Chemicals and oil processing Electrical and electronic Machine building and metal cutting Food, beverages and tobacco
40–50%:	Electricity and thermal energy

Source: National Statistics Institute, *Current Economic Business* June 1994.

are distributed differentially, in terms of occupational status, age, gender and region.[16]

In social terms the impacts of transition and modernisation have been highly differentiated. They have been differentiated through regionally uneven impacts, through socially uneven impacts, and in terms of differentiation within enterprises between categories of workers (particularly between core and non-core workers). Throughout the region, therefore, the winds of change have brought processes of creative destruction attended by massive capital transfers from public to private hands and from the mass of the population to certain sectors of it. Social differentiation and class formation are occurring rapidly, although often as a result of spontaneous privatisation, appropriation of public resources, primitive accumulation and speculation.[17] Processes of regional differentiation and spatially uneven development are deepening, particularly as a result of massive capital and skill transfers to the metropolitan economies and regions.[18]

The collapse of production – while serious throughout the country – has taken a particularly heavy toll in the economically more marginal regions of the country. The regional allocation of production under central planning has given way since 1991 to widespread branch-plant closure as state industries seek to maintain benefits and labour in the core plants. The core plants of the formerly multi-plant state enterprises have closed branch plants, labour has been shed, and the semi-autonomous enterprises have withdrawn from the workshop economy of peripheral, low-wage, often female-based factories. These so-called 'children of the mother firm' generally received all their raw materials from the mother firm and the products of the workshops were shipped back to it.[19] In some cases the machinery remains, but in others assets have been stripped as 'machines have departed' to the mother firms.[20] The result has been devastating in the peripheral localities and regions of the branch-plant economies. Bulgarian unemployment rates currently run about 18 per cent, but plant closures in the south, for example, have resulted in unemployment rates as high as 90–95 per cent.[21]

Neftochim

Between 1989 and 1993 Neftochim experienced severe crisis: supplies of crude oil from the former Soviet Union ceased and then later were re-established on a hard currency basis; heavy debts were incurred by oil shipments to Iraq that were not paid for as a result of the Gulf War; markets for petroleum and other products in COMECON were lost after 1989; and in 1992 the central government gave preferential tax breaks to foreign petroleum importers, severely disadvantaging Neftochim in the Bulgarian market.[22] By 1993, Neftochim had a debt of US $109 million from a World Bank loan, production had dropped from a high of 7.5 million tonnes in 1990 to a low of 4.7 million tonnes in 1992 (two-thirds of which derived from second-party processing).[23] The effect was efforts to shed workers in 1991 and 1992.

In 1992, under pressure from workers and surrounding communities (whose economic base was tied almost entirely to employment in the plant), Neftochim agreed to protect jobs at the cost of wage reductions for workers (although these reductions are small and wages for workers in Neftochim are very high by comparison with other jobs) (Interview, July 1992). Decreases in employment until May 1992 seem to have occurred predominantly in managerial, technical and supervisory positions, while the number of production workers has remained stable.[24]

Since 1989, Neftochim has been at the centre of major struggles over its financial resources. Most of its profits are redirected into central state coffers, and little is returned for reinvestment.[25] The vested interests in this honey-pot are numerous and widely spread, and they include individuals and groups at the central, regional and local levels, as well as managers and others within the enterprise.

The economic conditions and political forces within which Neftochim is currently embedded narrow the possibilities for effective technological restructuring and environmental reconstruction, and make highly unlikely any effective response to the current chronic situation. The political struggles over Neftochim as a 'Swiss Cow' (or cash cow) and the weak social basis for union and community action limit the possible response to restructuring and technological reconstruction.

Managers in the enterprise are not synonymous with managers in a capitalist enterprise. First, the firm is a large *combinat* comprising several discrete (albeit linked) production units (plants). Serving each of these plants are other units, such as maintenance, transport, research, accounting, etc. Each of these has its own structure of management, with its own director who sits on the executive board of Neftochim.

The possibility that the *combinat* will be privatised, or broken up and privatised, means that managers are forced to think very carefully about their own positions in the near future. The possibility of being in place to benefit from the privatisation of one of Bulgaria's most profitable and wealthy enterprises has created a massive inducement towards 'careful steps'. It also means that the position of managers towards restructuring alternatives tends to be very conservative. Moreover, the uncertainty surrounding privatisation has encouraged a position of 'wait and see', rather than a vigorous upgrading, capitalisation and internationalising of the enterprise with vigorous retraining programmes for workers and managers alike (BBN 1993). In sum, the management of the enterprise is both cautious and conservative, and seems unwilling to attempt to make policy under these conditions of uncertainty. Managers may actually have an incentive to run down the plant so that it privatises at lower cost, thus increasing their own chances of participating in its purchase. The situation of caution and conservatism is further exacerbated by the fact that the directors of the enterprise (appointed by the central state) have been changed five times in the past three years. The struggle over control and ownership of the enterprise has severely undermined the effectiveness of any real restructuring.

Workers are also in a highly ambiguous position. They are, of necessity, caught in the struggles over the future ownership of the enterprise, and the trade unions are adamant that there will be no privatisation without their participation. On the other hand, their social and political bases of action outside the enterprise are rather weak. Neftochim workers receive among the highest wages in the country, with many direct and indirect benefits (including extended holidays with pay, enterprise rest homes, health care, cheap food and access to special prophylactic foods such as yoghurt). The ability to mobilise the members and the neighbouring community in support of long-term labour actions against the enterprise – for example, to close down the plant and prevent its operation – is severely weakened by the economic conditions in which workers and their families find themselves in the contemporary transition, and the lack of popular support from the community for what local residents feel are overpaid workers.

Moreover, for historical and contemporary reasons, the four unions in the plant adopt similar views towards the enterprise: all fear the prospects of economic decline and job-loss, and all support management struggles to protect the industry. Tripartite committees work together to challenge the domination of the central state, and seek to redirect profits back into the enterprise and to change national policies towards taxation, non-loyal buyers, export restrictions and prices at the pump. The effect is a largely non-confrontational unionism, whose ability to address health and safety issues in the plant is limited to after-the-fact worker benefits, and whose ability to deal with broader community issues is very limited indeed.

On the other hand, the political and economic struggles over the plant have over the past four years generated a stasis in policy, in which sensible but thoroughgoing reinvestment strategies are not being considered. The state depends for a large proportion of its budget on the profits of the *combinat*, and many individuals along the way have strong vested interests in the continued short-term flow of profits to the central state. The consequence is that little modernisation of equipment is occurring and ongoing environmental and health effects continue to be externalised to workers and residents of nearby communities, as insufficient funds are returned to the *combinat* for reinvestment. The central state seems incapable of changing this situation and is unwilling to forgo the short-term financial benefits gained from running down the plant in favour of longer-term benefits of substantial reinvestment.

The groups most impacted by pollution from the *combinat*, besides workers in the plants, are the communities close to the plant and in the path of the prevailing winds. These communities have historically been 'captured' by the power of the large state *combinat* and, because of the ways in which the *combinat* has contributed to their social and infrastructural budgets (albeit in a minor fashion), the possibilities for effective municipal government have been undermined.

Consequently, the local councils of the villages, towns and county (*obstina*) centres remain relatively powerless to effect any real changes in the technological and environmental policy of the *combinat* (as they have been over the past forty years). These levels of influence are not uniform, however, but depend upon social power, i.e., on personal contacts and/or the size of the settlement and the political power it can wield: county governments (*obstini*) and larger town councils have closer contacts with Sofia and have been able to establish stronger regulatory control and informal financial linkages with the enterprise than have the smaller towns and villages, despite the often closer proximity of the villages to the plant and corresponding higher levels of environmental and health impacts for village residents. One effect of this policy has been that community leaders in smaller settlements have been forced to work with Neftochim management to 'beg for crumbs', as one local mayor described it, to support much-needed infrastructure in the village.

Transition to which capitalism?

Since state enterprises have still not been privatised, a great deal of restructuring is aimed primarily at shoring up the power of particular groups within the enterprises in the event of privatisation, share ownership, or some other reorganisation of the individual enterprises. In some cases smaller units of the enterprise have been spun off and informally privatised by former managers, and now operate on contract servicing or supplying the main plant.[26] But for the most part, the current situation is typified by what a recent ILO report described as 'enterprise desertion' – enterprises operating with little central guidance and with only weak incentives to restructure or to operate more efficiently (ILO-CEET 1994). In this context, the rent-seeking that occurred under central planning continues and the horizontal linkages between plants and enterprises that protected them in an economy of shortage seem to be maintained to protect supplies and guarantee benefits in an economy of decline.

Under conditions of economic crisis and diminished state oversight, regional or national monopolies such as Neftochim seem to be drawing on their experience under central planning of operating as political bodies to protect budgets, subsidies and supplies and prevent entry of new producers. Now unfettered by the party and the state these regional monopolies begin to operate as huge trading companies: they strive to maintain (and increase) their lateral or horizontal exchanges and contacts (previously restricted by the ministry and the party), lease out facilities to second-party producers, and focus attention on maintaining supplies, labour and markets through barter if necessary. As a result, far from transforming the industrial workplace and rationalising production systems and the labour process, this form of 'modernisation' is actually resulting in both mass unemployment (particularly in the branch-plant economies) and, in enterprises like Neftochim, continued collusion between management and workers to protect their budgets, supplies and wages. Supervision by the party has collapsed and managers are more attentive to the problems of supply and barter, particularly as monopolies become stronger and supply shortages intensify. As state enterprise managers devote their 'managerial' efforts to protecting the subsidies and supplies of the enterprise, work continues to be 'put out' to worker collectives within the enterprises.

The cost associated with the social reproduction of the community – an important enterprise responsibility under central planning – has now become a luxury the enterprises seek to shed. With the collapse of state and party control over the enterprises, the mechanisms for coercing them to support their local communities have also substantially diminished. As a result, communities have lost many of the social subsidies they used to enjoy under state management, ranging from investment in roads, education and health to underwriting of public transport, sports facilities and public building projects (such as housing). Moreover, since some communities, such as the city of Bourgas, remain both powerful and politically important, the restructuring of enterprise–community relations has resulted in inter-municipality struggles over remaining enterprise subsidies, with benefits accruing to the larger urban centres. The result is a further weakening of smaller, often more heavily impacted communities.

Communities in marginal regions of the country have been particularly badly affected by these changes within enterprises. As managers have withdrawn from their former responsibilities for the provision of infrastructural investments and social benefits, one strategy has been to close down and asset-strip branch plants – shedding the costs of what used to be called 'social industries' in favour of protecting the jobs of core workers in the metropolitan plants.

Despite these adjustments, financial vitality within the large state enterprises is elusive. Investment capital to purchase such large enterprises cannot be accumulated by workers or managers and new Bulgarian owners will not have the necessary investment capital for needed technological restructuring. The only possible buyers are then foreign companies, but not all managers and workers see benefits from such a take-over. Moreover, the central government has not yet been willing to cede control over such a potentially profitable enterprise to non-Bulgarian owners. The resultant stasis deepens the tendencies described above, weakens the common interests that bound together different groups within the enterprise under central planning, and in their stead different fractions of management struggle with each other for the support of the workers with an eye to future changes in ownership and management.

Conclusion: economic transition, industrial restructuring and the question of democratisation

What lessons of transition emerge from this brief case study? At the very least, the certainties of much extant transition theory must, I think, be questioned. The case study of Neftochim Bourgas (and parallel studies in Russia (Burawoy and Krotov 1993, 1995), the Czech Republic (Pavlínek 1996), and Slovakia (Smith 1994, 1995a, 1995b)) raises the question as to whether what is happening is: (a) an extension and deepening of the horizontal networks and barter arrangements that were typical of the classical model of state socialism; (b) residual forms within inefficient industries that will, in time, give way to more formal market arrangements based on prices and economic rationality; or (c) new or hybrid forms that signal an alternative, emergent model of industrial organisation, what Michael Burawoy calls 'merchant capitalism'.

Extant liberal interpretations of transition present a particular reading of the current situation of stasis, entrenchment and collaboration between workers, management and government officials. In this reading, at the enterprise level marketisation and competitive relations will eventually force managers to adopt demand-driven investment decisions and competitive market relations will provide the necessary stimuli for managers to take on the difficult political task of imposing careful planning procedures in production. Privatisation of state enterprises and the emergence of entrepreneurialism under competitive market conditions will lead to the rationalisation of production, the reorganisation of work practices, and increasing technical composition of production – resulting in labour shedding and investment in technologically efficient production systems. The effect will be a rationalised and more highly skilled, monitored and disciplined workforce working with more cost-effective and environmentally efficient production technologies. Profit-maximising behaviour will result in the introduction of rational accounting systems in which production inefficiencies will be counted as real costs, excess labour will be seen as unnecessary input cost, and fines and public sanctions will be assessed as negative costs, encouraging industries to become more sensitive to health and safety issues and to strive to reduce pollution on economic and public relations grounds, or enterprises will remain non-competitive and eventually close down.

At the community level, democratisation of the political process will force greater accountability of political actors and bureaucrats, and empowerment of individuals and communities. Decentralisation of decision-making to local and regional levels of government will permit increased accountability to be translated into stricter and more informed regulatory enforcement. Regional environmental and health regulatory agencies will be supplemented by activist *obstini* environmental agencies which will see in anti-pollution measures, besides an urgent practical necessity, a valuable source of political and monetary capital.

De-communisation of the public sphere will encourage increased co-operation between state institutions at the central, regional and local levels and should, therefore, foster more effective regulatory procedures that are more responsive to the needs of the population and less concerned with the need to maintain production at any cost.

Indeed, all these trends are evident within the Bulgarian case study. But is this the whole picture? Evidence from Bulgaria suggests that, in removing the state and the party from the enterprise, political reform and privatisation have enhanced (albeit in more difficult circumstances) the relative power of managers to foster relations of barter. The power of *some* worker collectives and brigades within the enterprise have also increased. Core workers, especially in national industries such as oil refining and chemicals, have been able to consolidate their control at the point of production to such an extent that the independent trade union affiliate (KNCB-Neftochim) in Neftochim now operates largely independently, and often at odds with the national union federation (KNCB). Moreover, de-communisation has led to a proliferation of trade union bodies within the enterprise, each motivated by the specific needs and goals of its membership. Managers compete to represent the interests of these worker groups in Neftochim, and workers thus become aligned with particular parts of management. No longer under the constraints of the plan, worker groups and segments of management compete with each other in an attempt to maintain and guarantee wages, subsidies, supplies and continued loose management control over production on the shop-floor. The extension of the regional monopoly power of the enterprise, or, in the case of an enterprise like Neftochim, the extension of national monopoly power results in increased, not decreased, barriers to the entry of new firms.

Marketisation merely deepens the already existing system of barter arrangement between enterprises to ensure reciprocal supplies. Democratisation results in the strengthening of the political power of the actors who are already embedded in existing systems of power, weakening smaller communities and creating new barriers to entry into the political arena. Decentralisation results from, and in, the withering away of the power of the central state in the enterprise and in local and regional government responsibilities. The result is a weakening of state control over the economy at the very time that de-communisation of government and enterprises results in increased competition among actors over the surplus from enterprises, especially cash cows such as Neftochim (but now outside the control of the party mechanisms and to an increased degree beyond the control of the state).[27]

In this scenario, the efforts of state enterprises in the transitional period are directed to the struggle for short-term profit to enable extension of power within the enterprise in preparation for planned privatisation. The social and environmental consequences are severe. Little new investment can take place. Very little investment capital is available and this capital shortage is perpetuated and deepened by management policies directed towards political power, protecting core workers, maintaining short-term profits through barter and asset stripping, maintaining barriers to new entrants sometimes through price undercutting and through informal (at times under the table) privatisation of enterprise assets. A weak state is unable adequately to monitor these activities and is too weak to force reorganisation within the enterprise (as for that matter is current management). The industry is heavily indebted and carries with it uncertain environmental liabilities from past and present practices. As a result, international firms have been unwilling to enter into joint ventures and provide investment capital for enterprises without firm guarantees of their lack of environmental liability and without major overhauling of relations in production. Even the surrounding communities who might otherwise be able to mobilise support for enforcement of environmental regulations (as with the Bourgas city environmental industrial passport system) now find themselves fragmented politically as workers and their families constitute a major interest group which opposes stricter enforcement of health and safety regulations. Under the general rubric of liberal reform ideology, these interest groups argue for the freeing of entrepreneurial capital to foster a regional economic recovery which will provide the economic resources to deal with environmental and health problems later.

Of course, one might argue that the pathology of transition described here is merely 'transitional' – a legacy of the past that will be swept away by the emergence of a rational calculus of cost–benefits, market incentives and entrepreneurial drive. This may be. But little concrete evidence from the actually occurring transition of large state industries currently leads us to believe that this is, or will be, the case. Instead, these changes force us to take seriously the enterprise and community level of social struggles and the emerging hybrid forms that seem to be the contemporary legacy of central planning and the reform process. Initial indications point, instead, to a situation of crisis management that generates protective and cautious strategies, little actual technical or organisational restructuring, a severing of formerly paternalistic ties with smaller neighbouring communities, and a careful deepening of strategic alliances with larger municipalities to mobilise support at the national level for state subsidies, guaranteed supplies and protected markets. The industrial geographies that are emerging from this type of restructuring will have far-reaching consequences for the social and political life of the people of the region.

Acknowledgements

Research for this chapter was supported by National Science Foundation International Collaborative Programs grants INT-8703742 and INT-9021910, the John D. and Catherine T. MacArthur Foundation Program for Peace and Cooperation, The University of Kentucky Office of Research and Graduate Programs, The Regional Research Institute at the West Virginia University, and the Institute of Geography, Bulgarian Academy of Sciences. The chapter has benefited from discussion with and comments from Bob Begg, Margaret Fitzsimmons, Mieke Meurs, Adrian Smith, Stefan Velev and Michael Watts, as well as participants in the IGU Industrial Geography Group workshop on 'Industry and Environment', Budapest, August 1994 (particularly Gyorgyi Barta, Bill Berentsen, Sergio Conti, Bogdan Domański, Riitta Kosenen, Eike Schamp and Michael Taylor), and members of the Restructuring Group (Oliver Froehling, Carole Gallaher, Katherine Jones, Susan Mains, Eugene McCann, Petr Pavlínek, Jeff Popke and Chad Staddon). All errors of fact and interpretation are, of course, my own.

Notes

1 For an interesting discussion of the effects of this war economy on the regional and industrial geography of Czechoslovakia, see Pavlínek (1996).
2 This chapter is part of a larger project assessing the geographies of transition and

modernisation in Eastern Europe, specifically through a detailed case study of Bulgaria.

3 This term is taken from Rutland (1995: 2).

4 For a parallel assessment of this form of restructuring in African structural adjustment policies see Ould-Mey (1996).

5 See, for example, Doreen Massey (1984), Andrew Sayer and Richard Walker (1992), Storper and Walker (1989).

6 Notable exceptions are Barta (1994), Berentsen (1994), Burawoy (1985), Burawoy and Krotov (1993, 1995), Burawoy and Lukacs (1992), Clarke, Fairbrother, Burawoy and Krotov (1993), Domański and Matykowski (1994), Pavlínek (1996), Schamp (1994), Smith (1994, 1995a, 1995b) and Wyzan (1993).

7 For a critique of the equating of Stalinist production methods with Fordism see Smith (1994, 1995a, 1995b) and Pavlínek (1996).

8 For a similar transitional phenomena in South Africa, see Pickles (1992b).

9 This focusing of attention on production politics is not intended to signify a prioritising of economic explanation over the political, but to redress the absence of political economic analysis of production dynamics within contemporary transition theory.

10 A conservative variant of this 'millenialism' is Francis Fukuyama's (1992) much celebrated and/or berated *The end of history and the last man* (see also Fukuyama 1989, 1989/90; and some criticisms from the Left: Peet 1993a, 1993b; Rustin 1992; Miliband 1992). A provocative variant of this form of 'millenialism' is Felix Guattari and Toni Negri's *Communists like us*, where they argue that the revolutions of 1968 mark 'a serious challenge to the dead weight of bureaucratic capitalism with its "over-coded" and de-individualized individual' (p.17) and the beginning of the end of Integrated World Capitalism (IWC).

11 Whether these new articulations are transitional forms (as much liberal economic theory would hold) or may lead to stabilisations of new and different development models is currently an open question. I prefer to keep this an open question. See Lipietz (1987).

12 The Burgas plant has a land area of 1,300 hectares and floor space of 740,000 square metres.

13 Over 60 per cent of the equipment is more than twenty years old.

14 Under this regime, information was carefully controlled and little that was reliable was available to either the public or to local government officials. Even data collected from the local mayors' and administrators' offices were not returned to the communities by the central government. Medical practitioners and local officials, as well as local residents themselves, were, however, generally aware of the high levels of incidence of respiratory, heart and skin diseases among the population, and particularly among the young, aged and pregnant. Even the enterprises themselves at times admitted that the gas releases (both low-level daily emissions and major catastrophic releases) were dangerous and harmful to workers and local residents. For example, on some production lines workers are not allowed to work for more than five to seven years without being reassigned, early retirement is mandatory on the most polluting lines, and daily rations of yoghurt (a so-called prophylactic food) were issued to all workers to clean their systems of pollutants. The enterprises argued, however, that under

the system of mandatory state quotas which pushed production levels ever higher and yet returned little capital for reinvestment in environmentally safe or efficient production technologies, enterprise managers were powerless to effect the types of changes that would protect workers and local residents.

15 In many industries in Bulgaria the government failed to supply enterprise wages over several months in 1992 and 1993.

16 For an analysis of similar conditions elsewhere in Central and Eastern Europe see Crossette 1994; ILO 1993, 1994; Standing 1993a, 1993b; Whitney 1994; and Zaniewski 1992.

17 These 'public resources' were, of course, only nominally owned and/or controlled by the public under state socialism. Ownership of these privatised resources by individuals is new, but control of them by a small elite for their own purposes is not.

18 For a parallel situation in the former East Germany, see Grabher 1994.

19 In districts like Ruen, textile workshops were formerly distributed throughout the several villages in the *obstina*. Until 1994 about 1,000 villagers were employed in these workshop factories, but by August 1994 this number had dropped to 200.

20 Interview with KNCB (Confederation of Independent Trade Unions) representative Usef Mehmed, KNCB Office, Ruen, 1 August 1994.

21 Significantly, these high unemployment levels differentially impact the Turkish and Pomack regions of the south, introducing an important ethnic dimension to restructuring that has received little attention.

22 In 1994 the average utilisation of capacity in the chemical and oil processing industry was between 50 and 60 per cent, down from over 87 per cent in 1989 (National Statistical Institute, *Current Economic Business* June 1994).

23 Second-party processing involves the processing of a customer's oil for a fee, instead of the refinery buying and processing its own oil. Of the 4.7 million tonnes of this second-party production, 1.3 million was carried out under contract for the Marc Rich Company (ILO-CEET 1994). The story of the role played by Marc Rich, who faces numerous indictments for fraud in the US, during the transition in Central and Eastern Europe and the former Soviet Union is still to be written.

24 The loss of earnings, combined with a decrease in the number of female workers in the plant, has, however, had important consequences for the local economy.

25 Interviews with managers in Neftochim, 1991, 1992, 1993, 1994.

26 There have been few of these in the case of Neftochim, and they remain small (mainly research and technical service groups) partly because of the capital necessary to effect such a privatisation and partly because of the degree of over-sight exercised by central ministries.

27 Neftochim – as with most large state enterprises in Bulgaria – is still a state enterprise. On several occasions the state has exercised its power over management of the enterprise, in particular by replacing recalcitrant or corrupt managers. However, the fact remains that the devolution of decision-making to the enterprises removes the state from day-to-day management decisions, and gives to the enterprise a greater level of autonomy and freedom from state apparatuses.

Bibliography

Altvater, E. (1993) *The future of the market: an essay on the regulation of money and nature after the collapse of actually existing socialism*, London: Verso.

Barta, G. (1994) 'Different strategies of multinational companies in Hungary and their Hungarian reception', a paper presented at the IGU Commission on the Organisation of Industrial Space Conference, Budapest, 16–20 August.

BBN (168 Hours) (1993) 'Real privatisation has not started yet', *168 Hours BBN* 3, 38 (20–26 September): 1, 5.

Berentsen, W.H. (1994) 'The spatial structure of industrial employment and production in Eastern Germany from 1895 to 1989 and contemporary regional problems', a paper presented at the IGU Commission on the Organisation of Industrial Space, 16–20 August, Budapest.

Burawoy, M. (1985) *The politics of production*, London: Verso.

Burawoy, M. and Krotov, P. (1993) 'The economic basis of Russia's political crisis', *New Left Review* 198: 49–69.

Burawoy, M. and Krotov, P. (1995) 'Class struggle in the tundra: the fate of Russia's workers' movement', *Antipode* 27, 2 (April): 115–36.

Burawoy, M. and Lukacs, J. (1992) *The radiant past: ideology and reality in Hungary's road to capitalism*, Chicago and London: University of Chicago Press.

Campbell, J.L. (1992) 'The fiscal crisis of post-communist states', *Telos* 93 (Fall): 89–110.

Clarke, S., Fairbrother, P., Burawoy, M. and Krotov, P. (1993) *What about the workers? Workers and transition to capitalism in Russia*, London: Verso.

Crossette, B. (1994) 'U.N. study finds a free Eastern and Central Europe poorer and less healthy', *The New York Times* 6 October.

Domański, B. and Matykowski, R. (1994) 'A new emerging spatial order? Mechanisms and prospects of industrial change in postsocialist Poland', a paper presented at the IGU Commission on the Organisation of Industrial Space Conference – Industry and Environmental Challenge: Focus on Industrial Transition in Post-Socialist Countries, 16–20 August, Budapest.

Fukuyama, F. (1992) *The end of history and the last man*, New York: Avon Books.

Fukuyama, F. (1989/90) 'A reply to my critics', *The National Interest* 18 (Winter): 21–8.

Fukuyama, F. (1989) 'The end of history?', *The National Interest* 16 (Summer): 3–18.

Grabher, G. (1994) 'Instant capitalism: fragile investment in East German regions', in P. Dicken and M. Quevit (eds), *Transnational corporations and European regional restructuring*, Utrecht: The Royal Dutch Geographical Society: 109–30.

Guattari, F. and Negri, T. (1990) *Communists like us: new spaces of liberty, new lines of alliance*, trans. by M. Ryan, New York: Semiotexte.

ILO (International Labour Organisation) (1993) 'Labour market reforms in Bulgarian industry', *ILO-CEET Newsletter* 2 (June): 4–5.

ILO (International Labour Organisation) (1994) 'Bulgarian G24 meeting sets up co-ordinating group on social policy', *ILO-CEET Newsletter* 2 (June): 4.

ILO-CEET (1994) *The Bulgarian challenge: reforming labour market and social policy*, Budapest.

Kornai, J. (1992) *The socialist system: the political economy of communism*, Princeton: Princeton University Press.

Lipietz, A. (1987) *Mirages and miracles*, London: Verso.

Massey, D. (1984) *Spatial divisions of labour: social structures and the geography of production*, New York: Methuen.

Mikhova, D. and Pickles, J. (1994a) 'GIS in Bulgaria: development and perspectives', *International Journal of Geographical Information Systems* 8, 5 (May): 471–7.

Mikhova, D. and Pickles, J. (1994b) 'Environmental data and social change in Bulgaria: problems and prospects', *The Professional Geographer* 46, 2: 229–36.

Miliband, R. (1992) 'Fukuyama and the socialist alternative', *New Left Review* 193 (May/June): 108–13.

Ould-Mey, M. (1996) *The stick and the carrot: global adjustment and structural adjustment in Mauritania*, Lanham, Maryland: Littlefield Adams Books.

Pavlínek, P. (1996) 'Transition and the environment in the Czech Republic: democratization, economic restructuring and environmental management in the Most District after the collapse of state socialism', Unpublished Ph.D. Dissertation, Department of Geography, University of Kentucky.

Pavlínek, P., Pickles, J. and Staddon, C. (1994) 'Democratisation, economic restructuring and environment in Bulgaria and the Czech Republic: a comparative perspective', in P. Jordan and E. Tomasi (eds) *Zustand und Perspectiven der Umwelt im oestlichen Europa*, Frankfurt am Main: Peter Lang: 57–84.

Peet, R. (1993a) 'Reading Fukuyama: politics at the end of history', *Political Geography* 12, 1 (January): 64–78.

Peet, R. (1993b) 'The end of history and the first human', *Political Geography* 12, 1 (January): 91–5.

Perlez, J. (1994) 'Along Eastern Europe's capitalist road, fast and slow lanes', *The New York Times International* Friday, 7 October: A1 and A6.

Pickles, J. (1991) 'Democratisation and its others: politics, economy and environment in contemporary Bulgaria', a paper presented at the International Symposium on 'The Impact of Economic and Political Restructuring on the Bulgarian Environment', West Virginia University, Morgantown, 7 December.

Pickles, J. (1992a) 'Community and *combinat*: local responses to state industries in the Bourgas region of Bulgaria', a paper presented in the Special Session on Comparative Perspectives on Restructuring in Central and Eastern Europe, Annual Conference of the Association of American Geographers, Atlanta.

Pickles, J. (1993a) 'Environmental politics, democracy and economic restructuring in Bulgaria', in J. O'Loughlin and H. van Wusten (eds), *The new political geography of Europe*, London: Bellhaven Press: 167–85.

Pickles, J. (1993b) 'Environmental politics, democratisation, and civil society', an invited paper presented to the Department of Geography, University of Hawaii at Manoa, 22 April.

Pickles, J. (1994a) 'Control, negotiation, and competition: the role of environment in social and economic bargaining in the transition from central planning', a paper presented at the Second International Symposium on 'The Impact of Economic and Political Restructuring on the Bulgarian Environment', West Virginia University, Morgantown, 8 May.

Pickles, J. (1994b) 'Environmental reconstruction and theorising transition in state industries in Bulgaria', a paper presented at the IGU regional Conference on Environment and Quality of Life in Central Europe: Problems of Transition, Prague, 22–26 August.

Pickles, J. and Staddon, C. (1994a) *Summary report of the 1993 survey on governance and environment in the Bourgas region of Bulgaria*, Research Paper 9413, Regional Research Institute, West Virginia University, Morgantown.

Pickles, J. and Staddon, C. (1994b) *Summary report of the 1992 survey on environment, well-being, and governance on the Bourgas region of Bulgaria*, Research Report 9417, Regional Research Institute, West Virginia University, Morgantown.

Pickles, J. and Staddon, P. (1995) 'Democratisation and changing state form: a case study of environmental regulation in Bulgaria, 1991–1995', unpublished mimeo.

Rustin, M. (1992) 'No exit from capitalism?' *New Left Review* 193 (May/June): 96–107.

Rutland, P. (1995) 'On the road to capitalism: reflections on East Europe and the global economy', a paper prepared for the International Studies Association Annual Convention, Chicago, 24 February 1995, p. 2.

Sayer, A. and Walker, R. (1992) *The new social economy: reworking the division of labour*, Oxford and Cambridge, Mass.: Basil Blackwell.

Schamp, E.W. (1994) 'Industrial transition in the post-socialist countries: geographical perspectives', a paper presented to the IGU Commission on the Organisation of Industrial Space, Budapest Conference, 16–21 August.

Smith, A. (1994) 'Uneven development and the restructuring of the armaments industry in Slovakia', *Transactions of the Institute of British Geographers*, NS 19: 44–424.

Smith, A. (1995a) 'Regulation theory, strategies of enterprise integration, and the political economy of regional economic restructuring in Central and Eastern Europe', a paper presented at the Annual Conference of the Institute of British Geographers, University of Northumbria, Newcastle upon Tyne, January.

Smith, A. (1995b) 'Comparative regional development, industrial restructuring and the transition to capitalism in Slovakia', paper presented at the session on Prospects for the Regions of Eastern Europe and the Former Soviet Union, Annual Conference of the Institute of British Geographers, Newcastle upon Tyne, January.

Standing, G. (1993a) *Labour market developments in Eastern and Central Europe*, Budapest: International Labour Organisation Central and Eastern European Team Policy Paper 1, January.

Standing, G. (1993b) *Restructuring for distributive justice in Eastern Europe*, Budapest: International Labour Organisation Central and Eastern European Team Policy Paper 2, January.

Storper, M. and Walker, R. (1989) *The capitalist imperative: territory, technology, and industrial growth*, Oxford and Cambridge, Mass.: Basil Blackwell.

Whitney, C.R. (1994) 'East Europe's hard path to a new day', *The New York Times* 30 September: A1, A4, A5.

Wyzan, M. (1993) 'Economic transformation and regional inequality in Bulgaria: in search of a meaningful unit of analysis', a paper presented at the Annual

Conference of the American Association for the Advancement of Slavic Studies, Honolulu, 19–22 November.

Zaniewski, M. (1992) 'Regional inequalities in social well-being in Central and Eastern Europe', *Tijdschrift voor Economische en Sociale Geografie* 83, 5: 342–52.

8

THEORISING TRADE UNIONS IN TRANSITION

Andrew Herod

> In the socialist state under construction, trade unions are
> needed not to struggle for better working conditions – this is
> the task of the social and political organisation as a whole
> – but to organise the working class for the purpose of
> production: to educate, discipline, allocate, collect, attach
> individual categories and individual workers to their jobs for a
> set period: in a word, hand in hand with the government in an
> authoritative manner to bring workers into the framework of a
> single economic plan.
>
> (Trotsky, quoted in Pipes 1990: 710)

> [The trade union is] an organisation for recruiting, educating
> and training, a school for administration – a school of
> Communism.
>
> (Lenin, quoted in Hayenko 1965: 1)

> A union's primary function as an organisation is to represent
> workers' interests.
>
> (Kochan and Katz 1988: 149)

The transition occurring in Central and Eastern Europe (CEE) is clearly
having a dramatic impact on labour relations and labour markets. The shift
from a centrally planned system of production and allocation to a market
economy is fundamentally changing the relationship between enterprise
managers, the workforce, the state and the trade unions. Privatisation and
restructuring augur nothing less than a complete transformation of civic and
work-life in the region. In the West the dominant discourse concerning
the transition – perhaps most clearly captured by Francis Fukuyama's (1992)
notion of the 'end of history' – sees current events in terms of a process of
modernisation along western lines through the birth of a new civil society
conceived out of the union of a market economy and political democracy. It
is assumed by many that, with the withering of the central state planning

197

and control apparatus and the flourishing of economic privatisation, the societies of the region will, given time, become just like those of Western Europe or North America (e.g., Fleissner 1994; Targetti 1992; Freeman 1992; and Silverman and Yanowitch 1993).[1] Within these new societies, it is argued, old institutions which were 'deformed' under Soviet influence will finally be able to operate as they 'should', which is to say that they will take on the characteristics of their corresponding institutions in Western Europe or North America (e.g., Gordon and Klopov 1992; Clarke 1993; Fogel and Etcheverry 1994; Hanson 1997).[2]

The trade unions, many of which played significant roles in the events of 1989, are viewed within this discourse to be crucial elements in the new societies whose role it is to serve as guardians of many civil rights and to act as 'responsible' partners in the transition to capitalism (Egorov 1996; ICFTU 1996; FTUI 1994; Silverman and Yanowitch 1992, 1993). Hence, whereas under central planning the role of the official trade unions was to serve as the 'transmission belts' of the central planners by ensuring production quotas were met (Pravda and Ruble 1986), it is assumed that, once the transition is complete, they will act first and foremost as representatives of the workers, whose interests they will further and protect through the practice of free collective bargaining and participation in tripartite negotiations with government and employers over such things as pensions or minimum wage levels. Indeed, many western trade unions and labour organisations have spent large amounts of time and money running seminars and educational workshops in the region to train local trade unionists to operate more along western lines. For example, the International Confederation of Free Trade Unions has established offices in several CEE countries and run conferences concerning matters such as setting up pension plans, financial accounting, and running union elections (ICFTU 1996). The International Metalworkers' Federation trade secretariat has similarly run workshops for the new and/or newly democratising metal unions in the region (Herod 1998b; for more on the various international trade secretariats themselves, see Busch 1983 and Windmuller 1980), whilst the American Federation of Labor-Congress of Industrial Organizations' Free Trade Union Institute (FTUI) has provided organising assistance, union-to-union exchanges, and research and technical support to unions in the region (FTUI 1994; Herod 1998a).[3]

There are, however, a number of issues that need to be addressed in claims such as these, which tend to assume that CEE under 'capitalism' will be just like the West and that institutions will therefore be just the same and operate in just the same ways. These issues relate to the nature of the transition and the kind of capitalism it is bringing forth, the geography of the transition, the role of the unions in this process, and the kinds of regional political economies that are emerging. Most particularly, within this discourse the unions have been placed in somewhat of a paradoxical situation, for they are seen by many both as essential to the creation of a new democratic civil society that will ensure the success of free market capitalism, and yet also as potential impediments to the unleashing of those very market forces through their efforts to protect workers' standards of living. In this chapter I examine some of these issues concerning trade unions in the transition. The chapter is in two main parts. In the first section I contrast how the new model of trade unionism advocated as appropriate for, and a necessary component of, the market economy differs from that of the old Soviet-inspired model. In the second section I address a number of problematics which, I think, need to be considered when thinking about the transition's impact on, and the role played by, the trade unions in CEE.

Competing models of trade unionism

The Soviet model

The Soviet model of trade unionism which was subsequently imposed in various forms throughout CEE after the Second World War owes much to the writings of Lenin, writings which are themselves somewhat contradictory on the matter (for an extended view of Lenin's thoughts, see Lenin 1963). At one level Lenin was deeply suspicious of trade unions, which he saw as essentially reformist organisations interested more in improving wages and conditions and in seeking to abate the excesses of capitalism than in overthrowing it. Instead, he argued that it was the party which was the true embodiment of the proletariat's aspirations and that, consequently, the unions – which Lenin saw as representing merely narrow sectional or local interests – should be subordinated to the party, which was to represent all workers' interests (Nove 1969). At another level, however, Lenin believed that the unions did have some legitimate protective functions to perform on behalf of workers because, he recognised, the views of managers and those of workers would not always coalesce, even in a proletarian state (Lenin 1971). Such beliefs about trade unions combined to produce what has been called a 'dualist' Leninist model of trade unionism (Pravda and Ruble 1986). Emerging from the Tenth Soviet Communist Party Congress of 1921, this model attempted to accommodate two competing visions for the unions: one in which they would be subordinate to and, effectively, part of the state, the other in which the economy would be under union control (see Hayenko 1965 for more details). The result of this dual mandate was a contradictory situation wherein the unions were supposed both to encourage the growth of labour productivity but also to protect workers from management abuse, a contradiction which Pravda and Ruble (1986: 2) suggest 'reflected the fundamental ambivalence of a system where labour was no longer pitted against capital yet remained under managerial control in an effort to further the economic growth of what was only notionally a workers' state'.

Under the Soviet model of development, state control over the trade unions was vital if central planning were to succeed, for independent worker committees and trade unions might prove problematic to implementing party goals. Hence, the unions were to serve as the link between the cadre of the party and the mass of the workers. Over time this link increasingly became a one-way street as the unions took on the role of 'transmission belts', tools for ensuring that the wishes of the central planners were fulfilled on the factory floor. As Hayenko (1965) has suggested, such a structural position meant that the role played by unions under the Soviet model was not determined by the unions themselves on the basis of members' needs but was, instead, fixed by the state on the basis of the Communist Party's economic requirements. This, of course, led to multiple contradictions in the unions' position in society. In official party rhetoric unions were seen as defenders of the proletariat's interests yet, in practice, they were usually denied the ability to use the one weapon which unions have traditionally used to defend workers' interests – the strike – because to do so would disrupt production at plants that were now 'the people's' and would thereby be 'counter-revolutionary'.[4] As a consequence of their role in fulfilling the goals of central planning, unions in the USSR and, subsequently, throughout CEE were legislatively required to participate in efforts at economic development, particularly through mobilising workers to fulfil production quotas set by the central planning committees, to encourage 'socialist emulation' among workers, and to provide input in drafting legislation concerning industry, workers' issues and broader 'proletarian cultural activities'. A key characteristic of central planning was that issues of labour efficiency were less of a concern than were those of ensuring that production quotas were met. This led to a situation in which enterprises often hoarded labour to make certain they had sufficient workers on hand to meet production quotas without having to worry about whether this was an efficient use of labour, and concealed resources from the eyes of central planners for fear that knowledge of resource availability might lead to higher production targets being set for them (Clarke 1993). Loyalty to the system and to the unions was established not through successful bargaining for higher wages or protection from unemployment but, often, through the unions' abilities to reward favoured workers with access to trade union-run community centres, medical facilities and vacation resorts or with vouchers for consumer goods which are hard to get, such as TV sets, automobiles, clothing, etc.

The model described above, of course, is very much an 'ideal' type and most closely approximates developments in the Soviet Union. Although it is often taken to represent the norm for all the Soviet satellite countries, the underlying differences in the economic and political geographies and in trade union traditions in the countries of CEE upon which it was super-imposed and within which it developed after the Second World War mean that, whilst the general parameters were similar, there were, in fact, some significant differences in the ways in which trade unionism developed across the region. Hence, unions in countries such as Czechoslovakia and East Germany which had a lengthy pre-communist tradition of large, well-organised labour movements aligned with social democratic forces tended to maintain a non-dualist structure for a longer period of time after the Second World War than was the case in countries such as Poland, which had weaker and more fragmented union movements and labour traditions (Pravda and Ruble 1986). Equally, pre-war union traditions in Czechoslovakia and Hungary may well have accounted for the greater concern paid by union leaders in these countries to issues of union independence from the party than was the case in more rural countries such as Romania and Bulgaria which lacked long traditions of independent trade unionism and where the unions were used primarily as agents of modernisation and enforcers on the shop-floor of industrial discipline on workers more attuned to the rhythms of agricultural work. Similarly, whereas all countries had union organisations at the national, regional and local levels, those countries with a federal state system such as the USSR, Czechoslovakia and Yugoslavia also had an additional set of trade union structures at the level of individual republics.

Based upon such differences, Pravda and Ruble (1986: 9–10) suggest that the 'ideal'-type Soviet model must be seen not as monolithic but, actually, as a number of variations on a theme which differed both historically and geographically. There are four arenas, they argue, in which variation could be seen among the trade union systems of Sovietised CEE. First, the different underlying national traditions affected the way in which the superimposed Soviet model subsequently played out on the ground, resulting in a variety of models ranging from the more authoritarian practices of Bulgarian and Romanian unions to those in Czechoslovakia and Hungary, where unions played a more protective role, though still within the limits placed by party control. (We might add that such underlying traditions themselves also varied locally *within* national systems, again based on pre-existing conditions, political distance from the capital and its ministries, types of economic activity conducted in particular localities, etc. (e.g. see the discussion in Petkov and Thirkell 1991 concerning the practice of 'counter-planning' in several industrial sectors in Bulgaria in which different shop-floor proposals to modify centrally devised plans were incorporated differentially in various enterprises; see also Nagy 1984 concerning variations in collective agreements).) Second, trade union organisation was modified over time, particularly in response to popular protest against the regimes (e.g. in Poland and Hungary in the late 1950s, and in Czechoslovakia in the late 1960s). Hence, in Czechoslovakia during the Prague Spring, for example, democratically elected workers' councils emerged in several enterprises with long pre-war histories of labour dissent – such as at the ČKD engineering works in Prague and at the Škoda plant in Plzeň – and continued to operate for some time even after the Soviet invasion (Fišera 1978). Third, there were some efforts to break radically with the model, as was the case in Poland when Solidarity arose to press for the independence of the unions and the adoption of a more adversarial and protective form of unionism. Fourth, changes taking place in the political economy and cultural realms of some societies were reflected in changes in the trade unions. Hence, for example, as the Hungarian state introduced various reforms such as the New Economic Mechanism and became more corporatist in nature in the late 1960s, national union officials adopted positions which were increasingly critical of government policy and in which the unions were able to exercise influence over policy through consultative channels, even to the extent of winning some major victories against the government on issues such as labour redeployment.

The western model

There is, of course, no single western model of trade unionism.[5] Unions operate under a number of assumptions and in a variety of ways throughout 'western' market economies. In the United States, for example, trade unions are locally formed and focused entities designed to reflect the local interests of their members (cf. Clark 1989). Consequently, the national trade union chamber, the American Federation of Labor-Congress of Industrial Organizations, is a loose and voluntary organisation in which substantial sovereignty is maintained by the individual unions in different crafts and industries that make it up. Likewise, in Japan unions are also primarily constituted locally in various enterprises and national confederations are fairly loose entities (see Hanham and Banasick 1998). However, in contrast to the United States, in Japan there is no requirement for representational exclusivity nor are supervisors barred from belonging to the same union as production line workers. Whereas US unions are mostly craft or industrially based, Japanese unions are primarily enterprise-based. In Germany, by contrast, the model of trade unionism is one in which there is a highly centralised system of control and contract bargaining. This centralised bargaining structure operates in combination with a system of locally orientated autonomous works councils that are legally separate from the multi-industrial unions but which provide for worker representation in the individual plants (Thelen 1991).[6] In France it is the national confederations to which unions belong which play the most significant role in industrial relations, rather than the industrial unions of which individual workers are actually members. The central confederations tend to be ideologically orientated and historically it has not been unusual for a worker to belong to two unions simultaneously, one because it better represents the worker's own ideological position, the other because it is a more effective representative of workers at the workplace (Fairweather and Shaw 1972; Goetschy and Rozenblatt 1992). Such confederations operate under a highly centralised system of control and exercise considerable influence over the activities of affiliated unions, particularly concerning the signing of collective agreements or striking. Britain and the Scandanavian countries have still different models (on Britain, see Edwards *et al.* 1992; on Scandinavian countries, see Kjellberg 1992; Dølvik and Stokland 1992; Scheuer 1992; and Lilja 1992). In still other countries such as Mexico, the leading union confederation is separate from but dominated by the ruling political party and operates as part of a corporatist arrangement within the state. Despite such differences, one of the more unifying elements of these models of trade unionism is that the unions involved are supposed, first and foremost, to be representatives of the workers. Western commentators, such as Fossum (1985: 96), have suggested that the major goals of unions operating under market economy conditions are to organise an increasing number and share of the labour force and to provide representation services that enhance the well-being of their members. Fossum suggests that trade union members will evaluate whether they want continued membership of a particular union based upon the quality and quantity of services they receive, whereas union officials will often desire growth in the union's size to enhance their power and stability whilst seeking approval of their actions from the membership through re-election.

In terms of bargaining, the 'ideal' model of trade unionism is one in which unions enter into free collective bargaining with private employers or the state in an environment in which the state provides certain legal frameworks and guarantees but in which most issues are determined around the bargaining table in a process of give and take by the representatives of labour and management. Whereas under the Soviet model unions' economic positions tended to be related more to the political conservatism or liberalism of the various political establishments and the demands of central planners, in a market economy the bargaining position that unions and employers enjoy is often closely tied to the ups and downs of the business cycle. Furthermore, a key goal of unions in market economies is to take wages out of competition so that workers compete for work on the basis of things such as productivity rather than on the basis of who will work for the lowest wage. Finally, whereas unions operating under the Soviet model may sometimes have supported the introduction of new technologies because these allowed production quotas to be more readily met, unions in market economies often oppose such measures because they fear that new technologies may in fact result in higher unemployment.

Issues in theorising trade unions in transition

Although some western commentators have argued that, as a result of the transition, the trade unions of CEE will simply become 'westernised' without really specifying what a 'western' model of trade unionism is, in practice the geographical variation in economic and political conditions and traditions means that theorising the status and role of trade unions in the transition will not be a straightforward matter. Several important issues will affect both the nature and speed of change being felt by the unions.

First, arguments about 'westernisation' tend to assume, rather than to demonstrate, that the nature of the market economy emerging in CEE will in fact be a carbon copy of the West's and that, consequently, the role to be played by the unions will mirror that of their western counterparts. However, early evidence suggests that such an assumption is far too simplistic. Indeed, as Stark (1996) has argued with reference to what he calls 'recombinant property', it appears that a distinctly East European type of capitalism is in fact emerging which is as different from that of Western Europe or North America (if not more so) as is the type of capitalism at work in, say, Japan or South Korea. Thus, for example, in several countries of CEE numerous companies appear to be thoroughly bankrupt by western standards but are nevertheless being kept running by their creditors, many of whom are banks concerned about damaging their own asset base should they shut them down (Hawker 1993; *Prague Post* 1995a). Likewise, Piore (1992: 172) argues that even the term 'market' economy has different connotations in CEE from those it has in the West, suggesting that in CEE the term refers to the 'dispersion of *political* power, the break-up of the state monopoly over economic resources and the destruction of its capacity for central control through a totalitarian state which those resources confer', whereas in the West 'the market is a much more narrowly defined economic institution . . . for co-ordinating and directing economic activity in which resources are controlled by private individuals who interact freely with each other in a competitive marketplace'.

Although western industrial relations literature primarily argues that in a market economy the state will provide certain guidelines within which free collective bargaining will take place between the representatives of labour and of management, the situation on the ground in CEE is, then, in many cases quite distant from such 'ideals' (e.g., see Bamber and Peschanski 1996 for an account of the situation with regard to industrial relations in Russia). Even after almost a decade of restructuring, much of the economy in some countries still remains in state hands. The speed and methods of privatisation and the form and degree of restructuring have varied tremendously across the region (for more on such variations, see Rondinelli 1994; Frydman *et al.* 1993, and OECD 1993). This is having significant impacts on the development of the trade unions. Whereas those unions in the private sector have had to deal with employers who themselves must respond to the pressures of market forces, those unions still in the state sector do not face the same kinds of issues in terms of negotiating collective bargains. The result has been a situation in which labour relations in privatised enterprises are often more conflictual in nature and more focused on issues of work redesign and reorganisation and workers' participation in enterprise management than is the case in enterprises that are still state-owned (Egorov 1996). Still further, the influx of much foreign capital to some countries (particularly into Poland, the Czech Republic and Hungary) and regions (especially core industrial or capital cities) has meant that many unions are now dealing with foreign employers who are often able to pay higher wages than can local concerns, especially since such foreign firms are frequently less interested in making a profit and thus in bargaining the lowest wages and conditions they can secure than they are in establishing market presence in anticipation of future growth in CEE's economies and consumers' purchasing power (IMF 1992; Murphy 1992; Michalak 1993; OECD 1995; Buckwalter 1995).

Second, many such accounts frequently pay little attention to geographical variations in the nature of the transition, particularly with regard to pre-existing patterns of economic development, industrial history and political tradition. Yet, given that the trade unions in CEE are developing in quite different political, cultural and economic contexts, different groups of workers in different countries or regions are developing different styles and structures of trade unionism. For example, Romanian, Russian and Bulgarian unions have often engaged in large-scale strikes, while Czech and Slovak unions have been more quiescent until recently.[7] Likewise, within countries the ways in which the transition is playing out in different locales are shaping the ways in which the unions are developing. In the Czech Republic, for example, economic restructuring and lay-offs are hitting coal mining and textile regions in northern Moravia more heavily than they are Prague, where the large influx of foreign investment and the growth of tourism have helped keep unemployment fairly low. One result of this has been the development of much more radical local unions and/or continued support for the old communist unions in the coal and steel towns of northern Moravia which are experiencing relatively high levels of unemployment, whereas in Prague the unions and workers have generally been more supportive of the transition and the process of economic restructuring.[8] Equally, whereas the coal, steel and textile unions from high unemployment areas have focused their demands upon issues related to job security, workers and unions in the Prague area have been more interested in issues related to the higher costs of living associated with marketisation.

Third, as conditions change over time, structures of trade unionism are also likely to change quite considerably. In the Czech Republic, for instance, the model of trade unionism envisaged for the country in 1990 when the old Communist Revolutionary Trade Union Movement (Revoluèni Odborové Hnutí (ROH)) was disbanded was one based upon the structure of the US AFL-CIO in which unions were affiliated to the national

Czechoslovakian Chamber of Trade Unions (ČSKOS) and, after 1992, with the Czech-Moravian Chamber of Trade Unions (Českomoravská Komora Odborových Svazů (ČMKOS)) in 'a very loose confederation'.[9] However, the failure of the unions to develop comprehensive national bargaining structures – itself largely a result of the highly decentralised structure of unionism which has developed since 1990 in which local unions jealously guard their sovereignty *vis-à-vis* national unions and the ČMKOS itself – has led many within the national Chamber and in the headquarters of some national branch unions such as the metalworkers to suggest abandoning the AFL-CIO model in favour of a more centralised organisational form fashioned after that of the German trades unions.[10] Some Czech unions are also beginning to adopt more confrontational stances towards the Klaus government than was the case earlier in the transition and there have been a number of strikes in the rail, mining and medical sectors (*Prague Post* 1992, 1995b, 1995c, 1995d), the result, according to ČMKOS President Richard Falbr, of the fact that 'people are getting increasingly tired of the transition and the loss of their living standards'.[11]

Fourth, in much of this discourse the unions are placed in a seemingly contradictory position with regard to the transition. Many of the new and reformed unions have been at the forefront of calls for active struggle to bring about the privatisation of enterprises, market reforms, the emergence of an entrepreneurial system and culture, and the restructuring of enterprises along profitable grounds as a means to stimulate economic development, increase wages, ensure greater production of, and access to, consumer goods, improve workers' living standards, and, perhaps most importantly, break the political control of the old *nomenklatura* (Volkov 1992; see Ost 1989, 1995 for an account of the 'neo-liberalisation' of Solidarity in Poland; see De Luce 1993 on complaints by the Czech power workers' union over the Klaus government's failure to dismantle the energy monopoly České Energetické Závody).[12] Capturing the flavour of this discourse, Paul Somogyi, Executive Director of the AFL-CIO's FTUI, has suggested that free trade unions are 'basic pillars of democracy around the world' and that 'Democracy, civil society, *and market reform* [will] all falter without their most forceful advocate, the free trade unions' (FTUI 1994: 6–7, emphasis added). Yet, given how labour hoarding operated under central planning, such economic restructuring invariably entails the loss of (often large) numbers of jobs and the reduction in real incomes as prices for food, rent and other living expenses dramatically increase to their market-determined 'proper' levels. In the dominant discourse on the transition, then, the unions are expected by many reformers to advocate marketisation as a means of ensuring democ-ratisation (which are seen as two sides of the same coin), whilst in practice their members often look to them to minimise the job losses and declines in standards of living resulting from marketisation and its associated enterprise restructuring and price rises. Union leaders are thus often put in the position of having to support economic positions which favour restructuring and privatising enterprises as a necessary part of the transition but which are invariably unpopular with those directly affected by redundancy. Such a position is, perhaps, eminently illustrated by the comments of Jiří Rusnok (1993), an adviser to the Czech ČMKOS, who suggested that, in fact, unemployment in the early 1990s in the Czech Republic was too low and that 'An increase [in unemployment] would be healthy' since this would bring higher productivity and, presumably, higher wages for those workers still in employment. However, now that many of the unions are functioning in more democratic ways, leaders favouring market reforms often become vulnerable to being voted out of office. This can itself lead to problems of high turnover among union officials, a situation which can make it difficult for individual leaders to gain the experience that only comes with lengthy tenure of office but which is necessary to engage effectively in collective bargaining with employers. Furthermore, high levels of turnover of some elected officials – as were particularly evident in several countries' unions during the early 1990s[13] – also make it more difficult for unions to foster international collaboration because personal contacts between unions in different countries are often lost as officials are voted out of office and so the slow process of building links with workers abroad who may be dealing with the same problems or same transnational corporations must continually start from scratch.

Fifth, whereas under the 'ideal' western trade union model the principal role played by unions is supposed to be to protect the interests of their members in negotiations with the employers, in many CEE countries the primary conflicts are not between the representatives of labour and those of capital, but between new and old unions – that is to say, inter-union conflicts – or between the unions and the old political *nomenklatura* (see Temkina 1992 for an account of the situation in the former USSR during the early 1990s). Again, focusing on the Czech unions as an example, there has been a great deal of inter-union conflict in the Czech Republic concerning the possible creation of German-style plant-based labour-management 'works councils' to which workers who are not necessarily members of the plant's trade union might be elected to address certain issues related to production and working conditions. The majority of the thirty-four national branch unions which make up the ČMKOS are opposed to this development because they fear the councils will simply be used to bypass their local trade union organisations in the plants.[14] On the other hand, the Communist Trade Union Association of Bohemia and Moravia (formed from the remnants of the old ROH confederation) and its leader, who is a Deputy in the Parliament, have been vociferous supporters of such works councils. With very little presence at the national level but with well-organised plant-level structures in northern Moravia, the communist unions would easily be able to dominate such councils in the region.

In most of the countries of CEE there exists the situation of dual or even triple unionism between the 'new' unions and confederations, the old 'reformed' communist unions and confederations, and even some old 'unreformed' communist unions and confederations (see ICFTU 1996 for more details).[15] Thus, in Poland many unions have disaffiliated with the old communist OPZZ confederation and become members of Solidarity. In Bulgaria there has been antagonism between the unions affiliated with the reformed Confederation of Independent Trade Unions of Bulgaria (CITUB) and those affiliated with the Podkrepa confederation which was established as part of the opposition to the communist regime. In Romania, there was much conflict and rivalry for members between two new confederations (CNSLR and Fratia) and, although these rivals merged in 1993, there was subsequently rivalry between the CNSLR-Fratia and a new federation CSDR (Democratic Trade Union Confederation of Romania) which formed in 1994. Likewise, in Russia the member unions of the Federation of Independent Trade Unions of Russia (most of whose leaders are former members of the old *nomenklatura*) has found itself in competition with newly formed unions and the rival Association of Social Trade Unions (*Sotsprof*) (Bamber and Peschanski 1996). Furthermore, throughout the region many unions have affiliated with western-orientated trade union organisations such as the International Confederation of Free Trade Unions and its related International Trade Secretariats (to which unions in particular industrial sectors belong) whereas others remain affiliated with the old Soviet-backed World Federation of Trade Unions and its industrial organisations (see Herod 1998a for more details).[16]

The situation of inter-union rivalry has been exacerbated by new laws in several countries which have encouraged the creation of often quite small unions. In Romania, for example, a 1993 law allowed as few as fifteen workers to create a new union while in the Czech Republic as few as three workers may do so. The rivalry between the different unions has often played itself out, particularly in the still state-controlled sector but also in the newly privatised sector, in the form of competition to gain access to resources and benefits. The principal purpose of such activity is to attract more members than rivals do and therefore to be able to exert greater political influence in shaping the privatisation and restructuring process or in pressuring the government to provide subsidies to particular enterprises or industries. In such cases, negotiating contracts with employers for the purposes of improving current members' living standards is frequently something of a secondary concern (Kirichenko and Koudyukin 1993). Indeed, so intense have been these inter-union conflicts that many of the western trade union organisations (such as the International Metalworkers' Federation) now working in CEE to train unionists about collective bargaining procedures in a market economy have had to run separate courses for officials belonging to rival unions (Herod 1998b).[17] Such conflicts are important because they highlight the limitations in simply assuming that western industrial relations literature and sociological theories, with their assumptions that in market economies the principal social conflict will be that between capital and labour, can be used unproblematically to analyse the economies in transition.

Finally, it has been the case with many of the new and/or reformed unions in CEE that 'intellectuals' have played a much more significant leadership and agenda-setting role than in comparable unions in western market economies. Using the example of Solidarity in Poland, Ost (1996: 30) argues that the reason for this stems from the nature of labour relations during the communist period in which the working class paradoxically acted as a class-*for*-itself but not as a class-*in*-itself, a situation which was itself the result of the development of a political and economic system in which

> when a self-proclaimed working-class party comes to power and nationalizes all industry . . . not only is everyone in some sense objectively a member of the working class (as employees of the state-owner), but everyone claims subjective membership as well, as 'working class' becomes an honorific, the equivalent of the liberal democratic 'citizen'.

In such a situation, 'workers appeared in politics not as workers but as citizens . . . condemned to be a "universal class" [and so] lost the sense of their own particularity [such that the] working class acted "for-itself," but not "in-itself"'. Ost suggests that such a universalist notion of class swamped a more particularist notion of class and allowed various 'intellectuals' and professionals to take over the workers' movement and to speak legitimately as worker leaders, particularly since many such professionals – consigned to jobs they did not like – thought of themselves as no different from any other worker. The result, Ost argues, has been the 'colonisation' of many workers' movements by activists who in western market economies would be considered middle-class professionals and who are often much more favourably inclined towards marketisation and more concerned about guaranteeing political pluralism than are working-class trade union members who are seeing their standards of living decline precipitously and are in many cases more concerned about bread-and-butter issues. This, of course, raises questions about the nature of the future development of the trade unions and broader workers' movements and the agendas that they might pursue.

Conclusion

In this chapter I have attempted to outline a number of issues I think important when considering the impact of the transition upon the trade unions and, equally, the role played by the trade unions themselves in shaping the process of transition. Clearly, one of the key issues to consider is the great geographical variation of political, economic and cultural conditions and traditions across CEE. The transition is not playing out in a uniform manner across the region, and unions in different parts of CEE face different challenges and are likely to develop in different ways. Consequently, any attempt to theorise the transition must be sensitive to the geographical context within which it is playing out. Such geographic variations greatly complicate efforts both to generalise the nature of development and labour relations' practices prior to 1989 (e.g., see Godson 1981; Schapiro and Godson 1981; Nagy 1984; Pravda and Ruble 1986; Petkov and Thirkell 1991) and, equally, to argue that throughout the region a process of modernisation is occurring which is uniformly making societies and economies more 'western'. In other words, the uneven geography of the transition begs the question of whether there is a single regional transition occurring or, instead, a multitude of local transitions based in different historical and contemporary conditions. Despite assumptions by many that CEE will eventually look just like other Western European or North American economies, it is questionable whether this is in fact the direction of change these economies are taking. If it is the case that a distinct type of capitalism is emerging differentially across CEE, then this will have important implications for the types of trade unions and labour relations processes and structures that will develop in the region.

Three further points also need consideration. First, given recent political events in CEE and the resurgence of former Communist Parties in several countries, the question of a reversal and/or slow-down in the trend towards privatisation, marketisation and restructuring is not as fanciful as it may once have seemed in the early 1990s. This, of course, would have significant implications for workers and for the future development of the trade unions. Specifically, given the manner in which the transition is playing out unevenly over space, workers in regions disproportionately hit by unemployment and the negative consequences of restructuring may be more favourably disposed to greater state intervention in the economy than those in regions where unemployment is less of a concern. If this is the case, we might expect to see quite different types of unionism and union demands emanating from different locales even within single countries or, relatedly, growing antagonism between local and national union structures depending upon the political affiliations of the parties involved. This latter in particular could spark intense political struggles over the issue of the geographical scale (national, regional, or local) at which power will be exercised within particular unions.

Second, although the legal relationships between the state and the trade unions may have changed a great deal in the countries of CEE, in many cases the class relationship between union members and enterprise managers/bureaucrats remains essentially the same. Indeed, many of the old *nomenklatura* have been able to convert their status from managers and party officials to owners of enterprises, since they are some of the few who have access to the resources and contacts necessary to acquire newly privatised enterprises. Thus, in many cases neither the individual actors involved in the operation of various enterprises nor their relationship to one another have significantly changed with privatisation.

Third, the question must be asked at what point the transition is considered to be complete. If the transition is deemed to have been completed when CEE's post-communist institutions all function exactly as do their western counterparts, then it may take a very long time, if it is ever completed. On the other hand, if the standard of completion is deemed to be simply the privatisation of former state-run institutions, then the transition may be seen as being completed in a much shorter time period. This is an important issue, not only intellectually – since it goes to the heart of how we use terms such as 'transition' – but also politically, for it raises questions about the direction and speed of social transformation and the possibility that new styles of trade unionism unique to CEE are emerging.

Acknowledgements

Fieldwork for this chapter was funded by two Faculty Research Grants from the University of Georgia Research Foundation. I would like to thank the many trade union officials who gave me their time and agreed to be interviewed, including: Richard Falbr, President, Czech Moravian Chamber of Trade Unions (ČMKOS), Prague, Czech Republic; Vlastimil Beran, Head, International Department, ČMKOS, Prague, Czech Republic; Jan Uhlíř, National President of the Czech Metalworkers' Federation (OS KOVO), Prague, Czech Republic; Lucy Zábranská, International Department, OS KOVO, Prague, Czech Republic; Dana Sakařová, International Department, OS KOVO, Prague, Czech Republic; Anne-Marie Mureau, former Co-ordinator for Central and Eastern Europe, International Metalworkers' Federation, Geneva, Switzerland; and Poul-Erik Olsen, Director, Education and Working Environment Group, Eastern Europe, International Metalworkers' Federation, Geneva, Switzerland.

Notes

1 Fleissner (1994: 4), for instance, suggests that the 'goal of [transformation] is a pluralistic and democratic political system comparable to Western European countries'.

2 For example, Gordon and Klopov (1992: 30) argue that 'there is no doubt that the transformation that began at the end of the 1980s, if not stopped by a violent return to the past, is [leading] our country [the USSR] onto the path of *normal* development' (emphasis added).

3 For several decades the AFL-CIO has had several regional organisations which operate in various parts of the world (Africa, Latin America and the Caribbean,

Asia). The Free Trade Union Institute is the Federation's European organisation. At the time of writing, the four centres are scheduled to be merged into a single international organisation.

4 Although this, of course, did not stop Solidarity in Poland making effective use of the strike weapon to wrest concessions from the state.

5 As Poul-Erik Olsen, Central and Eastern European co-ordinator for the International Metalworkers' Federation has commented: 'they [many of the CEE trade unionists participating in metalworkers' seminars] thought that in Western Europe you had a uniform system [of labour relations] and they didn't realise that actually there was the Canadian model, the German model, the French model, the British model, which are not easy to compare' (interview with the author, Geneva, 24 September 1995).

6 Such works councils are a central feature of German labour relations and are joint management–labour councils elected at individual plants with the power to control certain issues related to production and the way in which the plant is run. Works councils are separate from the unions and worker representatives do not have to be union members, though usually they are (see Thelen 1991 for more details). The European Union is currently discussing a provision that would create works councils in member states.

7 For more on different types of trade unionism and labour relations emerging in CEE, see Gill 1990; Hughes 1992; Széll 1992; Smith and Thompson 1992; Clarke *et al.* 1993, 1995; Egorov 1996; Ost 1996; Hegewisch *et al.* 1996; Kiuranov 1992; for more on variations in economic conditions across the region, see the collection of reports put out by the International Labour Organisation's Central and Eastern European Team (e.g. Vaughan-Whitehead 1993) and the series of *Transition Brief* newsletters put out by the OECD's Centre for Co-operation with the Economies in Transition in Paris.

8 Interview by author with Vlastimil Beran, Head, International Department, ČMKOS, Prague, Czech Republic, 21 August 1996.

9 Interview by author with Richard Falbr, President, ČMKOS, Prague, Czech Republic, 20 August 1996.

10 Interviews by author with Richard Falbr, President, ČMKOS, Prague, Czech Republic, and Jan Uhlíř, National President of the Czech Metalworkers' Federation (OS KOVO), Prague, Czech Republic, 20 August 1996.

11 Interview by author with Richard Falbr, President, ČMKOS, Prague, Czech Republic, 20 August 1996.

12 Interestingly, Czech Prime Minister Vaclav Klaus alleged the union (the Trade Union of North-western Energy Workers) was keen to break up the monopoly and privatise five power plants as part of an effort to place its officials on the administrative boards of the plants, something which the anti-union Klaus government opposed.

13 Interview by author with Anne-Marie Mureau, Co-ordinator for Central and Eastern Europe, International Metalworkers' Federation, Geneva, Switzerland, 30 August 1994.

14 Interview by author with Vlastimil Beran, Head, International Department, ČMKOS, Prague, Czech Republic, 21 August 1996.

15 In part, this conflict relates to the different ways in which the old official unions have been challenged during the transition and to struggles over who should

now control the assets of the old communist trade unions. In some countries the old unions were simply taken over by the new elements, many of whom had been involved in the various strike committees which formed and were active in the overthrow of the communist regimes (the 'Czech model'). In others, the forces opposed to the regimes instead created new unions to challenge directly the old (the 'Polish model'). This has meant that in some parts of CEE the transformation of the unions has involved working within the structures of the old communist entities (at least initially) whereas in others workers have been freer to develop completely new structures from the ground up, arguably less constrained by old organisational structures.

16 For most of the post-Second World War period the international trade union movement was split between the western-orientated International Confederation of Free Trade Unions and the Moscow-backed World Federation of Trade Unions (see Windmuller 1980 for more details).

17 Interview by author with Anne-Marie Mureau, Co-ordinator for Central and Eastern Europe, International Metalworkers' Federation, Geneva, Switzerland, 30 August 1994.

Bibliography

Bamber, G.J. and Peschanski, V. (1996) 'Transforming industrial relations in Russia: a case of convergence with industrialised market economies?' *Industrial Relations Journal* 27(1): 74–88.

Buckwalter, D.W. (1995) 'Spatial inequality, foreign investment, and economic transition in Bulgaria', *Professional Geographer* 47(3): 288–98.

Busch, G.K. (1983) *The Political Role of International Trades Unions*, New York: St Martin's Press.

Clark, G.L. (1989) *Unions and Communities under Siege: American Communities and the Crisis of Organised Labour*, Cambridge: Cambridge University Press.

Clarke, S. (1993) 'The contradictions of "state socialism"', in S. Clarke, P. Fairbrother, M. Burawoy and P. Krotov (eds) *What About the Workers? Workers and the Transition to Capitalism in Russia*, New York: Verso: 5–29.

Clarke, S., Fairbrother, P. and Borisov, V. (1995) *The Workers' Movement in Russia*, Aldershot: Edward Elgar.

Clarke, S., Fairbrother, P., Burawoy, M. and Krotov, P. (1993) *What About the Workers? Workers and the Transition to Capitalism in Russia*, New York: Verso.

De Luce, D. (1993) 'Premier charges extortion by union', *Prague Post* 17–23 February: 9.

Dølvik, J.E. and Stokland, D. (1992) 'Norway: the "Norwegian model" in transition', in A. Ferner and R. Hyman (eds) *Industrial Relations in the New Europe*, Oxford: Basil Blackwell: 143–67.

Edwards, P., Hall, M., Hyman, R., Marginson, P., Sisson, K., Waddington, J. and Winchester, D. (1992) 'Great Britain: still muddling through', in A. Ferner and R. Hyman (eds) *Industrial Relations in the New Europe*, Oxford: Basil Blackwell: 1–68.

Egorov, V. (1996) 'Privatisation and labour relations in the countries of Central and Eastern Europe', *Industrial Relations Journal* 27(1): 89–100.

Fairweather, O. and Shaw, L.C. (1972) *Labor Relations and the Law in France and the United States*, Ann Arbor: Graduate School of Business Administration, University of Michigan.

Fišera, V.C. (ed.) (1978) *Workers' Councils in Czechoslovakia 1968–9*, New York: St. Martin's Press.

Fleissner, P. (ed.) (1994) *The Transformation of Slovakia: The Dynamics of Her Economy, Environment, and Demography*, Hamburg: Verlag Dr. Kovač.

Fogel, D.S. and Etcheverry, S. (1994) 'Reforming the economies of Central and Eastern Europe', in D.S. Fogel (ed.) *Managing in Emerging Market Economies: Cases from the Czech and Slovak Republics*, Boulder: Westview Press: 3–33.

Fossum, J.A. (1985) *Labor Relations: Development, Structure, Process*, Plano, TX: Business Publications, Inc.

Freeman, R.B. (1992) 'Getting here from there: labor in the transition to a market economy', in B. Silverman, R. Vogt and M. Yanowitch (eds) *Labor and Democracy in the Transition to a Market System*, Armonk, NY: M.E. Sharpe: 139–57.

Frydman, R., Rapaczynski, A., Earle, J.S. *et al.* (1993) *The Privatization Process in Central Europe*, Budapest: Central European University Press.

FTUI (Free Trade Union Institute) (1994) *1994 Annual Report of the Free Trade Union Institute*, Washington, DC: FTUI.

Fukuyama, F. (1992) *The End of History and the Last Man*, New York: Free Press.

Gill, C. (1990) 'The new independent trade unionism in Hungary', *Industrial Relations Journal* 21(1): 14–25.

Godson, J. (1981) 'The role of the trade unions', in L. Schapiro and J. Godson (eds) *The Soviet Worker: From Lenin to Andropov*, New York: St. Martin's Press: 108–34.

Goetschy, J. and Rozenblatt, P. (1992) 'France: the industrial relations system at a turning point?' in A. Ferner and R. Hyman (eds) *Industrial Relations in the New Europe*, Oxford: Basil Blackwell: 404–44.

Gordon, L.A. and Klopov, E.V. (1992) 'The workers' movement in a postsocialist perspective', in B. Silverman, R. Vogt and M. Yanowitch (eds) *Labor and Democracy in the Transition to a Market System*, Armonk, NY: M.E. Sharpe: 27–52.

Hanham, R.Q. and Banasick, S. (1998) 'Japanese labor and the production of the space-economy in an era of globalization', in A. Herod (ed.) *Organizing the Landscape: Geographical Perspectives on Labor Unionism*, Minneapolis: University of Minnesota Press.

Hanson, S.E. (1997) *Time and Revolution: Marxism and the Design of Soviet Institutions*, Chapel Hill: University of North Carolina Press.

Hawker, A. (1993) 'Low unemployment perplexes officials', *Prague Post* 4–10 August: 5.

Hayenko, F.S. (1965) *Trade Unions and Labor in the Soviet Union*, Munich: Institute for the Study of the USSR.

Hegewisch, A., Brewster, C. and Koubek, J. (1996) 'Different roads: changes in industrial and employee relations in the Czech Republic and East Germany since 1989', *Industrial Relations Journal* 27(1): 50–64.

Herod, A. (1997) 'Back to the future in labor relations: from the New Deal to Newt's Deal', in L. Staeheli, J. Kodras and C. Flint (eds) *State Devolution in America: Implications for a Diverse Society*, Sage: Thousand Oaks, CA: 161–80.

—— (1998a) 'Of blocs, flows and networks: the end of the Cold War, cyberspace, and the geo-economics of organized labor at the *fin de millénaire*', in A. Herod, G. Ó Tuathail and S. Roberts (eds) *An Unruly World? Globalization, Governance and Geography*, London: Routledge: 162–95.

—— (1998b) 'The geostrategics of labor in post-Cold War Eastern Europe: an examination of the activities of the International Metalworkers' Federation', in A. Herod (ed.) *Organizing the Landscape: Geographical Perspectives on Labor Unionism*, Minneapolis: University of Minnesota Press.

Hughes, S. (1992) 'Living with the past: trade unionism in Hungary since political pluralism', *Industrial Relations Journal* 23(4): 293–303.

ICFTU (International Confederation of Free Trade Unions) (1996) *Report on Activities of the Confederation and Financial Reports, 1991–1994*, Brussels: ICFTU.

IMF (International Metalworkers' Federation) (1992) *Investissements étrangers et droits syndicaux en Europe centrale et orientale*, Geneva: International Metalworkers' Federation.

Kirichenko, O.S. and Koudyukin, P.M. (1993) 'Social partnership in Russia: the first steps', *Economic and Industrial Democracy* 14: 43–54.

Kiuranov, C. (1992) 'Incompatible: Bulgaria – from managed self-management to managerial management', in G. Széll (ed.) *Labour Relations in Transition in Eastern Europe*, Berlin: Walter de Gruyter: 139–46.

Kjellberg, A. (1992) 'Sweden: can the model survive?' in A. Ferner and R. Hyman (eds) *Industrial Relations in the New Europe*, Oxford: Basil Blackwell: 88–142.

Kochan, T.A. and Katz, H.C. (1988) *Collective Bargaining and Industrial Relations*, Homewood, IL: Irwin.

Lenin, V.I. (1963) *What Is to be Done?*, Oxford: Oxford University Press.

—— (1971) 'The role and function of the trade unions under New Economic Policy', in V.I. Lenin, *Selected Works in Three Volumes*, London: Lawrence and Wishart: 656–66.

Lilja, K. (1992) 'Finland: no longer the Nordic exception', in A. Ferner and R. Hyman (eds) *Industrial Relations in the New Europe*, Oxford: Basil Blackwell: 198–217.

Michalak, W.Z. (1993) 'Foreign direct investment and joint ventures in East-Central Europe: a geographical perspective', *Environment and Planning A* 25: 1573–91.

Murphy, A.B. (1992) 'Western investment in east-central Europe: emerging patterns and implications for state stability', *Professional Geographer* 44(3): 249–59.

Nagy, L. (1984) *The Socialist Collective Agreement*, Budapest: Akadémiai Kiadó.

Nove, A. (1969) *An Economic History of the U.S.S.R.*, Harmondsworth: Penguin.

OECD (1993) *Methods of Privatising Large Enterprises*, Paris: Organisation for Economic Co-operation and Development.

—— (1995) *Taxation and Foreign Direct Investment: The Experience of the Economies in Transition*, Paris: Organisation for Economic Co-operation and Development.

Ost, D. (1989) 'The transformation of Solidarity and the future of Central Europe', *Telos* 79 (Spring): 69–94.

—— (1995) 'Labor, class, and democracy: shaping political antagonisms in post-communist society', in B. Crawford (ed.) *Markets, States, and Democracy: The*

Political Economy of Post-Communist Transformation, Boulder: Westview Press: 177–203.

—— (1996) 'Polish labor before and after Solidarity', *International Labor and Working-Class History* 50 (Fall): 29–43.

Petkov, K. and Thirkell, J.E.M. (1991) *Labour Relations in Eastern Europe: Organisational Design and Dynamics*, London: Routledge.

Piore, M.J. (1992) 'The limits of the market and the transformation of socialism', in B. Silverman, R. Vogt and M. Yanowitch (eds) *Labor and Democracy in the Transition to a Market System*, Armonk, NY: M.E. Sharpe: 171–82.

Pipes, R. (1990) *The Russian Revolution*, New York: Vintage Books.

Prague Post (1992) 'Union mining for labor strategy', 10–16 November: 3.

—— (1995a) 'Economists question "miracle" of nation's low unemployment rate', 20–26 December: 1.

—— (1995b) 'Unions say it's time to speak up', 5–11 April: 4.

—— (1995c) 'Unions threaten chaos for Czech rail system', 21–27 June: 1.

—— (1995d) 'North Moravia's miners labor under strike alert', 21–27 June: 3.

Pravda, A. and Ruble, B.A. (eds) (1986) *Trade Unions in Communist States*, Boston: Allen and Unwin.

Rondinelli, D.A. (1994) *Privatization and Economic Reform in Central Europe: The Changing Business Climate*, Westport, CT: Quorum Books.

Rusnok, J. (1993) Statement by Jiří Rusnok, adviser to ČMKOS, quoted in A. Hawker 'Low unemployment perplexes officials', *Prague Post* 4–10 August: 5.

Schapiro, L. and Godson, J. (1981) *The Soviet Worker: From Lenin to Andropov*, New York: St Martin's Press.

Scheuer, S. (1992) 'Denmark: return to decentralization', in A. Ferner and R. Hyman (eds) *Industrial Relations in the New Europe*, Oxford: Basil Blackwell: 168–97.

Silverman, B. and Yanowitch, M. (1992) 'Introduction', in B. Silverman, R. Vogt and M. Yanowitch (eds) *Labor and Democracy in the Transition to a Market System*, Armonk, NY: M.E. Sharpe: ix–xxii.

—— (1993) 'Introduction: the transformation of work in postsocialist and post-industrial societies', in B. Silverman, R. Vogt and M. Yanowitch (eds) *Double Shift: Transforming Work in Postsocialist and Postindustrial Societies*, Armonk, NY: M.E. Sharpe: ix–xxvi

Smith, C. and Thompson, P. (1992) *Labour in Transition: The Labour Process in Eastern Europe and China*, London: Routledge.

Stark, D. (1996) 'Recombinant property in East European capitalism', *American Journal of Sociology* 101, 4: 993–1027.

Széll, G. (ed.) (1992) *Labour Relations in Transition in Eastern Europe*, Berlin: Walter de Gruyter.

Targetti, F. (ed.) (1992) *Privatization in Europe: West and East Experiences*, Aldershot: Dartmouth.

Temkina, A.A. (1992) 'The social base of economic reforms', in B. Silverman, R. Vogt and M. Yanowitch (eds) *Labor and Democracy in the Transition to a Market System*, Armonk, NY: M.E. Sharpe: 13–24.

Thelen, K.A. (1991) *Union of Parts: Labor Politics in Postwar Germany*, Ithaca: Cornell University Press.

Vaughan-Whitehead, D. (1993) *Minimum Wage in Central and Eastern Europe:*

Slippage of the Anchor, Budapest: International Labour Organisation Central and Eastern European Team.

Volkov, I. (1992) 'The transition to a mixed economy and the prospects for the labor and trade-union movement', in B. Silverman, R. Vogt and M. Yanowitch (eds) *Labor and Democracy in the Transition to a Market System*, Armonk, NY: M.E. Sharpe: 53–67.

Windmuller, J.P. (1980) *The International Trade Union Movement*, Deventer, Netherlands: Kluwer.

9

PRIVATISATION AND THE REGIONAL RESTRUCTURING OF COAL MINING IN THE CZECH REPUBLIC AFTER THE COLLAPSE OF STATE SOCIALISM[1]

Petr Pavlínek

Coal mining traditionally played a crucial role in the state socialist development model. The extensive regime of accumulation in which there was a continued process of expansion in the means of production (Pavlínek 1998; Smith 1998) relied on ever-increasing production of electricity, largely based on coal, to satisfy the growing energy demands of energy inefficient state socialist economies. The former Czechoslovakia was no exception. In 1989, the country derived 55 per cent of its energy needs and 78 per cent of electricity from brown coal (lignite) (Statistical Yearbook 1989: 394). The largest brown coal deposits are located in the North Bohemian Coal Basin, which accounted for 74 per cent of Czechoslovak brown coal production in 1990. In turn, more than 50 per cent of brown coal produced in northern Bohemia came from the Most region, which has been a centre of coal mining in northern Bohemia for more than one hundred years (Figure 9.1). Mining (along with the chemical industry) constituted the heart of what I will call the 'regional regime of accumulation', by which I refer to the character of the long-term development in the relations of production and consumption within a regional economy based upon a particular sectoral structure, which in turn is embedded within the national economy with its own forms of aggregated production–consumption dynamics. Coal and chemicals were therefore critical elements in the state socialist transformation of the Most region; together they accounted for more than 90 per cent of industrial production and more than 80 per cent of industrial employment in the 1980s.

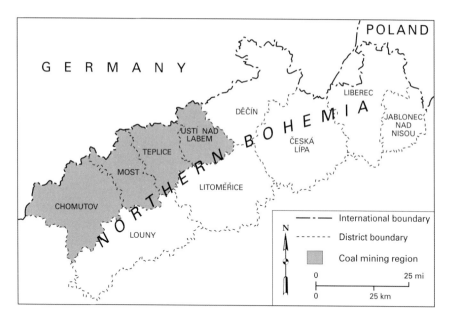

Figure 9.1 Northern Bohemia coal mining region

Both coal mining and the chemical industry have had a devastating effect on the environment in the region. Most of the brown coal has been mined in opencast mines causing large-scale landscape devastation and leading to the demolition of thirty-two villages located in the district, as well as the old city of Most. The burning of low-quality brown coal with a high sulphur and ash content in local power plants has resulted in high levels of air pollution and the devastation of forests in the nearby Ore Mountains. The Chemopetrol chemical complex has also contributed to air and water pollution in the district. Consequently, the state socialist development model has transformed the Most District into one of the most environmentally devastated regions in Europe and this extremely polluted environment has had negative effects on the health of the local population (see Pavlínek 1998).

The Most District is a typical example of state socialist extensive accumulation which has become unsustainable: the political power to relocate settlements in order to extract shallow and low-grade coal is no longer available, the economics of high sulphur coal production have become untenable, and the health and environmental costs of pollution from brown coals and chemicals have placed severe restrictions on what industry can do. Industrial restructuring after 1989 in this region therefore raises several important questions about how we understand the current economic transformation and its regional implications.

The aim of this chapter is to examine the implications of the collapse of state socialism and the 'transition to capitalism' for the mining industry and, consequently, for the regional economy of the Most region. First, I will demonstrate that the privatisation and restructuring of coal mining is likely to lead to the gradual closure of coal production, and this has a number of implications for the economy and environment of the region. The chapter thus provides an examination of the impacts of de-industrialisation on regional economic and environmental life.

Second, the privatisation and restructuring of coal mining in the Most region demonstrates the *contested nature* of the transition and the struggle over its outcomes. One result of the collapse of state socialism is the re-emergence of an open class struggle over the direction of change in the Czech Republic, individual sectors of the economy and in individual regions. The changes in coal mining therefore challenge the liberal notion of simple and linear transition in CEE (see also Grabher and Stark, and Smith and Swain in this volume).

Third, I will argue that, although there are a number of actors involved in the privatisation and restructuring of coal mining, the role of the *central state* has so far been critical because it has initiated and controlled the privatisation process. The central state has designed the strategy of economic transition from the centrally planned to a market economy, and this has profoundly affected the performance of coal mining in the Most region. Central state policy has so far also been the most important dynamic of social change in the Most District. This situation challenges a widely held perception of the Czech state as neo-liberal and indicates that its role in industrial restructuring is much more complicated and complex than neo-liberal models assume.

Restructuring of coal mining in the Czech Republic

The governmental restructuring strategies for coal mining

Following the collapse of state socialism in 1989 and the approval of the government's transition strategy in September 1990, there was an expectation that the coal industry would be privatised and restructured. However, the price of coal was not freed and no plan to privatise coal mining or to create a state 'coal policy' was in place. At the same time, the analyses of Czech and Slovak coal mining prepared by Czech, Slovak and western coal specialists showed that it would be impossible to sustain the very high coal consumption levels that had been achieved under the state socialist development model (Stružka 1992; Gheyselinck 1992; Kopečný 1992; Pěgřímek 1992; Formánek 1992; Cibulka 1993). The emerging problem of a long-term decline in demand for coal and its corresponding overproduction consequently opened the issue of mine closures. It was also becoming obvious that future coal mining would have to consider environmental concerns and that newly proposed ecological legislation including the 'ecological limits of mining' and clean air legislation would seriously influence coal mining (Stružka 1992).

For these and other reasons, the transformation of coal mining from the state-owned and centrally planned system to some form of private ownership and market regulation was understood as a very complicated task – one compounded by the fact that tens of thousands of highly paid jobs in the industry would be lost, thereby creating serious social problems in coal mining regions. The role of the state in the entire process was ambiguous, however. While the state wanted to withdraw from its direct role in coal mining as soon as possible, the coal mining enterprises expected the state to play a major role in their restructuring and to provide financial guarantees.

By the end of 1991, plans for the restructuring and privatisation of the coal industry were based on the three key components of governmental reform strategy: liberalisation of prices, the end of state subsidies for industrial operations and the privatisation of industry. Gheyselinck (1992), who was head of mining policy in the Czech government, for example, argued that in order to follow the government's transition strategy it was necessary to abolish the coal industry's old organisational and institutional structure inherited from the centrally planned economy and create a new structure based on commercial coal companies and a competitive environment. Setting up new rules that would govern relations of commercial coal companies with the state was also necessary. The new organisational structure would – he argued – have to be set up *before* the privatisation of coal mining could take place.

The restructuring strategy aimed to achieve three goals: first, to create commercial coal mining companies with at least a medium-term perspective, with a similar structure and the same relationship with the state as currently existed; second, to minimise the social, psychological and political impacts of restructuring; and third, to minimise state expenditures in the medium term (Ministry for Economic Policy and Development 1992).

Gheyselinck's plan classified existing coal mining enterprises into three groups: (1) those which were currently unprofitable; (2) those which were currently viable but endangered by the growth of wages in the medium term; and (3) those that would be competitive over the long term (Gheyselinck 1992; Ministry for Economic Policy and Development 1992). Based on this classification, Gheyselinck proposed to create five new coal mining companies, three of them based on brown coal and located in north-western Bohemia. He also argued that before the new coal companies could be privatised it was necessary to end the redistribution by the state of revenues produced in coal enterprises, replace it by direct financial relations between the state and new commercial coal companies, and free coal prices. Unprofitable enterprises were to be liquidated (Gheyselinck 1992; Ministry for Economic Policy and Development 1992). The Czech government opted to minimise state expenditure and merged these enterprises with larger profitable commercial companies.

However, there were three main criticisms of the governmental restructuring strategy of coal mining based on Gheyselinck's proposal. First, according to the coal mining trade unions and mine management, the creation of competing coal mining companies in a single coal mining district was not in the best long-term interests of the district or industry. Mine managers argued that the strategy was 'naïve'. The task of creating several competing commercial coal companies would not be enhanced when profitable mines had to finance and supervise the closure of unprofitable mines.

Second, the transfer of *past* state obligations associated with coal mining, such as recultivation and other environmental damage, to newly created brown coal commercial companies would dramatically influence the future profitability of these firms. Coal companies are obliged to finance from their current profits the closure of exploited mines and to recultivate any damaged landscapes where that damage took place in the past fifty years. The managers argued that because the state had redistributed profits made in coal mining and had not created any financial reserves for recultivation over the past forty years it should now take responsibility. Currently, the financial obligations of newly created coal companies are larger than their total capital.

Third, the government did not seriously consider any alternatives to Gheyselinck's plan, either from the coal mining unions or from the management. While negotiations with trade unions resulted in some small changes in the original restructuring plan, and mine directors had a chance to comment on these changes, the original restructuring strategy remained largely unchanged.

Changing forms of class struggle in coal mining after the collapse of state socialism

Although the labour unions were largely discredited among workers after more than forty years of symbolic existence as a 'transmission belt' of the Communist Party (see Herod, this volume), the coal mining unions were the first unions to protest the government's restructuring policy. Initially, neither the government nor management of the mines took labour unions seriously and did not expect them to participate in the discussions over coal mining strategies. However, mining unions and especially the unions from northern Moravia (the region most affected by proposed restructuring and privatisation strategy) put increasing pressure on the government to consider their concerns. North Moravian mine unions opposed the restructuring strategy and claimed that it was not viable but would result in a 'gradual liquidation of mines' (*Lidové noviny* (LN) 1992a: 1, 8).[2] They argued that the government was not concerned with the region's mining crisis and that its social programme was ineffective (Hawker 1993). Coal miners presented several demands to the Czech government regarding the restructuring strategy: protection of the Czech coal market from outside competition; price liberalisation of coal; and social protections for displaced workers. Coal miners from northern Moravia were also supported by coal miners from northern Bohemia (LN 1992b: 3).

The dispute between the government and trade unions over the restructuring strategy for coal mining escalated into an open conflict in the autumn of 1992. The Czech government negotiated with the unions and made some concessions, but only after demonstrations by about 5,000 coal miners in Prague. Yet this did not drastically change the original restructuring strategy. This demonstration was the first serious conflict between labour and the government after 1989 and is evidence of a shift towards open class struggle in the Czech Republic after the collapse of state socialism and Communist Party hegemony.

Restructuring of coal mining in the Most District

Pre-privatisation agony in coal mining in the Most District

The North Bohemian Brown Coal Mines (NBBCM) state company, which included all mines in the North Bohemian Coal Basin, was abolished following the rapid disintegration of central planning after 1989. Six independent coal mining enterprises were established to replace the NBBCM, three of them in the Most District. Because of this organisational fragmentation, no common coal mining strategy existed in the basin after 1989. The enterprises instead waited passively for organisational restructuring and subsequent privatisation without any substantial changes in their internal organisation and production strategies. Such behaviour by large enterprises 'locked into a preexisting system of economic relations' (Burawoy and Krotov 1992: 17) can be described as 'pre-privatisation agony' (Šulc 1993: 326). In such a situation, these state enterprises were not interested in or able to make any major investment or restructuring decisions. As a result, no substantial changes in production took place (Šulc 1993; OECD 1994). The symptoms of 'pre-privatisation agony' included: (1) passive approaches within the enterprises towards their restructuring; (2) attempts to keep the old system of production practices until privatisation was launched; and (3) efforts to achieve maximum profit for the management and employees at all costs, by using the assets of an enterprise and by increasing its debts. Enterprises strove to maintain employment levels and increase wages irrespective of productivity levels or the market realisation of their products. In doing so, they attempted to maximise their inputs and minimise the outputs – typical enterprise behaviour under state socialist command planning. Thus, surviving practices from the centrally planned economy and inertia resulted in minimal changes in production (Šulc 1993).

However, when privatisation of state-owned enterprises occurred this also does not necessarily lead to changes in production and overall rationalisation, a finding similar to that of Burawoy and Krotov (1993) in Russia. This situation also suggests that the changes in the external economic environment (collapse of central planning and introduction of the market) are much faster than the changes in the internal structure of the enterprises and especially in the social relations of production. Clarke *et al.* (1994) argue that such change in the internal structure of enterprises requires restructuring of both management and the labour force, including working and production practices, which is very difficult to achieve. It is also a potential source of conflict between workers and management, within the labour force and within management. This situation points towards the unevenness, complexity and contested nature of the transition from state socialism to capitalism.

Establishment and privatisation of the Most Coal Company

The Most Coal Company (MCC) is composed of the three coal mining enterprises located on the territory of the Most District and was established by the National Property Fund (NPF) on 1 November 1993, eleven months after it was planned. The NPF originally owned 100 per cent of MCC shares. Its ownership declined to 34 per cent after the second wave of voucher privatisation in 1994 in which 42 per cent of the shares were offered for sale to privatisation investment funds and individual investors. On 1 November 1993, 9 per cent of the shares were given away free of charge to the cities and villages directly influenced by the company's coal mining, with the ostensible aim of allowing cities and villages to influence the activities of MCC, including protection of the landscape and the environment (Privatisation Project 1992: 76).

As a result of privatisation a number of organisational changes occurred. MCC was organised into a three-tier system: plant, division and company headquarters. Three divisions identical with the former state enterprises were established. They had the same managerial staff and used the same management strategies. The only change was the creation of MCC headquarters that supervised the divisions and replaced the former General Headquarters of the NBBCM from the period of central planning.

In April 1994, four months after the establishment of the company, the three-tier managerial structure of the MCC was replaced by a two-tier organisation in which the divisions were abolished. According to the general director and chair of MCC board of directors, the major reason for the change was the need to lower the operating costs of the company to enable it to stay competitive (*Hospodářské noviny* (HN) 1994a: 7). The plants with their administrative offices are therefore now directly supervised and managed by headquarters of MCC.

The new organisational structure, however, did not result in any radical reduction in employment or any reorganisation of production, suggesting that restructuring and rationalisation lag behind organisational changes and changes in ownership. Although employment in the enterprises that formed MCC has declined in the past ten years, and especially after 1990, production declined much faster (Table 9.1). Consequently, the productivity of coal mining in the Most District declined.[3] Surprisingly, however, the coal enterprises remained profitable in spite of declining productivity, an excess workforce, and an absence of rationalised production methods.

Slower decline in employment than in output is a result of three main processes. First, the state (as the owner of MCC) did not put pressure on the company to rationalise its management. It is also doubtful, however, whether its future ownership structure, which will include many small individual shareholders and investment privatisation funds, will be able to exert such pressure either. Second, the state was not interested in a rapid decline in employment because it would necessarily have led to higher unemployment rates and potential social conflict in the region. Third, the state reimposed wage controls on non-financial enterprises with twenty-five or more employees in July 1993 and this curbed the growth in wages and allowed the state-owned enterprises to keep 'unnecessary workers' as a means of ensuring social peace.

Coal production declined rapidly after 1989, and especially after 1991, because electricity production declined as industrial production collapsed in

Table 9.1 Decline in the coal production and coal mining employment in the Most District

Year	Coal production			Employment		
	Thousands of tons	*1984=100*	*1991=100*	*Employees*	*1984=100*	*1991=100*
1984	38,562	100.0	–	18,988	100.0	–
1991	36,147	93.7	100.0	18,039	95.0	100.0
1993	26,471	68.6	73.2	16,373	86.2	90.8
1994	22,700	58.9	62.8	15,100	79.5	83.7
1995	21,800	56.5	60.3	13,500	71.1	74.8
1996	19,500	50.6	53.9	10,700	56.4	59.3
1998	18,100	46.9	50.1	n.a.	n.a.	n.a.
1999	16,300	42.3	45.1	10,400	54.8	57.7
2005	15,400	39.9	42.6	n.a.	n.a.	n.a.

Sources: Data compiled from OOČSÚ 1986; Privatisation Project 1992; Schreiber and Štáva 1994; LN 1994a: VIII; HN 1995: 7, 1996: 6.
Notes:
The 1996–2005 data reflect the expected situation; 1996 employment is as of June 1996.

some sectors and declined generally with economic transition;[4] environmental legislation after 1989 led some industrial enterprises to convert their heating systems from coal to natural gas or other cleaner energy sources; and competition from other brown coal companies in the Czech Republic increased.

The Most Coal Company and the marketisation of brown coal mining

The future of coal mining and its environmental consequences in the Most District will depend, *inter alia*, on the ability of MCC to sell the brown coal it produces. What does the collapse of state socialism and emerging capitalist competition mean for the future of brown coal and coal mining in the district? MCC now competes with both domestic and foreign firms. Its domestic competitors include not only other coal companies but also other producers of heat and energy in the natural gas and nuclear power industries. In this respect, the future of coal mining in the Most District depends on the government's energy policy.

MCC managers complain, however, that brown coal has lost state socialist privileges, such as subsidies for unprofitable mines and the ability to strip-mine virtually without restraint. They also complain that coal mining has become responsible for all its production costs, including environmental devastation, and even the costs of dealing with past environmental destruction. Many managers feel, however, that the nuclear energy sector still enjoys preferential treatment from the state in many respects. For example, the production costs of electricity produced in nuclear power plants do not include the cost of a permanent storage facility for used nuclear fuel. It also does not include comprehensive insurance for a possible nuclear disaster. Also, the energy sector as a whole underwent much less radical organisational restructuring in comparison with coal mining after 1989. The Czech Energy Works company, for example, was not fragmented by the government into several competing companies, but it remained a single company monopoly in energy production. Consequently, it remains one of the largest and most powerful industrial companies in the Czech Republic.

MCC, however, wants to compete on the energy market using several different strategies. First, it plans to focus on the production of high-quality sorted brown coal with a higher heating value and a lower sulphur content. Second, MCC competes with its price of coal. While inflation reached around 10 per cent annually in the Czech Republic during the 1990s, coal prices grew by only 2–3 per cent annually. Third, MCC plans to spin off and privatise some production services, such as the provision of spare parts for coal mining technology, repair and reconstruction, which are today part of the company. Company managers believe that this would rationalise and improve the efficiency of production. Finally, in 1994, a decision was made to close several high-cost production mines to make MCC more competitive.

Despite these efforts, MCC expected to be forced to steadily lower its coal production for several reasons: entire communities are gradually converting from brown coal to natural gas as a heating source; electricity is increasingly being used for direct heating; ecological legislation, local and regional efforts limit coal mining; and brown coal from other areas is cheap. The sale of brown coal in the Czech Republic will also be affected by the launching of the Temelín nuclear power plant in the late 1990s and the Clean Air Act that will come into effect in 1998 (LN 1994a: VIII). By the end of the century, the company plans to have cut coal production by 55 per cent from its 1990 level.

Struggles over the environment and coal mining in the Most District

The struggle over the environment has become one of the most important factors that influences restructuring and the future development of coal mining in northern Bohemia and in the Most District in particular. Such struggles illustrate the contested nature of the transition as local people, local governments, central government, district authorities and coal mining companies struggle over the future of coal mining. This section also illustrates the importance of democratisation at the local scale for local environmental struggles. The discussion also points towards the importance of environmental issues in local efforts to mobilise against any further expansion of coal mining.

Governmental ecological limits to coal mining and its impacts on MCC

A number of government resolutions have been passed dealing with environmental problems in northern Bohemia which limit the further development of opencast coal mining in the region.[5] The most important of these are the 'ecological mining limits'. These function as territorial boundaries drawn around communities endangered by coal mining beyond which mining is prohibited. They were imposed by the government in 1991 to protect northern Bohemian communities from further demolition because of coal mining. The mining limits were implemented as a result of popular resistance by citizens and opposition from local governments in several towns and villages that had been slated for demolition to make way for opencast coal mining.[6] Out of three new coal mining companies formed in the region, MCC is probably the most affected by these resolutions. About 70 per cent of the 6.1 billion tons of exploitable coal deposits in the north Bohemian coal district cannot now be exploited (Pěgřímek 1992).

Two opencast mines are the most affected by the 'ecological mining limits'. First, the Chabařovice Mine in the Teplice District (slated for closure by 1999) was not included in a commercial coal mining company and was not privatised. The Chabařovice Mine produces coal with the lowest sulphur content in northern Bohemia (0.4–0.6 per cent). The mine was supposed to operate until the year 2015. However, its further operation was only possible through the liquidation of the town of Chabařovice – with 2,500 inhabitants today but almost 8,000 before the Second World War – and two other villages (Roudníky and Hrbovice). One hundred million tons of high-quality brown coal are located beneath these settlements. The town and villages were saved because of the efforts of new local governments elected after 1989 and as a result of popular opposition. These efforts succeeded despite strong resistance by the Chabařovice Mine and the proposals of an international investment consortium to raze Chabařovice and build a new town elsewhere (see Bystrov 1993; Bensman 1992a, 1992b).

Second, the Čs. Armády Mine, the second largest opencast mine of MCC, is even more affected. The government's 'ecological mining limits' will freeze the mining of 1.5 billion tons of the best quality brown coal in the central portion of the North Bohemian Brown Coal Basin. The ecological limits include the Čs. Armády Mine, which must be closed by the year 2010, forty years earlier than originally projected. The miners, not surprisingly, see this decision as 'fatal' to the mine and their livelihoods (Pěgřímek 1992: 153).

The government imposed the ecological mining limits for the Čs. Armády Mine specifically to protect the village of Horní Jiřetín (Figure 9.2).[7] The early closure of the mine will seriously influence the production of coal by MCC. For example, if the Čs. Armády Mine is closed by the year 2010, as currently planned, annual coal production of MCC will decline from its current level of about 23.4 million tons to 10–11 million tons. The managers of MCC have also complained that the imposition of ecological limits on mining would reduce the competitiveness of the company compared with the North Bohemian Mines, which are affected to a much lesser extent by the ecological limits. Managers at MCC thus felt that the economic interests and future profitability of their company were being endangered by a form of democratisation in which local communities are allowed to stand up against the devastation caused by coal mining. Apparently, the relationship between the coal enterprises and the local governments (including rural and urban municipalities and the District Office – a local arm of the central state) has profoundly changed following 1989. According to officials in both settings, both coal mines and local governments felt 'controlled' by each other during the state socialist period, resulting in part from the way in which Communist Party hegemony operated in both. After 1989, relations between coal enterprises and the local governments were constructed on a new legal basis and began to change. For

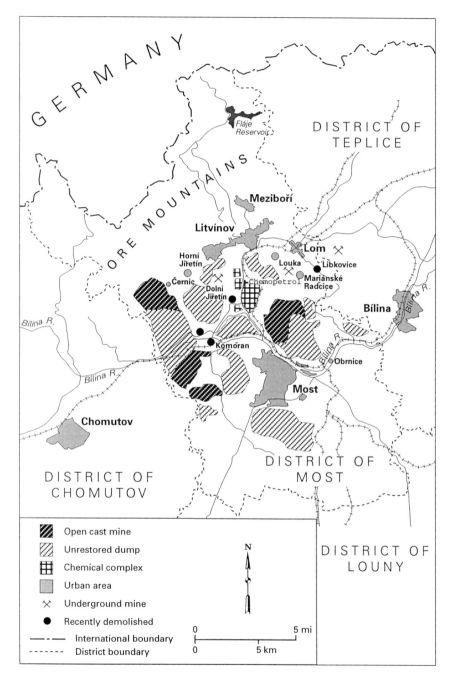

Figure 9.2 Most District, Czech Republic

example, in 1993 MCC paid about 100 million crowns to municipalities in the Most Basin for the use of their territories and the exploitation of coal deposits. According to a manager at MCC:

> I can say that we have built a new relationship with municipalities [after 1989]. I want to say that it is not a bad [relationship]. We do not throw our arms around each other's necks, but these are mutually very respectful relations and we also talk about many unpleasant issues. . . . We do not have conflictual relations with the existing villages except Horní Jiřetín, but I would not like to dramatise it. We have agreed that when we had something to talk about we would meet in peace at the round table negotiations and we would discuss the entire problem. We would attempt to find a mutual solution without using power and strong words from any side. We try to stick to this approach and I have to say that the other side does the same.[8]

Such a view represents a dramatic change in the approach of coal mining enterprises to the local communities. State socialist power relations between coal mining enterprises and local communities were based on the total subordination of local interests to coal mining interests. This situation resulted in the demolition of thirty-two villages and the city of Most. As discussed below, coal mining enterprises have been increasingly forced to recognise the new situation in which, after 1989, power relations between them and local communities began to change.

Struggles between local communities and coal mining enterprises in the Most District after 1989

Although the managers of MCC argued that they respect the new limits on coal mining, struggles exist between MCC and some villages. As one local mayor has argued, 'The coal mines do everything they can to control the situation here. Their behaviour did not change. The same people are sitting at their headquarters as before [1989] and they always express the megalomania of coal mining.'[9]

Although the government imposed the ecological limits for coal mining and decided that the village of Horní Jiřetín would not be razed, the future of the village remains uncertain. In 1991, the Czech Mining Authority was charged with the responsibility of delimiting the territory that could be mined and reducing the coal deposits that can be exploited in the area. Yet this had still not been done by August 1993, when the mayor of Horní Jiřetín complained that the government was not acting on its resolutions. As a result, life in the village is paralysed. For example, individuals wishing to rebuild a house or other building have to acknowledge that they must demolish the building at their own expense if the village is liquidated as a result of coal mining.

Illner's (1992) classification describing the paternalistic relationships under state socialism between coal and chemical enterprises on the one side and cities, small towns and villages on the other is useful here. He characterised the relations between enterprises and larger cities as a type of 'landlord' relationship, while the relations between the enterprises and small towns and villages represented forms of 'parasitic' and 'antagonistic' relations. According to Illner (1992), situations in which enterprises are 'partners' are typified by shared responsibility of enterprises and communities for the future development of communities. Where enterprises are 'neighbours', enterprises ignore the problems of communities and are not interested in participating in community development. Are these relationships changing during the 'transition to capitalism'?

Evidence from interviews with the mayors of cities and villages in the Most basin suggests that these relations *are* changing, but that the change is uneven. Enterprise–city relations are changing from enterprises as 'landlords' to enterprises as 'partners' or 'neighbours'. However, some in city government perceive the relationship of the city and the coal mining enterprises as a 'confrontational' one. For these people there was virtually no co-operation between the city and the mines because the coal mining enterprises did not know what was going to happen to them in the future and consequently behaved accordingly. Under these conditions, coal mining enterprises preferred 'neighbourly' relations to 'partner' relations because this allowed them to ignore the problems of the city of Most. The city of Most is, however, interested in 'partner' relations, not in 'neighbourly' relations with the coal mining enterprises.

According to managers of MCC, the coal mining enterprises are also interested in the 'partner' type of relations with those communities negatively influenced by the coal mining activities. They attempt to compensate cities, villages and the Most District Office financially for their environmentally harmful activities, to improve mutual relations, and to secure local support for their mining activities. For example, the company donated 38.2 million crowns in 1992 and 19.7 million crowns in 1993 to local governments in the Most District. Much of this has gone to the Most District Office and the city of Most. However, this support is not guaranteed and may quickly evaporate, particularly if the profits of MCC continue to decline rapidly. In 1994, for example, MCC planned to donate only 10 million crowns.

The parasitic and antagonistic relations between the coal mining enterprises and the villages that were typical of state socialism are slow to change. MCC is forced by national regulations, such as the ecological coal mining limits, gradually to abandon its antagonistic approach towards villages, although this type of relation continues in several cases. In contrast to the cities, where there are mutual attempts to establish 'partner' relations with coal mining enterprises, the relations between the coal mining enterprises and villages are either moving towards a type of 'neighbour' relation or the coal mines remain parasitical on the village.

Mayors of the villages in the Most basin are critical about the behaviour of coal mining enterprises towards their communities. Although the coal mining enterprises operate on municipal territory, they do not compensate the villages in any way. These activities have adversely influenced the environment in the villages and this situation has not changed much from the state socialist period. For example, the mayor of Mariánské Radčice argued that the only contributions his village received from the coal mining enterprises under state socialism were two wreaths and two members of the people's militia sent for important communist anniversaries. 'What has changed since is only that wreaths are not laid and members of the militia are not sent here.'[10]

However, although the coal mining enterprises were reluctant to compensate these communities or contribute to their development, mutual relations have changed. This change took place in the perceived and actual domination of local communities by coal mining:

> In the past, the coal mines behaved as if they were in power, because they were in charge. The industry ruled here. We were subordinated. The relationship between the village and coal mines has changed a little. [The mines] improved their approach towards our village. I would say that the relations are more equal now. Today, we negotiate as equal partners if there is a meeting between the village officials and coal mines. It is different from the past when the National Committee was always inferior.[11]

Tensions between communities and large industrial enterprises therefore point to the changing nature of power relations, and the attempts of both sides to benefit from this change. Both communities and MCC feel that they are no longer subordinated to each other and can make independent decisions regarding their mutual relations. MCC seeks local support for coal mining while cities such as Most understand the economic importance of coal mining in the area and seek financial support from MCC for their development. As I have already argued, the changes in the behaviour of coal mining enterprises differ depending on whether they are dealing with the cities or small villages. While both the cities and coal mining enterprises are interested in developing 'partner' types of relations for mutual benefit, the enterprises tend to ignore the needs of small communities even as they use their resources.

These accounts of change in the relations between the coal mining enterprises and villages suggest that there is no clean break between the state socialist past and the post-socialist present. Rather, a restructuring of state socialist relations is gradually taking place. Furthermore, the struggle between old and new social practices results in a complex mixture of old state socialist and new post-state socialist relations (see Smith 1997).

Class struggle, coal mining and regional restructuring in the Most District

Open class struggle after the collapse of state socialism influenced the government's transition strategy for coal mining. What was its role in the restructuring of coal mining in the Most District? The role of trade unions in the restructuring of industry and the region was influenced by the gradual transformation of unions from being subordinated to the Communist Party to being independent organisations defending miners' interests.

Coal mining trade unions and the Economic and Social Council of the Basin Region

Overall, Czech trade unions were divided over many issues and were looking for new forms of organisation that could articulate workers' interests (see also Herod, this volume). One of the major organisational problems was that the trade unions refused a centralised system of organisation after the disintegration of the old unions. As a result, the unions are troubled by internal divisions and are unable to act as a united social force (cf. Clarke and Fairbrother 1993).

However, the north Bohemian coal mining unions are playing an important role in the unfolding industrial and regional restructuring in the region far beyond the individual coal mining companies. The coal mining trade unions initiated the establishment of the Economic and Social Council of the Basin Region (ESCBR), which includes the trade unions, municipalities, state administration and enterprises.[12] The most important goals of this organisation, which co-ordinates the activities of its members, are the revitalisation of the region, industrial restructuring and the improvement of the quality of the environment. The ESCBR is now a recognised co-ordinating partner of the Czech government and co-operates with several governmental ministries, such as the Ministry of the Environment, the Ministry of Work and Social Affairs, and the Ministry of Health Care.

Trade unions chose to pursue their demands through the more corporatist ESCBR, instead of pursuing their individual interests, and this provides a mechanism for them to communicate with the Czech government and Parliament. The first goals of the ESCBR were to put through emergency measures for northern Bohemia such as a Clean Air Act, desulphurisation of coal-burning power stations, lower taxes for private entrepreneurs to support the development of the private sector in the region, subsidised housing for poor people, and a state subsidy of 15,000 crowns for the conversion of individual households from coal to natural gas or electric heating.[13] The government has eventually accepted all these demands in some form, although in 1991 they thought that they were 'unrealistic'.

The existence of the ESCBR also allows the trade unions to communicate effectively with municipalities and to a lesser extent with the state administration at the district level. Furthermore, the unions participated in the establishment of the Most Region Foundation, which plans to support the development of small and medium-sized private entrepreneurs in the Most region. Together with the German trade unions of Saxony, the ESCBR has established the inter-regional union council 'Saxony', the first of this kind in the former state socialist countries of Central and Eastern Europe.

Thus, the coal mining unions were able to transform themselves relatively quickly from the 'transmission belt of the Communist Party' to independent organisations attempting to defend the interests of their workers. However, in this new role they are restricted in several ways. They are plagued by organisational and institutional problems which result in internal divisions. Their role in the internal organisational restructuring of coal mining enterprises has been limited because of their focus on collective bargaining issues such as wages, employment and working conditions, and their decision not to challenge general issues of enterprise privatisation and restructuring (see also Clarke and Fairbrother 1993). The unions do play an active role in the regional restructuring of the entire North Bohemian Brown Coal Basin, including the Most District, through their active participation in the Economic and Social Council of the Basin Region. In this setting, the unions, in close co-operation with other important actors in the region, contribute to the development of a common approach towards regional restructuring and how this is presented to the Czech government and the Czech Parliament. The unions also participate in inter-regional and international co-operative efforts to alleviate some of the economic, environmental and social problems associated with transition and restructuring in similarly environmentally devastated regions within Germany and Poland.

Conclusion

The restructuring of coal mining in the Most District illustrates the complexity and contested nature of the 'transition to capitalism' in the Czech Republic. Changes in ownership and the organisation have only gradually changed mining activities and the relations of coal mining enterprises with the surrounding communities in the district. In this respect, what we are witnessing is not a clean break with the state socialist past and a linear and smooth transition to capitalism, but a gradual restructuring of the state socialist regional accumulation regime and its related social relations (Smith 1994; Clarke *et al.* 1994; Clarke 1993, 1992).

Although the central state was the decisive actor in the preparation of the restructuring strategy for coal mining, its final version resulted from the struggle between the central state, labour and coal mining management. Furthermore, the future of coal mining in the Most District will result from the local struggles between MCC and its individual enterprises with local communities and labour – MCC can continue coal mining at its already reduced levels only through the further demolition of additional communities in the region. It will also be influenced by the contested national strategies of accumulation and regulation and their impacts on the coal mining industry in the Most region. Restructuring of the coal mining industry and changes at the national level have direct implications for the quality of the environment in the Most District. The environment is becoming an important factor limiting the further development of coal mining. The quality of the environment seems to benefit not only from the drop in industrial and energy production, but also from new ecological legislation enacted after the collapse of state socialism, including the ecological limits of coal mining imposed by the central state. However, the future of the environment and communities endangered by coal mining will depend on the abilities of local communities to defend these policies at the local and national levels.

Notes

1 This chapter is based upon intensive research carried out on the dynamics of industrial restructuring and environmental degradation in northern Bohemia between 1992 and 1996. It is a revised version of part of a forthcoming book on this issue, *Economic Restructuring and Local Environmental Management in the Czech Republic*, 1998, Lewiston, NY: Edwin Mellen Press.

2 In northern Moravia, the centre of bituminous coal mining, 55,431 coal mining jobs out of 108,613 (51 per cent) were lost between 1989 and 1993 and the workforce was expected to drop to 20,000 jobs by 1996 (Schreiber and Štáva 1994; Hawker 1993). A Ministry of Industry and Trade official said in 1993 that 'all the mines in the Ostrava area will eventually have to be shut down' (Hawker 1993). Overall, the Czech coal mining industry shed more than 100,000 jobs between 1990 and April 1996. The industry employed 186,000 workers in 1990 and 83,000 in April 1996 (MF Dnes 1996:14).

3 Between 1989 and 1991 the employment of miners and other manual workers declined by 8.8 per cent (from 17,342 to 15,833), while the administrative white collar employment declined by 6.5 per cent (from 2,359 to 2,206) in the enterprises which later formed the MCC (Privatisation Project 1992: 11).

4 Industrial production of the Czech Republic declined by 22.3 per cent in 1991, 7.1 per cent in 1992 and 5.3 per cent in 1993 (HN 1994b: 7). It was expected to grow by 3 per cent in 1994 (LN 1994b: 10). Production of electricity and heat declined by 2.8 per cent in 1991 and 3.4 per cent in 1992 (ČSÚ 1993). The production of electricity declined by 3.6 per cent in 1991 but grew by 1 per cent in 1992 and by 3.1 per cent in 1993. The production of electricity

in non-nuclear power plants of the Czech Republic declined by 3.2 per cent in 1991, 2.8 per cent in 1992 and 1.7 per cent in 1993 (ČSÚ 1994).

5 (1) Resolution no. 287 was passed on 2 November 1990 – the set of measures to restore the environment in northern Bohemia; (2) resolution no. 166 was passed on 5 May 1991 – the review report about the implementation of tasks from resolution no. 287; (3) resolution no. 331 was passed on 11 September 1991 about the further development of the Chabarovice opencast mine; and (4) resolution no. 44 was passed on 10 October 1991 about the report dealing with the territorial limits of coal mining and energy production in the North Bohemian Brown Coal Basin (Pěgřímek 1992).

6 For example, the local government of the town of Chabařovice (located outside the Most District) was elected in the 1990 free local elections to save the town (Bystrov 1993). In the case of the village of Horní Jiřetín, the civic resistance against the plans of coal mining enterprises to raze the village culminated with the organisation of a meeting in April 1991 which was attended by the Minister of Environment and his deputy, first deputy of the Minister for Economic Policy, official from the Ministry of Culture, and several members of the Parliament. At this meeting, the inhabitants of the village and the local government voiced their concerns over plans to raze their village and determination to fight such plans.

7 Horní Jiřetín had 1,860 inhabitants in 1990 (9,076 in 1930 and 7,346 in 1950) and 504 houses in 1990 (1,005 in 1970) (OSS 1992). The village of Dolní Jiřetín, a part of Horní Jiřetín, was razed between 1980 and 1983. Similarly, the village of Jezeří, which used to be part of Horní Jiřetín, was demolished after 1988.

8 Interview with ing. Richter, manager of the Most Coal Company, Most, 18 August 1994.

9 Interview with Miroslav Štýbr, mayor of the village of Horní Jiřetín, 13 August 1993.

10 Interview with Jiří Kicl, mayor of the village of Mariánské Radčice, 3 August 1993.

11 Interview with Jiří Kicl, mayor of the village of Mariánské Radčice, 3 August 1993.

12 The members of the ESCBR are elected representatives of the individual participants. The Municipalities are represented by the mayors of the cities, towns and villages located in them. Enterprises are represented by the North Bohemian Economic Union, which organises sixty industrial enterprises in the region, and the state administration is represented by the heads of the district authorities and district employment offices. Trade unions are represented by their elected representatives (the Chamber of Trade Unions and NGOs). The ESCBR was established in 1991 after a visit by the Czech President Václav Havel and after negotiations with the then Prime Minister Pithart.

13 The Clean Air Act (Law No. 309/91) will come into effect in 1998 imposing strict governmental emission limits compatible with Western European standards on the amount of pollution released from the brown coal steam power plants and other industrial sources of pollution. By that time, the Czech Energy Works plans to install desulphurisation equipment and scrubbers in all of its thirty-one operating power plants in northern Bohemia and to close the

remaining ones. So far the company has closed eleven units. Desulphurisation equipment has been installed into the first two units of the Počerady power plant and launched in October and November 1994. The programme of desulphurisation of brown coal power plants should result in the decline of their SO_2 emissions from 722,000 tons in 1993 to 80,000 tons in 1999 (LN 1994c: III; Ministry of Environment 1993).

Bibliography

Bensman, T. (1992a) 'Coal mining plan threatens town's existence', *Prague Post* 2, 31: 1, 5, 8.

—— (1992b) 'Town's residents blast American mining proposals', *Prague Post* 2, 36: 7.

Burawoy, M. and Krotov, P. (1992) 'The Soviet transition from socialism to capitalism: worker control and economic bargaining in the wood industry', *American Sociological Review* 57, 1: 16–38.

—— (1993) 'The economic basis of Russia's political crisis', *New Left Review* 198: 49–69.

Bystrov, V. (1993) 'Chabařovice: město s nadějí' (Chabarovice: the town with hope), *Lidové noviny* 6, 229: Nedělní LN/VI–VII. 2 October: 2.

Cibulka, J. (1993) 'Východiska ke zlepšení životního prostředí v pánevní *oblasti* severních Čech v kontextu s rozvojovými záměry energetiky a těžby', Report prepared for the Economic and Social Council of North Bohemian Basin Region, Most, April.

Clarke, S. (1992) 'Privatization and the development of capitalism in Russia', *New Left Review* 196: 3–27.

—— (1993) 'The contradictions of "state socialism"', in S. Clarke, P. Fairbrother, M. Burawoy, and P. Krotov (eds) *What About the Workers? Workers and the Transition to Capitalism in Russia*, London and New York: Verso.

Clarke, S. and Fairbrother, P. (1993) 'Beyond the mines: the politics of the new workers' movement', in S. Clarke, P. Fairbrother, M. Burawoy and P. Krotov (eds) *What About the Workers? Workers and the Transition to Capitalism in Russia*, London and New York: Verso.

Clarke, S., Fairbrother, P., Borisov, V. and Bizyukov, P. (1994) 'The privatization of industrial enterprises in Russia: four case-studies', *Europe–Asia Studies* 46, 2: 179–214.

ČSÚ (1993) *Přehled ukazatelů sociálního a ekonomického rozvoje ČR 93/2* (Survey of social and economic development indicators in the Czech Republic 93/2), Prague: Czech Statistical Office.

—— (1994) *Přehled ukazatelů sociálního a ekonomického* rozvoje CR 1993 (Survey of social and economic development indicators in the Czech Republic 93), Prague: Czech Statistical Office.

Formánek, V. (1992) 'Úloha a perspektiva uhlí v ČSFR' (The role and perspective of coal in Czechoslovakia), *Uhlí-Rudy* 1, 1: 2–6.

Gheyselinck, T.O.J. (1992) 'Restrukturalizace uhelného průmyslu v České Republice' (Restructuring of coal mining industry in the Czech Republic), *Uhlí-Rudy* 1, 6: 183–6.

Hawker, A. (1993) 'Coal industry slips deeper into the pit', *Prague Post* 3, 37: 9.

Hospodářské noviny (1994a) 'Mostecká uhelná se snaží udržet ceny paliv', 23 February: 7.

—— (1994b) 'K trendům indikátorů ekonomického a finančního vývoje', 10 June: 7.

—— (1995) 'Trvalý pokles těžby hnědého uhlí se zřejmě zastaví až koncem století', 10 May: 7.

—— (1996) 'Mostecká uhelná zabrzdila pokles těžby', 1 July: 6.

Illner, M. (1992) 'Municipalities and industrial paternalism in a "real socialist" society', in P. Dostál, M. Illner, J. Kára and M. Barlow (eds) *Changing Territorial Administration in Czechoslovakia: International Viewpoints*, Amsterdam: Department of Human Geography, Faculty of Environmental Sciences, University of Amsterdam.

Kopečný, K. (1992) 'Postavení uhelného hornictví v palivo-energetické politice České republiky' (The place of coal mining in the fuel and energy policy of the Czech Republic), *Uhlí-Rudy* 1, 6: 187–8.

Lidové noviny (1992a) 'Horníci na Prahu', 4 November: 1 and 8.

—— (1992b) 'Půjdou horníci na Prahu?' 29 October: 3.

—— (1994a) 'Těžba uhlí se bude snižovat', 22 December: KORUNA LN/VIII.

—— (1994b) 'Zahranicní obchod ČR se letos propadl do mírného pasíva', 29 December: 10.

—— (1994c) 'Čisté počeradské mraky Made in ČEZ', 8 December: KORUNA LN/III.

Ministry for Economic Policy and Development (1992) 'Návrh na restrukturalizaci uhelného průmyslu v ČR v souvislosti s jeho privatizací' (Proposal for restructuring of coal mining industry in the Czech Republic in relation to its privatisation), Report for deliberations of the Economic Council of the Czech government, Prague, 28 April.

Ministry of the Environment (1993) 'Preliminary report on emissions of sulfur dioxide and nitrogen oxides in the Czech Republic', prepared for Working Group on Strategies Geneva, 30 August – 3 September, Prague: Ministry of the Environment of the Czech Republic.

Mladá Fronta Dnes (1996) 'Kvůli útlumu odešlo z dolů sto tisíc lidí', 5 August: 14.

OECD (1994) *OECD Economic Surveys: The Czech and Slovak Republics*, Paris: OECD.

OOČSÚ (Okresní oddělení českého statistického úřadu v Moste) (1986) *Statistická rocenka, okres Most 1985*, Most: OOCSÚ.

OSS (Okresní statistická správa Most) (1992) *Sčítání lidu, domů a bytů 1991, okres Most*, Most: OSS.

Pavlínek, P. (1998) *Economic Restructuring and Local Environmental Management in the Czech Republic*, Lewiston, NY: The Edwin Mellen Press.

Pěgřímek, R. (1992) 'Perspektiva a podmínky dalšího rozvoje těžební činnosti v SHR ve světle probíhající společenské a ekonomické transformace' (Perspective and conditions of further development of mining activities in the North Bohemian Brown Coal Basin in the light of unfolding social and economic transformation), *Uhlí-Rudy* 1, 5: 152–5.

Privatization Project (1992) *Privatization Project of the Most Coal Company*, Most: Most Coal Company.

Schreiber, P. and Štáva, J. (1994) 'Uhelný průmysl České republiky' (Coal mining industry in the Czech Republic), *Uhlí-Rudy, Geologicky pruzkum* 1, 6: 217–20.

Smith, A. (1994) 'Uneven development and the restructuring of the armaments industry in Slovakia', *Transactions of the Institute of British Geographers*, New Series 19: 404–24.

—— (1997) 'Breaking the old and constructing the new? Geographies of uneven development in Central and Eastern Europe', in R. Lee and J. Wills (eds) *Geographies of Economy*, London: Arnold.

—— (1998) *Constructing Capitalism? Industrial Restructuring and Regional Development in Slovakia*, London: Edward Elgar.

Statistical Year book of the Czechoslovak Socialist Republic (1989) Prague: SNTL.

Struzka, Z. (1992) 'Zásadní otázky postavení a struktury uhelného průmyslu v ČSSR' (Principal questions of the role and structure of coal mining industry in Czechoslovakia), *Uhlí-Rudy* 1, 5: 145–7.

Šulc, Z. (1993) 'K některým teoretickým otázkám transformace' (Some theoretical questions of transformation), *Politická ekonomie* 41, 3: 319–22.

Part III

SOCIAL AND POLITICAL MOVEMENTS AND THE POLITICS OF AGRARIAN TRANSITION

10

PATH DEPENDENCE IN BULGARIAN AGRICULTURE

Mieke Meurs and Robert Begg

Introduction

One of the few nations to re-elect the Communist Party (re-named the Bulgarian Socialist Party or BSP) in its first free elections, Bulgaria lags behind other East European nations in the pace of liberalisation and its attendant economic restructuring (Wyzan 1993). By early 1994, only one major state firm had been privatised by the National Privatisation Agency and a number of laws essential to the functioning of a market economy remained to be passed. At the same time, many elements of the reform programme which *have* been implemented, such as attempts to rapidly restructure trading relations and forcibly decentralise production have also worked against agricultural recovery. Bulgarian economic reforms have apparently gone both 'too fast' and 'too slowly' for the agricultural sector, squeezing productivity and economic viability from both sides. Yet, despite extremely difficult conditions, those dependent on the agricultural sector for survival have created a variety of local solutions to their problems and have continued producing. This has prevented even greater declines in agriculture. With some improvements in the policy environment, these diverse local initiatives may even provide the basis for a viable agricultural sector.

However, by the mid 1990s the outlines of Bulgarian agriculture had changed little from November 1989, when Communist Party leader Todor Zhivkov was ousted. Only 14 per cent of agricultural land had been distributed to new private owners by May 1994 (Ministry of Agriculture 1994), although in some areas of the country this occurred more quickly than in other areas. The majority of agricultural land continued to be farmed in large, collectively held farms. But the importance of the slow pace of reform is easily overstated. Both the new private and co-operative sectors posted declining yields and output through the mid-1990s, and we will argue that this poor agricultural performance is linked mainly to the more general problems of reforming and rebuilding the Bulgarian economy.

In this chapter, we try to illustrate the roles of both history and recent reform shocks in creating the current dynamics of change in Bulgarian agriculture. We will use the concept of 'path dependency' to capture the role of history in influencing current outcomes. But to say that current outcomes are 'path dependent' is to say more than that history matters. Theorists who have developed the concept of path dependency in recent years have high-lighted *specific* factors and historical moments which set events on a given path. In examining the causes of differential development in the north and south of Italy, for example, Robert Putnam (1994) argues that distinct sets of social institutions developed in the two regions, one of which facilitated the development of economic efficiency and one which did not. He points to the period around 1100 as the period when institutional structures diverged in such a way as to set the two regions on different paths. David Stark's (1991) examination of various strategies of economic transition adopted across East Central Europe points to the distinctive structures of class or power which developed in each country under socialism, structures which he argues were central to determining the form reform took after 1989. Other authors point to other variables as determinants of a historical path: W. Brian Arthur (1994), for example, argues that technological factors played an important role in guiding the direction of economic development, while Douglass North (1990) has pointed to the role played by property rights, social norms and the strength of the state.

To say that certain variables set a social system on a particular path does not imply that history is linear or predetermined. Instead, the concept of path dependency rather suggests that certain variables play a particularly important role in facilitating some outcomes while making others much more difficult. In Stark's (1992) analysis, for example, he argues that the power of Solidarity in Poland in 1989 facilitated a privatisation scheme which favoured workers, whereas in Hungary, where unions were not strong, no special advantages were conferred on workers and privatisation took place mainly through sales.

A path-dependency analysis thus suggests that traditional discussions of economic efficiency must be re-examined. While a particular technology or institution may be the 'most efficient' in an abstract sense, it may not be the optimal choice under the historically given conditions or for the particular actors making the decision. Certain outcomes may not be attainable. Solutions must instead be locally and historically appropriate.

In our examination of Bulgarian agriculture we suggest that – given the knowledge base of the Bulgarian population, the country's economic structure and the technical characteristics of Bulgarian agriculture – massive privatisation may not provide the best or even a feasible solution. The uneven privatisation of Bulgarian agriculture, therefore, may not be the root of recent weak performance. Instead, we argue that distinctive paths of restructuring provide solutions to the distinctive problems of different regions, and we examine the impact of economic restructuring and government policy to illustrate the role local pathways play in current agricultural performance in both private and co-operative sectors. While aggregate data are used to illustrate some general trends, Bulgarian data are seldom as consistent and complete as one would like (Mikhova and Pickles 1994; Wyzan 1993). Consequently, anecdotal and case study information are used to capture the impact on agricultural performance of a web of individuals, local, national and international institutions, and ethnic and ideological communities.

Historical 'paths' to agricultural transition

Rural Bulgaria retains vestiges of institutions dating back at least to the Ottoman Empire (1396–1878), including land fragmentation, collectivism and communal labour. These antecedents were altered during collectivisation, but they were often incorporated into the new system, rather than being eradicated. The transformed institutions and practices play important roles in the current transformation.

Ottoman legacies

As Ottoman control of Bulgaria eroded after 1858, property rights in land were extended to Bulgarians, creating a highly egalitarian system of landholding based on the extended family (*zadruga*).[1] Inheritance rights in land were extended to all children. As the *zadruga* holdings were broken up in the face of increased commercialisation of agriculture, family lands were subdivided and by 1934 the average land-holding was 6.8 hectares held in eleven scattered parcels. Eighty-four per cent of land-holders farmed fewer than 10 hectares (Dobreva and Meurs 1992). The fragmentation of land held back modernisation of the agricultural sector. At the end of the Second World War most villagers were still using a wooden or steel plough pulled by draft animals.

When anthropologist Irwin Sanders travelled in 1934 to Dragalevsky (a village then a few miles from the capital city of Sofia), village life was still dominated by family and collective labour: 'family members grow or make for themselves most of the things they need, they depend only to a minor degree upon stores, factories, and other commercial agencies' (Sanders 1949: 144). Some land remained under collective management by the village and labour was still performed collectively on this land. Sanders (1949: 59) described collective haying:

> The captains, appointed by the mayor to control the activities of the people from the respective neighbourhoods, stirred up the spirit of competition, and the stacks grew higher and higher. . . . The

towering stacks stood motionless, symbolic of that co-operation which is the heart of communal life everywhere.

But for the enthusiasm of the workers, this scene might well have been that of a collective farm brigade in 1984.

In the early 1900s, the state took on an important role in guiding the development of agriculture, channelling investment funds into agriculture from 1903 to 1944. Government banks loaned funds to local credit co-operatives, while guiding their lending policy. After the early 1930s, the state also exercised a monopoly on the purchase of wheat and rye, later extended to cotton, rose oil, tobacco and other products, setting prices and controlling both domestic and export marketing (Stoyanova 1993; Logio 1936: 190).

Communism and the collectivisation of agriculture

Bulgarian agriculture changed radically with the ascendancy of the Bulgarian Communist Party (BCP) to state power in 1948. Over 96 per cent of total arable land was consolidated into state and collective farms, and by the mid-1970s a structure of very large-scale, state-controlled farms (averaging 2,400 ha in 1976) was established, on which production was highly mechanised and specialised. This basic structure persisted until 1989 although farm size was reduced somewhat in the 1980s, and disinvestment in agriculture after the mid-1970s reduced available machinery.

Alongside the highly centralised and mechanised collective production, small-scale, low-input, individual agriculture also persisted. Collective farm (TKZS) members were allowed to farm plots of under 0.5 hectares and keep a few animals. Output from these plots grew to 44 per cent of total livestock production and 25 per cent of vegetables by 1989 (Dobreva and Meurs 1992). This 'personal' agriculture was highly articulated with the TKZS, however, which provided seed, chemical inputs and machinery to family plots, and purchased surplus final products. After 1982, quasi-private agriculture was extended, as TKZSs subcontracted production to groups of workers or families, providing inputs and production technology to the subcontracting groups. By 1989, 62 per cent of grain, 80 per cent of vegetables, and 70 per cent of meat were produced by contract groups under state guidance (Killian 1990).

These legacies complicate a transition to private, market-oriented agriculture. Collective labour and paternalism are long-standing forms of rural organisation; individual entrepreneurial initiative, commercial orientation and risk-taking, on the other hand, have not been important rural traditions. Fragmented pre-war land ownership, which is the basis for the restitution of agricultural land (see below), and the large-scale, centralised nature of agricultural capital may together undermine the formation of viable private farms in the current period, as may the large-scale, centralised nature of agricultural capital.

Politics and the legal basis of agricultural transformation

The agricultural policies of the last five years, both those of the Bulgarian Socialist Party (BSP) and the opposition coalition (Union of Democratic Forces – UDF), have been shaped more by political considerations than by any productive or economic rationale. In June 1990, a Grand National Assembly was elected to implement fundamental reforms. The BSP won a majority of the votes but only 47 per cent of the seats in the severely divided legislative body. Division slowed reform, but in February 1991 the main agricultural reform law, the Land Act, was passed.

The Land Act was of particular importance to the BSP because the party's support lay heavily in rural areas where it distributed resources and favours through the TKZSs. The 1991 Land Act attempted to protect this source of BSP power. The restitution of land to its pre-war owners was mandated and was widely popular because of the highly egalitarian (if fragmented) pre-war distribution. But the law also retained limits on private property. Land did not have to be distributed to individuals; groups could claim land collectively by registering as a new co-operative. Holdings were limited to 30 ha (20 in regions of intensive production). Ownership could not be transferred for three years, and then could land be sold only to neighbours, relatives, lease-holders or the state. Foreigners or firms with foreign partners were prohibited from purchasing land, and agricultural land could not be used for non-agricultural purposes. Machinery and other property remained under the collective farms' control, and the future of each TKZS was to be determined by its General Assembly. TKZSs could register as 'new' co-operatives, retaining the assets of the TKZS and claiming the land of those TKZS members who chose to remain in the collective.

Clearly, this law sought to protect the existing farm structure. In principle, villagers might claim land to farm privately, dissolving the TKZS. But recipients faced with potential ownership of a tiny plot, little appropriate machinery, and limited experience with independent farming could also choose to leave the land in a collective, to be farmed as part of a larger unit. With the TKZS controlling existing farm machinery and other assets, managers could pressure households to leave their land in the collective.

By late 1991, however, a reformist wave swept the country. New parliamentary elections gave the opposition UDF 34.5 per cent of the vote and a majority in Parliament. Alliance with the Movement for Rights and Freedoms (MRF),[2] the only smaller party to win the 4 per cent of votes necessary to secure parliamentary representation, gave the UDF a slim ruling majority within Parliament. In March 1992, the new government passed a number of important amendments to the Land Act, aimed at smashing the

TKZSs as a source of BSP power in the countryside. The amendments facilitated consolidation of private holdings into larger, commercial plots. The limitations on owned plot size, the three-year prohibition on sale, and the prohibition on non-agricultural use were all removed. Firms with minority (up to 49 per cent) foreign ownership would be permitted to own land. In addition, the law stipulated that land had to be returned according to its real boundaries, not as a share of a new co-operative. To block the direct transformation of old TKZSs into new production units, the law required that all assets of each TKZS be liquidated and all land distributed to individuals before new co-operatives could be formed. District officials, mainly appointed by the UDF, were to appoint Liquidation Committees for that purpose. TKZS members would receive share coupons at liquidation, which they could use to bid for farm assets.

This new version of the Land Act dealt a severe blow to the BSP and the existing farm structures. The changes transferred control of farm machinery from powerful (usually socialist) villagers to UDF-appointed officials. In addition, the new framework greatly complicated the organisation of new co-operatives on the basis of the TKZS.

The legal changes also complicated the implementation of the reform process, however. Restituting the millions of tiny plots according to their real (and often contested) boundaries proved to be a logistical nightmare. In addition, restitution only to individuals would result in land fragmentation even greater than in the pre-war period. According to the National Land Council, over 90 per cent of claims were for plots of under 1 ha (Table 10.1), and many of those claims would later be divided among several heirs to the original holder. Negotiating with a multitude of owners necessary to consolidate a viable plot would be difficult.

Another complication arose from the strategy for liquidating farm assets. Farm assets such as grain dryers, storage sheds and large machinery cannot be easily dispersed among, and used by, small-holders. Farm managers expecting rapid, fire-sale liquidation could not expect to benefit from capital maintenance, and many installations deteriorated or were simply

Table 10.1 Land claims under restitution, 1 July 1992

Grouping	No. of households	% arable land
0.1–1 ha	1,783,390	90.81
1.1–5.0 ha	171,394	8.72
5.1–10.0 ha	8,608	0.44
>10.0 ha	580	0.03
Total	1,963,972	100.00

Source: 168 Hours, 18 October 1992: 7.

abandoned. Nor could large livestock herds be easily dispersed among small-holders, and entire breeding stocks were sold for slaughter.

Finally, the imposition of the UDF-supported law on BSP supporters resulted in a feeling that the reforms were being imposed on the rural population by outside, urban forces. One village woman argued: 'First the communists made us give up our land, and now the UDF is making us take it back. It's like getting slapped on both sides of your face' (Creed 1994: 37). This situation was exacerbated by the fact that many farm workers, especially technicians, Turks, Gypsies and other formerly landless peasants were not eligible to claim land, as their families did not own land in 1946. For example, on the Bourgas TKZS (in the south-eastern part of the country), only 40 per cent of employees could claim land (Yarnal 1994). These problems created substantial resistance to the reform, as will be seen below.

By August 1992, over 1,705,000 applications for restitution had been filed (Dinkov 1993), but actual restitution of land and liquidation of farm assets proceeded slowly. By June 1994, owners of approximately 46 per cent of agricultural land had been told where their land would be, but only 14 per cent of land had been titled (Ministry of Agriculture 1994: 3). Without titles, farmers could neither legally register in new co-operatives nor obtain secured bank credit. By December 1993, only 2.5 per cent of the TKZS had liquidated their assets, so machinery remained concentrated on the old TKZSs (*Current Economic Business* December 1993: 30).

This inability to quickly establish private farming discredited the UDF government among its supporters. At the same time, de-collectivisation proceeded exceedingly quickly in the mountainous tobacco regions, where mechanisation had never been possible and most TKZSs were highly subsidised. When state support ended after 1991, those TKZSs that were not economically viable collapsed. Plots which had never really been unified were quickly identified and reclaimed by owners. This caused massive economic disruptions, both for previously landless producers and for new private tobacco producers suddenly working without state support. Since many tobacco producers are of ethnic Turkish origin, the problems discredited the UDF's coalition partner – the MRF. A political crisis resulted in late 1992 which brought down the government.

The new government led by centrist Luiben Berov (a scholar of the Bulgarian agricultural economy) took a softer line on agricultural reform, amending the Land Act and Co-operative Law to facilitate the formation of new co-operatives while also retaining the Land Act amendments which facilitated the transfer and consolidation of private holdings (Creed 1994). The changes reflected increased official acceptance of co-operative farming and willingness to allow rural dwellers to define their own method of restructuring.

Overall, however, the legal manoeuvrings added to the difficulties in rebuilding the agricultural sector. Both the first and second versions of the Land Act sought to impose ideologically defined forms of organisation, limiting the ability of rural dwellers to create farms suited to local conditions and preferences. The third version of the Land Act encouraged varied forms of agricultural organisation, but the continuous changes in the legal context and slow clarification of ownership created enormous uncertainty for producers already facing a host of economic difficulties.

Economic constraints on agricultural restructuring

The economic transition policies of the various Bulgarian governments resulted in rising relative input prices and loss of markets, but did not immediately reduce monopoly power in agricultural markets. Particular government policies have exacerbated some of these problems. We will examine each of these in turn.

Rising relative input prices

Under central planning, input prices were heavily subsidised but output prices were also held down to subsidise consumers. Wheat, for example, cost $35 a ton in Bulgaria in 1989, but $150 on the world market. Overall, the World Bank estimates a net tax on agriculture of 25–35 per cent in the 1980s (World Bank 1991). With the end of central planning, this form of taxation was eliminated. Little relief resulted for the agricultural sector, however. By the end of 1992, output prices for most agricultural goods had increased five to ten times, while input prices had increased six to twenty-five times (Ministry of Agriculture 1993a: 27). Fertiliser, feed and pesticide prices adjusted rapidly (although not completely) towards world market prices, in part because of their high import content and in part because of the power of local monopolies (Ministry of Agriculture 1993a: 27). Water prices rose eight to fifteen times by 1994 (depending on use and location). Prices for agricultural products adjusted more slowly. Cattle prices, for example, rose to nearly three times their 1990 level by 1992. Pork and poultry prices increased somewhat faster, but still failed to keep pace with feed prices (Ministry of Agriculture 1993b: 28).

High nominal interest rates further increased costs. From an interest rate of 2.5 per cent in 1990,[3] nominal interest rates increased to around 50 per cent by late 1991 and hovered there until mid-1993 (*PlanEcon Report* 12 October 1993: 25). While real interest rates were slightly negative, the high nominal rates combined with short-term lending practices were a particular problem for agriculture because of the long lag between planting and harvest. In April 1993, preferential interest rates for agriculture were introduced to assure spring planting (*168 Hours* 14 November 1993: 7). Few subsidised loans were extended, however, as banks feared the state would not be able to cover the subsidy (Zagorska 1994: 6). The few subsidised loans which were extended went mainly to well-connected managers of newly registered co-operatives (Meurs and Spreeuw 1997).

Markets

The slow upward adjustment in agricultural prices is linked to the decline in both foreign and domestic markets. In 1989, 75 per cent of Bulgarian exports went to CMEA countries (NSI 1993b: 151–2). As the former CMEA economies collapsed after 1991, Bulgarian exports to CEE and CIS fell dramatically (Table 10.2). New markets, however, have been slow to develop. The share of Bulgarian exports going to the European Union rose to nearly 30 per cent in 1993 (NSI 1993b; *Current Economic Business* October 1993: 59–61; *PlanEcon Report* 18 July 1994: 36), while exports to the European Free Trade Area and the US remained approximately unchanged (Davidova and Sukova-Tosheva 1993: 22–7).

Export restrictions by the Bulgarian government have added to the impact of declining markets. From 1989 to February 1990, exports of certain foodstuffs, including meat, fodder grain, and wheat were prohibited (Cochrane *et al.* 1993). From June 1991 until at least September 1994, a combination of quotas, export taxes and export bans limited exports of wheat and feed grains in order to hold down food prices (Davidova 1993: 177; Meekhof, Penor and Schmitz 1993: 31).

Domestic markets also collapsed as real incomes in Bulgaria fell by about two-thirds from 1990 to 1992 (Buckwell *et al.* 1994: 115, 129). Demand for food products fell to an average of 59 per cent of 1989 levels in 1992 to 1993 (*PlanEcon Report* 12 October 1993: 20; 18 July 1994: 25). Consumption patterns also changed as consumers substituted bread for more expensive products (Buckwell *et al.* 1994: 122). While the government

Table 10.2 Bulgarian exports by selected trading partners, 1985, 1989–93 (per cent of exports)

	1985	1989	1990	1991	1992	1993
CEE	74.7	–	80.2	57.7	41.9	35
CIS	56.6	65.2	63.8	49.7	16.1	15.9
OECD	9.6	–	9.0	26.3	42.3	43.1
USA	0.2	0.6	1.7	3.4	3.2	3.3
EC	6.4	–	5.0	15.7	30.8	28.1
Arab nations	9.6	–	6.1	8.3	7.9	6.9

Sources: NSI 1993b: 151–2; *Current Economic Business* October 1993: 59–61; *PlanEcon Report* 18 July 1994: 36.

has tried to protect the weak domestic market somewhat by retaining certain tariff barriers and continuing some price supports (Gueorgiev and Bashikarov 1993), net transfers to agriculture were again negative from 1990 to 1992 (Ivanova 1993: 103, 109).

Monopoly power in agricultural markets

The pre-reform industrial structure was based on a system of specialisation in which each branch was dominated by one large conglomerate. The 1991 Law on Competition attempted to reduce the power of these large state firms by breaking up 100 large companies into about 800 firms (Economist Intelligence Unit 1991: 33). In many cases, however, firms have retained regional monopolies. Tobacco growers suffer from similar institutional contexts. Bulgar-Tabak, the national monopoly under central planning, was divided into twenty-two subsidiaries. Tobacco prices are still fixed by the central Bulgar-Tabac, however, and farmers' bargaining power is reduced by their continued dependence on the local subsidiary for inputs. Plans for privatisation involve the sale of most local monopolies as a unit and are unlikely to affect market structure. In any case, political stalemate has prevented the mass privatisation of large firms.

By 1994, entry of new firms had begun to increase competition in some markets. Markets for machine service and fodder had become increasingly competitive. In the fruit and vegetable sector, Greek and Turkish buyers competed actively with local firms. In some regions, dairy and meat processing and even grain purchasing also have a significant number of new entrants. However, even competition has not solved the problem of low prices. In 1992, tomatoes and other crops rotted in the fields as prices made harvesting unprofitable. And in many areas local monopolies continued to distort price, contributing to the rapid rise in input prices and low purchasing prices for output (Meurs and Spreeuw 1997).

The economic factors and government policies outlined here affect both the new co-operatives and the fledgling private sector, although not equally. Co-operatives appear to offer some advantages in negotiating with input suppliers, banks and processing firms, and in gaining access to foreign markets, and these considerations may partially explain the emergence of varied local responses. Where traditional production specialisations benefit from large-scale, capital-intensive technologies, co-operatives may offer a particular advantage. A similar kind of path-dependent dynamic may be at work in the restructuring of Russian agriculture, in which local organising traditions combine with current economic structures to make co-operative farming a preferred alternative to private farming in many places (Theisenhusen 1995).

Uneven local responses to the agricultural crisis

The implementation of the land reform legislation and the restructuring of agriculture have varied greatly across regions, depending on physical geography, local production specialisation and regional structures of power. Even in the days of the Communist Party, a great deal of accommodation occurred between central dictates and local and regional structures. Misha Glenny (1993: 169) explains this in terms of Bulgaria's pre-war traditions:

> Whereas other Eastern European countries were moving away from the semi-feudal structures bequeathed by the inter-war period, Bulgaria retained much of its feudal tradition. The most important consequence of this was the development of a system of patronage, which was administered by the regional barons of the party.

As noted above, these powerful local structures remain dominated by the BSP. Even in the few rural areas where UDF members hold elected positions, members of the old structures remain powerful in village politics. Often, these individuals are expected by the local population to carry on in their traditional role of maintaining production and economic security. In many places they have done so by limiting the UDF attack on collective farming, although the approach they take to this varies from hard-line communist, to technocratic, to entrepreneurial (Begg 1993). In other places, legislation or lack of popular support has effectively neutralised local power structures, and the old farms were broken up. A brief review of two cases provides insights into the range of dynamics involved in the complex mosaic of local response.

Kameno (Bourgas region) is a grain-growing village near the Black Sea. Here the appointed director of the Liquidation Commission had been mayor of the village in the 1980s. He remained committed to large-scale collective production after 1989, but also embraced capitalism and revealed extra-ordinary entrepreneurial abilities. As director of the liquidation, he dragged out the process while making a profit for the old TKZS. The collective farm began attracting new members from adjoining farms in the midst of the de-collectivisation process. While never actually defying liquidation directives outright, the director simply kept the farm going until it became possible to restructure as a co-operative after the Land Act amendment of the Berov government. Not all grain-growing regions managed such a smooth re-collectivisation of land however. Creed (1994) describes the case of Zamfirovo, in which villagers could not save their grain and livestock farm from UDF-appointed liquidators. But in the case of Kameno successful local resistance to the legislative assault on the collective farms was possible in part because local conditions offered advantages to co-operative farming and in part because of the particular skills of one administrator and his staff.

In contrast, the tobacco-growing regions where much of the Turkish population is concentrated experienced unusually rapid de-collectivisation. As noted above, this population supported the MRF, coalition partner of the anti-communist UDF from 1991 to 1992 (Figure 10.1). Regional officials were thus in a strong position to drive the old guard out of local affairs. In addition, the area is hilly and arable land is dispersed, which makes it difficult to consolidate fields under the TKZS. Much of collective tobacco production had actually been carried out on small plots by families even during the socialist period, and individual fields were relatively easy to restore to pre-war owners. Tobacco had been highly subsidised and remained lucrative throughout the 1980s. After 1991 many households were eager to reclaim tobacco land in the hope that private tobacco production would continue to be profitable.

However, in many cases the rapid restitution of land did little to change production relations or improve conditions. The local tobacco-purchasing monopoly simply replaced the TKZS as the organiser of production, supplying technology and inputs and purchasing the final product at agency-set prices. As state subsidies were ended in 1991, these prices quickly became less favourable to farmers. Many former tobacco workers, including

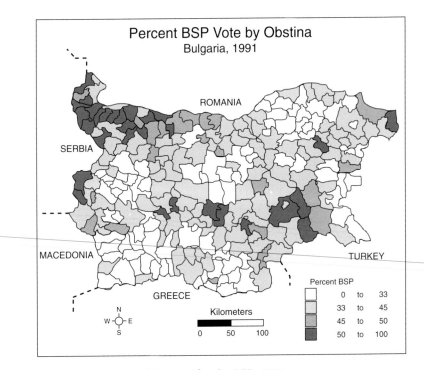

Figure 10.1 Percentage of the vote for the BSP, 1991

254

technicians, those whose families originated in other regions, and those from previously landless families had no rights to land at all and were completely displaced. Combined with the collapse of rural industry in this region, this resulted in unemployment rates ranging from 25 to 90 per cent depending on the locale. Rapid de-collectivisation was popular and feasible both politically and technically, but it failed to spur local economic growth. Unemployment and falling tobacco earnings contributed to the political backlash which brought down the UDF–MRF coalition government in October 1992.

As these cases illustrate, local geography and political structures have combined to influence the course of agricultural restructuring. Whereas the history of land fragmentation, collectivism and paternalism, plus the uneven reform effort complicate a transition to private, capitalist farming across Bulgaria, the impact of these factors is not uniform. Substantial amounts of de-collectivisation have occurred in regions where *small-scale* production is most viable. This includes the tobacco region described above, as well as the regions near Sofia and other large cities where market gardens and small livestock producers can sell directly to consumers. These areas have also tended to be dominated by anti-communist political parties which aid de-collectivisation efforts (Figure 10.1). In the grain-producing plains of northern Bulgaria, where economies of scale in production are more important, land fragmentation and slow market development have been more detrimental to private producers (Meurs and Spreeuw 1997). In these areas, historical experience with collective, centrally guided production has combined with continued communist control to facilitate the organisation of co-operatives (see Figure 10.1). These distinct patterns are also reflected in an uneven sectoral distribution of private production (Table 10.3). There has been a rapid emergence of private production in the fruit, vegetable and livestock sectors and a significantly slower transformation in grain crops, as well as sugar-beet and sunflowers, which are also highly mechanised and sold mainly to large state processing firms.

Poor agricultural performance

Agricultural production fell by almost 18 per cent in real terms from 1989 to 1993 (*PlanEcon Report* 18 July 1994: 20). The decline was felt most heavily in livestock production (which declined to 69 per cent of its 1989 level) while horticulture production in 1992 remained at 96 per cent of the 1989 level (NSI 1993b). Yields also fell in all crops. In the grain sector, which continued to be dominated by large co-operative farms, production fell substantially. Wheat production fell by 31 per cent between 1989 and 1993, barley by 39 per cent. In the quickly privatised tobacco and vegetable and fruit sectors, output also fell dramatically. Livestock production, also based heavily in the private sector before and after 1989, suffered a similar

Table 10.3 Percentage output on personal or private plots, 1989–93

	1989	1990	1991	1992	1993
Wheat	2.2	3.4	11.6	13.8	16.1
Barley	8.2	11.2	20.9	n.a.	20.1
Sunflowers	<0.1	<0.1	1.0	3.1	7.7
Tobacco	1.3	1.9	6.7	59.8	88.1
Sugar-beet	0.2	0.3	5.1	4.6	<0.1
Tomatoes	34.0	41.1	57.4	78.7	83.2
Grapes	48.8	50.4	52.1	57.8	48.6
Pigs	18.9	20.0	24.6	26.1	46.3
Cattle	19.1	17.9	26.7	31.9	62.7
Sheep	31.4	31.4	40.0	48.6	84.1
Poultry	42.4	38.2	44.6	49.2	72.3
Milk	24.1	28.3	38.8	51.3	70.4
Eggs	48.5	46.9	53.3	54.0	58.4

Sources: NSI 1992: 70; 1993b: 130, 132; *Current Economic Business* October 1993.

fate (NSI 1992, 1993b). The big drop in production is related to large declines in yields. Wheat yields in 1993 were 40 per cent below 1989 levels and barley was almost 33 per cent below.[4] Tomato yields declined 38 per cent and grape yields fell 15 per cent. Only tobacco yields remained steady (NSI, various years).

The change in relative prices played an important role in declining yields and output. In response to the profit squeeze, between 1989 and 1992 average national use of pesticides fell by 62 per cent while chemical fertiliser use fell by 74 per cent (NSI, various years). On the Kameno farm in Bourgas, for example, the area planted increased from 1991 to 1993, but nitrogen fertiliser was applied to 200 fewer hectares, at 78 per cent of the previous concentration, but at nearly 4 times the cost (Begg 1993). These data may slightly exaggerate the problem of declining fertilisation, since organic material can substitute for commercial nitrogen to some extent, but even this substitution is limited by the price of fuel for manure spreading. Between 1992 and 1994, farm land was covered with a profusion of broad-leaf weeds, and many fields were thinly covered with small, poorly developed plants.

While the degree of privatisation does not appear to have played a role in the output decline (both privatised and collectivised sectors have suffered similarly), the legal uncertainty caused by the slow pace of restitution does appear to be having a negative impact. Many private farmers interviewed in 1993 and 1994 used no fertiliser at all, citing both cost and uncertainty about their tenancy as reasons. Irrigation systems on former TKZS land, in which neither liquidators nor potential land recipients have a clear future stake, have fallen into disrepair or been stripped by looters. On the Zamfirovo farm (Mikhailovgrad), 600 ha were irrigated in 1985, but by

1993 the system was unusable and no land was irrigated (Creed 1994). Nationally, only 300,000 ha were irrigated during 1993, compared to the 1990 capacity of 1.3 million ha (*168 Hours* 16 May 1993, 24 October 1993). Clearly, rising water prices have also contributed to this decline, but field interviews highlight the important role of uncertain ownership in the rapid deterioration of infrastructure.

A similar problem has arisen with perennial plantings. Claimants of land covered by orchards must pay for the 'improvements' to their land since collectivisation, and many have been unwilling or unable to do so. Much of the orchard land depends on irrigation, and with future ownership unclear no one is willing to pay for this expensive upkeep. Over the summers of 1993 and 1994, miles of roadway were lined with abandoned and dying fruit trees.

The price squeeze and legal uncertainty described above thus appear to be undermining recovery in both privatised and co-operative sectors. At the same time, they are creating a legacy of abandoned and deteriorating capital stock, mined agricultural land, and general disinvestment in agriculture which will continue to slow the agricultural recovery long after the restitution process and economic reforms have been completed.

Conclusions

The particular traditions in Bulgarian agriculture militate against a seamless transition to private capitalist agriculture. Pre-collectivisation and collectivisation experiences in Bulgarian agriculture have prepared the rural population for collective production, small-scale self-sufficiency, and state-guided market production. But history offers few precedents for large-scale, entrepreneurial commercial farming. The distribution of resources resulting from pre-war and socialist ownership patterns also complicates the emergence of such capitalist farms. Land-holdings are fragmented into plots too small for viable farms, while existing machinery and buildings are too large for all but the largest farms. This legacy makes the 'path' of commercial capitalist farming a difficult one in some places and sectors.

Certain aspects of the transition period also complicate the emergence of private, commercial farming. Legal uncertainty, low profit margins and high short-term interest rates make individuals hesitant to make investments in private production, while restitution complicates the consolidation of plots into commercially viable enterprises.

Clearly, the impact of these problems varies by region and production specialisation. Where products can be produced economically on a small scale, private entrepreneurs can easily consolidate viable farms and begin production with little investment or risk. But where economies of scale are significant or substantial investments are needed to begin production, individuals currently face significant barriers to entry. Co-operatives, which can use old political networks and traditional forms of organisation to maintain the consolidation of contiguous plots and retain access to large machinery and installations, offer organisational advantages to land owners. By December 1993, 1,292 agricultural co-operatives had been legally registered (*Current Economic Business* September 1993, December 1993), incorporating 41 per cent of land (Dinkov 1993: 5).

There is little evidence that a more rapid development of private farming would improve the performance of the agricultural sector. The loss of markets and rapid increase in relative prices of agricultural inputs inherent in the transition process affect both private and co-operative sectors. The resulting profit squeeze underlies falling yields and output in both sectors. Indeed, certain aspects of government policy exacerbate the price squeeze. The rapid opening of the economy has resulted in significant imports of fruits, vegetables and livestock products from Greece, Turkey and Western Europe, and these products – often produced with government subsidies – compete with local producers. But potential Bulgarian exporters (of grain) fail to benefit from the lifting of trade barriers on account of government-imposed export restrictions. The inability of government to address regional monopolies also contributes to the price squeeze, as decentralised farms face centralised firms both when buying inputs and selling production.

Rural producers can find few alternatives to farming in the current period. Unlike the de-collectivisation of Chinese agriculture which began during a period of high aggregate economic growth, offering a wide range of employment opportunities to former farmers, de-collectivisation of Bulgarian agriculture followed a period of stagnation and then economic collapse. Output in East Central Europe has fallen farther and faster than during the Great Depression of the 1930s in the West. Consumer demand has declined dramatically and government spending has been slashed. There is little incentive to investment in new industries which could absorb workers fleeing agriculture.

Seeking to maintain their livelihoods, farmers instead continue to experiment with various forms of reorganisation and adjustment of agricultural production. In doing so, they have faced a number of ideologically based constraints from politicians from both 'neo-socialist' and 'neo-liberal' political forces seeking to impose a single, ideologically driven solution on the countryside. These constraints have often been successfully evaded by local producers, however, and varied forms of production have persisted. Increased government recognition of the local nature of organisational rationality after 1992 provided support for such experimentation.[5]

The prognosis for Bulgarian agricultural organisation is now one of mixed organisational forms, with private farmers emerging and coexisting with larger co-operative farms. The acceptance of local solutions to restitution, including the restitution of land in collectively owned blocks, should make it easier for agricultural producers to find quick, locally appropriate solutions to the use of existing capital stock, land and labour. This will reduce uncertainty and encourage a lengthening of time horizons.

Yet it remains clear that eventual completion of the restitution process will not turn the agricultural sector into an engine of economic growth. The majority of the causes of the decline in agriculture are unrelated to property form or the process of restitution. Loss of markets, price and export controls, high interest rates and input prices, and monopoly power in the agricultural markets are the most significant problems. Agriculture cannot serve as the engine of growth until it has itself recovered, and the continued taxation of agriculture to support the rest of the economy now risks undermining long-run growth. A more even implementation of the reform, which supports competition in supply and purchasing networks alongside the decentralisation of farming will contribute to both the viability of private farming and incentives for overall recovery of production. A more even application of trade policy, which consistently puts Bulgarian producers on an even footing with foreign agricultural competitors will also contribute to these goals. If the overall context in which producers find themselves is improved in these ways, producers themselves are likely to solve the problems of appropriate farm organisation.

Acknowledgements

The authors wish to thank the John D. and Catherine T. MacArthur Foundation, the National Council for Soviet and East European Research, and the National Science Foundation for supporting the research on which this chapter is based.

Notes

1 The *zadruga* was a Slavic, patriarchal system of joint farming, in which all property was owned in common by extended families and labour was performed collectively.
2 A party loosely representing the Turkish minority. Parties purporting to represent the interests of one or more ethnic groups are formally illegal. The Movement for Rights and Freedoms therefore defines itself more broadly, while most of its constituency belongs to the Turkish ethnic minority.
3 The central bank rate was 4.5 per cent, but agriculture enjoyed a preferential rate.
4 The 1989 data reflect yields in the state sector only.
5 The BSP government elected in 1995 attempted once again to constrain local choices, but without much success. Legal challenges prevented further significant amendments to the Land Act.

Bibliography

Arthur, W. Brian (1990) 'Positive feedbacks in the economy', *Scientific American* February: 92–9.

Begg, R. (1993) 'Political influences on the de-collectivisation of Bulgarian agriculture', paper presented at the Association American Geographers Annual Meetings, Atlanta.

Buckwell, A., Davis, J., Balcombe, K. and Davidova, S. (1994) 'Food consumption during economic transformation in Bulgaria', Sofia: Ministry of Agriculture, PHARE Program Economic Analysis Unit.

Cochrane, N., Koopman, R.B., Lamb, J.M., Lundell, M.R., deSouze, M. and Sremac, D. (1993) 'Agricultural policies and performance in Central and Eastern Europe, 1989–1992', Washington DC: USDA Economic Research Service, Foreign Agricultural Economic Report No. 247.

Creed, G. (1994) 'An old song in a new voice: decollectivization in Bulgaria', in D. Kideckel (ed.) *East-Central European Communities: The Struggle for Balance in Turbulent Times*, Boulder: Westview: 25–42.

CSO (Central Statistical Office) (various years) *Statistical Yearbook of the People's Republic of Bulgaria*, Sofia.

Current Economic Business (1991, 1992), Sofia: National Statistics Institute.

Davidova, S. (1993) 'Agricultural policy and trade development: review 1992–1993', Sofia: Ministry of Agriculture, PHARE Program Economic Analysis Unit.

Davidova, S. and Sukova-Tosheva, A. (1993) 'Trade and welfare implications of the European agreement between Bulgaria and the EC on agriculture', Sofia: Ministry of Agriculture, PHARE Program Economic Analysis Unit.

Dinkov, D.G. (1993) 'Progress in the agrarian reform to August 8, 1993', mimeo, Sofia: Ministry of Agriculture.

Dobreva, S. and Meurs, M. (1992) 'Clients, prols, and entrepreneurs: state policy and private agriculture under central planning', Washington, DC: American University Department of Economics Working Paper No. 93–25.

Economist Intelligence Unit (1991) *Country Report: Romania, Bulgaria, Albania*, London: Economists Publications.

Glenny, M. (1993) *The Rebirth of History*, London: Penguin Books.

Gueorgiev, N. and Bashikarov, G. (1993) 'Foreign trade and foreign trade policy in Bulgaria, 1991–1992', Sofia: Centre for Democracy.

Ivanova, N. (1993) 'Measuring the effects of government transfers from agriculture in Bulgaria: calculation of producer subsidy equivalents', Sofia: Ministry of Agriculture, PHARE Program Agricultural Analysis Unit.

Killian, M. (1990) 'The accord form of organization in the interests of agriculture', Ph.D. dissertation, Institute of Sociology, Sofia.

Logio, G. (1936) *Bulgaria: Past and Present*, Manchester: Sherratt and Hughes.

Meekhof, R., Penov, I. and Schmitz, A. (1993) 'Transition in the Bulgarian grain sector', Sofia: USDA Liaison Office.

Meurs, M. and Spreeuw, D. (1997) 'The evolution of agrarian institutions in Bulgaria: markets, cooperatives and private farming 1991–1994', in D. Jones and J. Miller (eds) *The Bulgarian Economy: Lessons from Reform during Early Transition*, Aldershot, UK: Ashgate.

260

Mikhova, D. and Pickles, J. (1994) 'Environmental data and social change in Bulgaria: problems and prospects of data availability and future research', *The Professional Geographer* 46 (2): 229–36.

Ministry of Agriculture (1993a) *Informatsionen Buletin: Zerneni kulturi*, Sofia: Ministry of Agriculture, April.

Ministry of Agriculture (1993b) *Informatsionen Buletin: Grozde, Plodove, i Zelenchutsi*, Sofia: Ministry of Agriculture, April.

Ministry of Agriculture (1993c) *Analiz na Sustoyanieto i Tendensiite v Sektor Zhivotnovudstvo*, 1: Sofia.

North, Douglass (1990) *Institutions, Institutional Change, and Economic Performance*, New York: Cambridge University Press.

NSI (National Statistics Institute) (1992) *Statistical Reference Book of Bulgaria*, Sofia: National Statistics Institute.

NSI (National Statistics Institute) (1993a) *Statistical Reference Book of Bulgaria*, Sofia: National Statistics Institute.

NSI (National Statistics Institute) (1993b) *Statisticheski Spravochnik*, Sofia: National Statistics Institute.

168 Hours (1993), Sofia: Bulgarian Business News.

PlanEcon Report (1993), Washington, DC: PlanEcon Inc.

PlanEcon Report (1994), Washington, DC: PlanEcon, Inc.

Putnam, Robert (1993) *Making Democracy Work*, Princeton, NJ: Princeton University Press.

Sanders, I. (1949) *Balkan Village*, Lexington: University of Kentucky Press.

Stark, David (1992) 'Path dependence and privatization strategies in East and Central Europe', *East European Politics and Society* 6 (1) Winter: 17–54.

Stoyanova, R. (1993) 'State policy toward agricultural co-operatives in Bulgarian from the end of the 19th century to 1944', mimeo, Sofia: Institute of History, Bulgarian Academy of Sciences.

Theisenhusen, William (1995) 'Landed property in capitalist and socialist countries: the Russian transition', in Gene Wunderlich (ed.) *Agricultural Landownership in Transitional Economies*, Lanham, MD: University Press of America.

World Bank (1991) *Bulgaria: Crisis and Transition to a Market Economy vols I, II*, Washington, DC: World Bank.

Wyzan, M. (1993) 'Economic transformation and regional inequality in Bulgaria: in search of a meaningful unit of analysis', paper presented at the American Association for the Advancement of Slavic Studies, Honolulu, Hawaii, 19–22 November.

Yarnal, B. (1994) 'De-collectivisation of Bulgarian agriculture', *Land Use Policy* 11 (January): 67–70.

Zagorska, M. (1994) 'Issues concerning the crediting of agriculture and the related activities in Bulgaria', Sofia: Ministry of Agriculture, PHARE Program Agriculture Analysis Unit.

11

DEFENDING CLASS INTERESTS

Polish peasants in the first years of transformation

Krzysztof Gorlach and Patrick H. Mooney

Introduction

Poland was atypical of countries under communism insofar as it retained at least two relatively autonomous enclaves: the Roman Catholic Church and the peasant family farm. In 1989, as the first democratic elections since the Second World War were held, the private sector contained slightly more than 2 million family farms, comprising about 76 per cent of Polish farmland. At this time, the socialised sector consisted of almost 1,000 state farms (*Panstwowe Gospodarstwa Rolne*) and about 2,200 collective farms (*spoldzielnie*). State farms held about 20 per cent of the available arable land and collective farms about 4 per cent. The private and socialised farms differed a great deal in size. Most of the family/private farm-holdings (about 35 per cent) were between 2 and 5 hectares and only 6 per cent had more than 15 hectares of arable land. The average size of family farms was about 7 hectares while those of state and collective farms were, respectively, slightly more than 4,000 hectares and about 350 hectares.

Under communism, the state had favoured state-owned and collective farms by both supplying the means of production and offering tax breaks, cheap credits, etc. (Kuczynski 1981; Szurek 1987; Gorlach 1989). Family farms were forced to counterbalance the disadvantages of state policy by intensive labour. In this sense, Polish agriculture not only deviated from most other communist countries in the extent to which land-holding remained private, but its labour intensity also rendered it unique in comparison with Western European and North American agricultural production systems.

At first glance, these 'family farmers' might be expected to be the most likely supporters of the movement towards an economic system based on private ownership and free markets. This large community of more than 2 million land-owning households seemed a natural ally for a government seeking support in the early stages of the transformation and appeared to be the best prepared to meet the demands of the private market. This was not a hope expressed only by politicians. Social scientists (e.g., Cohen 1992) also reasoned in the same way as they pointed to privately held farm land, a rudimentary agricultural market infrastructure and a collective memory of entrepreneurship.

Actual developments, however, were quite different. Already by 22 July 1990 the English language weekly *The Warsaw Voice* reported: 'Straggling the roads were over 29,000 tractors, carts, supply vehicles, combine harvesters and another agricultural machines. Police Headquarters noted 927 instances of roads being blocked.' The two-hour action organised by farmers represented the peak of a furious protest wave that had started almost a month earlier. This was the first 'heart attack' faced by the government led by Tadeusz Mazowiecki, the former Solidarity adviser and key negotiator during the round table talks in 1989 and the first non-communist prime minister in Poland since the Second World War. Who could foresee a greater paradox? The largest group of private owners were protesting the policy of a government which was supposed to be leading the country out of communism.

That these developments appear as a paradox is a reflection of the theoretical poverty associated with explaining this transition. Claus Offe (1992: 11) recently characterised the contemporary transformation of East Central Europe as a

> revolution without a historical model and a revolution without a revolutionary theory. Its most conspicuous distinguishing characteristic is indeed the lack of any elaborated theoretical assumptions and normative arguments addressing the questions who is to carry out which actions under which circumstances and with what aims, which dilemmas are to be expected along the road, and how the new synthesis of a post-revolutionary order ought to be constituted, and what meaning should be assigned to the notion of 'progress'.

Thus, Offe (1992: 11–12) argues that 'the rapid flow of events' is 'not guided by any premeditated sequence, nor by proven principles and interests about which the participants would be clear'. David Ost (1991: 4) makes a similar point: 'Social groups in post-communist society do not have a clear sense of what is in their interest and what is not.' One consequence of this is that 'few societal groups have as yet formed strong interest associations with a clear sense of program, and nowhere have they exerted a dominant influence on political life' (Ost 1991: 3).

Ost's point may not hold as well for Poland's farmers as for the larger population. Their experiences under communism and in the Solidarity years suggest that Polish farmers may have had a comparatively well-developed sense of their interests as agricultural producers. This identity, derived partially from peasant tradition and culture, functioned as a mechanism of resistance to communist collectivisation efforts which were all but abandoned in the 1950s. The survival of private property in agricultural production itself indicates the power of the Polish peasantry even under the old regime. Wilkin (1988: 8) writes: 'the determined struggle of the peasants to survive, as also the weakness of collective forms of agriculture, enabled the peasants to force through greater changes in the socialist economic system than other social groups.'

This argument is supported by Stark (1992: 1), who has pointed out that, regardless of intentions, the attempt to implement the 'free market' as a comprehensive blueprint for societal change misses the 'more pertinent comparative lesson' that the 'failure of socialism rested precisely in the attempt to organize all economic processes according to a grand design'. Rather than duplicating this 'rationalist fallacy', Stark argues that effective reconstruction must consider the institutional legacies of the old order and of the transition itself. These societies are not 'starting from scratch' but, to draw on Stark's (1992: 4) metaphor of collapse, 'it is in the ruins that these societies will find the materials with which to build a new order'.

The resource mobilisation tradition in social movement analysis provides a useful means of considering the extent to which such 'materials' or resources are available to the Polish peasantry in their attempt to 'build a new order'. Among these materials are the networks established by pre-existing organisations that provide the linkages among this category of the population that are the basis for representing interests (Tilly 1978). Polish peasants had a distinct party during the communist era and distinct organisations within the Solidarity movement. These institutions provided some of the materials that Polish peasants brought to the post-communist era that facilitated their capacity to represent their particular interests more clearly. Nor were Polish peasants operating in a theoretical vacuum, as suggested by Offe. These organisations, perhaps especially in the 1980s, developed a set of cognitive frameworks with which peasants could analyse the changing conditions. Indeed, in this sense, the Polish peasantry existed outside the grand design that constituted the 'rationalist fallacy' of planning that Stark points to as a cause of the failure of communism. To the extent that the Polish peasantry escaped this logic, they were better prepared to meet the challenges of the transition, even if, ironically, one of their tasks was to challenge the transition itself. Their daily lived experience reflected the conditions of private owners, who, despite a largely state-directed and controlled economy, still had to defend their own private economic interests.

Further, the Polish peasantry had an additional political resource insofar as election of political representation from their place of residence reflected the interests of a comparatively homogeneous occupational category in contrast to the more diversified urban and suburban populations. Thus, agricultural interests and the attendant internal class-specific interests could be more readily articulated in political discourse than among urban populations whose representation consisted of heterogeneous occupational categories and interests.

Thus, while transitional processes surrounded the Polish farm in the 'freeing' of both political processes as well as upstream and downstream markets, Polish agriculture was not subject to the severe 'shock' of privatisation in the means of production. In this sense, the 'collapse' is not as total as in the larger economy. However, the post-communist experience suggests that the opening of both economic markets and the political sphere has severely challenged peasant unity derived from the communist period.

This chapter briefly reviews Polish agriculture under communism and shows that there was already an emergent fragmentation of interests forming in the 1980s under Solidarity. The bulk of the chapter addresses the deeper schisms reflected in subsequent protests and mobilisations. These cleavages emerged with the diversification of networks and associations that themselves reflect the various specific interests in the 'peasant community' and with the representation of those interests in the political and economic spheres. Following this, we turn to a brief description of peasant politics in the first years of transitional socio-economic programmes that were developed by the five most influential peasant organisations at that time and examine the conditions associated with their emergence.

Defending peasants' interests under communism

The political process of collectivisation in the late 1940s and early 1950s was linked with certain measures intended to hasten the disintegration of the peasantry by setting various groups with differing socio-economic interests against each other. However, all these attempts failed:

> [The] attempt to break up the peasantry from within, by emphasizing differences in socio-economic interests between various class groups within the peasantry . . . overlooked the extent of peasant solidarity and the mistrust which originated before the war, during the Nazi occupation and the difficult post-war years.
> (Golebiowski and Hemmerling 1982: 28)

The shaky foundation of the collective farms was clearly revealed by the almost instantaneous break-up of most collective farms after October 1956 in the first phase of de-Stalinisation. Eight thousand of the 10,000 collective farms in existence before the October 1956 seventh plenary assembly of the Central Committee of the Polish United Workers' Party (PZPR) quickly ceased to exist. Most of the remaining 2,000 had been broken up by 1971. Only those collective farms which had been independently established by peasants themselves, and were economically viable, survived.

The vacillations and ambiguities in the agricultural policy of the 1960s and 1970s are easier to understand as the result of two contradictory forces. First, authorities needed to assure food supply and this could not be managed if collective farming was fully enforced. On the other hand, the authorities never abandoned their vision of collective farming. The outcome was a policy of 'increasing production in the peasant sector, without the development of the sector' (Kuczynski 1981: 47). By the late 1970s an interesting form of peasant protest emerged. Two small, local organisations appeared in the Lublin area and the Warsaw region calling themselves, respectively: Defence Committee of Believers and Peasant Self-Defence Committee of Grójec Region (Figure 11.1). The first group focused on broader issues concerning the right of believers to express their faith openly and to build churches. The second focused primarily on peasant economic interests. These organisations emerged as a result of the influence and activity of democratic opposition groups that increased in Poland after the 1976 crisis (Bernhard 1993). These first communist-era independent peasant organisations already showed a complex multi-dimensional character, combining occupational as well as national and religious interests. Peasant identity was grounded not only in farm ownership but also other identities, such as being Polish and Catholic.

This reveals a change in the forms and tactics of peasant protests. In a highly oppressive system (i.e., during the Stalinist collectivisation period), passive forms of resistance had been dominant. In the less oppressive conditions of the 1970s more 'active' methods were introduced in the peasant component of the democratic opposition movement, i.e. forming independent (not controlled by the state) organisations and in taking such actions as dumping apples on the roads in the Grojec region in late 1979. This period also consolidated a bridging of nationalist and religious commitments among the peasantry while reproducing a deep mistrust of the state. Religious symbols played an important role in peasant protest in both the 1960s and 1970s (see Gorlach 1992). To draw upon Scott's (1990) concept of 'hidden' and 'public' transcripts: religious values and activities became a kind of 'substitutive public transcript'. Basic elements of the transcript containing religious, nationalist as well as peasants' occupational discourses shaped peasant consciousness as Solidarity emerged in 1980–1. When the events of summer 1980 opened the public space for a relatively free political discourse, this led to the formation of a rural Solidarity programme in which national and religious problems blended with specifically agricultural issues.

Farmers' Solidarity was founded in the autumn of 1980 as part of the larger national movement challenging the communist system in Poland. The significance of a distinct peasant identity is reflected by the fact that farmers were the only occupational group that founded a separate 'Solidarity' organisation. Peasants' demands in 1980–1 followed those of workers in

Figure 11.1 Selected Polish towns and cities

covering general social and national issues, such as the recognition by the state of the role played by the Church in social life, the recognition of a Christian national tradition, and freedom of religious beliefs. However, there was also a set of demands pertaining to the condition of the peasantry. These demands included:

1 The recognition of peasant agriculture as a permanent and equal element of the national economy, along with securing appropriate conditions for its further development, and blocking the mechanisms of land turnover, which tended to destroy a peasant economy.

2 The securing of favourable management for the peasant economy, its further development through stable supplies of the means of production, and a proper level of income for the peasants.
3 The equalisation of individual peasants with other socio-occupational groups with regard to social benefits.
4 The creation of conditions for the functioning of local administrative and economic structures based on the principle of self-government.

(Halamska 1988: 150)

The most important feature that distinguished the peasant movement was an organisational division inside the movement different from that of 'Workers' Solidarity'. One might say that three different peasants' Solidarities existed in 1980–1, namely Independent Self-Governing Union for Individual Farmers 'Solidarity', 'Peasant Solidarity' and 'Rural Solidarity'. Each group concentrated on different aspects of peasants' demands. The Independent Self-Governing Trade Union for Individual Farmers 'Solidarity' (NSZZ RI 'Solidarnosc') stressed 'entrepreneurial' interests. Peasant Solidarity stressed political action and regional or even locale-specific issues. Rural Solidarity, whose members represented various groups of rural inhabitants (including farm workers on the state and co-operative farms), mainly concentrated on equalising rural living standards with other social and occupational groups.

Each group was particularly influential in different parts of the country. Independent Self-Governing Trade Union for Individual Farmers 'Solidarity' (NSZZ RI 'Solidarnosc') was especially strong in northern and western Poland as well as in the Warsaw region. This reflected the interests of larger and market-oriented private farms in these areas. Rural and Peasant Solidarity groups were present mostly in the eastern and southern parts of the country where smaller and more traditional farms predominate. Both Rural and Peasant Solidarity were more oriented than Farmers' Solidarity towards the peasant political tradition. Their presence was significant in areas of Poland where the first peasant political organisations emerged at the end of the nineteenth century. It was, for example, in the town of Rzeszów that the first peasant political protest in the Solidarity era led to the signing of an agreement with state authorities in February 1981, and it had been there that the Congress that led to establishing the first peasant political party in the history of Poland was held in 1895. Rzeszów thus symbolised a tradition of peasant political participation and highlights the significance of symbols of power and the power of symbols (Kubik 1994). Indeed, one of the cultural resources the Polish peasantry has been able to draw on has been its own role as a symbol of the Polish nation, and this constituted a valuable symbolic resource which the peasantry brought to its mobilisation against both communism and again, ironically, against aspects of the transition.

In May 1981, the Independent Self-Governing Union of Private Farmers 'Solidarity' (NSZZ RI 'Solidarnosc') was formed in Poznan. This organisation was dominated by those forces which were inclined to follow 'the farmerisation path'. As Szurek (1987: 248) stresses:

> When martial law was proclaimed in December 1981 Farmers' Solidarity activities were banned as were many other organisations. However, social and political activity did not cease. A significant part of this activity shifted to semi-legal local structures linked to local Church institutions (parishes) and concentrated on detailed, local affairs and the problems of running farms in hard economic times.

The Jaruzelski regime's policy towards peasant agriculture was 'of a carrot and stick variety' (Korbonski 1990: 274), what Gorlach (1989) has described as an instance of 'repressive tolerance'. On the one hand, martial law required peasant support against strong resistance from workers and the intelligentsia and 'to ensure at least a minimum supply of foodstuffs as part of its attempts to stabilize the explosive political situation and to start the process of normalization' (Korbonski 1990: 276). Elsewhere, Korbonski (1990: 273) adds:

> The record shows that starting with the declaration of martial law, General Jaruzelski has engaged in open flirtation with the peasants. Faced with open hostility on the part of the workers and the intelligentsia, the new leader had little choice but to turn to the one segment of Polish society which did not exhibit outright enmity to his policies. Thus, in the course of 1982 and 1983 the government's ire was directed primarily against the workers and intellectuals, and the regime clearly tried to curry favor with the peasants.

Hence, Polish communists in the early 1980s tried to follow the way of their predecessors who carried out agrarian reform in 1944–5 to deprive their political opponents of peasant support (Gorlach 1989). Authors considering this period of peasant–state relations stress various characteristics. Korbonski (1990: 220) points to indicators showing 'processes which strengthened the relative position of the private sector'. There was a slow but observable growth in the number of holdings above 15 hectares (large farms in Polish terms) as well as an increase in farms between 7 and 15 hectares. Moreover, there was a sharp increase in the amount of farmland transferred from the State Land Fund to the peasant sector. Another sign was the growing mechanisation of individual farms. Politically, the communist-dominated organisation ZSL (United Peasant Party), formed in 1949 as the only official political representation of peasants, took on a new and more visible role.

ZSL influenced the communist recognition of peasant farming in the 1984 constitutional amendment. The state-sponsored Agricultural Circles (KZKiOR) also played a role in the process of price negotiations with government. As Lonc (1992: 221) writes: 'A traditional co-operative style peasant farmers' organisation, Agricultural Circles, was entrusted with the role of a trade union and as farmers' representative in price and income negotiations with government agencies.'

However, some trends also ran counter to this 'farmerisation' path. In 1984 and 1985 the amount of farmland transferred from the State Land Fund to the peasant sector decreased significantly. While Korbonski (1990: 270) claims that the reasons for this 'are not entirely clear', they in fact lay in the larger context of communist policy towards peasant agriculture and its 'stick and carrot' or 'repressive tolerance' character. Much less intense use of fertilisers on individual farms as well as smaller deliveries of quality feedstuffs to the peasant sector by the state agencies were also a part of the agricultural picture in mid-1980s. Despite the visible role of some official peasants' organisations like ZSL or Agricultural Circles (KZKiOR) there were no fully autonomous official peasant structures involved in politics: 'the peasants had no authoritative spokesman of their own and remained largely at the mercy of the bureaucracy which controlled just about every aspect of agricultural production and distribution' (Korbonski 1990: 271).

In sum, anti-peasant attitudes dominated communist agricultural policy throughout the 1970s and 1980s. This was so despite moments of pro-peasant policy and public claims about the importance of peasant farms. These minor concessions and empty proclamations were both designed to pre-empt more serious peasant opposition. As a result, peasant farms remained underprivileged with respect to access to the means of production since both official organisations, ZSL and KZKOiR, were dominated by communists and had relatively little autonomy to respond to peasant needs or market signals.

Peasant politics under the transformation

After the elections of 4 June 1989 the situation changed rapidly. Tables 11.1–11.4 outline the conditions of the post-1989 Polish peasantry. These were the effects of the systemic economic reform of 1990 known as the 'Balcerowicz-Sachs Plan'. First, with respect to 'terms of trade': prices of agricultural commodities increased 3.8 times in 1990, while prices of products bought by peasants increased 7.6 times (Dzun 1993: 41–2). Hard times for peasants were also visible through the declining value of products sold both by private as well as co-operative and state farms (Table 11.1).

The declining value of products sold directly resulted in declining peasant income (Table 11.2).

Table 11.1 Prices of products sold by farms, 1989–92

Types of farms	1989	1990	1991	1992	
	Previous year =100			1985 = 100	1990 = 100
State	94.4	93.1	80.8	54.4	59.1
Co-operative	97.9	82.5	81.0	55.8	69.4
Industrial	91.6	85.1	83.4	65.1	79.9
Total	92.5	86.8	82.7	62.6	74.8

Source: *Glowny Urzad Statystyczny* (Statistical Yearbook) 1993: Table 45 (p. 457).

Table 11.2 Individual (peasant) farm income, 1989–92

Type of allocation	1989	1990	1991	1992	
	Previous year = 100			1985 = 100	1990 = 100
Consumption	114.8	48.5	82.9	58.6	95.9
Modernisation of housing	88.1	91.0	64.2	48.0	56.8
Modernisation of farms	87.5	70.9	34.9	18.7	27.6
Total	113.6	48.6	73.9	44.7	75.1

Source: *Glowny Urzad Statystyczny* (Statistical Yearbook) 1993: Table 28 (p. 440).

These figures indicate that since 1990 farming families (in private farms) experienced a dramatic decline in income. As a result, the whole pattern of income allocation in peasant households has changed. In 1990 this was especially visible in terms of consumption. The increase in 'consumption income' in 1991 and 1992 resulted in a strong decline in income allocated to the modernisation of both housing and farms. Thus, the struggle for survival impeded the process of farm modernisation demanded by the new economic conditions. Peasant dissatisfaction and frustration that led to the first waves of protests in 1990–3 are rooted in this declining income.

Tables 11.3 and 11.4 show more clearly the changing levels of income among various groups of Polish society and its significant decline in the years 1990–2.

In this new situation, *political* and administrative pressure on individual and group initiative was severely reduced. However, producers now faced a different type of constraint in the form of *economic* competition and its demands for flexibility and adaptability. These latter pressures have been

Table 11.3 Changes in real income (per capita) in families

Year	Type of family			
	Workers	Peasants	Peasant-workers	Retired
1989	100	100	100	100
1990	66	75	71	72
1991	66	62	60	90
1992	66	57	52	73
1993[1]	67	64	59	87
1994[1]	77	72	70	101

Source: Dziurda (1993).
Note
1 Projections.

Table 11.4 Incomes in families (retired = 100)

Year	Type of family		
	Workers	Peasants	Peasant-workers
1989	164	168	171
1990	136	137	148
1991	109	87	102
1992	119	83	105
1993[1]	126	83	105
1994[1]	140	81	112

Source: Dziurda (1993).
Note
1 Projections.

increasingly important as the market permeates various areas of socio-economic life.

The second group of factors influencing recent changes in East Central European societies is grounded in the legacy of the past (Stark 1992). These changes are taking place within the context of a cultural heritage that contains the experiences, customs, habits and patterns of activity developed during the years spent under the logic of the communist system. These, too, are the 'materials' which peasants bring to bear on the construction of a new agricultural politics and economy.

The legacies of the declining communist system are discussed by Mokrzycki (1992). He stresses that communism left a clearly defined structure of material interests that functions as a serious barrier to economic and political change. Among the agents of these interests are workers as well as the intelligentsia and peasants. However much these major classes and/or strata of the former communist society opposed the state, they were, in fact, deeply dependent on its policies. Their material interests lay in state assistance to industry and agriculture, or in state sponsorship of arts and culture. As Mokrzycki (1992: 274) writes with respect to private farming:

> One of the goals of the reform, assumed and often formulated by the government in early 1990, was to activate a positive selection of farms. It was assumed that larger, stronger, more modern, and more productive farms, would better adjust themselves to the reform than would the small ones, which use a primitive technology and are marked by a low productivity. It has turned out, however, that so far just the reverse has proved true. The error of the assumptions was the underestimation of the connection between more developed farms and their socialist milieu. Unlike the autarkic primitive farms, the more developed and, in particular, specialised ones were adjusted to the cheap credits, to centrally fixed and relatively stable prices of supplies and agricultural produce, and above all to an economy marked by chronic shortages, which made the purchase of means of production extremely difficult but guaranteed the easy and centrally organised sale of all products, of virtually any quantity and in any amount.

These economic and political changes in Poland immediately brought the first of several waves of peasant protests. We turn now to a brief description of the more significant events between September 1989 (when the first non-communist government emerged) and September 1993 (the time of the parliamentary election which marked the emergence of the first government formed by post-communist forces, i.e., Social Democracy (former Communist Party) and Polish Peasant Party (PSL, the former United Peasant Party – ZSL).

By the beginning of 1990, the 'honeymoon' between peasants and the new communist government was already over. Peasants protested the liberal economic policy introduced by the Mazowiecki government (commonly known as the 'Balcerowicz-Sachs package') and the expected sharp increases in prices for machinery, chemicals and other means of production. In January 1991, peasants protested again, demanding minimal guaranteed prices for basic commodities and a system of cheap credit as well as protection of domestic producers against massive food imports.

In August 1991 in the historic town of Zamość, in eastern Poland, a new phase of peasant protest began. Peasants staged a spectacular form of sit-in, living in tents in the town centre to protest high interest rates and demanding postponement of debt repayment to a local bank. This protest was supported by PSL and NSZZ RI. On 4 October, this protest was extended to a national scale. The National Protest Committee on Indebted Farmers formed and began negotiations with the Minister of Agriculture. Searching for a compromise, the Minister of Agriculture sent a letter to the agricultural banks asking them to stop foreclosures on farms where owners had used their credit strictly in line with bank contracts. On 18 October, some of the peasants protesting in Warsaw established the Committee of Farmers' Defence, later called 'Self-Defence' (Samoobrona). Its leader Andrzej Lepper argued that 'We are neither a trade union nor a political party but a peasants' social movement emerging on the basis of peasants' discontent.'

In autumn 1992, the political mobilisation of farmers seemed to be declining. Farmers became less inclined to join protest actions and less receptive to calls for demonstrations. They seemed to be frustrated with the ineffectiveness and the impossibility of co-operation between NSZZ RI, KZKiOR and 'Self-Defence' (Samoobrona). Despite several waves of farmers' protests the economic situation showed no significant signs of improvement (see Tables 11.1–11.4). On the other hand, an increasing sense of the efficacy of political parties was visible. According to a national survey in November 1992, the majority of farmers (54 per cent) recognised peasant parties and politicians as the best chance to improve their situation. Only 18 per cent of peasants pointed to the spectacular protests and not more than 5 per cent saw the violent actions as a way of improving the farmers' situation (CBOS 1993b: 12). This added to the increasing significance of the Polish Peasant Party (PSL) and, at the same time, forced the existing trade union-type organisations in the countryside to form 'their own' political parties. Former leaders of NSZZ RI formed the Peasant-Christian Party (SL-Ch) and the Polish Peasant Party 'People's Agreement' (PSL-PL) while Samoobrona established the National Committee of Self-Defence. The aim was to organise under one umbrella the social and political forces which rejected the existing state, Parliament and main political parties, both Solidarity-rooted as well as post-communist ones. In turn, KZKiOR began to co-operate closely with the Social Democratic Party (SdRP), i.e., the former communists (PZPR). We turn now to a brief description of these organisations and their programmes.

Major peasant organisations and their programmes for transition, 1989–93

The Polish Peasant Party (PSL, formerly the United Peasant Party) stresses that in the process of integration with Western Europe, Poland should protect its own economic interests. It advocates an active state role in the economy, protecting domestic markets against foreign producers and products by means of customs barriers, quotas, etc. The PSL would use the state to guarantee minimal prices for agricultural commodities to protect

Polish farms as they face rapid changes in the market economy. However, these prices should be linked to many other instruments of state agricultural policy.

The legitimacy of PSL has been problematic as a result of its association with 'those' who collaborated with the communists under the previous regime. However, PSL has some significant organisational advantages in its capacity to mobilise farmers. In each of Poland's forty-nine voivodships, a party structure exists.[1] Many PSL members are involved in local self-government structures or in local economic institutions and organisations. Their political opponents point to this fact in referring to them as 'rural *nomenklatura*'. Moreover, PSL still has about 200,000 members, far more than any other peasant organisation.

The second organisation, Polish Peasant Party-Peasant Agreement (PSL-PL) is the primary structural remnant of Solidarity in the peasant political movement. PSL-PL shares many views with PSL, its post-communist adversary. It was formed as a political branch of the Self-Governing Independent Trade Union for Individual Farmers 'Solidarity' (NSZZ RI). Its leader, Gabriel Janowski, entered the Cabinet in 1992 as Minister of Agriculture in the government led by Hanna Suchocka. Like PSL, leaders of PSL-PL also stress the necessity of protecting Polish economic interests before integration with Western Europe begins. They also contend that the weaknesses of many Polish family farms (small acreage, low level of mechanisation, traditional methods of farming) should become their strength. As Janowski stressed in a television broadcast: 'The taste of home-smoked ham is much better than a processed one.' Polish farmers, according to him, should compete on the international market selling high-quality food, produced in more expensive but natural ways.

There are no reliable data on the number of members belonging to the PSL-PL. According to the party, it has about 400,000 members. However, it was unable to mobilise them in the September 1993 parliamentary elections, collecting less than 2 per cent of votes and showing a lack of support even in rural areas. It seems to be merely a small group of relatively ineffective activists or what Poles call a 'sofa' party.

The third organisation, the Christian Peasant Party (SL-Ch), stresses the necessity of integration with the countries of the European Union. In their view, the Polish state should not be overprotective and the import of foreign commodities should be encouraged to force Polish farmers to become more competitive. State agricultural policy should facilitate change in Polish agriculture and not protect individual farmers at any price. However, leaders of the party argue that a minimal prices programme could cause sharp increases in food prices that would reinforce anti-peasant attitudes among consumers. Further, the state should also encourage privatisation in the food processing industry. This solution is seen as best for farmers, especially for those who are interested in enlarging and modernising their farms. This

275

would also create new jobs in rural communities, though not in agriculture. However, lack of mass support in rural communities is also visible in the case of the Christian Peasant Party. In 1991 it had about 15,000 members.

The Farmers' Trade Union 'Self-Defence' (Samoobrona) officially registered in January 1992. Samoobrona argues that Poland should reach the economic level of Western European countries before integration proceeds. Polish agriculture should be rejuvenated after the first four years of Solidarity governments and liberal economic policy. Samoobrona activists introduced the idea of 'profitable prices' for agricultural commodities into political discourse. This means that the state should guarantee profit for every farmer in Poland.

Samoobrona reflects the strong populist component in the contemporary peasant movement in Poland. The leader of Samoobrona, former Communist Party (PZPR) member Andrzej Lepper, graduated from agricultural high school, worked on a state farm, and then operated a comparatively large but debt-ridden farm in Zielnowo, Pomerania. Most leaders of Samoobrona are also heavily indebted to banks. Between autumn 1991 and May 1993 Lepper and his colleagues organised ten protest campaigns: the hunger-strike in front of the Sejm (Polish Parliament) building in November 1991, three sit-in strikes in the Ministry of Agriculture, three road blockages, and three demonstrations in Warsaw. In March 1992, during the first Samoobrona actions in Warsaw they organised 'anti-execution' or anti-foreclosure groups in order to defend farmers whose farms were offered for sale by auction because of bank debt. Lepper claimed that Samoobrona is an organisation for people being harmed and claims a membership of about 300,000 people. 'Samoobrona's roots are the same as Solidarity's. The cause of Samoobrona's emergence lies in the arrogance of the power elite,' Lepper told *Gazeta Wyborcza* (15 August 1992). Some violent actions organised by Samoobrona are similar to those of radical French farmers' organisations.

The National Union of Agricultural Circles and Agricultural Organisations (Krajowy Zwiazek Kolek i Organizacji Rolniczych: KZKiOR) was formed by the communist bureaucracy in the 1970s as an umbrella for controlling social and economic organisations. In 1980–1 it was presented by communist authorities as a trade union alternative to the Farmers' Solidarity movement. Under martial law and later in the 1980s KZKiOR became part of a trade union movement called National Agreement of Trade Unions (Ogolnopolskie Porozumienie Zwiazkow Zawodowych or OPZZ), a trade union that was accepted by communist authorities. At the same time Farmers' 'Solidarity' as well as mainstream 'Solidarity' were dissolved by the new law on trade unions in late 1982. In fact, KZKiOR became an organisation for the 'agricultural *nomenklatura*', i.e., mainly for activists and staff of various state-sponsored agricultural organisations. After 1989, it still remained the partner of OPZZ, urging protection of Polish agriculture in the face of the increasing competition and hardship that resulted from the development of markets.

Each farm organisation discussed so far can be characterised in terms of concepts of 'distributive politics' and 'symbolic politics' (Molotch 1976) (see Table 11.5). Distributive politics refers to the socio-economic programme of each organisation while symbolic politics refers to its political tradition or ethos. Three types of 'distributive politics' can be distinguished: 'farmerisation', 'third way' and 'fossilisation'. Farmerisation refers to the idea of large private farms as the dominant form of agriculture and minimal involvement of the state in markets. Third way refers to private farms as the basic form of agriculture, although other forms such as co-operatives and state farms are also acceptable. This includes an active state policy promoting the quality of agricultural commodities and protection of smaller and more traditional farms in the face of market competition. Fossilisation represents the idea of state support for any farm at any cost in the name of social justice and people's rights.

Symbolic politics also may be divided into three types. The first represents the political tradition of Solidarity as carried out by the organisations that resulted from its fragmentation. The second reflects the political tradition of 'People's Poland' carried out by stigmatised organisations that took part in official political life under communism. Finally, post-Solidarity politics are carried out by organisations which emerged in protest against both Solidarity and post-communist politics in the period of rapid social and political transformation.

Discussion

Peasant protests in the years 1989–93 can be divided into two periods that we could call 'Solidarity' (1989–91) and 'post-Solidarity' (1992–3). Both these periods are characterised by local as well as national protests of

Table 11.5 Peasant organisations

Symbolic politics (ethos)	Distributive politics (socio-economic programme)		
	Farmerisation	'Third way'	Fossilisation
Solidarity	SL-Ch	PSL-PL NSZZ RI	x
Post-communist	x	PSL	KZKiOR
Post-Solidarity	x	x	Samoobrona

farmers. However, they were slightly different in terms of the types of claims made, the methods of protest used, and the frames of meaning deployed. In the first period, in which NSZZ RI 'Solidarnosc' was the leading force, farmers became more focused on changes in agricultural policy with respect to creating better conditions for farms. They advocated cheap and accessible credits, tax breaks, fixed prices for agricultural commodities, etc. However, in 1992, as Samoobrona became the leading force of peasant protests, the number of demands concerning material compensation, clearance of farm debts and favourable conditions for every farmer rapidly increased. Demands for change in general economic policies also increased in that year, as a reflection of Samoobrona's concentration on material compensation and economic equality for all 'hard working people'.

In the first period, various forms of non-violent actions such as marches, demonstrations, sit-in strikes and road blockades constituted the peasants' repertoire of collective action. In the second period, however, hunger strikes were added, as were confrontational tactics that moved beyond the traditional non-violent Solidarity ethos: clashes with police, violent attacks on court officers, Molotov cocktails, etc.

This constituted a 'frame transformation' (Snow *et al.* 1986) and a shift in the dominant collective identity associated with the protest actions. In the first period the Solidarity ethos dominated. This meant that farmers were determined to avoid violence during their protests and that they also pointed to the need for social justice. This need shaped the demand for favourable farming conditions, while stressing deep nationalist traditions and religious sentiments long associated with peasant movements after the Second World War. In the second period, however, populist attitudes towards the state and rejection of the Solidarity ethos dominated. Samoobrona developed the idea that both former communist apparatchiks and Solidarity leaders were united 'against the common people', and the Solidarity government's advocacy of free market policies was described as a legacy of communist government tricking citizens and neglecting their needs.

These identities are also seen by the character of their demands. Most demands were made directly in the name of protesting groups or in the name of farmers. This reflects a strongly formed consciousness concerning the specificity of peasant 'class' interest, especially visible in the first period of protests. In the second period, however, there was an increase in demands made in the name of 'nation' and/or 'society', as Samoobrona's populist and 'post-Solidarity' strategy to gather all the 'deprived', 'dominated' and 'tricked' 'common people' under its umbrella emerged.

The second period of peasant protests might be described as a microcosm of the whole struggle conducted by democratic 'opposition forces' against the communist state. The rejection of both the state and the idea of 'an alternative society' (the Samoobrona concept developed at the very end of 1992) is reminiscent of the 'Self-Defence Committee' (KSS 'KOR') programme in the late 1970s. Even the name Samoobrona 'bridges' (Snow *et al.* 1986) these two organisations.

However, the second period of peasant protest, despite its radical and sometimes furious moments, also saw diminishing peasant protest. This is confirmed in a sociological survey in which the majority of farmers showed no interest in protest actions (CBOS 1993a: 12), but were mostly interested in the functioning of local self-government (CBOS 1993b: 6). In the 1993 parliamentary elections peasant Solidarity organisations as well as Samoobrona were soundly defeated by the PSL. Thus, peasant protests faded as their claims were institutionalised within democratic political processes.

Peasant collective identity in public space between 1989 and 1993 revolved around a polarised 'master frame' (Snow *et al.* 1986) that included a *legitimate ethos* and *illegitimate institution.* The legitimate ethos consisted of 'Solidarity values' developed in the process of challenging the communist system and struggling for 'freedom', 'human rights', 'democracy', against 'evil', etc. This ethos was carried by the 'Solidarity' trade union and political groupings, like: Peasants' 'Solidarity' (NSZZ RI), Polish Peasant Party-'Peasant Agreement' (PSL-PL), and Peasant-Christian Party (SL-Ch). Peasant protests were usually organised spontaneously by informal and local groups of farmers and by trade unionist peasant organisations. National protests like road blockades or mass demonstrations in Warsaw were led by NSZZ RI 'Solidarnosc' or Samoobrona or even KZKiOR. However, despite some efforts in 1993, peasants were not inclined to co-operate on the national level. This state of affairs reflected a 'symbolic politics' that was more important among leadership than among rank-and-file members. A symbolic politics that stresses tradition and ideology has been particularly useful to leadership whose interest is in building collective identities.

Despite the differences among the organisations under consideration, all of them stressed that they were a 'trade union' rather than a 'political party'. Even Samoobrona, which evolved to political party status by the end of 1992 following protests in the spring of that year, denied accusations by the Minister of Agriculture that it was a party-like institution reaching out for power. This phenomenon reveals a very important aspect of Polish public discourse in which there has been a clear distinction between trade unionist and party politics, a legacy of Solidarity political culture as well as evidence of 'anti-politics' in public discourse (Ost 1991).

For these reasons, peasant political organisations did not take part, at least formally, in the protests that took place between 1989 and 1993. They saw themselves as a political subject integrated into institutionalised struggles for power through democratic procedures. PSL seems to be the only example among the peasant organisations which defined itself as a political party from the beginning of the period under consideration. Other parties, like PSL-PL and SL-Ch, which were formed during this period, while supporting some peasant protests, decided not to be officially involved. Instead, they sought to institutionalise peasant protests by putting protesters' claims on the parliamentary agenda. This was especially true of PSL, which stood in opposition to the governments led by former Solidarity advisers through the years 1989–93. PSL used peasant dissatisfaction and demands to build support for its local structure in the September 1993 parliamentary elections and establish legitimacy despite its 'communist past', a strategy that proved successful as the results of the elections have shown. PSL was able, together with the post-communist Social Democratic Party (SdRP) and other post-communist organisations (mainly OPZZ and KZKiOR), to form a stable parliamentary majority and a new government.

Conclusion

We have argued that these events showed that the Polish transition is neither smooth nor simply the result of a 'grand design' shaped by International Monetary Fund and World Bank experts. Polish peasants reacted quickly to the economic package known as the 'Balcerowicz Plan' introduced in January 1990 by the first non-communist government. This indicated that the transition would follow a painful path that derived from both govern- ment policies and responses from various social groups. The question asked by Ost (1991) is whether these groups have clear visions of their preferences and interests that are able to guide their behaviour in response to economic, social and political changes.

This analysis challenges Ost's and Offe's statements on the lack of clear interests or groups in post-communist society. Polish peasants, at least, have shown clear interest in preserving their farms. They have also exerted a dominant influence on political life. The strength of PSL's new legitimacy was manifest clearly during the 1993 parliamentary elections and consolidated in the 1995 presidential election. This revealed that the majority of Polish peasants, holding relatively small and traditionally run farms, could mobilise politically to vote for the organisation which in their opinions best represented their collective interest.

The ideological and political changes in the peasant movement seem to be signs of a search for the best defence of their material interests. Despite some dramatic performances, especially by Samoobrona, mainstream peasant politics aimed at establishing protection of family farms. The fragmentation of Solidarity in the countryside and its declining significance (populist Samoobrona succeeded NSZZ RI 'Solidarnosc' in leading peasant protest) has been one dimension of this change. At the same time the growing significance of post-communist PSL confirms that the majority of peasants effectively sought and finally found the institutional framework to express and defend their basic interests. But the fragmentation of the Solidarity movement in the countryside also confirms this trend in a slightly different way. The split in the NSZZ RI between the SL-Ch and PSL-PL was based on material interests. SL-Ch became the institutional framework for a minor group of large, successful and entrepreneurial family farmers while the PSL-PL represents the interest of small-holders. All this indicates that peasant politics under the transformation has been, quite contrary to Ost's claims, directed by clear material interests of peasants. Further, peasants had existing organisational 'materials', networks and experiences with which to engage in the rebuilding of the Polish state and society. Some of these derived from the long period of communist domination and some derived from the Solidarity era. Contrary to Offe's claim of an ideological vacuum, these peasants drew on a variety of ideological frameworks that enabled them to diagnose the conditions they confronted and to frame their interests politically according to a diversity of specific material interests, as indicated by the rather rapid coalescence of peasant groups around different political organisations. As the process of economic transformation continues to make the stratification system in the countryside more complex, it may be expected that these political formations will become even more firmly established on the Polish political stage.

Note

1 Voivodship (*wojewodztwo*) is a middle level administrative entity. Poland is divided into forty-nine voivodships. The head of each is appointed by the Prime Minister and acts as a representative of the national government. Each voivodship consists of a certain number of 'counties' (*gmina*). Each *gmina*'s council is elected every four years by local residents. They, in turn, appoint the *gmina*'s administration.

Bibliography

Bernhard, M. H. (1993) *The Origins of Democratization in Poland. Workers, Intellectuals, and Oppositional Politics, 1976–1980*, New York: Columbia University Press.

CBOS (Centre for Public Opinion Research) (1993a) 'Strategie adaptacyjne rolnikow' (Farmers' adaptation strategies), Warsaw: mimeo.

—— (1993b) 'Rolnicy indywidualni o instytucjach zycia spolecznego i politycznego wsi polskiej' (Individual farmers on institutions of social and political life in rural Poland), Warsaw: mimeo.

Cohen, S. (1992) 'Staggering toward democracy. Russia's future is far from certain. A roundtable discussion', *Harvard International Review* 15 (2): 14–17.

Dziurda, M. (1993) 'Dochod na glowe – na leb, na szyje' (Per capita income – the rapid decline), *Tygodnik Solidarnosc* 27 (July): 2.

Dzun, W. (1993) 'Spoleczno-ekonomiczne aspekty rozwoju rolnictwa w warunkach gospodarki rynkowej' (Socio-economic aspects of agricultural development in market economy), in M. Wieruszewska (ed.), *Wies i rolnictwo na rozdrozu* (Agriculture and countryside at the crossroads), Warsaw: IRWiR PAN.

Gazeta Wyborcza (1992) 15 August.

Glowny Urzad Statystyczny (GUS) (Central Statistical Office) (1993) *Rocznik Statystyczny* (Statistical Yearbook).

Golebiowski, B. and Z. Hemmerling (1982) 'Chlopi wobec kryzysow w Polsce Ludowej' (Peasants in the crises of the People's [Republic of] Poland), *Kultura i Spoleczenstwo* 2: 21–150.

Gorlach, K. (1989) 'On repressive tolerance: state and peasant farm in Poland', *Sociologia Ruralis* 26 (1): 23–33.

—— (1992) 'The vanishing point: the religious components of peasant politics of protest in Poland, 1970–1980', in B. Misztal and Anson Schupe (eds), *Religion and Politics in Comparative Perspective: Revival of Religious Fundamentalism in East and West*, Westport and London: Praeger.

Halamska, M. (1988) 'Peasant movements in Poland, 1980–1981: state socialist economy and the mobilization of individual farmers', in L. Kriesberg, B. Misztal and J. Mucha (eds), *Social Movements as a Factor of Change in the Contemporary World. Research in Social Movements, Conflicts and Change. A Research Annual*, vol. 10, Greenwich, CT: JAI Press.

Korbonski, A. (1990) 'Soldiers and peasants: Polish agriculture after martial law', in Karl-Eugen Wadekin (ed.), *Communist Agriculture: Farming in the Soviet Union and Eastern Europe*, London and New York: Routledge.

Kubik, Jan (1994) *The Power of Symbols against the Symbols of Power*, State College, PA: Pennsylvania State University Press.

Kuczynski, W. (1981) *Po wielkim skoku* (After a great leap), Warsaw: PWE.

Lonc, T. (1992) 'Social and economic adjustment and policies in a period of crisis and emergence from it. The experience of Poland in the 1980s', in A. Hakkinen (ed.), *Just a Sack of Potatoes? Crisis Experience in European Societies, Past and Present*, Helsinki: SHS.

Mokrzycki, E. (1992) 'The legacy of real socialism, group interests, and the search for a new utopia', in W. D. Connor and P. Ploszajski (eds), *Escape from Socialism: The Polish Route*, Warsaw: IFiS Publishers.

Molotch, H. (1976) 'The city as a growth machine: toward a political economy of place', *American Journal of Sociology* 82: 309–32.

Offe, Claus (1992) 'Democratisation, privatisation, constitutionalisation', paper presented at Post-Socialism: Problems and Prospects Conference at Charlotte Mason College, Ambleside, Cumbria, July.

Ost, David (1991) *Solidarity and the Politics of Antipolitics: Opposition and Reform in Poland since 1968*, Philadelphia: Temple.

Scott, J. C. (1990) *Domination and the Arts of Resistance. Hidden Transcripts*, New Haven: Yale University Press.

Snow, David A., Rochard, E. Burke Jr, Warden, Steven K. and Benford, Robert D. (1986) 'Frame alignment processes, micromobilization, and movement participation', *American Sociological Review* 51: 464–81.

Stark, David (1992) 'Path dependence and privatization strategies in East Central Europe', *East European Politics and Society* 6: 17–54.

Szurek, J.-Ch. (1987) 'Family farms in Polish agricultural policy: 1945–1985', *Eastern European Politics and Societies* 1: 2.

Tilly, Charles (1978) *From Mobilization to Revolution*, Reading, MA: Addison-Wesley.

Warsaw Voice (1990) 22 July.

Wilkin, J. (ed.) (1988) *Gospodarka chłopska w systemie gospodarki socjalistycznej* (Peasant economy in socialist state), Warsaw: Warsaw University Press.

12

RURALITY AND THE CONSTRUCTION OF NATION IN ESTONIA

Tim Unwin

Transition and uncertainty

This chapter explores the intersection of political and economic interests as they have found their expression in a particular kind of place and time. The place is a farm-yard. But it is a nested hierarchy of farms; it is situated in a rural landscape; it is in Estonia; in the Baltic region; is becoming/has become part of Europe. The theoretical and the empirical nestle together in a forest bed. The world is there; globalisation impinges; capitalism seeks to penetrate; rhetorics of liberal democracy and free market dominate the listened-to spoken worlds of ministries and bars. But it is a world of peat bogs, of dry dusty fields, of fungi and berries, of empty cattle sheds, of foreign tourist voices, of home and away.

Amin (1997: 1) has succinctly captured the increasing confusion and uncertainty associated with differing theoretical positions on the idea of globalisation, when he suggests that 'The more we hear about globalisation from the mounting volume of literature on the topic, the less clear we seem to be about what it means and what it implies'. The same can be said for the mounting volume of literature on transition. However, one cannot begin to understand the idea of transition in Eastern Europe without some comprehension of the extensive array of arguments concerning the global spread of capitalism, its linkage with so-called liberal democracy, and the growing dominance of the principles of the free market. Neither can one begin to understand the dramatic changes that have taken place in the former Soviet Union and Eastern Europe since the mid-1980s without recognising that interpretations of transition reveal as much about those making the claims as they do about the events themselves. To address, for the moment, just the ideas of democracy and revolution, Eisenstadt (1992: 21) has argued that the collapse of the communist regimes of Eastern Europe 'has been one of the most dramatic events in the history of mankind'. However, taking a longer-term perspective of revolution, Tilly (1995: 234) suggests that 'In the perspective of 500 revolutionary years, the collapse of Eastern European regimes loses some of its close-up magnitude'. Similarly, Markoff (1996) suggests that they reflect merely the latest of a series of waves of democracy which have transformed the global political system since the 1780s.

Two key observations can be made in seeking to synthesise these differing interpretations of change. First, the meanings of the words that we use to describe them are themselves constantly changing. As both Tilly (1995) and Markoff (1996) emphasise, core concepts like *revolution* and *democracy* have altered significantly in meaning throughout history, as individuals have sought to transform their social conditions. The events of the last decade have therefore been of critical importance in helping to shape our own under-standings of such processes. However, second, it is important to reflect on the relational and interdependent aspects of these changes. While some may see transformations of government as being essentially political in character, it is important to be cognisant of other social, economic and ideological processes if we are to grasp the full complexity and significance of such change. This chapter therefore focuses on the complex interconnections between political and economic change, by highlighting both the complexity of transition, and also the contested ways in which ideas of nationalism and capitalism have been worked through in the specific arena of Estonia.

Theorising the nation and economy in transition

The concept of *transition* is highly problematic. Transition generally implies change from one known state to another. Even if some agreement can be reached on the character of regimes in Eastern Europe and the former Soviet Union prior to 1990, itself a contentious issue, the final outcome of many of the processes currently under way remains far from clear. More-over, the rhetorics used to describe processes of change owe as much to the wishful thinking of their advocates as to any realities of political or economic practice (Callinicos 1991; Miliband 1991; Burawoy 1992). Whatever the variant perspective, though, most theoretical statements concerning transition make specific claims about the *end characteristics* and *outcomes* of the processes involved. These are widely seen as being the introduction of a so-called liberal democratic political structure and a free-market economy. These twin elements of 'transition' owe much to the shock-therapy model proposed by Sachs (1990, 1995), which, as Gowan (1995: 9) has argued, sought to create a very specific kind of political economy in the countries of Eastern Europe, with 'a state as open as it is possible for it to be to the forces of international economic operators: a state with a globalized institutional structure, through which the resources of what he [Sachs] calls "the global mainstream economy" can flow'.

The wish-fulfilling connections between theoretical discourse and political practice are particularly well expressed in the neo-conservative arguments of Francis Fukuyama (1992). In his book *The End of History and the Last Man*, Fukuyama (1992) employs Hegelian categories to explore the historical victory of capitalist democracy over communism (Pensky 1994). For Fukuyama, as for other officials of capitalist governments and international agencies such as the International Monetary Fund (IMF), it has been important to portray the collapse of communism as being the direct result of capitalism's economic superiority, in order to encourage the newly emerging political regimes of Eastern Europe and the former Soviet Union to adopt the political and economic underpinnings of the capitalist way of life. Significant elements in this package were the economic reviews undertaken by the World Bank and IMF, and the structural adjustment programmes subsequently initiated with their help (for the Estonian example, see Odling-Smee 1992; World Bank 1993; see also Ould-Mey in this volume and Sidaway and Power in this volume). These have been designed to incorporate the economies of Eastern and Central European countries ever more closely into the expanding capitalist world economy.

The empirical experiences of political and economic change in Eastern Europe and the former Soviet Union over the last decade, however, emphasise that the model of transition adopted within this reworked modernisation theory has four significant flaws: first, different states have undergone vastly different political and economic trajectories since 1991; second, it is not yet certain that all states in the region will necessarily become successful capitalist economies, however these are defined; third, the concept of transition fails satisfactorily to account for the contested nature of the changes that have occurred; and fourth, it fails to acknowledge the fundamental tension that exists between the concepts of capitalist modernisation and political freedom. Paradoxically, as Habermas (1994: 85) has argued,

> the more democratic the framework is, the harder it is to carry out the kind of socially and economically painful reform programs that are demanded by the deconstruction of the systems of social benefits and subsidies, monopolistic state industries, the stranglehold of the managerial elites, and so on.

This chapter seeks to explore the tensions between the political and economic change in Eastern Europe through the specific lense of Estonia. However, it grounds this interpretation within three theoretical frameworks: first, those of Habermas (1994) concerning the relationships between capitalist modernisation and political freedom (see also Held 1995); second, debates over the relative roles of national and global processes and institutions (see, for example, Ohmae 1990, 1995; Hirst and Thompson 1996a, 1996b);

and third, the conflicts of interest in the re-emergence of nationalisms in Eastern Europe and their association with capitalist interests (Žižek 1990; Gowan 1995).

Modernisation and political freedom

There is a fundamental tension between the ideas of *capitalist modernisation* and *political freedom* so central to models of transition in Eastern Europe. Habermas (1994: 84), for example, argues cogently that 'the evidence continues to mount that there is no automatic relationship between capitalist modernization and political freedom'. He goes on to point out that in many countries without 'the cultural peripheral conditions for European modernity, capitalist developmental poverty, ecological devastation, political repression, and cultural disintegration braid together into an increasingly desperate feedback loop' (Habermas 1994: 84). This argument raises the fundamental question of whether the countries of the former Soviet Union and Eastern Europe do indeed possess the requisite cultural conditions for positive, rather than negative, feedback between the political and economic structures emerging through their reintegration into the capitalist world economy. Precisely what such cultural conditions are, though, also remains problematic.

These issues have been explored at length by David Held (1995) in his analysis of the relationships between different kinds of democracy and the global order. In this context, his argument that the links between national sovereignty and liberal democracy have generally been taken for granted is highly pertinent. Likewise, his suggestion that 'national communities by no means exclusively "programme" the actions, decisions and policies of their governments and the latter by no means simply determine what is right or appropriate for their own citizens alone' (Held 1991: 142) emphasises that there are important linkages between global interconnectedness and issues of national identity and systems of political representation at the state level. In particular, Held's (1991: 148) claim that 'while there has been rapid expansion of intergovernmental and transnational links, among other things, the age of the nation-state is by no means exhausted' is one that will be explored in this chapter by examining the way in which Estonian national identity is being forged in the context of the aspirations of its ruling elite to integrate the state within international bodies such as the European Union (EU) and the North Atlantic Treaty Organisation (NATO).

What Habermas (1994) emphasises, though, is the need to understand and distinguish between economic and political interests and processes. This is tied up historically with the relationship between different forms of political control and economic activity. It is thus no coincidence that the rise of the nation-state in the nineteenth century was fundamentally connected with the emergent strength of industrial and commercial capitalism at that time. The parallels between processes operating then and the close linkage between *national identity* and free-market *capitalism* in Eastern Europe today are striking. In both cases it is significant to note that rhetorics associated with human 'freedoms' in the political sense were associated with the creation of new class structures in the economic arena, which were not necessarily any more equal or 'free' than those which previously existed. There are resonances here with Théret's (1994) exploration of the articulation of economic, political and domestic orders and levels of practices within Europe, which draws upon regulation theory (see Smith and Swain in this volume; Dunford in this volume). Théret's emphasis that the economic order is essentially concerned with the accumulation of 'things', and the political order with the accumulation of 'power', is thus one that coincides with the central claims of this chapter on the political economy of Estonian transition.

National and global processes and institutions

A second theoretical arena engaged by this chapter concerns the 'hollowing out' of the nation-state; while global and local processes and institutions are gaining in power, the role of the nation-state is declining. As Amin (1997) has emphasised, this widely accepted argument is one that has recently come under considerable criticism. Hirst and Thompson (1996a, 1996b) thus argue strongly that national-level economic processes remain central to the global economy, and that nation-states have a significant role to play in creating and sustaining the governance necessary for the successful functioning of that economy. In their words,

> [t]he central function of the nation state is that of distributing and rendering accountable powers of governance, upwards towards international agencies and trade blocs like the European Union, and downwards towards regional and other sub-national agencies of economic co-ordination and regulation.
>
> (Hirst and Thompson 1996a: 408)

Amin (1997: 4–5), though, remains sceptical of their critique of globalisation, suggesting that it is both 'a denial based on a caricature of the term' and that in their formulation '[s]ight is lost of anything between globalisation as an international trading system and globalisation as the end of the territorial state'. In his critique of Hirst and Thompson's (1996a, 1996b) arguments, Amin (1997) stresses the importance of two phenomena discussed by Held (1995) in his analysis of democracy and globalisation: the increasing global linkages between economic, social, cultural and political activity; and the intensification of connectivities between different states and societies. Whereas Amin (1997) concentrates essentially on the implications of these arguments for urban places and societies, this chapter seeks to explore their resonance among rural peoples and places.

In the Estonian case, what is interesting is the way in which the drive for acceptance within the international order in general, and the European Union in particular, articulated through both economic and political vehicles, has had significant differential influences on people and places within its territorial limits. More specifically, it is argued that the drive for international economic and political acceptance, and the pursuit of rapid integration within the global political economy, has had damaging short-term implications for the viability of Estonia's rural economy and society. This reinforces Gowan's (1995) arguments that the powerful role of western influence in Eastern Europe, while creating so-called liberal democracies and free markets, has also denied the newly emerging states the public subsidies, forms of protection and regulation that exist in most western states.

Nationalisms and capitalist interest

The third theoretical arena engaged by this chapter concerns the types of nationalisms that are emerging in Eastern Europe. Žižek (1990: 51) has thus emphasised 'the dark side of the processes current in Eastern Europe' with 'the gradual retreat of the liberal-democratic tendency in the face of the growth of corporate national populism'.

The debate concerning the role of nationalism in Estonia, and the creation of a so-called ethnic democracy, has been hotly argued (see for example Smith 1994, 1996), and it is not the purpose to extend that debate here. Rather, the chapter seeks to explore some of the cultural resonances that have been used to construct Estonia's national identity, and to illustrate how different groups within Estonian society have sought to emphasise particular themes in that identity in pursuit of their political and economic interests. Žižek (1990: 60) has suggested that 'the re-emergence of national chauvinism in Eastern Europe' is 'a kind of "shock absorber" against the sudden exposure to capitalist openness and imbalance'. This claim re-emphasises the connectivities between the emergence of capitalism and nationalism that have already been alluded to, but it inverts their relationship. Thus, rather than seeing national identity and capitalism as emerging hand in hand, as happened in the nineteenth century, he suggests that vehement nationalism is in some ways a reaction to the introduction of capitalism.

The political economy of Estonia's transition

One of the difficulties in generalising about 'transition' in Eastern Europe is that the paths followed by individual countries, while having many things in common, also have many unique characteristics. Estonia provides an excellent case study of the connectivities between political and economic processes for at least three main reasons. First, it has followed one of the most determined paths towards capitalist modernisation of all the states of the former Soviet Union and Eastern Europe (Lugus and Hachey 1995). The social, political and economic effects of this strategy, and the way that they have led to increasing social and spatial differentiation, are already becoming starkly apparent. Second, Estonia provides a good, and well studied, example of the ways in which a specific new national identity is being created (Lieven 1994; Smith 1996). With its substantial minority population of Russian speakers, the methods through which the Estonian government has sought to introduce a specific kind of democracy to the country are particularly interesting (see Smith 1994, 1996). Third, Estonian politicians have sought to move as rapidly as possible towards integration with the European Union for both political and economic reasons. The implications of such a policy for local social and economic change within Estonia have been of considerable significance in shaping the lives of people in different parts of the country.

In order to understand the contested nature of Estonia's political economy in the 1990s, it is important to situate the discussion in the context of the main political and economic events associated with its independence from the Soviet Union in 1991 (for a full account see Lieven 1994; for an introduction to the Baltic economies see Van Arkadie and Karlsson 1992). This section therefore summarises key aspects of these changes, before their wider historical context is explored in the ensuing section.

The politics of Estonian independence

The reforms of the Soviet political system introduced by Gorbachev in the 1980s provided the immediate context within which the Estonian independence movement was able to vocalise its opposition to Soviet rule. This created an opportunity in which increasing political activity by those opposed to rule from Moscow could develop. In August 1988, the Estonian National Independence Party (ENIP) was formed, with a Popular Front being established two months later. In November 1988, the Supreme Soviet of the Estonian SSR adopted a declaration of national sovereignty, and during the following year the Popular Front's aspiration to declare independence became increasingly widely supported. In February 1990, elections were held to the Congress of Estonia, which was designed to be 'a national parliament untainted by any Soviet origins' (Taagepera 1993: 170). However, during 1990 tensions increased between the Estonian Supreme Soviet and the Congress of Estonia over the precise path by which independence was to be achieved. It was not until the attempted coup against Gorbachev in August 1991, which threatened military intervention in the Baltics, that Estonia's leaders were precipitated into an actual declaration of independence, which was eventually recognised by the Soviet Union on 6 September 1991.

While there was enormous variation in the demands of the different groups seeking Estonia's independence, two central aspirations came to dominate their rhetorics: the creation of a political *democracy*, and a *free-market* economy. The main legal instruments by which these were to be achieved were implemented in 1992: monetary reform and departure from the rouble zone; the approval of a new 'democratic' constitution; and parliamentary and presidential elections. During this period, though, there was apparently negligible recognition among the mass of the population seeking freedom from the Soviet Union of the contradictions that this would involve. Nationalism was not specifically equated with capitalism, and there was little conscious realisation of the problems which submission to globalised capitalism might involve. A common sentiment was expressed as follows: 'We have learnt how to cope with Moscow; we will therefore certainly be able to handle Brussels!'

The precise nature of the 'democracy' upheld in the 1992 Constitution has been the subject of considerable discussion and debate, because of the way in which it addresses the ethnic diversity of people living within the state's boundaries (Park 1994; Smith 1994, 1996). This relates closely to Žižek's (1990) arguments about the characteristics of Eastern European nationalisms noted above. In 1989, according to the census, there were some 121 different nationalities among the 1.5 million people in Estonia. Only 61.5 per cent of the population was Estonian, with 30.3 per cent claiming Russian, 3.1 per cent Ukrainian and 1.8 per cent Belorussian nationality. Article 1 of the 1992 Constitution states that 'Estonia is an independent and sovereign democratic republic wherein the supreme power of the state is held by the people'. Here, quite clearly, the founders of the new state make their claim for democracy.

However, the 1992 Constitution emphasises that only those who were citizens of the inter-war republic or their descendants have an automatic right to citizenship; for others, a period of residency, Estonian linguistic competence and an oath of loyalty are required (Smith 1996). Most of the Russian-speaking population were thus excluded from automatic citizenship of the new state. This has led Smith (1996) to adopt Smooha and Hanf's (1992) term 'ethnic democracy' to describe the Estonian polity.

One of the central political issues facing the new Estonian state was thus the problem of how to incorporate a substantial minority population into its polity in the context of rising claims for an 'exclusive' national identity. Moreover, the tension was heightened because this Russian-speaking minority had previously been the politically dominant force. Estonian governments since 1991 have placed great emphasis on Article 9 of the 1992 Constitution, which states unequivocally that 'The rights, liberties and duties of everyone and all persons, as listed in the Constitution, shall be equal for Estonian citizens as well as for citizens of foreign states and stateless persons who are present in Estonia'. The primary difficulty with this claim is that Article 57 states that 'The right to vote shall belong to every Estonian citizen who has attained the age of eighteen'. Thus, while there is a claim that basic human rights apply to all people within the Estonian polity, a substantial number of Russian-speaking people, who, for whatever reason, have not taken Estonian citizenship, have been excluded from voting in national elections.

Partly as a result of international pressure in the context of Estonia's desire to join the European Union, the Estonian Parliament therefore passed a new Law on Citizenship in January 1995, which was designed to clarify the situation and make it easier for those Russian speakers who so desired to become Estonian citizens. Moreover, the Law on Local Government Elections, while not permitting voting in national elections, does allow resident non-Estonians to vote in local elections, which the central government sees as being the main service providers to the republic's peoples.

Any understanding of the creation of an Estonian national identity following independence in 1991 must therefore take into consideration the tensions between the Russian-speaking population who migrated into the state from the 1940s onwards and Estonians claiming citizenship from the 1930s. However, the intention of this chapter is not so much to dwell on the creation of an Estonian national identity in opposition to an externally imposed Soviet or Russian identity, but rather to understand the way in which this identity has been forged internally from a historical and largely rural past. This highlights a further tension, between those forces supporting largely urban-centred capitalist modernisation and those seeking to emphasise the republic's rural cultural identity. To understand this tension, it is essential briefly to consider some of the key economic policies adopted by Estonian governments since independence.

The economic framework of independence

Economic restructuring since independence has been based on three principles: a stable currency, a balanced budget and liberal foreign trade (*Eesti Pank Bulletin* 2 1996). This policy aims to promote a rapid transformation to a liberal free-market economy, and has been developed in close association with the IMF, once again reflecting the importance of globalised regulatory frameworks in Eastern Europe's economic and social restructuring.

Estonia was the first of the countries in the former rouble zone to implement currency reform, and in June 1992 it pegged its new Kroon to the Deutschmark at a rate of 8:1. Significantly, this rate has never been changed, despite a banking crisis in late 1992 and early 1993 when several of the weaker banks were allowed to fail. A conservative fiscal and monetary policy has thus been one of the cornerstones of Estonia's economic stance since independence. Other central planks of this policy have been the privatisation of state enterprises by international tender, the liberalisation of trade through an emphasis on open international competition, the restructuring of its economy to enhance new private enterprise, and the development of a legislative base to support such enterprises. Underlying such policies are a strict adherence to the principles of competition and government non-intervention in the economy. As a Ministry of Foreign Affairs report has summarised,

> [t]he government's position is that free access to Western markets and the gradual growth of foreign investments will produce more viable long-term results than passively received aid could. State support is seen as not encouraging local economic initiative or guaranteeing the irreversibility of the economic and political reforms of Estonia.

(Ministry of Foreign Affairs, Estonia 1995)

Estonia's dogged determination to follow free-market principles has resulted in considerable flows of foreign direct investment, and a rapid transformation of the economy. However, this has not been without painful costs, and these have been particularly seriously felt in the rural agrarian sector (see Lugus and Hachey 1995; Venesaar and Hachey 1995; CCET 1996). After four years of economic decline, 1995 saw Estonia's first post-reform growth in GDP with an increase of 2.9 per cent over the previous year. This was led mainly by the service sector with a 59.4 per cent share of GDP, but manufacturing also increased its share over the previous year by 0.2 per cent to 19.1 per cent (*Prime Minister's Economic Report* 1996). In contrast, agriculture continued its downward slide to a 9.8 per cent share of output compared with 11.0 per cent in 1993, and 14.9 per cent in 1991. While some indicators suggest that Estonia has turned the corner, and is indeed successfully making the 'transition' to a 'modern' capitalist economy as espoused by the IMF, not all interpretations are so optimistic. Lugus and Hachey (1995) thus highlight five basic problems that have been encountered during the first four years of independence.

First, as with most other states in the region, there has been a dramatic collapse in production across all sectors of the economy as Estonia has sought to reorient its links from East to West (see Dunford in this volume). Second, the agricultural sector has been particularly hard hit as a result of the slowness with which agrarian reform has proceeded, and also because of the competition it has faced as a result of the tariff barriers and other farming subsidies enjoyed by its European competitors (Unwin 1994, 1997). Until very recently, Estonia's post-independence governments have consistently and proudly refused to support the agrarian sector with subsidies, arguing that this would be inconsistent with their avowed policy of maximising competition and minimising state interference in the economy. Third, Lugus and Hachey (1995) point to the rapid turnover in the number of small businesses created since independence, and the considerable difficulties faced in privatising large, technically obsolete state enterprises. Fourth, they emphasise that the privatisation of living space has encountered numerous barriers, largely because of the poor quality and energy inefficiency of housing built during the Soviet era. Finally, they stress that growing income disparities, both between different population groups and between different parts of the country, are generating increasing social problems and tensions.

The remainder of this chapter explores some of the implications of these tensions, particularly as they express themselves in the conflicts of economic and political interest between people living in rural and urban areas, and also in the formulation of Estonia's post-independence national identity. The following section examines the intersection of political and economic interests as reflected in the March 1995 elections to the Estonian Parliament. It highlights the dominance of a liberal-market discourse, even when popular support for parties shifts, particularly towards a coalition superficially representing the initial 'losers' from the economic and social transformation associated with independence. The chapter then moves on to explore oppositions to this globalised political economy through the dynamics of ruralism and national identity.

Elections and the political economy of independence

Estonian politics since independence have been characterised by a large number of parties grouped together in varying electoral coalitions. Following the adoption of the new Constitution, 38 parties contested the September 1992 elections to the 101-seat Parliament (Riigikogu). These elections attracted a turn-out of 67 per cent and were won by the Fatherland (Isamaa) Party with 29 seats, followed by the Secure Home (Kindel Kodu) Party with 17 seats, and the Popular Front with 15 seats. This election victory proved to be highly influential in strengthening the neo-liberal agenda begun at independence, and in aligning the economy ever more closely towards the principles of a free market and minimal government intervention. Lieven (1994: 285) has thus noted that 'Fatherland had stood on a mixture of free-market economics, restitutionalism, and moderate nationalism, under the overall slogan "cleaning house", represented by an election poster depicting a man with a broom'.

The second elections, held in March 1995, suggest, as elsewhere in Eastern Europe, that there were varying levels of support for such policies. The total number of parties contesting these elections fell to thirty. While the Reform Party-Liberals (Reformierakond) gained the greatest number of seats (19) of any individual party, the winners of the election were the Coalition Party and Rural Union (Koonderakond ja Maarahva Ühendus) (41 seats), combining the forces of the Coalition Party (Koonderakond), the

Rural Union (Maaliit), the Country People's Party (Maarahva Erakond) and Farmers' Assembly (Põllumeeste Kogu), and the Pensioners' and Families' League (Pensionäride ja Perede Liit).

The new government appeared at first sight to represent many of the groups, such as pensioners and farmers, who suffered adversely from the economic restructuring programme introduced by the previous government, and made promises to introduce economic and social policies that would be to their benefit. However, the collapse of the government in October 1995, as a result of the involvement of the Interior Minister in a 'phone-tapping scandal, threw these agendas into turmoil. Discussions between the KMU and the Reform Party created a new coalition government which once again reinforced the continuing dominance of free-market economic principles. In part this reflects the widespread acceptance in Estonia of the need to adopt an economic and political strategy that will enable the republic to join the European Union and other international bodies, but it also reflects the personal interests and beliefs of the political leadership. The coalition agreement, for example, emphasised the need to maintain a balanced budget and a stable currency, to remain an open economy, to accelerate privatisation of state-owned property, and to implement land reform. It also specifically stated that '[w]hile developing the state's economic policy, protective tariffs, subsidies, donations or other direct support will not be used' (*Government Coalition Agreement* October 1995). Given this emphasis, disillusionment among many rural voters and the elderly, who had given their support to the KMU earlier in the year, was high.

Figure 12.1 illustrates the spatial distribution of support for the two main adversaries in the March 1995 elections, the Coalition Party and Rural Union (KMU) and the Reform Party (R). Clearly, there are wide differences in levels of support for the two sides in different parts of the country. In general terms, the KMU attracted its highest relative support from the most rural regions, reaching its maximum in Järvamaa and Viljandimaa at 47.2 per cent of the vote, whereas the Reform Party attracted most support from the capital Tallinn.

Such an electoral geography is in part an expression of the direct links between economic interest and political practice that have manifested themselves in Estonia since independence. Those who have benefited most from the economic changes since 1991 have tended to be people living in urban areas and working in the service sector, particularly in the capital region of Tallinn. It is here that support for the Reform Party has been highest. It is not possible, though, simply to equate urban residence with support for economic and political reform. While the rapid spread of producer and financial services, closely integrated with a globalised economy, has been most apparent within the Tallinn metropolitan area, and while the urban-located informal sector has also benefited directly from inward investment, large numbers of poor and unemployed people also live there. There is some

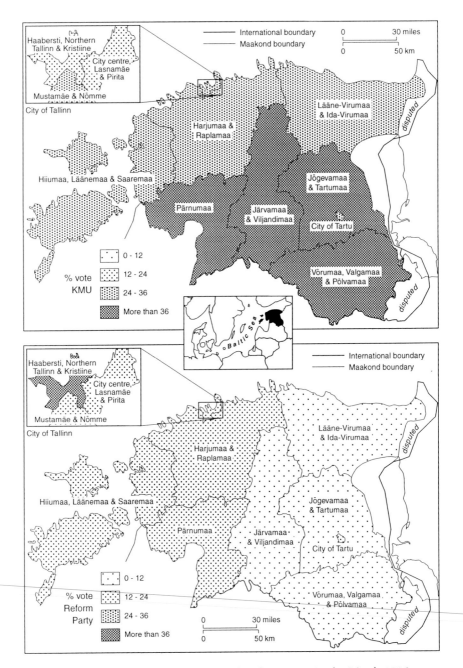

Figure 12.1 Percentage vote for KMU and Reform Party in the March 1995 parliamentary elections

evidence to suggest that class divisions have therefore emerged much more rapidly in urban communities than in rural areas. This also raises doubts as to whether reform has led to real democratisation, or whether it has actually replaced one form of class power with another.

Urban–rural conflicts of interest in the construction of Estonia's national identity

The growing divide between the winners and losers of 'transition' is not only represented in the political and economic arenas. The complex interplay between social, political and economic processes associated with the republic's independence from the Soviet Union has also been powerfully reflected in the contested symbolism of Estonia's emerging national identity. In short, while rural landscapes and environmental imagery played an important part in the shaping of Estonia's past national identity, these have increasingly been swept aside in the drive for modernisation and integration with the global capitalist economy.

During the latter 1980s, the independence movement relied heavily on past symbols and conceptualisations of Estonia's former identity (for a comparative example of Latvia, see Bunkse 1990). These placed considerable emphasis on the country's rural heritage and its Finno-Ugric linguistic context (for a discussion of the broader regional context of Estonia's national identity, see Buttimer 1994). Moreover, the initial spark of environmental protest in 1987 similarly reflected a concern with Estonia's rural heritage, and opposition to what was seen as Soviet, and more specifically Russian, despoliation of the state's 'natural' environment. Consequently, much of the earliest post-independence legislation sought to address crucial issues of concern to people living in rural areas, focusing in particular on land reform and farm reform (Abrahams and Kahk 1994; Unwin 1994, 1997).

The main post-independence Land Reform Law in Estonia was passed in October 1991 (Abrahams and Kahk 1994; Unwin 1997) and was based on the central principle that anyone who was a citizen of the Republic of Estonia on 16 June 1940, or who was the heir of such a person, should have the right to demand return, substitution or compensation for land that was unlawfully alienated during the period of Soviet rule (Maide 1995). There was little conscious examination of the implications that this might have for the creation of class structures in rural areas, and the rapidity with which the legislation was passed reflects the deep-seated cultural aspirations and ideologies of those still living in Estonia who wanted their land back. The tenurial structure of the new Estonia was to be based in essence on what had existed during the First Republic in the 1930s. The reworking of Estonian national identity therefore needs to be understood in the context of complex historical and agrarian legacies.

During the nineteenth century, Estonia was ruled by Russia, with the dominant landowning class being the German Baltic barons. It was not until 1919 that Estonia gained its independence and, as with its second independence in 1991, one of the first acts of the government was to pass a Land Reform Law. This created a substantial number of small independent farm-holdings, which played a fundamental role in shaping the economic, social and cultural identity of the new republic (White 1994). The 1922 census, for example, records that 76 per cent of the population was then rural, with 59 per cent employed in agriculture (Pullerits 1928). Even at the end of the 1930s, some two-thirds of Estonia's population was classified as rural, with approximately three-quarters of all farms being run on a single family ownership basis. It is essentially this structure, based on small family-owned farms, that the 1991 Land Reform Law has sought to replicate. The importance of such an agrarian structure and its associated landscape elements for contemporary Estonian national identity owes much to the 'national awakening' of the nineteenth century.

Abrahams and Kahk (1994; see also Kahk 1982; Taagepera 1993) have shown convincingly how the emergence of a specifically Estonian national identity first occurred during the quarter century following 1860. Although initially led by the indigenous middle class, this was essentially a rural and agrarian movement, closely associated with technical improvements in agriculture and concerned with improving the position of the Estonian peasantry. Expressed through Estonian language publications, and particularly through agricultural periodicals, this movement provided the basis for the specific rural national identity promulgated during the First Republic. Although Estonia had a relatively developed industrial base in the early part of the twentieth century, the character of the country remained fundamentally rural throughout the 1930s. Just under two-thirds of the state's area was agricultural land, a fifth forest, and around 14 per cent marsh, bog or waste. Agriculture provided more than half of the country's foreign currency earnings, and these were heavily based on dairy produce, with butter accounting for more than 80 per cent of all agricultural exports.

However, this rural identity, which portrayed Estonia as being a country with plentiful livestock grazing lush pastures, set amidst extensive areas of marsh and forest (Pullerits 1928), was not undifferentiated, nor was it exclusively Estonian. The country's history of domination by foreign powers, first the Germans and Danes, then the Swedes, and finally the Russians, meant that the landscape had been heavily shaped by other cultural influences. Furthermore, while most of the country is indeed relatively low lying, there is considerable subtle variation in the topography and soil quality, and the landscapes of the coastal area and islands are markedly different from those of the interior (Varep and Maavara 1984; Peil 1994).

During the period of Soviet rule, several forces tended to coalesce to enhance the rural character of an Estonian national identity, in opposition to the urban–industrial identity of the Russian-speaking immigrants. Initial resistance to the Soviet terror of 1940–1 and the deportations of Estonians, mainly farmers, in 1949 was situated in the countryside (Taagepera 1993). Here, the appropriately named Forest Brothers (Metsavennad) sought to wage a guerrilla warfare campaign against the Soviet occupation. Likewise, despite the widespread urbanisation which occurred from the 1950s onwards, many Estonians who moved to the main cities retained close links with the countryside, frequently visiting their old farms and cottages. Moreover, relatively few Russian speakers ever settled in the countryside, so that by 1989, although Estonian speakers accounted for only 49 per cent of the urban population, they represented some 87.5 per cent of the rural population of the country (Marksoo 1992). Furthermore, despite Soviet industrialisation and urbanisation, agriculture remained a central element of Estonia's economy throughout the period 1945–91. Thus, as late as 1980, about 14 per cent of the labour force was employed in agriculture and forestry, and 30 per cent of the population was still classified as rural.

At independence in 1991, there were therefore strong forces within Estonia promoting an essentially rural national identity and cultural heritage. This has subsequently found clear expression within a range of publications directed at the overseas tourist market (Unwin 1996), which typically emphasise the country's rural heritage in terms such as the following:

> The beauty of animate and inanimate nature with its wide variety, and the preservation of naturalness – this is the greatest treasure of Estonia. Land and sea, jagged coastline and more than a thousand islands, high bank and picturesque dunes, stone-covered areas side by side with patches of fertile soil, hills and primeval valleys, springs and a thousand lakes, karst areas and marshland, extensive territories covered with forest, flora extremely rich in species, habitats of water-fowl, landscapes long forgotten in many European countries . . .
>
> (Huma 1996)

Estonia's national identity is nevertheless a highly contested one. For example, the Council of Europe's (1995: 24) review of Estonia's cultural policy emphasises that

> while the arts, culture and cultural initiatives showed the way during the crucial stages of the liberation process . . . they had to recede to the backstage when the crucial stage of liberation was over and the practical work for organising political institutions and the transition to market economy began in the Independent Estonia.

Culture's initial oppositional position is now seen as having to make way for the transition to a *market* economy, closely linked to the creation of new political institutions. Even the Council of Europe, which is widely seen as being more culturally sensitive and aware than bodies such as the European Union, is thus propagating a political and economic model combining the forces of the free market and democracy.

The recent marginalisation of rurality in both cultural and political discourse reflects the further polarisation of urban and rural interests. Employment in agriculture fell to only 4.5 per cent of total employment in 1995, from a level of around 14 per cent in 1990; large areas of land are being abandoned, and the vast majority of external funding (both foreign direct investment and EU funding) is still being directed to the urban core around Tallinn (Unwin 1997). While there was an extraordinary sense of optimism among people establishing their own farms in the early years of independence (Unwin 1994), this is rapidly withering away. As elsewhere in Eastern Europe, rural depopulation continues apace, with few young people being willing to live in the countryside, preferring instead the employment possibilities and social life to be found in Tallinn. This is a process that governments have done little to prevent, and is a major consequence of the dominance of a globalised and urban–industrial discourse. The measures announced in the summer of 1996 to regulate the import prices of agricultural products have done little to help many people in the agrarian sector. Of the collective and state farms existing at independence, it is estimated that one-third have collapsed entirely, with one-third functioning more or less on subsistence level, and only one-third having been transformed into anything like commercially viable units (Unwin 1997).

European Estonia

This chapter has argued that, while a rural national identity was central to the Estonian independence movement in the late 1980s, it has rapidly been replaced by the urban-industrial and commercial imperatives of the capitalist system. For many Estonians, what matters most is to be modern Europeans. This implies taking on board entirely new labour practices and providing the conditions under which market competition can flourish as freely as possible.

A key element in this newly emerging Estonian identity is the aspiration for the country to become a full member of the EU and NATO as rapidly as possible (Friedrich Naumann Foundation 1994). At the beginning of 1995, a Free Trade Agreement with the EU came into force, and then in June 1995 Estonia signed an Association Agreement with the EU. The conditions laid down in this agreement are playing an increasingly prominent role in influencing government policy. In the Government Coalition Agreement signed in October 1995, for example, it was explicitly stated that the Association Agreements signed with the EU would provide the basis for the new Law on Supporting Rural Life due to be drafted during 1996. While accession to the EU and NATO is sought primarily for political security and anticipated economic advantages, the wider social and ideological influences of accession have not yet been fully thought through. Given the extent of rural and agricultural problems that already exist within the EU, Estonia's hoped-for accession does not bode well for the future of those people who sought to recreate their farm-holdings from their families' properties dating from the 1930s.

Conclusions

This chapter has sought to highlight the complex interplay between political and economic factors in the changes that have taken place in Estonia since 1991. In particular, it has focused on the effects of these changes in rural parts of the country, and their implications for the contested nature of Estonia's national identity.

The Estonian example has important ramifications for the theoretical debates which form the central focus of this book. With reference to debates over modernisation and political freedom, this chapter supports Held's (1991, 1995) view that the links between liberal democracy and national sovereignty should not merely be taken for granted. The emergence of a new national identity in Estonia owes much to the globalised character and rhetoric of so-called liberal democracy and the free market. The historical Estonian national identity which found its expression in the First Republic of the inter-war years, and which was derived from the country's national awakening in the nineteenth century, briefly re-emerged in the early 1990s. However, this has rapidly been swamped by a new identity, based on rhetorics of a free market and a liberal democracy, which owe much to the external influences of bodies such as the IMF, the EU and the Council of Europe.

The chapter has also suggested that there exists an extremely complex set of political and economic processes intersecting across many different scales from the global to the local. The simplified hollowing-out model, whereby national-level politics and economic decisions are deemed to be declining in importance, finds little support from the Estonian example. While international agencies and discourses have played a significant role in shaping Estonia's economic and political agendas since 1990, these have been strongly mediated by the republic's elected governments. Moreover, decisions made by these governments have had a powerful influence on the structure of the economy and on the daily lives of the country's population. If, for example, decisions had been made to restrict imports and to subsidise farmers, the rapid decline in agrarian production might well have been reduced. However, the medium-term relative success of the urban-commerical sector would nevertheless have been dented by such a policy, and the longer-term aspirations to join the European Union would likewise have suffered. This reinforces arguments, such as those of Gowan (1995: 54), that for the countries of Eastern Europe 'the costs of [shock therapy] have been far in excess of what was, from an economic point of view, necessary', and that from the European Union's point of view the policy has been a remarkable success. The evidence from Estonia thus emphasises Gowan's (1995: 60) conclusion that

> in Eastern Europe, the death of communism had led the West to try to stamp out economic nationalism in favour of its own national and collective interests in the region. But this does not so much suggest a new era on the globe as something rather old-fashioned which, in the days of communism, used to be called imperialism.

Third, although there are arguments such as those of Smith (1996) to suggest that there is indeed a dark side to Estonian nationalism, as propounded in general for Eastern Europe by Žižek (1990), there is no strong evidence to suggest (as Žižek does) that this is a specific reaction to the introduction of capitalism. Rather, it would appear that Estonian nationalism owes more to a longing for a past identity that was strongly suppressed by Russian conquest and domination. Moreover, the period of harking back to the past in the immediate post-independence situation of the early 1990s was relatively short-lived, and has now been replaced by a nationalism which is marching hand in hand with capitalist interests towards a new future in Europe. This shift reflects the contested character of national identity, as different classes and groups of people seek to create and impose their own political and economic interests on the character of the state.

Three more general conclusions concerning the character of 'transition' can also be drawn from the Estonian example. First, the changes that have taken place have been highly contested. This applies as much to the economic and political path that has actually been followed as it does to the images of the republic's emerging national identity. Second, it is apparent that those advocating the combined creation of a free market and a political democracy have so far been able to dominate the arena. This has been at a cost of increasing social differentiation, and of widening variations in levels of economic activity across the country, which have been particularly severe in rural areas. Although the ideas of a *free market* and *democracy* are still mentioned in the same breath, there is increasing evidence to support Habermas' claim that there is a very real tension between the two. The free market has clearly benefited a small group of people who are beginning to constitute the top end of an emerging class structure, and political instruments need to be developed to restrain the excessive expropriation of profit both by foreign investors and also by certain sections of the population within Estonia. Otherwise, it is likely that social tensions will have increased so much that by the end of the decade many of the aspirations of those founding the new republic may lie shattered. This is of particular importance in Estonia, where the Russian-speaking population provides a substantial minority which already feels threatened by the new political context within which it finds itself. If the economic position of this important segment of the republic's population is not improved then the risk of social and political unrest will become heightened.

Finally, the Estonian example illustrates that the path of change does not always lead in a straight direction. In the late 1980s and early 1990s, the political and ideological policies of independence led to the introduction of land reform legislation which looked backwards to Estonia's land-holding structure in the 1930s and to its national awakening in the latter part of the nineteenth century. As Abrahams (1994: 157) has so appropriately commented, this was a policy of 'back to the future', in which the republic's new identity was to be forged from a long tradition in which the family farm was seen as both socially desirable and economically effective. Within a couple of years of independence, though, this model is now seen by many Estonians as being out of step with the realities of the 'modern' world. The drive towards a free-market economy has revealed all too clearly that an Estonian national identity based on the republic's rural past, dominated by small family farms, has no place in the Europe of the twenty-first century. What that future will be, however, is still contested. Which of the several competing Estonian national identities will eventually become dominant in the political and economic context of the new Europe remains to be seen.

Acknowledgements

I am immensely grateful to Adrian Smith and John Pickles for their patience and for their extensive comments on an earlier draft of this chapter. The research upon which it was based was kindly funded by the British Academy and the Estonian Academy of Sciences. I am also very thankful to Anton Laur, Reet Karukäpp and Jaan-Mati Punning for their help and advice during my visits to Estonia.

Bibliography

Abrahams, R. (1994) 'Conclusion', in R. Abrahams and J. Kahk, *Barons and Farmers: Continuity and Transformation in Rural Estonia (1816–1994)*, Göteborg: Faculty of Arts, University of Göteborg.

Abrahams, R. and Kahk, J. (1994) *Barons and Farmers: Continuity and Transformation in Rural Estonia (1816–1994)*, Göteborg: Faculty of Arts, University of Göteborg.

Amin, A. (1997) 'Placing globalisation', paper presented to the Royal Geographical

Society (with the Institute of British Geographers' Annual Conference, Exeter, 7–9 January 1997.

Bunkse, E.V. (1990) 'Landscape symbolism in the Latvian drive for independence', *Geografiska Noticer*, 4: 170–8.

Burawoy, M. (1992) 'The end of Sovietology and the renaissance of modernization theory', *Contemporary Sociology*, 21, 6: 774–85.

Buttimer, A. (1994) 'Edgar Kant and Balto-Skandia: *Heimatkunde* and regional identity', in D. Hooson (ed.) *Geography and National Identity*, Oxford: Blackwell.

Callinicos, A. (1991) *The Revenge of History: Marxism and the East European Revolutions*, Cambridge: Polity Press.

CCET (Centre for Co-operation with the Economies in Transition) (1996) *Investment Guide for Estonia*, Paris: OECD.

Council of Europe (1995) *Cultural Policy in Estonia: Interim Report of a European Group of Experts*, Strasbourg: Council for Cultural Co-operation, Council of Europe [CC-CUL (95) 20 B].

Eesti Pank Bulletin (1996), number 2, Tallin: Eesti Pank.

Eisenstadt, S.N. (1992) 'The breakdown of communist regimes', *Daedalus*, 121, 2: 21–42.

Friedrich Naumann Foundation (1994) *Estonia and the European Union*, Tallinn: Friedrich Naumann Foundation.

Fukuyama, F. (1992) *The End of History and the Last Man*, New York: Free Press.

Government Coalition Agreement (1995).

Gowan, P. (1995) 'Neo-liberal theory and practice for eastern Europe', *New Left Review*, 213: 3–60.

Habermas, J. (1994) 'Europe's second chance', in M. Pensky (ed.) *The Past as Future: Jürgen Habermas Interviewed by Michael Haller*, Cambridge: Polity Press.

Held, D. (1991) 'Democracy, the nation-state and the global system', *Economy and Society*, 20: 138–72.

Held, D. (1995) *Democracy and Global Order*, Cambridge: Polity Press.

Hirst, P. and Thompson, G. (1996a) 'Globalisation: ten frequently asked questions and some surprising answers', *Soundings*, 4: 47–66.

Hirst, P. and Thompson, G. (1996b) *Globalization in Question*, Cambridge: Polity Press.

Huma (1996) *Estonia for Tourists*, Tallinn: Huma Publishing.

Kahk, J. (1982) *Peasant and Lord in the Process of Transition from Feudalism to Capitalism in the Baltics*, Tallinn: Eesti Raamat.

Lieven, A. (1994) *The Baltic Revolution: Estonia, Latvia, Lithuania and the Path to Independence*, New Haven and London: Yale University Press.

Lugus, O. and Hachey, G.A. Jr (eds) (1995) *Transforming the Estonian Economy*, Tallinn: International Center for Economic Growth.

Maide, H. (1995) 'Transformation of agriculture', in O. Lugus and G.A. Hachey Jr (eds) *Transforming the Estonian Economy*, Tallinn: International Center for Economic Growth.

Markoff, J. (1996) *Waves of Democracy: Social Movements and Political Change*, London: Sage.

Marksoo, A. (1992) 'Dynamics of rural population in Estonia in the 1980s', in

Estonia: Man and Nature, Tallinn: Estonian Academy of Sciences and Estonian Geographical Society.

Miliband, R. (1991) 'What comes after communist regimes?', *Socialist Register 1991*: 375–89.

Ministry of Foreign Affairs, Estonia (1995) *Estonia Today: Estonia's Economy*, 14 February, Tallin: Ministry of Foreign Affairs.

Odling-Smee, J. (ed.) (1992) *Economic Review: Estonia*, Washington DC: International Monetary Fund.

Ohmae, K. (1990) *The Borderless World*, London: Collins.

Ohmae, K. (1995) *The End of the Nation State*, New York: Free Press.

Park, A. (1994) 'Ethnicity and independence: the case of Estonia in comparative perspective', *Europe–Asia Studies*, 46, 1: 69–87.

Peil, T. (1994) 'Estonian islets: a historical-geographical study', in J. Bethemont (ed.) *L'Avenir des paysages ruraux européens*, Lyons: COMCO Edition.

Pensky, M. (ed.) (1994) *The Past as Future: Jürgen Habermas Interviewed by Michael Haller*, Cambridge: Polity Press.

Prime Minister's Economic Report (1996).

Pullerits, A. (ed.) (1928) *Eesti Põllumajandus Statistiline Album*, Tallinn: Tallina Eesti Kirjastus-Ühistus trükikoda.

Sachs, J. (1990) 'What is to be done?', *The Economist*, 13 January 1990.

Sachs, J. (1995) 'Consolidating capitalism', *Foreign Policy*, 98: 50–64.

Smith, G. (ed.) (1994) *The Baltic States: the National Self-determination of Estonia, Latvia and Lithuania*, London: Macmillan.

Smith, G. (1996) 'When nations challenge and nations rule: Estonia and Latvia as ethnic democracies', *Coexistence*, 33: 25–41.

Smooha, S. and Hanf, T. (1992) 'The diverse modes of conflict regulation in deeply divided societies', in A. Smith (ed.) *Ethnicity and Nationalism*, Leiden: E.J. Brill.

Taagepera, R. (1993) *Estonia: Return to Independence*, Boulder: Westview Press.

Théret, B. (1994) 'To have or to be: on the problem of the interaction between state and economy and its "solidarist" mode of regulation', *Economy and Society*, 23: 1–46.

Tilly, C. (1995) *European Revolutions 1492–1992*, Oxford: Blackwell.

Unwin, T. (1994) 'Structural change in Estonian agriculture: from command economy to privatisation', *Geography*, 79, 3: 246–61.

Unwin, T. (1996) 'Tourist development in Estonia: images, sustainability, and integrated rural development', *Tourism Management*, 17, 4: 265–76.

Unwin, T. (1997) 'Agricultual restructuring and integrated rural development in Estonia', *Journal of Rural Studies*, 13, 1: 93–112.

Van Arkadie, B. and Karlsson, M. (1992) *Economic Survey of the Baltic States: the Reform Process in Estonia, Latvia and Lithuania*, London: Pinter.

Varep, E. and Maavara, V. (1984) *Eesti Maastikud*, Tallinn: Eesti Raamat.

Venesaar, U. and Hachey, G.A. Jr (eds) (1995) *Economic and Social Changes in the Baltic States in 1992–1994*, Tallinn: Institute of Economics, Estonian Academy of Sciences.

White, J.D. (1994) 'Nationalism and socialism in historical perspective', in G. Smith (ed.) *The Baltic States: the National Self-determination of Estonia, Latvia and Lithuania*, London: Macmillan.

World Bank (1993) *Estonia: the Transition to a Market Economy*, Washington DC: World Bank.

Žižek, S. (1990) 'Eastern Europe's Republics of Gilead', *New Left Review*, 183: 50–62.

Part IV

SOCIAL TRANSFORMATION AND THE RECONSTRUCTION OF IDENTITIES

13

'THE POLITICAL' AND ITS MEANING FOR WOMEN

Transition politics in Poland

Joanna Regulska

Introduction

The debate on the transition in Central and Eastern Europe has focused predominantly on the interplay of political and economic factors, known as the question of simultaneity; on party politics or lack of it; on the social and economic implications of regime change; on privatisation and a decreasing role for the state in public service delivery; and on democracy and democratisation (Offe 1991; Ekiert 1996; Bryant and Mokrzycki 1994; Armijo, Biersteker and Lowenthal 1994; Waller 1994).

Within these theoretical frameworks a wide range of social actors in the region have attempted to understand the processes of political change, the transformation to market economies and the role of civil society during the transitional period (Rychard 1993; Lewis 1994; Nelson 1994; Balcerowicz 1994; Diamond 1994; Staar 1997). The policy agenda has been preoccupied with economic issues, and the expectations that their outcomes will resolve social policy needs (Adamski 1992; Kowalik 1993; Vinton 1993). Finally, through training and development, non-governmental organisations have concentrated on the practical realisation of democracy and on the establishment of new democratic practices of participation (Siegel and Yancey 1992; Quigley 1993; Les 1994). In Poland, the majority of these analyses, regardless of their theoretical or policy dimension, have been carried out chiefly at the national level.

While these achievements are impressive, other aspects of the transition have been given at best only lip service and have often been marginalised or entirely ignored in public debates. In particular, two areas call for attention: (1) the local dimensions of transition, and (2) the gender dimensions of the transitional politics. While the latter has been addressed in the practical and policy sense, the former, except for its practical dimension, has been virtually ignored (Bryant and Mokrzycki 1994; Ekiert 1991; Kurczewski

309

1993; Funk and Mueller 1993; Regulska 1997; Kubik 1994; Wolchik 1994; Staar 1997). Indeed, the preoccupation with economic restructuring (understood as the need for liberalisation, privatisation and deregulation) and political transformation (defined predominantly in terms of national and party politics) resulted in the omission of a local dimension in transition debates. The strong conviction that successful transition can be achieved by virtually ignoring input from citizens and institutions at the local level has *de facto* slowed the progress of reforms and hampered the building of a strong civil society.

From the perspective of gender, the fact that these dimensions of transition are ignored is not entirely unexpected. Their inclusion would require the sharing of power with women by those who do not wish to do so, namely the national state and recently empowered dissident men who occupy high-ranking positions (Fuszara 1997; Graham and Regulska 1997). This chapter argues that for Central and East European women, the exclusion of local and gender dimensions means a double political marginalisation and a *de facto* prohibition of women from exercising their political rights. The evident lack of gender issues in transitional debates strengthens the notion that the democratisation process is gendered. As women's political activities are concentrated predominantly at the local level, outside formal politics, the general lack of attention to the local scale – in both national debates and theoretical analysis – reinforces the structural rejection of women's contributions and shows how political change under transition is exclusionary.

This chapter aims to advance the discussion of the gendered political in the context of democratic transition in Poland. While an expansion of the idea of 'the political' and the actual ways to exercise political rights are crucial for democratic citizenship in general, the spaces where women are now developing political culture in Poland have not been much noticed. They have not been the subject of theoretical debates or the focus of extensive enquiries into why women are acting politically as they do and why they are encountering such serious problems. The goal is therefore to make visible the ways in which women are currently doing political work and to argue for a public redefinition of politics to include that work. The chapter argues that a narrowly defined concept of political rights helps produce exclusionary practices that marginalise women's political participation. It recognises the complexity of the processes women are using to construct political space and it discusses the forces that contribute to the way in which politics is currently defined, perceived and practised by women in Poland. The chapter concludes that the redefinition of politics beyond the commonly accepted forms based on male experiences and practised through formal political institutions is one necessary component of the changes needed in order for women to assert their developing political identities. Yet before we can discuss strategies that Polish women have employed, the question of the 'political' and the process through which it becomes constructed should be discussed.

Feminists in western democracies have attempted to explain the usual lack of women's political visibility by seeing the roots of oppression in patriarchy and patriarchal culture. They have argued about the role the state plays in perpetuating that exclusion and over the state's motives. They have asserted that the public–private divide relegates women to the private sphere and that therefore the separation of the private sphere from the public has impinged on women's capacities to exercise their political power (Pateman 1989, 1991; Phillips 1992, 1993). They have insisted that the roots of this unequal political identity have their foundation in sexual inequality.

At the same time, arguments have been advanced that the issue is not to dissolve the distinction between public and private but rather to re-examine the notion of this division and to revitalise the public (Phillips 1993: 13). Fraser's conceptualisation of *subaltern counter-publics* as contesting forces with dominant discourses focuses the debate precisely on the significance of constructing alternative spaces. By withdrawal and regrouping (function of space), and creation of training grounds (formation of base), social groups marginalised along gender, class and ethnic lines are challenging formations and structures that reinforce dominant power (Fraser 1993: 15). The public sphere becomes the terrain of contestation not by retaining the homogeneity of singular dominance but rather by creating a plurality of publics that battles that dominance. It is the 'multiplicity of publics' and the dialectic of spaces and bases that 'enables subaltern counter-publics partially to offset, although not wholly to eradicate, the unjust participatory privileges' (Fraser 1993: 15).

But this partial termination presents a problem, in both theoretical and practical terms. If indeed subaltern counter-publics have a limited power of contestation, then their vitality is in question. What will happen when they achieve their goal of partial change? Does a new public emerge and, if so, what happens to the old one? Does it dissolve itself, reconfigure, change an agenda or continue to contest? How do different subaltern counter-publics interconnect among themselves and with dominant power? Since women are currently more likely to be involved in the creation of these new alternative spaces than they are likely to be integrated into existing male-dominated structures, then these questions require attention if we are interested in both recognising and increasing women's participation.

The limited ability of conventional political structures, especially at the national level, to provide for adequate representation of marginalised groups has been observed by a number of feminist scholars (see, for example, Phillips 1993: 104, 131). Instead, they see decentralised formations (e.g., labour unions, local organisations and even local governments) as more inclusive and participatory (Tetreault 1994). In fact, it is at the local scale that in many western countries the conventional formations began to be open to women. Women run in larger numbers for local offices than for national, and they are successful in gaining local seats. Women's infiltration into the existing formal political structures is making – and will increasingly make – a difference (Carroll 1994; Kathelene 1994; Wadstein 1996).

Simultaneously, women are creating alternative spaces and are engaging in neighbourhood organisations, projects and in forming women's groups (Cohen 1989; Waylen 1992, 1994; Jacquette 1994). Political theorists stress that indeed 'the domain of "democracy" is now more likely to extend (and increasingly does extend) to institutions and practices outside of institutionalised politics' (Warren 1996: 250). Women's engagement in local activism – addressing issues of housing, education, the social safety net or a clean environment – represents a new form of the political, one that is less concerned with the struggle over power, but puts more emphasis on addressing citizens' needs. Through local initiatives and activism women are affecting the formation of social and economic relations and the redistribution of resources. They see political empowerment as inseparable from access to resources. As citizens of the community, they claim their rights but also deliver their obligations.

Some scholars have argued, however, that the impact of such forms of participation and mobilisation is negligible, because they take place outside what is recognised as the policy-making process, and therefore 'these concerns are less likely to be reflected in social programs' (Hernes 1988; Orloff 1993). It is at such points that we fail to recognise the distinct meaning of the political as women local activists conceive of it. By beginning the long-term process of giving new meaning to the political that is not narrowly founded on masculine definition and experiences, women's political actions have the potential to create 'subaltern counter-publics'. They contest the status quo and engage in construction of new political spaces. Moreover, what is important at this stage is not so much the extent of women's engagement in local activism, but rather how they and others assign meaning to these activities. When women's activism is publicly named as feeble and located outside power structures, women are disempowered. At the same time, there is much evidence that this currently 'unrecognised' power of women is beginning to challenge the established regime practices at various locations.

The process will be – and needs to be accepted as – a slow one, since politics in all senses of the term is often in conflict with fast-paced economic developments. In the West (for example, the USA or in Nordic countries) the local corporatist world still remains largely closed to political influence (Jezierski 1994; Wadstein 1996). This perpetuates the separation of capital production from socially controlled redistributive mechanisms, including the kinds of social activism women are developing. The power of the economic over the political sphere doubly excludes women who have only begun to gain access to political power and need that new political power to influence economic decision-making.

This chapter is composed of two sections. The first section explores some ways in which women's political invisibility is maintained. The second section points to some of the ways in which women's current activities in Poland have the potential to expand the public perception of the 'political' and women's participation in it.

The invisibility of gender

How are the naming and mapping of the 'political' in Poland invisibly gendered so as to limit women's participation? So far little attention has been given to questions in this form. Rather, women's participation in politics has been the focus of traditional political science enquiry through the study of party structure, party development, party membership, participation in elections and representation in elected offices (Siemienska 1994). While these data locate women in the realm of Polish politics, they tend to see women as conforming to existing political spaces, rather than constructing alternative ones. Few studies have questioned the conditions under which the narrow public definition of politics is being reproduced in post-1989 Poland (Fuszara 1993, 1997; Bellows 1996; Graham 1997; Graham and Regulska 1997). These few have shown the kinds of resistance that exists to seeing women as political actors, but they have also delineated new locations where alternative political spaces are being created. Drawing on these analyses we can argue that the political silence of women in Poland is primarily conditioned by at least four sets of forces: (1) past legacies of socialism and current adoption of liberal thought; (2) rejection by a large proportion of women of formal political structures and simultaneous lack of women's political identity and collective agency; (3) unwillingness of male-dominated formal political structures to share political power with women; and (4) the ability of the state to readjust itself and to sustain control over the political sphere. Each of these points is briefly discussed below.

The socialist past and liberal future

There is no question that Polish women do share experiences of oppression and inequality with women in other parts of the world, yet the specific context of this striving – the simultaneous struggle to dismantle the socialist, unequal past and to construct a new market-driven future – introduces contradictions that are difficult for women to resolve (see Regulska 1997 for a more detailed discussion). Take as an example the tension between individualism and collectivism. The endorsement of liberalism by Polish political elites brought prominence to 'the individual'. This is a reaction to the past four decades, that 'worked for the "abstract" human through the systematic destruction of the "concrete" human being: the individual' (Miriou 1994: 107). Women in Poland began to accept and develop this

313

new identity based on individualism. Groups and individuals often first assert what they are not, and how different they are from others. They build their identity through the construction of 'walls of difference'. Yet what is required to challenge the power of dominant male elites is women's collective agency and action. This collective consciousness is unlikely to develop rapidly, since in order for it to emerge it will need in the first place to ease the collective memory of socialism. Women are then caught in a double bind: what they require the most is what they forcefully reject: the collective. What they embrace – the individual – is bound to produce new forms of oppression.

A second dimension that points to the struggle between the past and the present is the question of the morality of politics. Convincing arguments have been put forward that extensive involvement by Polish society in the civic organising and protests of the 1980s was derived from the moral rather than the political sphere (Marody 1990). People were opposing authoritarian regimes on moral rather than political grounds and, as a consequence, the outburst of civic protest was stripped of its political power. As communist politics was dirty, immoral and corrupt, and thus the subject of contestation and not of participation, the oppositional politics that emerged appropriated a moral rhetoric forming an anti-politics. The new apolitical politics 'was comprehended as a social activity, not dealing with the question of power, but demanding less politics' (Jalusic 1994: 8). Under the new political system, the philosophy that political leaders have 'dirty hands' and the ideal of an apolitical identity remain culturally important. These values clash with new needs for greater political involvement. The rhetoric of anti-politics has been embraced by some women but it has also been appropriated by populist parties that have aimed to reinforce patriarchal values. As a result, the lack of women's political identity is effectively used by the dominant political powers to reproduce women's exclusion from the political.

The construction of politics and of women's political identity is therefore shaped by behaviours, values and institutions that still exhibit the imprint of the socialist past. At the same time the arrival of liberalism, with its emphasis on individuals and their sole responsibilities for exercising their rights, reinforces women's isolation from each other and weakens their ability to challenge exclusionary practices. In a way little has changed. As Jalusic argues in the case of Slovenia: 'The interpretation of the political has not changed. Before, political space was understood as a (socialist) state and its institutions. Now the notion is the same, only the regime has changed' (Jalusic 1994: 10).

Women's political culture or lack of it

How do women themselves see the impact of these changes on their political identity? While there have not been any systematic studies regarding women's attitudes towards politics in the post-1989 regime, the scattered data point to several forces that shape women's reluctance to participate in formal political institutions and their ambivalence in acknowledging their activism as political. Patriarchal values and the lack of recognition that equal representation is a basic right result in a low priority given to that entitlement by all citizens (Nowakowska 1995). The unwillingness of political leadership to create opportunities for women's participation in public and political spheres *de facto* reinforces the persistence of traditional values and attitudes, and augments women's role as mother and homemaker (Kalinowska 1995). The lack of serious public debate about women's roles in political and public life accentuates these problems (Fuszara 1994; Kuratowska 1996; Regulska 1995).

Women are also unwilling to tie themselves too closely to any one party, as they feel constricted by the party structures and they feel that they do not fit into the party framework. Therefore they feel ambivalent about joining party ranks, participating in party politics, or even becoming a party member (Fuszara 1994; Kuratowska 1996). For many, the term 'party' has negative connotations. As a result many women in Poland lack the motivation to join formal political institutions (Regulska 1995). The lack of drive to associate themselves with political institutions is further compounded by a political culture that downgrades women's experiences and sees them as of lesser importance. As recent studies conducted by Graham indicated: 'women's own interests and community needs . . . [are often] not placed on the local agenda' (Graham and Regulska 1997). Hence they cannot see their own interests represented as public interests.

Nevertheless, some women do enter formal structures, especially when they believe such representation is necessary and that they themselves can influence decision-making. When directly asked what prevents women from entering formal political structures, women officials newly elected in 1990 identified five significant barriers: (1) lack of time and the feeling of being overwhelmed by the responsibilities of paid employment and work at home (47 per cent); (2) women's lack of trust in women who hold public office, and the assumption that men can do those jobs better (46 per cent); (3) women's belief that they are not assertive enough, that they lack trust in themselves and lack self-esteem (14 per cent); (4) the lack of a tradition of active participation in public life among Polish women (9.5 per cent); and (5) women's lack of interest in participating (5 per cent) (Regulska 1994). Although these results were obtained early in the transition (1993) and are based on a specific sample of women who already occupy leadership positions in the local government structures (mayors, city president, chair of city council), they are still indicative of the weight that different barriers to participation in politics are given by women respondents. These women politicians don't see women's lack of interest as of primary concern. Rather the barriers are the difficult conditions under which women's participation takes place. Of course, women's exhaustion is an old and well-known issue, and not surprisingly it emerged as the most important problem in this survey. Women's fear of participation and their lack of political culture are more hidden and more recent ideas, although they may come to be recognised as equally important impediments.

Formal political structures

So far I have shown how past and present ideologies constrict women's political experiences and the degree to which women generally perceive themselves as less suitable political actors. These behaviours and perceptions are not, however, formed in a vacuum but rather are subject to constant pressures from political institutions. The transformation of the Polish state has altered the ways in which the conventional political system operates and the degree of access that women can have through formal channels. What is interesting in this transformation is the fact that the regime changes which appear to enhance women's access to political institutions in fact have narrowed down these opportunities. The recent legislative changes of the electoral law illustrate the point. The newly adopted electoral laws are primarily based on the proportional system and permit non-party groups to nominate candidates.[1] In other countries both of these changes have proved to enhance women's chances of winning elections (Fabian 1995; Osterdahl 1996; Wadstein 1996). Indeed, the analysis of the first round of elections confirms similar trends in Poland, where women in the local elections of 1990 and the 1991 parliamentary elections were slightly more likely to run and be elected under the proportional system (GUS 1995).[2] The fact that a diverse number of citizens' groups, organisations and associations can nominate candidates in local elections has also represented an unquestionable asset for women. As the results of the 1990 local elections indicated, over 81 per cent of elected women were indeed nominated by non-party groups (GUS 1995).

The increased party control of the parliamentary elections is an obvious cause for alarm, especially as women are more likely to avoid formal structures and resist joining political parties. This may in fact diminish women's chances of political participation through formal channels. This obstacle is further compounded by the fact that Polish political parties are not in a rush to increase women's membership. As there is no tradition to reveal party membership data, especially broken down by gender, it is difficult to present systematic analysis. The informal survey conducted in the autumn of 1994 by Senator Zofia Kuratowska reinforced the fact that parties continue to be uninterested in women both as members and candidates. Only one party, the Freedom Union, responded to the survey, although its data were limited. None of the other parties bothered to do so, though several gave as an excuse that they do not have any data (Kuratowska

1996). More recent estimates provided by the Main Statistical Office put women's membership in parties at about 17 per cent (BSE 1997). There is also some indication that party ideology is not irrelevant for women as more progressive parties (those to the left and centre-left such as the Freedom Union) seem to have more women as members (30 per cent) than peasant parties (6 per cent) and conservative parties (6 per cent). Centre-left parties also tend to have more women candidates on their lists. These trends are indirectly confirmed by the data showing women's inclusion in the parties' leadership. While the Social-Democratic Alliance have one woman for every thirteen men and the Freedom Union has two for every eleven, the Peasants' Party ratio is one for every fifteen (BSE 1997). Clearly these numbers are very low and inadequately represent more than half of Polish citizens (women represent 51 per cent of population).

A slight change in the attitudes of political parties may, however, be under way. At the time of the 1993 parliamentary elections women deputies began to complain publicly: 'Politics is male dominated. Colleagues in the party usually do not encourage women's access to party leadership, and during elections, place them at the lower end of the list – automatically restricting their chances of winning' (Olczyk 1993). Although national politics in 1993 still remained male dominated, more women were placed on the electoral lists and, once they were included as candidates, they had better chances of winning than men. The current campaign for the autumn 1997 parliamentary elections indicates that several parties to the left and centre-left will attempt to put forward women candidates. Furthermore, the Labour Union was the first party to break the ranks and implement quotas for women candidates of at least 30 per cent. This may be only smoke surrounding campaign fevers, but it could also be an indication of a slight change in attitudes towards women's political participation.

State responses

The state in Poland plays a significant role in the formation of new political spaces through shaping social processes, political participation and economic development. Through the control of its institutions, the state reinforces cultural values and shapes in its own way the construction of women's political identity. Through the creation of a legislative framework it defines its philosophy and political interests and it creates and applies the law that has regulated women's sexuality and their reproductive rights (e.g., abortion legislation); reinforced family values (e.g., lack of housing for non-nuclear families); and regulated women's access to economic opportunities (e.g., education, protective legislation). Through the extensive use of other traditional tools of central control – administrative, judiciary and police – the centre can regulate distribution of resources across social groups and geographical scales.

In the post-1989 political framework, socially and economically dis-advantaged groups found themselves in a new political location. The past rhetoric of perceived equality became translated by the state apparatus and its bureaucracy either into a paternalistic attitude towards women or into the state's withdrawal of many social benefits. What makes the state's current positions most different from those of the past is its loss of interest in those aspects of society which do not affect the pursuit of free-market economics. Thus while in the past the socialist state needed women to legitimise its rhetoric and power and virtually forbade private-market activities, in the new free-market era the repositioned state vigorously supports entrepreneurship and business development and is only selectively interested in the well-being of its citizens.

Simultaneously, the state has maintained its control over the political structures and institutions that continue to shape political spaces. For example, during the past few years the state attempted several times to tax the activities of NGOs.[3] Because a large proportion of NGO funding comes in the form of grants and is awarded by foreign funders rather than earned domestically, such taxes would have eliminated overnight many of the NGOs in Poland. While this move has not explicitly targeted women's groups, the importance of women in NGOs and the still fragile status of many organisations would have severely damaged many women's initiatives. The elasticity with which the state has seized the opportunity to reassert its power shows how it reinforces cultural values and shapes in its own way the construction of women's political identity.

Making visible the invisible

Women's experiences in local and often non-party politics call for an enlarge-ment of our understanding of political rights that would include new ways of accounting for and measuring political participation. But despite a wealth of literature documenting difficulties encountered by women in claiming the political, the need for redefinition of the notion of political rights has not been addressed (Dolling 1991; Einhorn 1991; Heinen 1992; Janova and Sineau 1992). I would argue that the conventional definition of political rights is too limiting and engages only narrowly defined participatory experiences: these definitions relegate women's political participation at the local level to the non-essential and the apolitical; they see women as located outside the traditional political structures and do not account for the diversity of their engagement in politics over their entire life course.

The new definition of political rights needs, then, to encompass both the right to *de jure* equal representation and to *de facto* women's participatory experiences. It should include women's local experiences and by doing so give to women's political culture needed legitimacy. Women's diverse experiences of participation in NGOs, in informal networks, neighbourhood organisations, and administrative and economic spheres need to be part of the new definition along with their participation in formal structures. The existence of women's agency at the local level (even though one should rightly argue that it is still fragmented) has already stretched the boundaries of conventional politics and reconfirmed the diversity of women's political experiences. Most importantly, however, a new notion of political rights would transform the current individual, fragmented experience of exercising political rights (e.g., through voting) into a collective and dynamic process that would make use of other means of political expression such as negotiation, bargaining, lobbying, and protesting in cases when citizens' needs are not met and their rights are threatened. Through mobilisation and collective actions that span local, national and international levels all citizens would be better able to access and utilise a variety of forms of political power.

At the community level, an expanded idea of political rights would mean that women could claim their rights not only through the right to vote, but also through access to services, through input into decision-making processes, and through the freedom to exercise pressures that would permit them to claim these rights (e.g., referenda or protests). At the national level, along with participation in electoral politics, under an expanded definition of political rights, women would have to have greater representation in appointed administrative bodies that prepare issues for decisions, set agendas and draw general guidelines (Regulska 1994). They would also then gain access to international politics where states act beyond borders, and where they are bound by international law.[4] The right to be elected provides an illustration of the limits that the current definition of political rights imposes on women's access to political power across all the above levels. Polish women gained their political rights in 1918. Under the past regime their political numerical representation was high (e.g., in 1978 it was 25.6 per cent in local government), as the ruling party needed to legitimise their claims of equality between women and men. In the post-1989 climate, as we have seen in the evidence from Poland, parties are not interested in women's issues and women are ambivalent about joining party ranks. In this context, the right to be elected becomes rather meaningless. If the political landscape is dominated by party politics and the only road to be elected is through party channels, then the technical right to be elected when one is outside the party structure is empty.

The call of Polish feminists and women's groups for widening the definition of what is political is consequentially linked with the reformulation of how a new sense of the political can be realised in practice. A review of strategies and activities now being employed by women in Poland reveals that the process of constructing this new politics takes place both at alternative sites and through linking these locations with the dominant power structures. Moreover, the majority of these initiatives are located at a level where the emergence of non-governmental organisations has provided new opportunities for an active process of constructing alternative civil identities. The recent legislative initiative to pass the Equal Status Act, undertaken by the Parliamentary Women's Group in Poland, but initiated by feminist groups, provides an interesting example of the latter case, while the unprecedented growth of NGOs represents the former. Both illustrate how claims for new political identity and actions are executed by women in practice.

The Equal Status legislation has been passed by a few countries in Europe, as they have become convinced that, without specific legislation advancing women's access to the decision-making process, their participation and representation cannot be realised. In some countries, it was a single general act that encompassed all spheres of life (Norway); in others these were specific single-sphere acts that addressed one particular area (employment in Denmark and Sweden; education in Great Britain). In Central and East European countries, as far as available information indicates, Poland is the only country where such legislation is being considered.

The first initiative to prepare the Equal Status Act was undertaken by Polish feminist groups in 1992. During the next five years through meetings, discussions and debates, many individuals and groups developed the Act's philosophical and legal framework. While some issues found places in other legislation, although not in a very satisfactory way (e.g., in the labour code), others were incorporated into Acts of Parliament (e.g., equal pay for equal work). In the end, the Parliamentary Women's Group collected over 160 signatures (ten times more then the required minimum: fifteen signatures are needed for legislation initiated by deputies).[5] The proposal was officially presented to the Sejm (Lower Chamber) in January 1997 and the Act awaits its first reading in Parliament. Because of the autumn 1997 parliamentary elections it was unlikely, however, that this equal status legislation would be enacted during this term of office. Considering also the immediate, negative responses after its first public announcement, from both male and female parliamentary deputies, a heated debate over its passage is expected.[6]

Despite the past and projected difficulties, the value of this initiative lies in the approach that the Act proposes to take. First, it argues that past experiences proved 'that a constitutional regulation concerning equality and the prohibition of discrimination alone, not supported with specific legislation on real possibilities of claiming rights, is not sufficient' (The Act Concerning the Equal Status of Women and Men: Draft 1996: 10). Second, the question of claiming rights is not simply an outcome of political representation but rather needs to be seen as interconnected with equal representation in all sectors of public life and at all scales, and needs to be reinforced by equal access to social and economic resources. In justification of the Act we read that:

special attention is given to the need for equal rights for women and men in political life and the principle of representation of both sexes in all public institutions. . . . The draft clearly defines equal rights in the areas of education and the protection of health. . . . There is also a catalogue of rights concerning employment. . . . The draft emphasises the right of women and men to the same employee benefits.

(The Act Concerning the Equal Status of Women and Men 1996: 10–11)

The approach employed by the authors makes two significant points. First, it recognises that, while theoretically the capacity to choose and be chosen is equal and guaranteed by law, the conditions under which that capacity may be exercised are not. This is indeed evident under current changes in Poland, as well as in the other countries of Central and Eastern Europe. While women do have their conventionally defined political rights, they also represent a group that has experienced the loss of social and economic benefits. If political equality is not seen in the context of social inequality, women, along with other socially and economically marginalised groups, are automatically denied the right of political participation. The political and social cannot be separated. Thus to equalise the freedom to choose requires equalisation of the actual access to material and social resources, which in turn should create equal conditions for exercising political rights. Second, the Act acknowledges the need for the inter-relationship between formal political institutions representing dominant power and 'subaltern counter-publics' that are being formed by women's and feminist groups. By initiating discussion and debate about the Equal Status Act outside the formal structures, and yet by recognising the need to legitimise the Act through state political institutions, Polish women are breaking new political ground, learning how to contest while at the same time utilising to their advantage dominant power structures.

The unprecedented growth of women's groups and local initiatives represents new forms of active participation by women in the formation of alternative public and political spaces. Indeed, since 1989 women have created a large number of women's groups and organisations. The Polish Committee of NGOs – Beijing 1995 – indicated that seventy such groups have been created since 1990 (the Polish Committee of NGOs 1995), and the 1996 survey done by the newly created National Information Centre on Women's Organisations and Initiatives in Poland claims that there are over 100 such groups. The Directory of Women's Organisations published by the Centre for the Advancement of Women puts the number at over 200 (but the directory also includes Women's Studies programmes at universities and organisations that existed before 1989) (Centre for the Advancement of Women 1995). The point is not so much the accuracy of the data, but the

321

sheer fact that such a rapid growth of women's active participation has taken place within a very few years.

The data gathered by the Main Statistical Office on women's participation in a variety of organisations (not only women's and not only NGOs) indicate that women constitute 31 per cent of the membership (Daszynska 1995). While the survey points out that women shy away from participating in political parties and political organisations *per se* (only 17 per cent in the latter), they are extensively involved in education (65.8 per cent), health and social welfare groups (62 per cent), and art and cultural groups (49 per cent). Their participation in scholarly (29 per cent) and professional associations (35 per cent) is lower but remains quite broad. It is also possible that in fact the numbers are much higher, as the rate at which organisations are being formed is very high and also many small groups and initiatives operate without being formally registered. This rapid growth of NGOs and other groups has taken place despite the lack of political traditions, skills and experiences and despite women's often exhibited unwillingness to work together.

Women's widespread involvement in diverse social and economic sectors speaks to the range of the issues that women who are actively involved consider as vital and requiring their public action. This diverse commitment indicates women's belief that multi-locational actions (not only in purely spatial but also in social terms) may bring the best outcomes, although these groups are predominantly located in urban areas, and therefore many women from smaller towns and villages still have limited access to them. Nevertheless, the increased interest on the part of women from smaller communities to participate in NGOs' activities is unquestionable. It represents a growing awareness by women that these alternative spaces do exist and that they may potentially be of some value to them (Graham and Regulska 1997; CRCEES 1996). The few examples of collective actions that Polish women have undertaken to assert their rights (e.g., The Federation for Women and Family Planning) or to mobilise their resources (e.g., The National Information Centre on Women's Organisations and Initiatives in Poland) symbolise the recognition on the part of women already engaged that the 'individual experience of male oppression and conflicting interests between women and men [need to be addressed through] collective action as women' (Eduards 1994: 181). This shift from individual to collective actions signifies also a switch from passive to active agency. Indeed, unless women themselves recognise that they have the capacity to question the isolation of the 'subaltern counter-publics', their collective political agency will remain silent.

There is yet another emerging threat to women's local political identity originating from the Polish Roman Catholic Church. In the 1980s the Church provided significant support to citizens in their fight for democracy. By doing so, however, it has attempted to establish its claim to the right, both at the national and local level, to interfere in political space. In the

1990s the Church has consciously chosen women's reproductive rights as its most visible target for asserting its political power over society. But this is by no means the only arena of influence. Women are being actively courted by local parishes to engage in meeting the social and charitable needs of the local population in their communities. While in countries where there is a clear separation of the state and church this would not appear to be unusual; in Poland these activities carry a political overtone and a clear but hidden alliance of the state and church in their pursuit to maintain patriarchal values. I would argue that the Church has focused so much attention on women because of its long-standing conviction regarding women's traditional role as mothers, wives and homemakers, and also that issues of women's rights and their public participation represent a convenient political 'wedge' allowing the Church to extend its sphere of dominance more widely without provoking much resistance.

Despite these negative cultural and ideological barriers, there seems to be an emerging recognition among some citizens that women are insufficiently visible in public and political life. A recently conducted public opinion survey indicated that at least one-third of citizens believe that women should be more represented in public life (between 31 and 44 per cent, depending on the type of institutional involvement) and that concrete action should take place to change the situation of women's exclusion (44 per cent) (CBOS 1997). While many citizens believe that a single constitutional guarantee would be sufficient to eradicate existing inequalities, the fact that for the first time the marginalisation of women in public life has been so widely acknowledged is a significant development (CBOS 1997; Fuszara 1997).

Clearly these few developments cannot be seen as sufficient to eradicate the marginal position of women as political actors who have a strong political voice. Changing existing power relations, the established institutions and structures and expanding women's collective agency will require sustained effort. Yet the activities initiated so far by different women's groups are making a difference in the way that the political is perceived, defined and practised.

Conclusions

In the Polish context the construction of the political is not only moulded by the scale at which the political is located and the agency that engages in the process of its reformulation, but is also an outcome of past histories and local cultures, and of the unclear future of liberal economies. Polish men and women have distinctive understandings of the nature of politics and these differences profoundly affect the process through which they engage in the political, the focus of their activities and the actual location of the political within institutional structures and hierarchies. Since 1989, women's engagement in politics has been located primarily at the local level in non-governmental organisations, local initiatives and in civic organising. These local institutions are less formalised, rigid and closed than conventional political structures. They operate within a decentralised framework that permits wider citizen participation and they are less guided by a hierarchical top-down climate typical in political and corporatist structures.

The emergence of women's local political identities (even though not yet equally visible in all different spheres) raises, at a minimum, three significant points: (1) it exposes the gendering process of social and economic relations that otherwise remained hidden, and therefore acknowledges their direct affinity with political empowerment; (2) it constructs these relationships as public discourses that are subject to scrutiny, questioning and contestation; and (3) it reaffirms that mobilisation on the part of women is essential, but that they will be even more empowered when their own interests are articulated with the diversity represented by class, race, ethnicity and locale.

The existence in Poland and other countries of Central and Eastern Europe of a contradiction between women's own perception of their activities as empowered and the failure to recognise those activities by the general public, ruling elites and scholars has had critical and disabling effects on the political identity of women: women are stripped of political agency despite their engagement in the formation of political space. They act, but these actions seem to have very limited influence. The fact that women's activism is seen as immaterial has yet another important implication for our understanding of the way in which 'subaltern counter-publics' relate to each other and interact with dominant power. For if they are indeed inconsequential, neither the existence of the linkages and interactions between them needs to be recognised, nor do the strategies for change that women develop need to be acknowledged. As women's collective actions and agency are at the heart of the formation of their political identity, to deny these is to disempower women anew.

Despite these mounting obstacles, women are motivated to engage politically: they desire to act and to achieve change. The increased ability of Polish women to work across different scales (local, national and global) attests to their commitment to redefine the political in a way that will be inclusive and flexible, and will permit choice of strategies that will contribute most to their empowerment. While this process is bound to be long and difficult, many Polish women have already embarked on this trajectory of change.

Acknowledgements

The author wishes to thank Ann Snitow, Julie Mostov, Eleonora Zielinska and Eva Vesinova for the comments on earlier drafts and suggestions.

Notes

1 The exceptions here are communities below 40,000 inhabitants, where a majority system operates.

2 Since 1989 the number of women elected to local government and Parliament has been growing, albeit very slowly. The 1990 local elections returned 10.4 per cent women and in 1994 13 per cent. The corresponding figures for parliamentary elections are: for 1991, 9.5 per cent (Sejm) and 8 per cent (Senat), and for 1993 13 per cent for both Sejm and Senat.

3 Similar attempts to tax NGO activities have been made by the Slovakian government. They have met, however, with strong opposition from the NGO community.

4 This can take place through the conventional, direct input by citizens via national parliamentary elections or through appointments of national delegations made by the political ruling elites. The first case is illustrated by current membership of the Polish parliamentary delegation in the European Parliament, an arm of the Council of Europe. The second case is illustrated by the United Nations with its extensive bureaucratic network, where delegates to various committees are selected by the governing political groups in each country.

5 The Parliamentary Women's Group was created in 1991 and currently brings together women deputies from five parties. It serves as a watchdog and a supporter of gender-related legislation (e.g., the Equal Status Act). On a regular basis it provides a forum where women representing different national and local elective offices, economic sectors as well as those engaged in NGOs, business, public administration or mass media can meet to discuss a variety of women's concerns.

6 When the completion of the Act was announced publicly in November 1996, many deputies had strong negative reactions. Men argued that 'discrimination of women has social and economic roots, but not legal' (Tomasz Nalecz, Labour Union) and they worried that this was an 'antidemocratic proposition' (Bogdan Pek, Peasant Party), and that 'there will not be enough active women to fill the 40 per cent threshold' (Jan Litynski, Freedom Union). Women deputies, on the other hand, reaffirmed the difference between women and men but rejected the Equal Status Act as unnecessary. Former Prime Minister Suchocka of the Freedom Union called it a 'crazy idea', and Alicja Grzeskowiak, a 'Solidarity' deputy, argued 'it's not a stereotype, but nature that results in women's and men's different roles' (*Zycie* 1996: 3).

Bibliography

Adamski, W. (1992) 'Privatization versus group interests of the working class and bureaucracy: the case of Poland', in P. Volten (ed.), *Bound to Change*, New York: Institute for East West Studies: 212–227.

Armijo, B.L., Biersteker, T.J. and Lowenthal, A.F. (1994) 'The problems of simultaneous transitions', *Journal of Democracy* 5, 4 (October): 161–175.

Balcerowicz, L. (1994) 'Understanding post-communist transitions', *Journal of Democracy* 5, 4 (October): 75–89.

Bellows, A.C. (1996) 'Where kitchen and laboratory meet: the tested food for Silesia program', in D. Rocheleleau, B. Thomas-Slater and E. Wangari (eds), *Feminist*

Political Ecology: Global Issues and Local Experiences, London and New York: Routledge: 251–270.

Bryant, C. and Mokrzycki, E. (1994) *The New Great Transformation?* London: Routledge.

BSE (Biuro Studiów i Ekspertyz Kancelarii Sejmu) (1997) 'Informacja o Sytuacji Kobiet w Polsce', Warsaw, February 24.

Carroll, S.J. (1994) *Women as Candidates in American Politics*, Bloomington: Indiana University Press.

CBOS (1997) 'Udzial Kobiet w Życiu Publicznym – Prawne Gwarancje Równosci Płci', Warsaw, February, BS/26/26/97.

Centre for the Advancement of Women (1995) *A Directory of Women's Organisations and Initiatives in Poland*, Warsaw.

Cohen, Y. (1989) 'The role of associations in democracy', in Y. Cohen (ed.), *Women and Counter Power*, Montreal, Quebec: Black Rose Books: 220–230.

CRCEES (Centre for Russian, Central and East European Studies), Graham, A. and Regulska, J. (1996) 'Advocacy and lobbying: women's strategies for empowerment in Central and Eastern Europe', New Brunswick, Rutgers University: CRCEES.

Daszynska, M. (1995) 'Udzial Kobiet w Życiu Publicznym w Świetle Danych Statystycznych', *Biuletyn Ośrodek Informacji i Dokumentacji Rady Europy* 5: 42–45.

Diamond, L. (1994) 'Rethinking civil society: toward democratic consolidation', *Journal of Democracy* 5, 3 (July): 4–17.

Dolling, I. (1991) 'Between hope and helplessness: women in the GDR after the turning point', *Shifting Territories: Feminism and Europe*, Special Issue of *Feminist Review* 39 (Winter): 3–15.

Eduards, M.L. (1994) 'Women's agency and collective action', *Women's Studies International Forum* 17, 2/3: 179–184.

Einhorn, B. (1991) 'Where have all the women gone? Women and the women's movement in East Central Europe', *Shifting Territories: Feminism and Europe*, Special Issue of *Feminist Review* 39 (Winter): 16–36.

Ekiert, G. (1991) 'Democratisation processes in East Central Europe: a theoretical reconsideration', *British Journal of Political Science* 21: 285–313.

Ekiert, G. (1996) *The State against Society: Political Crises and Their Aftermath in East Central Europe*, Princeton: Princeton University Press.

Fabian, K. (1995) 'Overview of women's interests articulation in Central and East Europe', unpublished paper.

Fraser, N. (1993) 'Rethinking the public sphere: a contribution to the critique of actually existing democracy', in B. Robbins (ed.), *The Phantom Public Sphere*, Minneapolis: University of Minnesota Press: 1–32.

Funk, N. and Mueller, M. (eds) (1993) *Gender Politics and Post-Communism: Reflections from Eastern Europe and the Former Soviet Union*, London: Routledge.

Fuszara, M. (1993) 'Women's legal rights in Poland in the process of transformation', *Beyond Law* 3, 8: 35–47.

Fuszara, M. (1994) 'Women in parliamentary elections 1993', unpublished paper.

Fuszara, M. (1997) 'Wizerunek Kobiet w Społeczeństwie Demokratycznym', address to the joint session of the Polish Parliament, Warsaw, 8 March.

Graham, A. (1997) 'Citizen participation in small and medium-sized Polish towns', unpublished Ph.D. dissertation, Rutgers University.

Graham, A. and Regulska, J. (1997) 'Expanding political space for women in Poland: an analysis of three communities', *Communist and Post-Communist Studies*. Forthcoming

GUS (1995) 'Udział Kobiet w Wyborach Parlamentarnych i Samorzdowych w Latach 1991 i 1993 oraz 1990 i 1994'.

Heinen, J. (1992) 'Polish democracy is a male democracy', *Women's Studies International Forum* 15, 1.

Hernes, H.M. (1988) 'The welfare state citizenship of Scandinavian women', in K.B. Jones and A.G. Jonasdottir (eds), *The Political Interests of Gender*, Thousand Oaks, CA: Sage Publications: 187–213.

Jalusic, V. (1994) 'Politics as a whore, women, public space and anti-politics in post-socialism', unpublished paper.

Janoa, M. and Sineau, M. (1992) 'Women's participation in political power in Europe: an essay in East–West comparison', *Women's Studies International Forum* 15, 1.

Jaquette, J. (ed.) (1994) *The Women's Movement in Latin America: Participation and Democracy*, Boston: Westview Press.

Jezierski, L. (1994) 'Women organizing their place in restructuring economies', in J. Garber and R. Turner (eds), *Gender in Urban Research*, Thousand Oaks, CA: Sage Publications: 60–76.

Kalinowska, E. (1995) 'Women in public life', in *The Situation of Women in Poland: The Report of NGOs' Committee*, Polish Committee of NGOs – Beijing, 1995, Warsaw, March: 25–31.

Kathelene, L. (1994) 'Power and influence in state legislative policy making: the interaction of gender and position in Committee Hearing debates', *American Political Science Review* 88, 3 (September): 560–576.

Kowalik, T. (1993) 'Can Poland afford the Swedish model? Choices in post-communist societies', *Dissent* 40 (Winter): 88–96.

Kubik, J. (1994) 'The role of decentralisation and cultural revival in post-communist transformations', *Communist and Post-Communist Studies* 27, 4: 1–25.

Kuratowska, Z. (1996) 'Polish civil society today and the situation for women', unpublished paper.

Kurczewski, J. (1993) *The Resurrection of Rights in Poland*, Oxford: Clarendon Press.

Les, E. (1994) *Voluntary Sector in Post-Communist East Central Europe*, Washington DC: Civicus.

Lewis, P.G. (1994) 'Political institutionalisation and party development in post-communist Poland', *Europe–Asia Studies* 46, 5: 779–799.

Marody, Mira (1990) 'Perception of politics in Polish society', *Social Research* 17, 2 (Summer): 254–274.

Miriou, M. (1994) 'Open space: from pseudo-power to lack of power', *The European Journal of Women's Studies* 1: 107–110.

Nelson, J. (1994) 'Linkages between politics and economics', *Journal of Democracy* 5, 4 (October): 49–62.

Nowakowska, U. (1995) 'Women's rights', in Polish Committee of NGOs –

Beijing, 1995 *The Situation of Women in Poland: The Report of NGOs' Committee*, Warsaw, March: 11–19.

Offe, C. (1991) 'Capitalism by democratic design? Democratic theory facing the triple transition in East Central Europe', *Social Research* 58, 4: 865–892.

Olczyk E. (1993) 'Każda Lista ma Swoj Kobiet: Nieliczne, ale Znane', *Rzeczpospolita*, Warsaw, 20 July: No. 167(3511).

Orloff, A.S. (1993) 'Gender and the social rights of citizenship: the comparative analysis of gender relations and welfare states', *American Sociological Review* 58: 303–328.

Osterdahl, I. (1996) 'Zgodność Demokracji Parytetowej ze Szwedzkim Prawem Konstytucyjnym i Wyborczym', *Biuletyn Ośrodek Informacji i Dokumentacji Rady Europy* 3: 85–96.

Parliamentary Women's Group (1996) *Act Concerning the Equal Status of Women and Men*, Warsaw: Sejm (draft).

Pateman, C. (1989) 'The disorder of women', in *Democracy, Feminism and Political Theory*, Cambridge: Polity Press.

Pateman, C. (1991) in M. Lyndon (ed.), *Feminist Interpretations and Political Theory*, University Park: Pennsylvania University Press.

Phillips, A. (1992) 'Universal pretensions in political thought', in M. Barrett and A. Phillips (eds), *Destabilizing Theory: Contemporary Feminist Debates*, Stanford: Stanford University Press: 10–31.

Phillips, A. (1993) *Democracy and Difference*, Cambridge: Polity Press.

The Polish Committee of NGOs – Beijing (1995) *The Situation of Women in Poland: The Report of NGOs' Committee*, Warsaw, March.

Quigley, K. (1993) 'Philanthropy's role in East Europe', *ORBIS* 37, 4 (Fall): 581–591.

Regulska, J. (1993) 'Transition to local democracy: do Polish women have a chance?', in M. Rueschemeyer (ed.), *Women in the Politics of Postcommunist Eastern Europe*, Armonk, NY: M.E. Sharpe: 35–62.

Regulska, J. (1994) 'Strategies for women's approaches in regional and urban planning: women's participation in decision-making bodies', in *The Challenges Facing European Society with the Approach of the Year 2000*, Council of Europe: 184–197.

Regulska, J. (1995) *Women's Participation in Political and Public Life*, Background Document for European Conference titled 'Equality and Democracy: Utopia or Challenge', Council of Europe, Strasbourg, February: EG/DEM (95)10, 20 pp.

Regulska, J. (1997) 'Decentralisation or (re)centralisation: struggle for political power in Poland', *Environment and Planning C: Government and Policy* 20, 3: 643–680.

Rychard, A. (1993) *Reforms, Adaptation and Breakthrough*, Warsaw: IFiS Publishers.

Siegel, D. and Yancey, J. (1992) *Rebirth of Civil Society: The Development of the Nonprofit Sector in East Central Europe and the Role of Western Assistance*, Report by Rockefeller Brothers Fund.

Siemienska, R. (1994) 'Polish women as the object and subject of politics during and after the communist period', in B.J. Nelson and N. Caudhuri (eds), *Women and Politics Worldwide*, New Haven and London: Yale University Press: 610–624.

328

Staar, R.F. (1997) *Transition to Democracy in Poland*, New York: St Martin's Press.

Tetreault, M.A. (1994) 'Political space for women', paper presented at the annual meeting of the International Studies Association, Washington DC.

Vinton, L. (1993) 'Poland's social safety net: an overview', *RFE/RL Research Report*, No. 17, 23 April: 3–11.

Wadstein, M. (1996) 'Udział Kobiet w Rozwoju Demokracji w Krajach Nordyckich', *Biuletyn Osrodek Informacji i Dokumentacji Rady Europy* 3: 50–59.

Waller, M. (1994) 'Voice, choice and loyalty: democratisation in Eastern Europe', in G. Parry and M. Moran (eds), *Democracy and Democratisation*, London: Routledge: 129–151.

Warren, M.E. (1996) 'What should we expect from more democracy? Radically democratic responses to politics', *Political Theory* 24, 2 (May): 241–270.

Watson, P. (1993) 'The rise of masculinism in Eastern Europe', *New Left Review* 198 (March/April): 71–82.

Waylen, G. (1992) 'Rethinking women's political participation and protest: Chile, 1970–1990', *Political Studies* 40: 299–314.

Waylen, G. (1994) 'Women and democratization: conceptualizing gender relations in transition politics', *World Politics* 46 (April): 327–354.

Wolchik, S. (1994) 'International trends in Central and Eastern Europe: women in transition in the Czech and Slovak Republics – the first three years', *Journal of Women's History* 5, 3 (Winter): 100–107.

Zycie (1996) 'Kobiety Chca Bye Równe pod każdym wzlęden', 21 November: 3. 11 November: 3.

14

IMAGINED AND IMAGINING EQUALITY IN EAST CENTRAL EUROPE

Gender and ethnic differences in the economic transformation of Bulgaria

Mieke Meurs

Introduction

Clearly, the transformation currently under way in East Central Europe (ECE) is offering very different opportunities to different groups of people. Some individuals have quickly and easily become fantastically rich, while others find themselves struggling to find any employment sufficient to pay for heating and bread. Patterns in the distribution of these opportunities continue to be an under-analysed aspect of the economic transformation. The emergence of classes has been little examined. A literature on the impact of the transformation on women has begun to emerge, although data limitations have made early conclusions somewhat tentative. Furthermore, the differential impact of restructuring across ethnic groups has only just begun to be examined.

In this chapter I use Bulgarian data to examine the relative economic status of men and women and of Bulgaria's three main ethnic groups (the majority ethnic Bulgarians, and members of the minority ethnic Turkish and Roma populations) just prior to the transformation, by examining patterns of occupational segmentation and earnings. Using information on the early dynamics of economic restructuring, I suggest how the transformation is affecting the relative position of the various groups. Finally, I contrast the political responses of two groups, women and the Turkish population, to their changing economic position and suggest ways in which economic and political equality might be expanded.

Prior to 1989, women across East Central Europe faced similar patterns of occupational segmentation to those faced by Bulgarian women, and since that time they have also experienced similar dynamics of economic restructuring. As a result, the findings presented here may generalise to women in other ECE countries (see, for example, the work of Monica Fong and Gillian Paull 1993). Data on the position of ethnic minorities in other ECE countries before and after 1989 is much more limited, however, and generalisations must therefore be approached more cautiously.

Background: equality and inequality prior to 1989

Prior to 1989, social groups in ECE coexisted in a kind of 'imagined equality'. On the one hand, women and ethnic minorities were offered greatly expanded opportunities for education, paid employment and participation in institutionalised politics. The state's need to fully employ all available labour in order to achieve industrialisation goals and Stalin's policy of dealing with the 'nationalities question' through co-optation ensured this expansion of opportunities.

It would be hard to defend this increase in opportunity as equality, however. Recently, authors examining socialist policy on women have argued that the state provided 'too much' formal equality to women, and in doing so undermined the achievement of real equality between women and men. In a 1993 essay, Zuzana Kisczkova and Etela Farkasova illustrate the problem metaphorically, describing the reactions of women who are permitted, even forced, to enter the hallowed (but male-constructed) buildings of employment, politics and the democratic centrist state. Having been excluded from the design of the buildings, however, these women found the buildings alien and uncomfortable structures. Unable to choose the conditions of their emancipation, women experienced formal equality not as emancipation so much as a change in the conditions of oppression (albeit in some cases this change did offer some improvements in quality of life).

Another critique of women's emancipation in ECE argues that policies of increased employment and political participation reinforced or even increased inequality, by ensuring women's access to the workplace without challenging essentialist conceptions of the 'correct' division of household labour (a traditional patriarchal division of domestic activity). What had previously been a division of labour became, for women, a double burden, which prevented them from really participating equally in the workplace or in the political realm (Einhorn 1993a).

In the case of minority ethnic groups, similar sorts of formal equality were offered in workplace and political organisation, with a similar lack of recognition of distinct needs for expression of the cultural minorities. Official policy also failed to challenge existing conceptions of cultural hierarchy, such as the superiority of western 'civilisation' over oriental 'backwardness'. This expansion of certain formal equalities of opportunity within a structure organised and controlled by ethnic Bulgarian men resulted in clear patterns of occupational segregation for women and minorities.

The data presented below illustrate these patterns. The data are drawn from a 1986 survey of 6,000 Bulgarian households carried out by the Institute of Sociology of the Bulgarian Academy of Sciences. The sample is representative for the nation as a whole, but only the data on the 4,350 rural households have been analysed here. The survey is unusual in that it allows for the identification of ethnicity through the question of which languages the interviewee speaks fluently (including Bulgarian, Turkish and Roma). Language, of course, captures only one facet of ethnic identity. Individuals could identify themselves as members of a minority group without speaking the language, or members of one ethnic group could speak another ethnic language fluently. To eliminate the latter source of error, fully bilingual individuals (twenty-six people) have been dropped from the sample. In addition, social prejudices against minority groups might encourage under-reporting of minority language skills. Surprisingly, 18 per cent of rural households responded that they did speak Turkish in the home, while 4 per cent claimed to speak Roma (Institute of Sociology 1986). These data are approximately consistent with official data on the ethnic makeup of the population, which listed Turks as making up approximately 20 per cent of the rural population in 1992 and Roma as about 5 per cent. Turkish and Roma households are concentrated in rural areas – nationally they make up 10 per cent and 3 per cent of the population respectively (NSI 1993: 92). The derivation of ethnicity from language data prevents us from examining the economic situation of the Pomak ethnic group, which consists of ethnic Bulgarians who have converted to Islam. Members of this group made up approximately 3.5 per cent of the population in 1989 (Ilchev and Perry 1993).

With truly equal access to education and employment, all ethnic groups and both sexes would be expected to be relatively equally distributed across sectors of the economy. As can be seen from Table 14.1, there was a significant clustering of women and ethnic minorities in certain economic sectors prior to 1989. Members of the Turkish ethnic group tended to be clustered in agriculture (42 per cent of Turks worked in this sector, compared to 27 per cent of the sample), in part due to their residence in rural areas. Roma also held slightly more than their share of jobs in this sector in 1989, for the same reason. Women were particularly overrepresented in the low-paying service and commercial sectors. In the government sector, which offered the highest average wage, both minority groups and women were significantly underrepresented. All three groups are also underrepresented in the high-paying transport and industrial sectors, although to varying degrees.

Within each sector, women and members of ethnic minorities tended to be concentrated in the jobs that do not demand high qualifications, with some noteworthy exceptions. While 4.2 per cent of the sample worked in jobs requiring high qualification, only 3.1 per cent of Turks and 1.2 per cent of Roma held these jobs (Table 14.2). In particular, women and ethnic

Table 14.1 Sphere of employment by language competence and gender (figures as percentages)

Sphere	Turkish	Roma	Female	Sample	Av. wage 1989
Industry	15.4	6.9	15.3	16.2	3,475
Construction	4.5	3.1	0.8	3.6	3,670
Agriculture	41.7	27.7	23.6	27.3	3,033
Commerce	2.3	3.1	3.9	3.1	2,788
Service	4.4	1.9	7.4	5.3	2,702
Science and culture	0.2	0.0	0.4	0.4	2,944
Government	1.3	0.6	1.2	2.7	3,767
Transport	2.7	2.5	0.5	2.9	3,580
Private	0.4	1.3	0.2	0.5	
Retired	27.0	52.8	45.8	38.0	

Sources: Institute of Sociology 1986; CSO 1988.

Table 14.2 Qualification level by language competence and gender (figures as percentages)

Qualification	Turkish	Roma	Female	Sample
Industrial workers				
Low	18.2	13.8	18.1	16.9
Medium	19.3	13.2	15.6	22.0
High	1.3	0.6	0.7	2.1
Agricultural workers				
Low	19.2	56.0	50.3	39.9
Medium	6.1	10.1	4.6	7.9
High	1.3	0.0	0.3	0.8
Service workers				
Low	1.4	0.0	2.5	2.7
Medium	2.7	5.7	6.6	6.3
High	0.5	0.6	1.3	1.3
Total				
Low	68.8	69.8	70.9	59.5
Medium	28.1	29.0	26.8	36.2
High	3.1	1.2	2.3	4.2

Source: Institute of Sociology 1986.

minorities were underrepresented in the category of highly qualified industrial workers, some of the most lucrative jobs under central planning. In the sectors where they made up more than their share of employees, however, women and minorities were slightly more likely to have jobs requiring more qualifications. In the service sector, where women were concentrated, women gained equal access to the highly qualified jobs. Likewise, members of the Turkish-speaking group controlled more of the highly qualified jobs in the agricultural sector, where they were concentrated. Roma speakers were absent from highly qualified jobs in all sectors, but did achieve significant representation among workers in agriculture with middle-range qualifications.

Finally, Table 14.3 illustrates the participation of the three groups in enterprise decision-making. In the firms which were considered important enough to be kept under central state control, rural women and ethnic minorities held no positions of responsibility, although they were employed by these firms. In municipally controlled firms, the three groups again held no positions of high authority, but some did hold positions of some responsibility. Most notably, Roma speakers made significant gains and were significantly overrepresented in municipal jobs with some responsibility. In locally controlled firms, women and ethnic minorities also held some positions of responsibility, but were still greatly underrepresented.

To some extent, these patterns of occupational segregation can be linked to differences in schooling. Among those with post-high school education, women and members of the Turkish and Roma-speaking minorities were

Table 14.3 Firm leaders by language competence and gender (figures as percentages)

Leadership level	Turkish	Roma	Female	Sample
Firm under national control				
Highest	0.0	0.0	0.0	0.0
Lower	0.0	0.0	0.0	0.2
No leadership responsibility	0.9	3.2	0.8	2.3
Firm under municipal control				
Highest	0.0	0.0	0.0	0.1
Lower	0.0	3.2	0.2	0.5
No leadership responsibility	2.6	1.6	0.9	2.7
Firm under local control				
Highest	0.5	0.0	0.3	8.0
Lower	1.5	3.2	1.3	4.8
No leadership responsibility	94.6	88.9	96.5	89.1

Source: Institute of Sociology 1986.

significantly underrepresented, although people with post-high school education made up only 5 per cent of the entire sample. Those having completed primary school (eight years of schooling) or less made up 72 per cent of the sample, but 75 per cent of women had only this level of education, as did 82 per cent of Turkish and 84 per cent of Roma speakers. At the middle levels of education, groups also differed somewhat: 23 per cent of the sample had some middle-school or vocational education, as did 20 per cent of women, 16 per cent of Turkish speakers and 13 per cent of Roma speakers (Institute of Sociology 1986).

Surprisingly, the combination of concentration in sectors of the economy with lower average wages and in the less skilled and less responsible jobs in these sectors did not result in significant differences in average wages between Bulgarian and Turkish speakers (145.5 and 143 leva/month respectively).[1] This may be explained by the high prices the government offered for tobacco, produced mainly by Turkish agricultural workers, and by high wages in the construction industry, where Turkish speakers were also concentrated. The pattern of occupational segregation also did not result in statistically significant differences between Bulgarian and Roma speakers, although this is probably due to the small number of Roma speakers in the sample (110 of the total 4,350), since the wage gap was quite large (145.5 versus 132 leva). Between men and women, however, the wage gap was large and statistically significant (175.5 versus 118.2 leva) (Institute of Sociology 1986). Overall household incomes followed the same pattern as average wages: an average of 5,362 leva per year for all sample households, 5,362 for Turkish speakers and 5,128 for Roma speakers (Institute of Sociology 1986).

Impact of the transformation

Given these distinct patterns of pre-transformation employment, the current economic restructuring is likely to have very different effects on the three social groups analysed here. One of the most significant impacts of the restructuring has been the rise in unemployment. In 1995, unemployment stood at 16 per cent, down from a peak of 20 per cent in 1994 (European Commission 1996).

Early predictions were that unemployment would have a greater impact on women than men, as the light industries and government-run services where women were concentrated would be among the first businesses to close (Ciechocinska 1993). In addition, women were expected to be dismissed in larger numbers than men and to be re-hired more slowly because of their continued legal rights to maternity leave, leave to care for sick children, and other benefits in the state sector which made women relatively more costly to employers (Einhorn 1993a; Fong and Paull 1993).

The unemployment rates of men and women have not differed significantly in Bulgaria, however. In 1993, 22 per cent of Bulgarian women were registered as unemployed compared to 21 per cent of men. In 1995, the difference remained small: 16 per cent for women, 15.5 per cent for men (European Commission 1996). Nor have women left the labour force in significantly larger numbers than men, unlike in the Czech Republic, where the low level of national unemployment can be attributed to the withdrawal of women from the labour market. Despite a certain amount of popular pressure to spend more time in their 'natural' role caring for their families after the experiment with socialist equality, and despite the rapid decline in the availability and quality of child-care services, labour force participation rates fell almost equally for women and men. Ninety per cent of both men and women participated in the labour force in 1989, while 78 per cent of men and 75 per cent of women did so in 1995 (European Commission 1996).

The unexpectedly equal impact of restructuring on men's and women's employment may be explained by the rapid growth of new service industries. Light industry and state-run services, which had employed the majority of women before 1989, did indeed experience collapse or privatisation more rapidly than the heavy industry, transport, or construction sectors where the majority of men worked. But at the same time, nearly all new private sector jobs have been concentrated in the service sector – banking, restaurants and cafés, and retail stores. The majority of these new jobs have gone to women, especially young and attractive women. So while men have been slower to lose their jobs, some women have more easily found new jobs in the growing private sector.

The impact of restructuring on the relative earnings of women and men is harder to judge. Overall, real wages fell almost 50 per cent in 1991, recovered somewhat in 1992 and 1993 (about 20 per cent), and then fell another approximately 20 per cent in 1994 (European Commission 1995: 40). Data are not currently published on wages in the newly emerging private service sector where many women have found employment, however. Given the high unemployment rates and lack of unionisation or labour regulation in the private sector, we might expect low wages in that sector. But wages in the state sector have also fallen, and many state firms pay wages only intermittently, when cash flow permits. The overall impact of restructuring on relative wages is thus difficult to gauge without further data.

The traditional division of household labour has meant that restructuring has another difficult-to-measure impact on women's work. As fiscal and foreign exchange crises have created shortages of medicines, high prices for food, and decline in child-care services, women have carried much of the burden of searching for alternatives or preparing home-produced substitutes. Freed from the burden of standing in long lines for goods in the socialist economy, they now face a new set of tasks which detract equally from their ability to advance in paid employment.

Comparing the impact of economic restructuring across ethnic groups is complicated by the fact that unemployment data are not published for individual ethnic groups. However, some aspects of the ethnic distribution of unemployment may be inferred from electoral support for the Movement for Rights and Freedoms Party in the 1991 parliamentary elections (Figure 5.3, Begg and Pickles in this volume). This party is supported mainly by members of the ethnic Turkish minority.[2] Comparing this map with Figure 5.2 (Begg and Pickles in this volume), which illustrates regions of highest unemployment, suggests that the Turkish-populated areas face levels of unemployment well above the 16 per cent reported nationally. Many areas face levels from 25 to 90 per cent.

In part, the high regional unemployment rates can be explained by the importance of agriculture in Turkish areas and the particularly rapid collapse of collective tobacco production, which Turkish speakers dominated. The small, disbursed tobacco plots had never been effectively collectivised in any case, and were quickly returned to their previous owners after 1989 (Meurs and Begg 1998). De-collectivisation eliminated many of the skilled jobs in management and technical support, while dispersed land was far from adequate to support the population in private farming. In addition, much of the industrial employment in the Turkish-speaking regions was subsidised by a government anxious to offer year-round employment to members of tobacco farming households, thereby ensuring the continued production of this foreign exchange-earning crop. After 1989, such subsidies were no longer available and much of the industrial production collapsed.

Comparing data from Smolyan, one of the counties (*okrug*) in the Turkish-populated regions for which data are published, with national data, indicates that nominal wages in this region have increased much more slowly than the national average (NSI 1995: 99). Combining this with the high unemployment rates has certainly caused average household incomes to fall significantly, although regional data are not available on this issue.

Political reorganisation

Women were not permitted to vote in Bulgaria until after the socialists took power in 1944, and thus had little pre-war experience in organising political parties. Middle-class urban women were, nevertheless, an organised political force in Bulgaria prior to the Second World War. The Bulgarian Women's Union was formed in 1901, and by 1931 the national organisation had 8,400 members (Todorova 1993: 34).

A certain level of political participation continued under socialism, where some participation in party life was often a requirement for job advancement. Women never attained the official target of 15 per cent of state and party jobs, however (Petrova 1993). Of rural women surveyed in 1986, 1.1 per cent had been elected to some party or state position, compared to 3.3

per cent of men (Institute of Sociology 1986). Instead, women enjoyed a token level of representation in most political organs and were encouraged to participate in the separate women's organisations, which advised women on how to manage their joint duties of socialist worker and socialist mother.

Political activity across ECE in the years leading up to 1989 gave some hope that women would play an important role in the reorganisation of society after state socialism collapsed. In the German Democratic Republic, feminist women's organisations played an important role in the green and anti-nuclear movements which challenged the socialist government in the 1980s (Einhorn 1993a). In Poland and Czechoslovakia women were active in anti-government movements, although they did not put forth a feminist agenda (Siklova 1993). In Bulgaria, women took to the streets of the northern city of Ruse in 1989 to protest environmental problems in a demonstration which signalled the beginning of the end for the socialist government. The ECE experience is similar to that seen in Latin America in the 1970s and 1980s, where women used the conservative, state-espoused ideology of women as mothers, subject to special protections, in order to challenge state actions (Noonan 1995).

In the GDR, the feminist Independent Women's Association (UFV) of the GDR did play an important role in 1990 in the organisation of the Roundtable, a grouping of socialist and opposition forces which governed the GDR in the period between the collapse of the socialist state and the first free elections. The UFV won the right for women to sit at the table as an interest group, alongside unions and political parties. Once seated, the UFV insisted on certain feminist democratic processes, such as shifting from majority voting to an increased emphasis on discussion, compromise and decentralisation (Bohm 1993: 156). With unification, however, the Roundtable was dissolved, and the UFV (along with most other East German political forces) lost much of its influence, although it did continue to support two representatives in Parliament in the mid-1990s.

In Bulgaria, like most other ECE countries, women disappeared from the political scene even more quickly. The reformed Communist Party women's organisation remains the only mass membership women's organisation, but continues to limit its focus to homemaking issues (Panova, Gavrilova and Merdzhanski 1993). Oppositional politics are mainly limited to mainstream struggles for parliamentary seats and cabinet appointments, where women are decreasingly represented. Whereas women made up approximately 21 per cent of national legislators in Bulgaria before 1989, in 1990 they made up only 8 per cent of the Great National Assembly (formed in 1990 to pass basic reform legislation) (Kostova 1993: 107) and in 1991 they made up less than 12 per cent of Parliament (Todorova 1993: 36). Unsupported by a strong women's group, the few women legislators continue to avoid 'women's issues', such as defence of equal employment opportunities, preferring to focus on the more widely accepted 'family issues', such as poverty relief. Women do turn out in approximately equal numbers with men when demonstrations are called by trade unions or other organisations, but explicitly women's organisations do not play a role in the organisation of such events.

The history of organising among the Turkish minority in Bulgaria is distinct from women's experience in a number of ways. Members of the minority remaining in Bulgaria at the turn of the century enjoyed full voting rights.[3] While a separate national Turkish organisation or political party did not exist, members of the Turkish minority participated actively in local and national politics. In 1923, there were ten (male) Turkish members of Parliament. This number declined to four in 1933, however, with the rise of fascism in Bulgaria, and Turkish political organisations were banned altogether by the fascist government in 1934 (Stoyanov 1994: 270).

From 1944 to 1956, separate Turkish organisations enjoyed renewed life with support from the new socialist government. A Turkish Department was created in the Central Committee of the Communist Party, Turkish language schools and newspapers were expanded through a policy of cultural autonomy under the guidance of the socialist state (Moutafchieva 1994: 32).

After 1956, however, state policy shifted. The very existence of separate ethnic groups was increasingly denied, and the separate organisations were gradually eliminated. This did not prevent the participation of Turkish speakers in politics, however. In 1986, 1.1 per cent of rural Turkish speakers interviewed reported that they had been elected to a local or national office. This level of participation equalled that reported by women, despite the fact that the Turkish-speaking population made up only about 10 per cent of the population. In contrast, none of the Roma speakers interviewed had been elected to any office.

In the period just prior to 1989, over 350,000 ethnic Turks (about 40 per cent of the ethnic Turkish population) were expelled from Bulgaria (Vasileva 1991) in an apparent effort by the government to distract the Bulgarian population from pressing economic problems. This fiasco may have contributed to the discrediting of the socialist government, but during this period ethnic Turkish organisations were not an active force in Bulgarian politics.

Despite this, the ethnic Turkish minority has emerged as a clear political voice in the post-1989 period. With the political opening of 1990, almost half of the expelled ethnic Turkish population returned to Bulgaria (Vasileva 1991), and one of the many new political parties formed that year was the Movement for Rights and Freedoms (MRF). This organisation won 12 per cent of seats in the Great National Assembly in 1990. In 1991, the MRF won 10 per cent of seats in a Parliament dominated by the two main political forces, the reformed Communist Party (BSP), which won 44 per cent of seats, and the neo-liberal coalition of opposition parties (UDF), which held 46 per cent of seats. In the 1994 parliamentary election, the MRF won 8 per cent of seats to the 29 per cent held by the UDF and 52 per cent held by the BSP (Koulov 1995). In both Parliaments, MRF delegates voted as a bloc under the tight control of the central party organisation, and this allowed the MRF rapidly to become the critical swing vote and a powerful voice in national politics.

Although the party has been very active in cultural and political issues at the local level, to date the MRF has not often exercised this parliamentary power to pursue issues of specific interest to the ethnic Turkish minority. Between 1990 and 1992, the MRF did set improvement of economic conditions among the Turkish and mixed populations as a condition for supporting the government (FBIS 21 October 1990: 12; 27 October 1992: 1). But in more recent years there is little evidence that such demands have been forcefully pursued.

The contrast between the experiences of women's and of Turkish organising is striking. Why have the political experiences of interest group organising been so different for women and members of the Turkish-speaking minority in Bulgaria? One source of the difference may lie in the perceived potential for interest group politics to address the problems of the transition. While economic devastation is clearly concentrated in the regions populated by ethnic Turks, asymmetric economic impacts on women are harder to capture. The biggest economic hardship for women may well be the increased time and effort needed to assure household reproduction. This hardship appears in the first instance to affect the family as a whole, however, and the concentration of its impact on women is hard to measure. Organising specifically as women may not immediately emerge as a solution to these problems, whereas organising under the banner of the MRF may appear a more direct response to the problems faced by Turkish speakers.

Another explanation may lie in the differences in the cultural resources available to the two groups for use in organisation in 1989. The pre-1989 attack on members of the ethnic Turkish minority served to heighten their own awareness of their distinct culture. In the period just following the collapse of state socialism, language and religious freedom became prominent issues in the local politics of Turkish regions. This cultural mobilisation was also aided by sympathetic Turkish and Islamic organisations abroad, which donated resources for the construction of mosques, religious schools and the printing of ethnic and religious publications. The increased cultural mobilisation contributed to, and was in turn fed by, successful political organising by the MRF.

Women, in their protests against the state in 1989, mobilised under the conservative, state-sponsored rubric of women as mothers. In Latin America, some similar movements managed to transform the conservative identity into a more feminist one in the process of successful protest (Noonan 1995). The independent political action of Bulgarian women lasted only a few months, however, and was too short-lived to achieve such an evolution. After 1989, the very identity which had served to mobilise women contributed

340

to their demobilisation, as women/mothers turned their attention to the overwhelming task of household reproduction in a time of economic crisis.

This reaction was reinforced by the ideology of East European dissident movements, which emphasised the importance of the unified family as the realm of anti-politics under state socialism. As the writings of Mikhail Gorbachev or Vaclav Havel on women illustrate, opposition movements conceptualised individual rights only in the public sphere and did not extend them to the household (Todorova 1993; Eisenstein 1993). In many places (the GDR is an interesting exception), this context undermined the development of women's consciousness of themselves as a group with distinctive interests. Examining the lack of political organising among Czech women, Hana Havelkova notes: '[w]henever the serious problems of women are debated in the Czech lands, someone, usually a woman, raises the question: "What about the problems of men?"' (1993: 62). Women continue to act mainly in the name of the family unit.

Another explanation for different levels of political response may lie in the form of political organisation which emerged after 1989. During the socialist period, women in East Central Europe mainly participated in mainstream politics predominantly at the local level (Einhorn 1993b: 56, 1993a), where they addressed concerns of immediate relevance to their families and daily lives – schools, transport, food distribution and environmental problems. Such local problems were also the basis of the anti-state organising of 1989, in which women participated quite actively. In the period immediately following 1989, a variety of forms of local participatory governance emerged spontaneously to fill the vacuum left by the collapse of the state. These forms of governance did not conform to the basic liberal conception of democracy based on hierarchical structures in a well-defined public sphere. Instead, they were decentralised and participatory (and sometimes chaotic). Local women often continued to participate in these structures.

The development of institutionalised ('normal') politics quickly eclipsed these experiments, however. True to the liberal conceptions which inspired the dissident movements, institutionalised politics was hierarchical and increasingly limited to 'public' issues. Local institutions were stripped of resources as changes in tax policy shifted the increasingly limited resources to the national level (Martinez-Vasques 1995), centralising institutionalised politics. This form of liberal politics may have contributed to emancipation from the all-controlling state, but it did little to promote women's expression of their common interests. Instead, post-socialist politics became a male domain. With women's opportunities to participate in the building of new (less alien) social institutions diminishing, they may have little motivation to remain in the hallowed 'buildings' of politics and the state.

This lack of political participation does not mean, of course, that women are not fighting back. Many struggle daily in their homes and communities to retain jobs, protect access to child care, or simply hold on to the idea that they have a right to participate as full citizens in the emerging economy. Under the right conditions, such unorganised efforts may provide the basis for more structured actions.

The new institutionalised politics, in which two monolithic parties were locked in a relative stalemate, was a much more hospitable context for organising by the MRF. Male organisers in the MRF easily exploited the existing hierarchical structures of the religious community and territorial government to mobilise voters behind a hierarchical and centralised political party. This form of mobilisation contributed to electoral successes and parliamentary power.

Ironically, however, this 'successful' strategy seems to have yielded little in terms of protecting the interests of the ethnic Turkish minority. While before 1989 members of the Turkish-speaking minority enjoyed little access to the best paid and most responsible jobs, government policies assured individual and household incomes nearly equal to those of ethnic Bulgarian households. In the post-1994 period, the MRF has given the Turkish minority a disproportionately important voice in Parliament, but members of the ethnic group have experienced the most severe and concentrated economic devastation in the country. The MRF has created strength as a voting bloc, but it has not used this strength to support policies to prevent or address this devastation. In part, the MRF may be trapped in its own rhetoric, in much the same way as Bulgarian women. While women organised under the conservative banner of wives and mothers, as a political party the MRF faced a legal ban on the organisation of ethnically based parties and thus organised under a universalist banner of internationally established human rights. This claim to universalism may now complicate MRF claims for special attention to the economic and cultural problems of the ethnic minority which supports it.

Imagining increased political and economic equality

To date, Bulgarian women's apparent preference for decentralised, local, political organising has contributed to their relative exclusion from political power. Ironically, however, this preference could hold the key to more effective political organising against the increasing economic hardships faced by Bulgarian women and ethnic minorities

The centralised, hierarchical political organisation of the MRF has not met the needs of the ethnic Turkish population which has supported it. On the contrary, the centrist, urban-focused political model which it supports would appear to have contributed to the economic devastation of regions with high concentrations of ethnic Turks. Local branch plants have been closed in the interests of the central firms, and tobacco prices have been driven to new lows in what appears to be a scheme to privatise the tobacco monopoly cheaply to government insiders (see Begg and Pickles in this volume). The Sofia-based MRF leadership has done little to resist these processes and has done little else to protect the economic interests of its rural supporters.

The potential of alternative forms of organising to better address local economic needs is illustrated by Hillary Wainwright in her study of successful forms of democratic governance in Western Europe. The organisations which Wainwright examined were like the early forms of post-socialist organising – decentralised, participatory, with decisions based on discussion and compromise. These organisations are examples of 'democratic management of public provision, co-ordination, and regulation of the economy', in which local organisations use local expertise to meet local needs (1994: 10).

Clearly, in 1989 much of the population of Eastern Europe was in no mood to experiment with radical new forms of democracy, despite the claims of leaders of the democracy movements, particularly in the GDR. The attitude was one of 'You rich countries experiment, give us something that works!' and what worked may seem to many to be liberal democracy in its traditional form.

Still, a revitalisation of decentralised, local forms of government might offer greatly expanded opportunities for expression by women and ethnic minorities, both of whom have a strong history of local political partici-pation. At the same time, by focusing political control at the local level, such organisation would provide an institutional base for expanded control of economic resources. Villagers across Bulgaria have used local political structures to co-ordinate the collective purchase of liquidated collective farm assets for co-operative use (see Meurs and Begg 1998 and in this volume). Expanded local governments in de-industrialising regions might co-ordinate collective purchase of other local assets slated for liquidation if the population were mobilised for activity at this level,[4] or women might use local political structures to co-ordinate continued child-care services. Such a change in the focus of political organising would not change the fundamental conditions of the current transformation. Unemployment levels would remain high and government spending greatly restricted. But a change in the form of organising might permit greater equality of political expression and greater control of policy by those most negatively affected by the transformation. Imagine!

Acknowledgements

The author wishes to thank the International Research Exchanges Board, the John D. and Catherine T. MacArthur Foundation, and the US National Science Foundation Social and Behavioral Research Program (Grant 9515244) for supporting the research on which this chapter is based.

Notes

1 The hypothesis that the population means are equal is not rejected at p = .01 using a separate-variance t-test. This same standard is used for other findings of significance reported here.
2 Parties purporting to represent the interests of separate ethnic groups are illegal. The MRF thus formally defines its goals in terms of national development and cultural understanding. Nearly all of the party's supporters are ethnic Turks, however.
3 Most of the Turkish population left Bulgaria when it was freed from Ottoman control in the late 1800s, but some ethnic Turks chose to remain.
4 These assets may be subject to liquidation by central firms, which perceive a negative rate of return on the assets. Local populations may include employment and other positive externalities in their valuation of potential returns, and thus find it economically beneficial to purchase these assets, which in any case are liquidated at fire sale prices.

Bibliography

Bohm, T. (1993) 'The women's question as a democratic question: in search of civil society', in N. Funk and M. Mueller (eds) *Gender Politics and Post-Communism: Reflections from Eastern Europe and the Former Soviet Union*, New York: Routledge: 151–159.

Ciechocinska, M. (1993) 'Gender aspects of dismantling the command economy in Eastern Europe: the case of Poland', in V. Moghadam (ed.) *Democratic Reform and the Position of Women in Transitional Economies*, Oxford: Clarendon Press: 302–326.

CSO (1988) *Statistical Yearbook of the People's Republic of Bulgaria*, Sofia: Central Statistical Office.

Einhorn, B. (1993a) *Cinderella Goes to Market: Gender, Citizenship, and Women's Movements in East Central Europe*, New York: Verso.

Einhorn, B. (1993b) 'Democratisation and women's movements in Central and Eastern Europe: concepts of women's rights', in V. Moghadam (ed.) *Democratic Reform and the Position of Women in Transitional Economies*, Oxford: Clarendon Press: 48–74.

Eisenstein, Z. (1993) 'Eastern European male democracies: a problem of unequal equality', in N. Funk and M. Mueller (eds) *Gender Politics and Post-Communism: Reflections from Eastern Europe and the Former Soviet Union*, New York: Routledge: 303–317.

European Commission (1995) *Employment Observatory: Central and Eastern Europe*, 7.

European Commission (1996) *Employment Observatory: Central and Eastern Europe*, 8.

FBIS (Foreign Broadcast Information Service) (1990, 1992, 1993) Daily Report. Washington DC.

Fong, M. and Paull, G. (1993) 'Women's economic status in the restructuring of Eastern Europe', in V. Moghadam (ed.) *Democratic Reform and the Position of Women in Transitional Economies*, Oxford: Clarendon Press: 217–247.

344

Hauser, E., Heyns, B. and Mansbridge, J. (1993) 'Feminism in the interstices of politics and culture: Poland in transition', in N. Funk and M. Mueller (eds) *Gender Politics and Post-Communism: Reflections from Eastern Europe and the Former Soviet Union*, New York: Routledge: 257–273.

Havelkova, H. (1993) 'A few pre-feminist thoughts', in 'Eastern European male democracies: a problem of unequal equality', in N. Funk and M. Mueller (eds) *Gender Politics and Post-Communism: Reflections from Eastern Europe and the Former Soviet Union*, New York: Routledge: 62–73.

Ilchev, I. and Perry, D. (1993) 'Bulgarian ethnic groups: politics and perceptions', *RFE/RL Research Report* 2, 12 (March): 35–41.

Institute of Sociology (1986) *Town and Village Study Survey Data*, Sofia: Bulgarian Academy of Sciences.

Kisczkova, Z. and Farkasova, E. (1993) 'The emancipation of women: a concept that failed', in N. Funk and M. Mueller (eds) *Gender Politics and Post-Communism: Reflections from Eastern Europe and the Former Soviet Union*, New York: Routledge: 84–94.

Kostova, D. (1993) 'The transition to democracy in Bulgaria: challenges and risks for women', in V. Moghadam (ed.) *Democratic Reform and the Position of Women in Transitional Economies*, Oxford: Clarendon Press: 92–109.

Koulov, B. (1995) 'Democratization "from above" and its limitations', Mimeo, Washington DC: American University.

Martinez-Vasques, J. (1995) 'Central and local tax relations', in Z. Bogetic and A. Hillman (eds) *Financing Government in the Transition: Bulgaria*, Washington DC: World Bank.

Meurs, M. and Begg, R. (1998) 'Path dependence in Bulgarian agriculture', in J. Pickles and A. Smith (eds) *Theorizing Transition: The Political Economy of Change in Central and Eastern Europe*, London: Routledge.

Moutafchieva, V. (1994) 'The Turk, the Gypsy and the Jew', in *Relations of Compatibility and Incompatibility between Christians and Muslims in Bulgaria*, Sofia: Intercultural Centre for Minority Studies and Intercultural Relations Foundation: 3–63.

Noonan, R. (1995) 'Women against the state: political opportunities and collective action frames in Chile's transition to democracy', *Sociological Forum* 10, 1: 81–111.

NSI (National Statistical Institute) (1993) *Demografska Kharakteristica na Bulgaria*, Sofia: National Statistical Institute.

NSI (National Statistical Institute) (1995) *Statistical Yearbook*, Sofia: National Statistical Institute.

Panova, R. Gavrilova, R. and Merdzhanski, C. (1993) 'Thinking gender: Bulgarian women's im/possibilities', in N. Funk and M. Mueller (eds) *Gender Politics and Post-Communism: Reflections from Eastern Europe and the Former Soviet Union*, New York: Routledge: 15–21.

Petrova, D. (1993) 'The winding road to emancipation in Bulgaria', in N. Funk and M. Mueller (eds) *Gender Politics and Post-Communism: Reflections from Eastern Europe and the Former Soviet Union*, New York: Routledge: 22–29.

Siklova, J. (1993) 'Are women in Central and Eastern Europe conservative?' in N. Funk and M. Mueller (eds) *Gender Politics and Post-Communism: Reflections from Eastern Europe and the Former Soviet Union*, New York: Routledge: 74–83.

Stoyanov, V. (1994) 'The Turks in Bulgaria', in *Relations of Compatibility and Incompatibility between Christians and Muslims in Bulgaria*, Sofia: Intercultural Centre for Minority Studies and Intercultural Relations Foundation: 268–271.

Territorial Statistical Bureau-Smolyan (1995) *Statistical Handbook Smolyan 1994*, Smolyan, Bulgaria: Statistical Publishing and Printing.

Todorova, M. (1993) 'The Bulgarian case: women's issues or feminist issues', in N. Funk and M. Mueller (eds) *Gender Politics and Post-Communism: Reflections from Eastern Europe and the Former Soviet Union*, New York: Routledge: 30–38.

Vasileva, D. (1991) 'Bulgarian Turkish emigration and return', *International Migration Review* 26, 2: 342–352.

Wainwright, H. (1994) *Arguments for a New Left: Answering the Free Market Right*, Oxford: Blackwell.

15

DEMOCRATISATION AND THE POLITICS OF WATER IN BULGARIA

Local protest and the 1994–5 Sofia water crisis

Caedmon Staddon

Introduction: geographies of water and power in Bulgaria

Storming the barricades at Sapareva-Bania, Bulgaria, 8 February 1995

In the pre-dawn hours of 8 February 1995 special troops of the Bulgarian Interior Ministry mustered in some fields near the picturesque resort town of Sapareva-Bania (Figure 15.1). A brigade of mounted police was put on alert in nearby Samokov. Their immediate objective was to break through a two-month-old barricade that local citizens had put across an access road into the Rila Mountains (Plate 15.1). This was deemed necessary in order to guarantee access to construction crews that were to construct a pipeline to divert water from the Skakavitsa River to the Djerman River, a distance of some 3 kilometres, and thence into the system that supplies the capital city, Sofia, with drinking water. Sofia, a city of 1.2 million people, was at that time experiencing a serious drinking water shortage resulting in an extensive water-rationing programme and significant impacts on regional industry and agriculture. Public health authorities also expressed concern about the effects of the lack of potable water on sanitation, and rumours about cholera and dysentery in various parts of the city were rampant (e.g., *Demokratsiya* 18 January 1995). Members of the National Parliament had even suggested that a large proportion of the population should be evacuated from the capital during the summer of 1995 if water supplies did not improve (*BTA Daily Bulletin* 2 February 1995). Notwithstanding these problems in the capital, those opposing the construction of Djerman–Skakavitsa demanded that their own long-standing water shortages be addressed. Indeed, a government

347

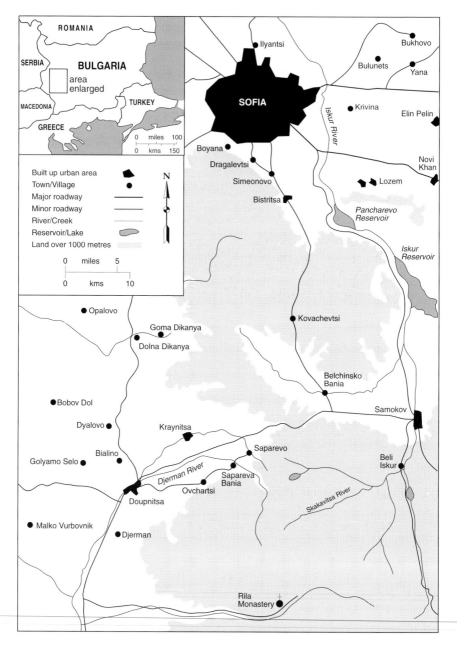

Figure 15.1 Bulgaria study area map

Plate 15.1 Barricades at Sapareva-Bania, 8 February 1995
(Source: Green Patrols Video, 1995, with permission)

decree (Decree No. 137) of late 1990 had already explicitly declared that no water was to be taken from the Rila communities until their own security of supply had been assured (*Zemedelsko Zname* December 1994, no. 224).

Shortly after dawn the troops moved in, supported by heavy trucks and wearing full riot gear. To inhibit any police sympathy or leniency towards demonstrators, the Bulgarian Interior Minister had shrewdly replaced the local police detachment with troops ordinarily based in Sofia. Their orders were to clear the barricades and secure the road for the passage of waiting construction vehicles (*BTA Daily Bulletin* 13 February 1995). Fearing just such an application of brute force, local activists had implored the protesters to be passive, to avoid any action that could be interpreted as aggression or incitement. Against the motley opposition of farmers, unemployed miners, mothers with children, retirees, environmental activists and the not inconsiderable retinue of spectators marched the disciplined ranks of grey-uniformed national policemen carrying clubs, rifles and the weight of central government (Plate 15.2). Videotape of the event shows many elderly Bulgarian women pleading with the soldiers to turn back, to understand and empathise with their situation, even as the military phalanxes removed the barricades and brusquely pushed people out of the way of the oncoming construction trucks. Nineteen people were arrested and about a dozen were injured, including one person who was seriously injured in what the Interior Minister later referred to as 'a textbook model of police work guided

Plate 15.2 Troops marching on protestors at Sapareva-Bania, 8 February 1995 (Source: Green Patrols Video, 1995, with permission)

by principles of democracy and respect for human rights' (*BTA Daily Bulletin* 13 February 1995). Additionally, in a move reminiscent of Soviet-era practices, several journalists had their cameras confiscated and their film destroyed by police (Staddon 1995, no. 23).[1] A large police force remained in the area through to the end of February 1995, guarding the road and harassing locals. Construction of the diversion was completed in late March 1995.

Political change in Central and Eastern Europe: reconfiguring the political imaginary

This chapter is about the conditions under which a particular political and environmental problem, scarcity of potable water, became a key fulcrum of political contestation between the rural communities of North Rila (Figure 15.1) and the Bulgarian central state as that nation continues to experience the paroxysms of post-state-socialist transition. More than just a straightforward struggle over the division of water resources, however, the protests against Djerman–Skakavitsa were fundamentally bound up with attempts by different social groups within Bulgaria to come to terms with the complexities and ambiguities of 'transition'. Indeed, the highly divisive issue of what normatively constitutes 'transition', and in particular 'successful' transition, is key to understanding the Sapareva-Bania protests. Under conditions of wholesale social and political transformation, in which the

350

informal 'politics of the street' played such an important role, the problem of what constitutes the 'democratisation' of political discourse and action looms large. Zygmunt Bauman (1992), for example, has argued that this current phase of the transition process is potentially a very dangerous one, where the inchoate tide of social transformation hurls itself against the twin buttresses of the equally entrenched *nomenklatura* and anti-communist classes. In this chapter I argue that careful examination of the political constellations created by the Djerman–Skakavitsa protests helps us to understand the realities of the post-1989 transition by highlighting the resources (cultural, political, social and economic) upon which those on both sides of the issue drew in developing their respective discursive representations of the crisis. Once these representations are fully appreciated it becomes clearer that what was ultimately at stake in this instance of local protest was the very reconceptualisation of the 'political *as such*; the political imaginary of post-communist "governmentality" (Foucault 1991).

The critical issue then involves how the political antagonists on the issue of Djerman–Skakavitsa both laid claim to the (western/westernising) discourse of democracy to legitimise diametrically opposed positions, including civil disobediance on the one hand, and the state's monopoly over the means of violence on the other. To these ends it has proved useful to employ Foucault's (1991: 93) concept of 'governmentality' as entailing a deeper sensitivity to people 'in their relations, their links, their imbrication with those other things which are wealth, means of subsistence, the territory with its specific qualities . . . customs, habits, ways of acting, etc. understanding what took place on 8 February 1995 at Sapareva-Bania requires a close examination of the relations between the subjects and objects of institutionalised *and institutionalising* power. The power to institutionalise subject–object relations here is the prize of achieving discursive hegemony, and the ability to destabilise such institionalising power is consequently the ultimate expression of local autonomy. In this chapter I argue that just such a set of destabilisations was constituted through three major discursive formations: first, those that addressed the experience of peripheralisation in North Rila within the context of post-1989 fast capitalism; second, the relations of political identity to local social structures; and third, the rootedness of local cultural identity in place. Overall I stress, following Pred and Watts (1992), Bauman (1992) and others, the point that social location and cultural process cannot be left out of our attempts to understand the contours of post-communist transition. Indeed Pred and Watts (1992: 18) have suggested that 'it is the *production of cultural difference within a structured system of global political economy* that is central to an understanding of both the territory-identity relation and to the reworking of modernity' (emphasis in original).

In the following sections I treat each of the major competing discourses about Djerman–Skakavitsa in turn. For both sides I claim that the emerging structures of 'fast capitalism' in Bulgaria are serving to perpetuate, rather than obviate, deeply rooted core–periphery and rural–urban cleavages. I therefore touch on some of the ways in which processes of market transition (de-statisation, marketisation, privatisation) are perpetuating geographically uneven development in Bulgaria, especially insofar as they helped constitute emergent discourses about the water shortage. The pro-diversion opinion broadly held in Sofia was constituted through a discourse of distanciated responsibility for the water crisis, a long-standing sense of socio-cultural primacy, and the political machinations of national politicians. I then turn more directly to the constitution of opposition to Djerman–Skakavitsa in Sofia as well as in Sapareva-Bania. I examine in more detail the internal constitution of the *social movements* involved in the Sapareva-Bania episode, the environmental NGOs, the Civil Initiative Committee for the Protection of Rila Waters and the other local institutions, particularly the Orthodox Church, the school and the community centre, that helped support the citizens' protest actions. Implicit in this treatment is a challenge to Habermas' celebration of social movements as transparently autonomous vehicles of a post-identitarian politics (Habermas 1987; Warren 1995). The Bulgarian case study reveals a very specific geographical differentiation in the structure and content of social movements which cuts across and greatly complicates the primary cleavage between the core and the periphery. In other words, in the case of Sapareva-Bania there was no easy alliance between environmental and local protest movements. In the concluding section I consider the implications of the specific expression of local autonomy at Sapareva-Bania and the subsequent police action it provoked for the restructuring of Bulgarian politics and for the theorisation of political change in Central and Eastern Europe.

The Sofia water crisis and Djerman–Skakavitsa: social discourse and politics

My aim in this section and the next is to elaborate on the ways in which social and economic change since 1989 has contributed to the constitution of competing discourses about the meaning of Djerman–Skakavitsa. Thus, I sketch out the competing water discourses of the central state and the locality, highlighting their divergent forms and their socio-epistemological structures of intelligibility. This part of the project could be presented as an attempt to delineate the contours of the 'representational spaces' inherent in normalised modes of discourse about transition, local–central relations, and Djerman–Skakavitsa (Lefebvre 1991). The distinction sought here is well expressed by de Certeau (1984: xiii):

> the presence and circulation of a representation . . . tell us nothing about what it is for its users. We must first analyse its manipulation by users who are not its makers . . . the tactics of consumption, *the*

352

ingenious ways in which the weak make use of the strong . . . lend a political dimension to everyday practices. (emphasis added)

Thus I will focus on the different ways in which residents of Sofia and North Rila structured their thinking and speaking about the water crisis and the sorts of spatial praxis that resulted. It will quickly become apparent that the two sides were constituted as antagonists not only on the basis of a manifest disagreement about the necessity of the Djerman–Skakavitsa diversion. Rather, the fundamental intransigence of the problem arises from the fact that they inhabited two radically different representational spaces in which democratisation, local autonomy and environmental management meant something different to each. Certainly these were a product of differing experiences of economic transition, but they were also configured through the continuation of long-standing divisions between core–periphery, urban–rural and the problem of Macedonian identity.

The 1994–5 Sofia water crisis

On 21 November 1994, the caretaker government of Renata Indjova imposed the first water regime on residents of the capital, with running water provided to private households for only twelve consecutive hours in every thirty-six-hour period (*Demokratsiya* 23 January 1995). Between that time and June 1995 when they were lifted, the regimes were intensified to 3:1 (eighteen hours on, fifty-four hours off) and expanded to include ever more areas of the city. By the end of April 1995, roughly two-thirds of the city was subject to water rationing (Staddon 1996). There was a general belief that, since Indjova's government was only a temporary bridge between the resignation of the BSP–MRF coalition government of Lyuben Berov in late September 1994 and the national elections scheduled for December 1994, it alone could afford to bear the political costs of taking such drastic measures.[2] Certainly the problem of timing is important since the prospect of water shortages had been the subject of popular political discourse since at least the early 1990s (*Standart* 8 January 1995). Knight, Valev and Staneva (1995) note that conditions for chronic water shortages in many Bulgarian towns and cities are the result of a decade-long drought, the very poor state of the physical infrastructure, and the complex nature of restructuring in the water sector. But the timing of the water crisis was fortuitous for all political parties, as each could play the issue to its advantage without having to assume any significant measure of responsibility. Thus the majority-BSP government of Zhan Videnov elected in the December elections found it quite expedient to continue the Indjova-imposed water regime while simultaneously denouncing past government inaction and noisily consti-tuting special 'expert' commissions to deal with the problem. Handling of issues such as the water crisis and the general economic crisis featured prominently in the White Paper issued by the BSP cabinet in March 1995 (BTA Special Report 1995).

Effects of the water regime were immediately apparent in the capital. Even those who were spared the water regime were compelled to procure mineral spring or creek water for drinking purposes, as the level of the Iskur Reservoir had sunk well below the level required for providing even 'conditionally clean water'. Everywhere in the city one saw citizens carrying reused soda bottles, large water jugs and even open pails for filling at one of several public taps located throughout the city (Plate 15.3). Fortunately, the Sofia basin is well supplied with mineral springs, so that alternative water sources were available even to those reliant upon public transport. In fact there is a long-standing tradition of utilising for domestic purposes the water from springs such as those outside the now closed Central Mineral Baths in downtown Sofia. Other springs at suburban locations such as Gorna Bania and Bankya were also favoured by many residents of the capital. Unfortunately, by mid-April 1995 many of the most centrally located taps had been closed by municipal authorities for reasons of 'bacteriological pollution', thereby closing off this critical avenue for augmenting domestic water supply. Intended water savings from the water regime were also blunted by the widespread practice of filling every available household vessel during the eighteen hours of water supply every three days. Bathtubs, jugs, soda bottles and even improvised cisterns were filled to cushion the subsequent two and a half days of water shut-off. All of this led some to doubt that the regimes resulted in any real conservation effects at all (Knight, personal communication July 1995; see also *Pari* 8 June 1994).

Plate 15.3 Sofians filling water bottles at public taps, Central Mineral Baths, Sofia (Photograph by the author, February 1995)

Fast capitalism in post-1989 Bulgaria

This atmosphere of crisis was greatly compounded by the dire condition of the Bulgarian economy and by the rapidly widening rift between the small class of the prosperous and the much larger class of the financially insecure. Seven years of post-1989 transition have been painful for Bulgaria as a whole, and for North Rila in particular. Table 15.1 shows that, while nominal GDP increased between 1989 and 1994, real GDP, denominated in US$, has fallen by almost 50 per cent. During 1995 and 1996, real GDP increased slightly, though the effects of these gains on an economy now plunged again into crisis are unclear (*BTA Daily Bulletin* January 1997). Official unemployment has risen to approximately 13 per cent of the working population, though this considerably underestimates realities. Worst of all for general standards of living, the consumer price inflation index has recently risen rapidly, with estimated inflation for 1996 at over 1,000 per cent (*BTA Daily Bulletin* January 1997). Frequent changes of government since 1989 have exacerbated the trials of economic transition. Having failed to negotiate the crisis within the state socialist development model that initiated the current period of reform and transition, Bulgaria has lapsed into a form of what Ben Aggar (1989) has called 'fast capitalism'. This has been typified by multiple and constantly changing articulations between formal and informal, legal and illegal, productive and non-productive, state and private spheres of economic activity. Pre-existent forms of social organisation and inequality have not been done away with, but rather have been grafted onto new conditions in a way analogous to Pred and Watts's (1992: 28) observations about Nigerian fast capitalism's 'articulation of precapitalist and capitalist institutions, and of new material practices with deeply sedimented cultural forms'. In the following paragraphs I investigate some of the main ways in which an emergent, though as yet unstable, mode of accumulation is being grafted onto a new politics of peripheralisation of North Rila.

Wyzan (1996) points out that Bulgaria's transition experience has been heavily conditioned by its relative lack of access to foreign capital investment and the convoluted nature of financial sector restructuring. With respect to the latter, it now appears that lax credit terms and extensive plundering of state assets by private holding companies such as Multigroup and Tron helped bring about the collapse of the banking sector in 1996 and 1997 (Angelov *et al.* 1994). In addition to deposit banking activities, these 'economic formations' have been heavily involved in 'cherry-picking' the most lucrative state enterprises, and mutualising the mass privatisation vouchers, thus compounding their nominal capital assets and simultaneously defeating the anti-monopolist impulses of Videnov's voucher privatisation scheme.[3] Indeed, the activities of these holding companies have been so pervasive and extensive that economists of the Bulgarian Academy of Sciences have commented:

Table 15.1 Bulgarian macroeconomic indicators, 1989–96[a]

	1989	1990	1991	1992	1993	1994	1995	1996
GDP (billion leva)	35.6	45.4	131.0	195.0	299.0	543.0	—	—
GDP (billion US$)	17.6	6.9	7.5	8.4	10.8	10.0	13.0	—
GDP per capita (US$)	1,957.0	769.0	836.0	990.0	1,280.0	1,184.0	1,546.9	—
GDP share of private sector	n.a.	9.1	11.8	15.3	19.4	28.0	—	50.0 (est.)
Unemployment (000s)	0.0	65.0	419.0	577.0	626.0	488.0	—	—
Registered unemployment (%)	0.0	1.6	10.8	15.5	16.4	12.8	10.8	12.5
Average monthly wage: leva	274.0	378.0	1,012.0	2,047.0	3,145.0	4708.0	—	12,290.0
US$	136.0	58.0	58.0	88.0	114.0	87.0	—	25.0
Annual % change in:								
Real GDP	−1.9	−9.1	−11.7	−5.7	−2.4	1.4	2.1	−9.0
Gross industrial production	−1.1	−16.8	−22.2	−15.9	−6.9	4.5	—	−6.0
Gross agricultural production	0.8	−6.0	0.0	−12.0	−18.2	0.8	—	—
Gross fixed investment	−0.5	−25.1	−15.6	−26.3	−29.7	n.a.	—	—
Consumer price inflation	10.0	72.5	473.7	79.6	64.0	121.9	33.0	311.0
Producer price inflation	n.a.	n.a.	284.0	24.9	15.3	91.5	—	257.1
Nominal average wage growth	8.7	38.0	167.7	102.3	53.6	51.4	—	161.0

Sources: Wyzan 1996: 82; Bulgarian Economic Review various dates; OECD 1997.
Note
a 1996 figures reflect end-of-year estimates taken from the bi-weekly Bulgarian Economic Review, published in Sofia by the Pari Publishing Group, and the OECD 1997 Country Report for Bulgaria.

> What exists in this country rather is a semi-criminal gangster economy where normal settlement mechanisms often give way to the arsenal of armed thugs. Some law enforcement authorities passively or actively help gangster formations. Instead of being marketised, the Bulgarian economy is being mobsterised.
>
> (Angelov *et al.* 1994: 10)

More worrisome still is the fact that these processes of 'shadow transition' have become identified in popular discourse with new political structures and the emergence of a new political elite (Crampton 1997).

The centre speaks: Sofia and the discourse of crisis

Seen in this political-economic context it is perhaps not surprising that ordinary Sofianites were by and large unsympathetic towards the mounting local opposition to VK-Rila in North Rila. Everyday conversations over coffee, in parks, around offices, tended to converge on a core set of representations of both the water crisis and on the local opposition movement against it. Central to the discourse promulgated in the centre was the commonly held notion that *Sofianites were not themselves responsible for the shortages.* A review of press reports, which covered more than three years of water issues coverage in the major daily newspapers, turned up few invocations of household profligacy as a significant contributing factor to the shortage (Staddon 1996). Over 50 per cent of newspaper articles discussing the water shortages blamed government 'mismanagement' or 'malfeasance' for the water crisis, with 30 per cent of articles indentifying Bulgaria's economic 'mafias' as the culprits (Staddon 1996). Interestingly, given the pronounced drought Bulgaria has experienced over the last decade (Knight *et al.* 1995), only a small handful of articles suggested that climate played more than a minor role. In some instances water sector experts made the case that household use patterns played a role and had to be altered (e.g., Paskalev 1994), but such considerations were not a common part of public discourse. In general it was held that the crisis was caused by some combination of meteorological/technical and social/political factors, drawing attention to the prolonged drought period and the insalubrious state of the urban water supply network. Technical evaluations have repeatedly pointed out that the 4,000 km of antiquated water mains that comprise the supply network lose more than 50 per cent of their load between reservoir and end user (Staddon 1996, no. 2; Knight *et al.* 1996). Even so, both technical specialists and political leaders tended to report the shortage in terms that sustained the view of the capital's blameless need for new water supplies to avert the impending crisis and its democratic right to take it from North Rila. In short, *continued extensive development* of the water sector was consistently proposed, rather than its intensification, upgrading and improvement.

Politicians were quick to pick up on this discourse of personal non-responsibility, redeploying it to underwrite solutions that combined immediate political advantage with a definitive solution to the Sofia water shortage. Former Sofia mayor Yanchoulev was a vociferous proponent for the reinstitution of the VK-Rila water diversion scheme (*Troud* 16 January 1995; *Demokratsiya* 27 January 1995). So, too, leading central government politicians such as Doncho Konakchiev (BSP Minister of Territorial Planning and Construction) and former Prime Minister Renata Indjova came to the meetings at Sapareva-Bania with a preset and unchangeable agenda, seeing them only as a vehicle for silencing local opposition and for posturing on the Sofia political stage. In one memorable episode Indjova postured outside a meeting hall near Sapareva-Bania rather shamelessly proclaiming for the assembled Sofia-based journalists that 'Sofia is dying of thirst! All good Bulgarians must do what they can!' To this hyperbole, a local loudly retorted, 'Come turn *my* taps and see what you get!' Once again playing to the gallery, former Prime Minister Indjova stormed out of one of the last meetings with her phalanx of 'advisers' loudly branding the local contingent as disloyal 'separatists' (Staddon 1995, no. 11; Green Patrols 1995). The interesting issue in this situation is not that the capital was at loggerheads with North Rila, but rather that all political factions appealed to the very same mode of governmentality, the very same mode of 'relations, links and imbrications' between people and water resources and between rural and urban populations. This incipient mode of governmentality received amplification through the symbolically powerful vocabulary of a 'democratic politics'.

This discourse of non-responsibility also fits well with the parallel discourse about the 'mafia-isation' of the Bulgarian economy. It is a generally accepted proposition that the Bulgarian economy is increasingly being controlled by a group of 'economic formations' known collectively as the 'G-13',[4] which includes the aforementioned companies Multigroup and Tron. Specific charges linking these groups to the water crisis have been alleged from various quarters. Perhaps most prominent is the claim that the giant Kremitkovtsi metallurgical works located just outside Sofia, which is partly controlled by Multigroup through a subsidiary holding company, had access to potable water for industrial purposes throughout the period of water rationing in the capital (*Standart* 20 January 1995; *Pari* 1 June 1994). Additionally there was the claim that the National Electricity Company (NEK) had been diverting millions of cubic metres of water from the Iskur Reservoir to produce hydroelectric power for export to Greece, as well as the claim that the infrastructure was being deliberately run down in order to facilitate a joint takeover of the Sofia municipal ViK by Bulgarian and foreign interests.[5] With extremely bad timing, the Bulgarian and Greek governments chose this moment to reopen treaty negotiations on the allocation of water resources from the Mesta and Strouma Rivers in southern

Bulgaria, thus fuelling conspiracy-oriented speculations in the yellow (tabloid) press.

These theories of the water crisis were inflected through the current 'highly bipolar' political situation: with two major political groupings vying for control of government in the December 1995 national elections, it was perhaps inevitable that each would attempt to use the water crisis for partisan advantage. For members of the opposition Union of Democratic Forces the implication of Multigroup in the Sofia water crisis is tantamount to clandestine BSP involvement. In this view, the BSP is manipulating the water crisis in order to curry favour with the overwhelmingly 'blue' Sofia electorate. According to an editorial in *Demokratsiya* (10 February 1995):

> The Red strategists estimate with cool heads and clean hands that the gain in dividends in Sofia is a goal that justifies the means. The goal is to stroke the naughty blue Sofian, tearing him away from the VDF by insinuating that the VDF does not defend the suffering citizens of the capital and even sets its face against their interests. In other words, the left hand reaches out for the right vote.

Not surprisingly, the BSP rejects this view out of hand, choosing instead to emphasise the technical factors contributing to the problem arising from problems of drought compounded by the incompetence of the pre-1989 regime and the governments of the 1990–4 period (i.e., the UDF). In March 1995 the newly elected BSP government published a White Paper outlining their diagnoses of the malfeasance of previous governments. With respect to the water problem, the White Paper stressed that the resource pricing system was 'ambiguous', that the infrastructure had not been maintained (Bulgarian Telegraph Agency 24 March 1995) and that the VK-Rila diversion system must be completed.

Popular analysis of the water crisis in Sofia was characterised by at least two key features. First, the dislocation of responsibility to government mismanagement and the economic 'mafias' allowed residents of the capital to represent themselves primarily as victims, as not responsible for the water crisis. Second, in the popular imagination any opposition to proposed solutions to the water crisis plaguing the capital was seen as either evidence of corruption or as anti-Bulgarian. After the establishment of the barricades at Sapareva-Bania on 21 December 1994, residents of the capital frequently referred perjoratively to North Rilans as obstreperous 'Macedonians', 'peasants' or as old-style communist revanchists. As I show in the next section, these appeals to nationalistic symbols are greatly complicated by the structuring of the discourse of opposition to Djerman–Skakavitsa in North Rila.

Opposition to Djerman–Skakavitsa: forging a discursive coalition

The return of the VK-Rila diversion scheme to public attention during late 1994 naturally enough caused something of a panic amongst those communities most directly affected. The problem was, to be sure, partly symbolic. Since popular opposition to VK-Rila had been one of the major rallying points for anti-government protest in the dying days of the Zhivkov regime in October 1989, its revival could reasonably be expected to stir public emotions. Reinstitution of this socialist-era project was therefore seen by many as a sign of incipient re-communisation, of the 'red hand' reaching out for the 'blue vote' to borrow a memorable metaphor, as well as a significant threat to the natural environment of the Rila Mountains. In North Rila, however, activists were much more concerned about the fact that the project was being imposed on the locality without any apparent regard for local needs or any consultation with local representatives. From this perspective, VK-Rila was thus a direct challenge to an as yet incipient democratic politics. Blockades had been erected on 21 December 1994 at two different places on the road running from the town to the site of the proposed diversion, and had garnered substantial support from other nearby localities and also from the major national environmental NGOs.

Notwithstanding the apparent unity of opposition to Djerman–Skakavitsa, political opposition was in fact constructed through the coalitional activities of two rather distinct constituencies: the local community organised through the locally based Citizens' Initiative Committee for the Protection of Rila Waters, and a number of Sofia-based environmental NGOs, including Green Patrols, Ecoglasnost, the Bulgarian Association for the Protection of Birds, and Borrowed Nature. Though the coalition functioned quite smoothly in practice, it is important not to lose sight of the fact that each constituency was in fact agitating against a *different face* of the diversion project. In the following paragraphs I discuss the manner in which these two groupings, with overlapping but also crucially divergent understandings, came together in a specific time and place to lobby against VK-Rila and the central state.

Peripheral vision: the view from Sapareva-Bania

While environmentalist discourse about the putative damages attending Djerman–Skakavitsa provided a powerful independent counter-argument against central government attempts to legitimise VK-Rila, local protesters tended to link the project with the lengthy history of material peripheralisation of the Bulgarian south-west. People in the locality repeatedly expressed anxiety about the level of economic exploitation and vulnerability that resulted from the state-socialist development model (Staddon 1996).

For them the Djerman–Skakavitsa project was material confirmation of their continued role in the political economy of transition literally as a primary resource-providing periphery: that is, as 'hewers of wood and drawers of water'.

Peripheral areas such as North Rila have more than shared in the overall decline in industrial production and the erosion of living standards that have plagued Bulgaria since 1989. Plants in the area have largely ceased operation, with residual industrial demand shifted to plants in larger towns and cities. Official statistics suggest that approximately 12 per cent of the south-west's working age population are among the so-called 'registered unemployed' (Natsionalen Statisticheski Institut 1995b), having been made redundant from the industrial *combinats* at Doupnitsa, Pernik or Radomir or the coal operations at Bobov Dol as successive waves of economic crisis overwhelmed the state sector after 1989. Markedly higher unemployment –perhaps as high as 25 per cent – in the smaller towns and villages of North Rila is a product of a relatively aged population as well as industrial decline (Table 15.2). Table 15.2 also shows that average incomes in smaller settlements such as Sapareva-Bania are well below the national and even regional averages. There is also an obvious urbanisation effect, as larger settlements including Blagoevgrad, Kyustendil and Doupnitsa tend to have higher average wages, reflecting the greater propensity of larger settlements to retain some level of industrial production, particularly in the state sector. Conversely, state disinvestment in industrial production has proceeded most quickly in smaller, more peripheral areas, where the political costs to the central state and the financial costs to the national economy are presumably lowest (OECD 1997; see also Begg and Pickles in this volume).

For these reasons, recent attempts at local economic development have depended heavily upon the natural and cultural resources of the local region. The Rila Mountains have been a well-known destination for nature tourism since at least the turn of the century (Kolaritov 1988). Currently hiking, camping, skiing and mountaineering are all pursued within the boundaries of the Rila Mountain Nature Conservation Area and nearby Pirin National Park (Staddon 1995, no. 20). There are considerable nature tourism resources already in place in the western Rila Mountains, including an extensive and very popular trail network, supported by a number of *hizhas* (mountain chalets), the Malyovitsa and Borovets mountaineering and skiing complexes, and the mineral baths that give Sapareva-Bania its name (Kolitarov 1988; Hristov, Vuchvarov and Loboutov 1972). Cultural importance is lent to the area through the region's strong historical associations with the struggle to preserve Bulgarian culture during the Ottoman period, symbolised foremost by the location of Rila Monastery right in the heart of the region (Figure 15.1). More than this, however, Rila, and the Bulgarian south-west more generally, is the region from which came many of the most important figures of the nineteenth-century *vuzhrazhdane* (national awakening), including

Table 15.2 Social statistics, Kyustendil and Blagoevgrad *okrugs*, 1992

			>65 yrs old as % of popn	Average income Leva	as % of		% income from state sector
	Total popn	% urban			national average	oblast average	
Boboshevo	4,620	39.9	44.9	12,495	53.2	56.7	29.8
Bobov Dol	13,655	56.1	26.7	38,200	162.7	173.4	75.8
Doupnitsa	55,737	74.3	22.5	27,499	117.1	124.8	30.6
Kocherinovo	7,500	40.5	38.7	15,902	67.7	72.2	29.6
Kjustendil	78,328	69.5	24.6	20,343	86.6	92.3	32.2
Nevestino	5,894	0.0	62.0	18,078	77	82.1	20.0
Rila	4,410	77.6	33.1	20,644	87.9	93.7	22.3
Sapareva-Bania	9,544	48.6	29.0	18,690	79.6	84.8	11.4
Trekliano	1,659	0.0	64.4	16,929	72.1	76.8	16.8
Blagoevgrad	78,810	90.7	14.8	21,631	92.1	98.2	35.5
Sofia *oblast*	986,253	59.7	23.7	22,032	100.0	100.0	30.1

Source: Natsionalen Statisticheski Institute 1993.

Yane Sandanski and Nikoli Vaptsarov (Bozhilov *et al.* 1994; Ilieva 1982). Together these cultural and natural resources create a considerable foundation for tourism and recreation development in the region, a potentially important basis for a region hard hit by eight years of transition-induced de-development.

The Djerman–Skakavitsa diversion poses direct and indirect threats to such rural development plans. For example, it is contended that removal of further forest cover and diversion of surface water will augment trends towards the aridification of the region (Green Patrols 1995). More obvious is the negative impact of road and diversion construction on the natural aesthetics of the area (Plate 15.4). Little is known about the long-term effects of large-scale water diversion on ground water aquifers in the area, and about the potential implications of reduced water availability for the development of agricultural and industrial production. The Environmental Impact Assessment (EIA) legally mandated for all such projects would have determined whether some of these fears were well founded. However, the legal process for the Djerman–Skakavitsa diversion was abrogated by the Videnov government in February 1995.

Politically, local activists thought their protest actions were an attempt to assert a fuller sense of local autonomy in the face of apparent recentralisation of allocative and authoritative power by the central state. Indeed, considerable pains were taken by organisers of the protests to show that the protests were not directed against any particular political formation, but rather at an emerging 'democratic centrism' in which all parties, from the UDF to the BSP, were participating. Thus, the blockades were deliberately established three days *after* the 18 December 1994 parliamentary elections, in which the

Plate 15.4 Segment of Djerman–Skakavitsa diversion
(Photograph by the author, June 1995)

BSP won a large majority, so that they would be less prone to being over-looked in the run-up to the elections, or (mis)interpreted in party-political terms. Similarly, abrogation of EIA requirements was widely interpreted as a turning away from post-Zhivkov attempts to establish a system of governance based on western democratic principles. Such manoeuvres were popularly understood in North Rila as harbingers of 're-communisation' in Bulgaria. Thus, the demand of the Djerman–Skakavitsa protesters for real political voice was referred to by one of their leaders as 'the only democratic experiment in Bulgaria right now' (Staddon 1995, no. 25).

The Civil Initiative Committee for the Protection of Rila Waters (CIC) was formed by citizens from Sapareva-Bania and Saparevo on 22 December 1994, after the first barricades had been established spontaneously by some local citizens in an unorganised attempt to block movement of construction equipment into the Rila Mountains. In registration documents filed with the government there were only thirty-one official members of the CIC, though meetings regularly drew hundreds of locals and the CIC was popularly recognised as the official leader and representative of the protest movement. Participation in the group transcended age, economic and party-political divisions as well as drawing participants from other nearby villages such as Bobov Dol and Krainitsa. Leaders of the CIC tended to occupy important social positions within the local community: one was a school principal, another a director of the community centre, and a third was a local entre-preneur and a member of Green Patrols.

Such economic and political grievances were, however, not the only factors giving shape to local opposition to Djerman–Skakavitsa. The response of the North Rila localities to the diversion project was also strongly conditioned by two major socio-cultural factors, both of which are highly contested: the twin problems of rural and of Macedonian identity. The way in which the local opposition was articulated with a celebration of these two aspects of identity sets the oppositionist discourse of the CIC clearly apart from that of the environmental NGOs. Like the central state, environmental NGOs such as Ecoglasnost tended to eschew the localist dimension to the struggle, emphasising its universalist components. Conversely, the central state attempted to subvert the deployment of both the rural and the Macedonian sense of identity in the discourse of the CIC. That these two factors exist as 'essentially contested concepts' in Bulgarian political culture further points to their centrality and ineluctability in analyses of social transformation in that country (Bates 1994; Creed 1993).

Environmental NGOs: the politics of environmentalism in Bulgaria

Also startled by the reappearance of the VK-Rila plan was the small community of Bulgarian environmental NGOs, many of whom had been involved in the protests against VK-Rila and other mega-projects during the October 1989 meetings of the Conference for Security and Cooperation in Europe. For them, its resurgence was doubtless a symbolic setback as well as an environmental one, for VK-Rila had come to represent for the nascent environmentalist movement the callous communist attitude towards nature (*Troud* 16 January 1995). These groups began to initiate different sorts of social and political action after the national government announced the restart of the Djerman–Skakavitsa diversion in November 1994, and intensified their activities after the police actions of 8 February 1995. The concerns of these groups related primarily to the implications of Djerman–Skakavitsa and VK-Rila for Bulgaria as a whole, rather than the North Rila local communities *per se*. They felt it necessary to move against an emerging regime of environmental governmentality at the national level which they saw as irresponsible and unsustainable. Thus for them, protest against Djerman–Skakavitsa was only one part of a larger struggle for better environmental management throughout the country (Staddon 1995, no. 18). The community of environmental NGOs was also especially incensed by the passage of the special legislation waiving the detailed EIA required for large projects under the Environmental Protection Act.

A number of rather different groups based in Sofia were actively involved in the protests against Djerman–Skakavitsa, including the national environmental-political party Ecoglasnost National Movement, the Green Patrols, Borrowed Nature, as well as prominent members of research institutions such as the Institute for Water Problems of the Bulgarian Academy of Sciences. Sofia-based Green Patrols is described by its national co-ordinator Amadeus Krastev as a 'rapid reaction environmental hit squad' (Staddon 1995, no. 10) which tries to deploy its resources to help publicise and organise activism around important local environmental issues (Penchovska 1993: 85). These activities have made Krastev a fairly well-known public figure, and he is referred to in some papers as the 'Green Führer' because of his preference for khaki military-style uniforms and his outspoken nature. The environmental organisation Borrowed Nature is in some ways an opposite of Green Patrols. Formed in March 1992 with twenty registered members, Borrowed Nature is primarily a 'behind-the-scenes' operator, preferring work on infrastructural, education and networking projects to Green Patrols' sensationalist approach (Penchovska 1993: 115). Ecoglasnost-National Movement (Ecoglasnost-NM) is one of the many direct descendants of the first environmental protest organisations in Bulgaria: the Committee for the Ecological Protection of Rousse, the Clubs for Perestroika and Glasnost, and the Independent Association Ecoglasnost (Mindjov 1995: 7 and 40).

The discourse of environmentalism deployed by these groups against VK-Rila concentrated on its potential impacts on the local and regional environments. Petr Slabakov, a prominent Green Party environmentalist, has

365

argued that the diversion will further desiccate the Rila Mountains region, a process that could eventually lead to the widespread desertification of the Bulgarian south-west with the attendant pedological, hydrological and meteorological effects (Green Patrols 1995). This line of argument also associated VK-Rila with communist-era anti-environmentalist planning practices. Because protests against VK-Rila had been an important part of the anti-communist movement in 1989, the return of the project was seen as a reassertion of a characteristically communist anti-environmentalist development agenda (Staddon 1995, no. 17). The return of the BSP to power in December 1994 further cemented this association, transforming for many the Djerman–Skakavitsa diversion project into a symbol of 're-communisation' in Bulgaria.

Thus, the opposition of North Rila locals to the water diversion project was not merely a product of heightened environmentalist sensibilities, though these were important. The strength of the coalition between the Civil Initiative Committee for the Protection of Rila Waters and national NGOs such as Ecoglasnost and Green Patrols depended on the fact that VK-Rila combined threats to both the natural environment and to local autonomy. Djerman–Skakavitsa was popularly perceived as yet another example of the exploitation of the south-west by a wholly self-interested capital city and a suppression of local community identity and political autonomy. Even local residents who seemed less sure about the technical environmental dimension of the issue invariably stressed that the action was also about democratic local self-organisation. Indeed, the Sapareva-Bania mayor who had originally sparked the protests by signing the construction orders was ousted from office in the October 1995 local elections, which at the same time returned the members of the local council who opposed him. There is some evidence of a recognition of these basic facts by the national NGOs, particularly in the perceptions and programme initiated by members of Borrowed Nature (Staddon 1996).

Water local autonomy and democratisation: a preliminary summary

'[T]he near order, that of the locality, and the far, that of the state, have of course long ceased to coincide: they either clash or are telescoped into one another' (Lefebvre 1991: 230).

It is tempting to conclude that the protests against the Djerman–Skakavitsa water diversion and the police crackdown that terminated them constitute a significant failure of democratisation after 1989. However, such a conclusion is only partially warranted. The citizens of Sapareva-Bania did fail spectacularly in their attempts to force the state to change its plans to construct the Djerman–Skakavitsa diversion. They even failed to get the state to follow its own legal requirements for Environmental Impact Assessment and community consultations. However, a more nuanced way of framing the issues would be in terms of the nature of power and the democratisation process, and the changing relations of political scale, between local and central states.

Normative models of political change in CEE would, I think, predispose us to interpret the events at the barricades in Sapareva-Bania as failures of democratisation. They would point not only to the application of military force against the citizenry, but also to the civil disobedience of the local citizens. This view secretly imports a western desire to dispense completely with the state-socialist past in ways that are at best naïve and at worst duplicitous. Established political structures are just that, *established structures*, and it will take considerably more than periodic elections to change them. Citizens of bureaucratic-authoritarian states developed many covert and informal means of expressing opposition to the state and it seems disingen-uous to expect them to simply give these up overnight just because USAID and foreign election-monitoring groupies are increasingly important in defining the nature of post-communist democratisation. This point is especially pertinent as the new politics is again foreclosing on the possibilities for formal, institutionalised local autonomy. Research in Sapareva-Bania, and elsewhere in Bulgaria, has highlighted the irony that in many ways localities are politically worse off now than they were before 1989. Prescriptive normative models, such as those offered by political scientists like Sam Huntington (1991) and Guiseppe di Palma (1993), are simply the wrong point of entry.

More sophisticated are the interventions of social theorists like Jürgen Habermas who want to recuperate the positive normative moment through appeals to transcendent conditions such as the 'ideal speech situation', and models of human moral-cognitive development. Consider for a moment the implications for local political autonomy of this way of thinking: if it is accepted that a key part of the struggle at the barricades in Sapareva-Bania took place on the level of discourses about culture and about political accom-modation, then we are not so far from a definition of local democratisation as the creation of 'an arena in which individuals participate in discussions about matters of common concern, in an atmosphere free of coercion or dependencies that would incline individuals towards acquiescence of silence' (Warren 1995: 171). This formulation in fact belongs to Habermas, whose project is oriented towards revealing the normative conditions necessary for a world characterised by this brand of what he calls 'discursive democracy' (in Warren 1995: 171). Attractive as such a formulation may be, empirical realities such as those expressed at Sapareva-Bania demonstrate just how difficult it would be to separate out the pure intersubjectivity demanded by Habermas from the noxious effects of power (i.e. the ideal speech situation will be exceedingly difficult to even recognise, let alone achieve).

More productive in my view is an engagement with the dialectics of social and political power, with its fluid and always contested nature. The case of local protest at Sapareva-Bania is interesting precisely because it exposed the ragged juncture between the overweening state and the ever-present possibility of local autonomy and resistance to the state. At the very least, the people of the local Civil Initiative Committee proved again that it is possible to resist the pressures of the state and to achieve some limited gains even if the bigger issues are lost. Local political mobilisation also proved sufficient to replace the mayor and some of the municipal council in the October 1995 local elections. To some extent political gains have been institutionalised and a local political identity was articulated that provides a positive model for residents of the Rila Region which is not rooted in any backward looking revanchisms.

The protests at Sapareva-Bania were considered by some on both sides to be a clash between different models of democratisation: one based on the adherence to central managed procedures and electoral politics, and another expressed through the articulation of local traditions with political-economic realities. In fact the formulation might better be thought of as an 'anti-model' implying that democratisation is not a process that can be preordained in the abstract, but rather one that must come from a basic commitment to empowering local agents to deal with local challenges. This is close to what Foucault meant when he characterised political modernity as a strategic intervention in the antagonism between the authoritarian and the polymorphous aspect of the will to power:

> A double movement, between state centralisation on the one hand, and dissidence on the other ... it is at the intersection of these processes that the problems of governance come to pose themselves with greatest intensity; of how to be ruled, how strictly, by whom, to what ends and with what methods.
>
> (Foucault 1991: 88)

Implied here is a model of political power that is not just instrumental, as in the use of military force, but is also social, cultural and economic, as in the attempts by Sapareva-Banians to define their actions in terms of cultural identity, material interests and the project of de-communisation.

Acknowledgements

I would like to thank Barbara Cellarius, Greg Knight, John Pickles, Adrian Smith and Dick Walker for helpful criticisms of earlier drafts of the chapter. Generous research support from the National Science Foundation (USA), the University of Kentucky Graduate School and the Social Sciences and Humanties Research Council (Canada) is also gratefully acknowledged.

Notes

1 References to research fieldnotes give the author, year and, where available, the interview number.
2 The three major political parties in national politics are the reformed communists, the Bulgarian Socialist Party (BSP), the anti-communist coalition the Union of Democratic Forces (UDF) and the Movement for Rights and Freedoms (MRF), which by and large represents Bulgaria's Turkish minority.
3 There is a considerable public mythology surrounding the 'mafia' formations. What is documentable is their involvement in a very selective process of industrial privatisation, the mutualisation of mass privatisation vouchers in 1996 and 1997, former Yugoslavia embargo-busting and property speculation, overwhelmingly in the capital. Less easily demonstrable is their purported involvement in the financial 'pyramid' schemes, the collapses of which have seriously destablised political structures in Bulgaria and elsewhere in south-eastern Europe.
4 This is a sardonic reference to the 'G-13' group of western industrialised nations.
5 In fact two French companies, SAUR and Compagne d'Eau, have made joint-venture offers to purchase the Sofia Water Supply and Sewerage Company ('ViK' in Bulgarian).

Bibliography

Aggar, B. (1989) *Fast Capitalism*, Urbana, IL: University of Illinois Press.
Angelov, I. *et al.* (1994) *Economic Outlook of Bulgaria 1995–1997*, Sofia: Institute of Economics, Bulgarian Academy of Sciences.
Bates, D. (1994) 'What's in a name? Minorities, identity and politics in Bulgaria', *Identities*, 1 (2–3): 201–25.
Bauman, Z. (1992) *Intimations of Postmodernity*, New York: Routledge.
Bozhilov, I. *et al.* (1994) *Istoria na Bulgaria (History of Bulgaria)*, Sofia: Printing House Hristo Botev.
Bulgarian Economic Review, a bi-weekly published by the Pari Publishing Group, Sofia, Bulgaria.
Bulgarian Telegraph Agency, *Daily News* (in English), various dates, Sofia.
—— *Daily Bulletin* (in Bulgarian), various dates, Sofia.
—— (1994) *Internal Information* (in Bulgarian), 25 January, Sofia.
—— (1995) *Special Report on the New Government* (in Bulgarian), 30 January, Sofia.
Carter, F. (1987) 'Bulgaria', in F. Carter and D. Turnock (eds) *Environmental Problems in Eastern Europe*, New York: Routledge, pp. 38–62.
de Certeau, M. (1984) *The Practice of Everyday Life*, Berkeley: University of California Press.
Crampton, R.J. (1987) *A Short History of Modern Bulgaria*, Cambridge: Cambridge University Press.
—— (1997) *A Concise History of Bulgaria*, Cambridge: Cambridge University Press.
Creed, G. (1991) 'Between economy and ideology: local-level perspectives on political and economic reform in Bulgaria', *Socialism and Democracy*, 13: 45–65.
—— (1993) 'Rural–urban oppositions in the Bulgarian political transition', *Sudosteuropa*, 42: 369–82.

de Certeau, M. (1984) *The Practice of Everyday Life*, Berkeley: University of California Press.

Demokratsiya (Bulgarian daily newspaper), various dates, Sofia.

Diamond, L. and Plattner, M. (1992) *The Global Resurgence of Democracy*, Baltimore, MD: Johns Hopkins University Press.

di Palma, G. (1993) 'Why democracy can work in Eastern Europe', in L. Diamond and M. Plattner (eds) *The Global Resurgence of Democracy*, Baltimore, MD: Johns Hopkins University Press, pp. 257–67.

Dobreva, S. and Kouzoundra, V. (1994) 'Some problems of the transformation in Bulgarian agriculture', in N. Genov (ed.) *Sociology in a Society in Transition*, Sofia: Regional and Global Development Institute, Bulgarian Academy of Sciences, pp. 77–102.

Donchev, D. and Karakashev, H. (1995) *Fisicheska i Sotsialno-ikonomicheska Geografia na Bulgaria* (Physical and socio-economic geography of Bulgaria), Veliko Turnovo: Slovo Publishers.

Douma (Bulgarian daily newspaper), various dates, Sofia.

Foucault, M. (1991) 'Governmentality', in G. Burchell, C. Gordon and P. Miller (eds) *The Foucault Effect: studies in governmentality*, Chicago, IL: University of Chicago Press, pp. 87–104.

Garnizov, V. (1995) *Konfliktut v Obshtina Sepereva Bania Shest Mesetsa Sled Negovia Pik* (The conflict in Sapareva Bania *obstina* six months after the demonstrations), Sofia: Centre for Social Practice.

Genov, N. (ed.) (1993) *Society and Environment in the Balkan Countries*, Sofia: Institute for Sociology, Bulgarian Academy of Sciences.

Gergov, G. (1991) 'The use and protection of water resources in Bulgaria', in J. DeBardeleben (ed.) *To Breathe Free: Eastern Europe's environmental crisis*, Baltimore, MD: Johns Hopkins University Press, pp. 159–73.

Green Patrols (1995) *Documentary about the Djerman–Skakavitsa Protests*, Sofia: Green Patrols.

Habermas, J. (1987) *The Philosophical Discourse of Modernity*, Cambridge, MA: MIT Press.

Hankiss, E. (1992) 'Grants and advice – thank you – but send us also a Durkheim and a Habermas', *East European Politics and Societies*, 6, 3: 359–63.

—— (1994) 'Our recent pasts: recent developments in East-Central Europe in the light of various social philosophies', *East European Politics and Societies*, 8, 3: 531–42.

Hristov, T., Vuchvarov, M. and Loboutov, G. (1972) *Economic-Geographical Characteristics of Kyustendil Okrug with a View of the Development of the Microregional Structure* (in Bulgarian), Sofia: Yearbook of the Sofia University Geological-Geographical Faculty, volume 64.

Huntington, S. (1991) *The Third Wave: democratisation at the end of the twentieth century*, Oklahoma City: University of Oklahoma Press.

Ilieva, M. (1982) 'Rilo-Pirinski Raion', in R. Naidenova *et al.* (eds) *Prirodniat i Ikonomicheskiat Potentsial na Planinite v Bulgaria* (The natural and economic potential of the mountains of Bulgaria, population and economics, vol. 2), Sofia: Bulgarian Academy of Sciences, pp. 163–76.

ING Bank (1994) *Bulgaria: political and economic assessment*, 14 December.

Institute for Geography (1989) *Geography of Bulgaria: physical-geographical and socio-geographical regionalisation of Bulgaria*, Sofia: Bulgarian Academy of Sciences.

Knight, C.G., Velev, S.B. and Staneva, M.P. (1995) 'The emerging water crisis in Bulgaria', *GeoJournal*, 35, 4: 415–23.

Kolitarov, V. (1988) *Sapareva-Bania: putevoditel* (Sapareva-Bania: tourist guide), Sofia: Meditsina i Fiskultura.

Kontinent (Bulgarian daily newspaper), various dates, Sofia.

Lefebvre, H. (1991) *The Production of Space*, Cambridge, MA: Blackwell.

Medjidiev, A. (1969) *Istoria na Grad Stanke Dimitrov (Doupnitsa) i pokrainata ot XIV Vek do 1912–1963* (History of Stanke Dimitrov and its region from the 19th century to 1963), Sofia: Izdatelstvo na Otechestvenia Front.

Mindjov, K. (1995) *Proceedings of ECO Parallel Conference 'Environment for Europe'*, Sofia, 20–25 October 1995, Sofia: Borrowed Nature Association.

Murphy, A. (1995) (ed.) *Geographic Approaches to Democratisation: a report to the NSF*, Washington, DC: National Science Foundation.

Natsionalen Statisticheski Institut (1993) *Oblastnite i Obshtinite v Republika Bulgaria, 1992* (*Oblasts* and *obstini* in the Republic of Bulgaria, 1992), Sofia: Statistical Publishing House.

—— (1995) *Oblastnite i Obshtinite v Republika Bulgaria 1994* (*Oblasts* and *obstini* in the Republic of Bulgaria, 1994), Sofia: Statistical Publishing House.

Natsionalen Statisticheski Institut (1993–5) *Statisticheski Godishnik* (Statistical Yearbooks, 1993–5), Sofia: Statistical Publishing House.

—— (1995) *Sotsialno i Ikonomicheska Razvitie na Republika Bulgaria, 1990–1994* (Social and economic development in Bulgaria, 1990–94), Sofia: Statistical Publishing House.

OECD (1997) OECD Economic Surveys: Bulgaria, Paris: Organisation for Economic Co-operation and Development.

Pari (Bulgarian daily newspaper specialising in financial news), various dates, Sofia.

Paskalev, A. (1994) 'Kak da predlojicm vodnata kriza v Sofia' (How to overcome the water crisis in Sofia), *Pari*, 15 September.

Penchovska, J. (1993) *Catalogue of Environmental NGOs in Bulgaria*, Sofia: Regional Environmental Centre for Central and Eastern Europe.

Pickles, J. (1991) 'Democratisation and its others: politics, economy and environment in contemporary Bulgaria', paper presented at the International Symposium on the Impact of Political and Economic Restructuring on the Bulgarian Environment, Morgantown, WV: West Virginia University, 7 December 1991.

Pred, A. and Watts, M. (1992) *Reworking Modernity: capitalisms and symbolic discontent*, New Brunswick, NJ: Rutgers University Press.

Staddon, C. (1995) Unpublished field notebooks, Lexington, KY: Department of Geography, University of Kentucky.

—— (1996) 'Democratisation, environmental management and the production of the new political geographies in Bulgaria: a case study of the 1994–95 Sofia water crisis', unpublished Ph.D. dissertation, University of Kentucky.

Standart (Bulgarian daily newspaper), various dates, Sofia.

Troud (Bulgarian daily newspaper), various dates, Sofia.

United Nations Development Programme (1995) *Bulgaria: human development report 1995*, New York: UNDP.

Warren, M. (1995) 'The self in discursive democracy', in S.K. White (ed.) *The Cambridge Companion to Habermas*, New York: Cambridge University Press, pp. 167–200.

Wolff, L. (1994) *Inventing Eastern Europe: the map of civilisation on the mind of the enlightenment*, Stanford, CA: Stanford University Press.

Wyzan, M. (1996) 'Stabilisation and anti-inflationary policy', in I. Zloch-Christy (ed.) *Bulgaria in a Time of Change: economic and political dimensions*, Aldershot: Avebury, pp. 77–106.

Zahariev, I. and Tankova, D. (1974) *Spetsialisatsia na ikonomicheskite raioni i okruzite v NR Bulgaria v Otraclite na Promishlenost* (Regional and district economic specialisations in Bulgaria by branch of production), Sofia: Bulgarian Academy of Sciences.

Zaharieva, M. (1993) 'Local environmental risks: perceptions and behavioural responses', in N. Genov (ed.) *Society and Environment in the Balkan Countries*, Sofia: Regional and Global Development Section, Institute for Sociology, Bulgarian Academy of Sciences, pp. 41–53.

Zemedelsko Zname (Agricultural Banner) no. 224, December 1994, Sofia.

24 Hours, various dates.

16

SOCIAL EXCLUSION AND THE ROMA IN TRANSITION

David Sibley

The Roma[1] in Central and Eastern Europe, as in other parts of Europe where they have settled during the past six hundred years, are commonly seen as a marginalised or excluded people. They have experienced a long history of oppression, at its worst during the Second World War, when they were the second-largest ethnic group to suffer Nazi genocide. In the post-Second World War period, in several socialist states they have been subject to forced resettlement. With the end of communism, the Roma have fitted uneasily with new conceptions of national identity and have been the frequent targets of racist attacks. Thus their marginal status has been confirmed under very different political regimes but, in this chapter, I will suggest that the concept of marginality in relation to the Roma is problematic. The label 'marginal' masks a complex relationship between the Roma and the larger society. This relationship needs to be examined more critically if we are to understand how the status of the Roma populations in Eastern and Central Europe is changing as both the demands and the supports provided by the state are removed and Gypsies adjust to the market economy. I begin by reviewing two contrasting models of social change and then, drawing on literature (principally from Hungary) and on my own observations, I will suggest how the Roma have been forced to the edge while, at the same time, retaining their ability to shape their own economic and social relationships with the outside.

Theories of social change

Unfolding models of change and cultures of assimilation

In order to provide a sense of the contradictory forces shaping the boundaries of Roma culture and society, it is useful to consider a set of arguments associated with the modern project of social transformation involving cultural assimilation. This is particularly so because of the teleological trajectories implied by programmes for social modernisation in Eastern Europe, whether they encompassed the Stalinist transformations of the 1930s or Ceauçescu's programmes for social change in Romania. Such programmes can be considered as 'unfolding models of social change' in which they treat 'social change as the progressive emergence of traits that a particular society is presumed to have within it from its inception' (Giddens 1979: 223). Increased social homogeneity was central to social progress and categories of 'deviants', such as the Roma, were therefore created among those who were slow or unwilling to take up their assigned roles in the dominant economy. In other words, there was no space for autonomy in which such 'deviant' groups could find social spaces for self-expression.

Such experiences have been mirrored in the West. Indigenous peoples in a variety of contexts have also been the subject of social transformations and marginalisation often associated with white settlement and internal colonialism. For example, the Inuit in Arctic and sub-Arctic Canada have been affected by such an unfolding model of change in which the moving of Native peoples into state housing schemes has been interpreted as 'an essential step in the *progress* of the Eskimo people', enabling them 'to adapt to the *more sophisticated* communities that will develop as the potential of the non-renewable resources of northern Canada [is] realized' (Yates 1970: 49, emphasis added). For some critical Marxists, however (e.g. Brody 1973, 1975), this process meant that indigenous peoples in developed societies would be dislocated, separated from their traditional economic supports, and incorporated into wider economic relations as members of a lumpen-proletariat. The result, from this perspective, was the breakdown of egalitarian arrangements for sharing resources and a shift to individual consumerism. The claims articulated by both Yates and Brody, while differing in their assessment of the impact of integration, insist, however, on the inevitability of change, involving some kind of assimilation into the dominant system.

Such assimilationist assumptions were also implicit from the mid-1960s to the mid-1970s in the work of Soviet social theorists. Gurvich (1978), for example, in his discussion of cultural change among ethnic minorities in Siberia, argued that the elimination of linguistic and other cultural differences and the incorporation of Arctic peoples in the dominant economic system were both necessary and desirable. Tipps (1973) – anticipating some of the claims of post-colonial theorists (Spivak 1987; Bhabha 1994) – has, however, criticised this kind of assimilationist thinking for its ethnocentrism, its roots in the dominant value system and its lack of sensitivity to the cultures of incorporated minorities. Indeed, one can perhaps begin to interpret the experience of the Roma in transition from a standpoint that examines cultures of survival rather than cultures of assimilation.

Cultures of survival

John Berger has provided a quite different argument regarding the transformation and continuity of social relations in the light of wider social change, such as 'modernisation'. In *Pig Earth* (1979: 196), Berger suggested that peasants, for example, have a life 'committed completely to survival' which 'is the only characteristic fully shared by peasants everywhere'. The peasant economy, Berger observed, 'was always an economy within an economy. This is what enabled it to survive global transformations of the larger economy – feudal, capitalist, even socialist.' A strong tradition, maintained through oral culture and reinforced by a history of exploitation, has given peasants the capacity to develop survival strategies, maintaining their existence in the face of external threats. While peasants have been involved proactively in a variety of revolutionary struggles throughout their histories, they have been largely focused on maintaining the material position of the peasantry. Many of these claims have their parallels with other groups who live on the edges of society, such as the Roma. However, this does not mean that nothing changes within these groups. In fact, change is vital if the culture is to stay the same. Without change and adaptation, or what might otherwise be described as the agency of the group, the culture would be transformed to make it accord with the values of the dominant society. Any discussion of 'cultures of survival' therefore involves defining a power relationship where the weaker group has some power and a capacity to act and mobilise in order to resist social transformation.

Peasants and the Roma are survivors in the sense that they have retained a cultural identity despite large-scale economic and political change. As Berger (1979: 203) notes:

> A class of survivors cannot afford to believe in an arrival point of assured security and well-being. The only but great future hope is survival. . . . The future path through future ambushes is a continuation of the old path by which the survivors from the past have come . . . the path is tradition handed down by instructions, example and commentary. To a peasant, the future is this future narrow path across an indeterminate expanse of known and unknown risks.

Cultures of survival therefore become 'a sequence of repeated acts. . . . Each act pushes a thread through the eye of a needle and the thread is tradition' (Berger 1979: 203–4). Those on the margins are not passive, rather there is agency in these acts, and such agency may range from individual and family survival strategies to collective forms of mobilisation.

Gypsies in Eastern and Western Europe (and in the Americas and Australasia) have maintained their identity through cultural tenacity and economic flexibility. The culture is reproduced through the maintenance of strong boundaries which distinguish the pure inner body, including the inner *social* body, that is, the Roma community, from the defiled outer body, which includes the defiled larger society, the *gaje*. An array of acts, objects and animals, and a system of ordering these things in space, inside and outside, determine the purity of the Roma and the defilement of the *gaje*, and this gives structure to Roma society. At the same time, the *gaje* have to be engaged. There is no Roma economy without the *gaje*, so it is imperative continually to seek out new niches in the fluid economic landscape of capitalism or (less fluid) socialism. This may involve the Roma presenting themselves in ways which are acceptable to their customers or employers – as respectable business people, as hard workers, or as 'scary' Gypsies – in order to gain maximum economic benefit, but without compromising their cultural boundaries. Thus, when begging on the streets of Budapest, a woman may accentuate her 'Gypsyness', conforming to a negative stereotype in order to unsettle the prospective donor.[2] To be willing to change, to be flexible and adaptable, are essential qualities for survival.

This commentary on two opposing conceptions of social change, an unfolding model embodied in the idea of a culture of progress with expanding horizons, and a model of survival and adaptation, provides the basis for the following account of relations between the Roma, the larger society and the dominant economy in socialist and transitional Central and Eastern Europe.

Work and settlement under socialism

Officially, the Roma did not exist under state socialism. They were not acknowledged as a nationality but they were required to work – primarily in factories or on state or collective farms. However, the range of employment in the formal sector was quite restricted. Ladányi and Szelényi (1996) point out that in Hungary the Roma were completely left out of the redistribution of land after the Second World War and, being landless, were not eligible to become members of the new co-operatives. As factory workers or farm labourers, they had to adapt to disciplined regimes and they clearly did not have the control over their economic destiny that they enjoyed in the informal sector. As Michael Stewart and Judit Szegö demonstrated in their 1988 Channel 4 documentary *Across the Tracks*, however, Vlach Gypsies in rural Hungary were able to combine compulsory factory work and other state employment with their own forms of production and exchange.[3]

In the Vlach settlement, the mainstay of the informal economy was horse-dealing, a traditional Roma occupation, supplemented by food production on small-holdings. One of the most successful horse dealers, who appeared to make a considerable amount of money by brokering deals between other Gypsies, was disqualified from state work by a 'dubious disability'. When asked why he did not work for the state, he replied: 'How could I join the

collective? ["He's ill", his wife interjected] I couldn't manage the hard work.'
Another illustration of the way in which the Vlach had adapted to the state
socialist system was the case of a woman who worked as a registered scrap
dealer, another traditional Gypsy occupation:

> I get steel, papers, batteries. I've been like this for thirty years. I earn
> more money than I would in a factory. There's more money in this.
> . . . It's dirty work but it pays well. What can I do about it? Nothing!

The same woman also scavenged food for a small number of pigs, and several
families in the settlement grew their own vegetables, activities typically
associated with the *gaje*. However, for the whole group, horse-dealing was
clearly the main source of income, providing considerably more wealth
than could be gained from factory or farm work and also continuity, a
connection with the pre-communist past. The film also indicated that
this community was able to extend the settlement to accommodate new
families by 'repairing' houses, that is, demolishing all but one wall and then
rebuilding on a larger scale.

This case of Vlach Gypsies in Hungary may not be representative, but it
does demonstrate the ability of some Roma to adapt to the state socialist
system, to retain some degree of economic autonomy and to maintain
cultural identity. There was little need to compromise even though, as
Ladányi and Szelény (1996: 3) argue, in late-communist Hungary rural
Roma were very much on the economic and spatial margins, despite their
incorporation in the formal economy:

> rural development was highly uneven. . . . Larger villages, nearer to
> urban workplaces and markets flourished while small villages,
> especially on the periphery of the regional system decayed. The
> more dynamic families moved out from these settlements. The
> 'no-hopers', Roma and non-Roma, were left behind.

Stewart and Szegö's research, however, suggests this may be an unduly
pessimistic view of the Roma under communism in Hungary. Settlements
from which the *gaje* were migrating provided opportunities for the Roma
to strengthen extended family and kin-group relationships; even though
material conditions were poor, informal economic activities could be
pursued out of sight of state officials in wholly Roma settlements, and they
did not have to endure the racism of the *gaje* on a daily basis.

Information on urban Roma populations during the socialist period is
rather limited, but studies by Kemény (1975) and Ladányi (1993) in
Budapest may provide some indications of what was happening elsewhere,
given the similarities in the structure of urban housing markets in Central
and Eastern Europe. In broad terms, the Budapest Roma comprised an

inner-city population, concentrated particularly in the poorest districts, but with a presence also in peripheral estates, self-built settlements and remnants of rural housing also on the urban fringe. The case of Budapest is instructive because it demonstrates the limited impact of the state on the urban settlement and underlines the problematic nature of marginality and exclusion in relation to the Roma.

In 1971, according to Kemény (cited in Ladányi 1993), 30 per cent of the Gypsy population in the city lived in 'Gypsy settlements', built with or without permission largely by the Roma. The housing conditions in these mostly peripheral settlements were very poor but they served an important function as a first destination for rural Roma migrating to the capital city. In this respect, they were equivalent to the *barriadas* and other spontaneous settlements in many Third World countries. From the mid-1970s onwards, the city authorities began to demolish these settlements so, by the time of Ladányi's survey in 1987, there were very few left. As Ladányi (1993) points out, however, the conditions encouraging rural–urban migration did not change during this period and the replacement of spontaneous Roma settlements was prevented only by vigorous police action. As in the capitalist West, the state was concerned with 'cleaning up' the city, 'sanitising' it by removing disorder – the Roma and their settlements – rather than addressing the problem of rural–urban inequalities. At the same time, the larger concentrations of Roma in the older tenement districts in Pest, particularly Jósefváros, were squeezed by slum clearance. Some families apparently moved into space sublet by other Roma. Others moved into peripheral estates, taking advantage of relocation policies which gave priority to larger families, but this only accounted for about 20 per cent of the total Gypsy population. In addition, there was a small element of the population which lived in scattered family houses from which they ran private businesses, similar to the Kalderas Gypsies in Paris (Williams 1982). Indeed, small businesses were increasingly encouraged by the state as a part of the move to a market economy during the 1980s. While many of the urban Roma were employed in poorly paid, unskilled factory jobs, city life simultaneously provided opportunities for work in the informal economy.

As Kovács (1990) demonstrated, the population of Budapest was already highly polarised in terms of income and housing conditions before the end of communism and the pattern of Roma settlement would have contributed to this polarisation, particularly because the majority of Roma were settled in the poorest tenement districts of the city. Superficially, there was a close similarity between Gypsy space in Budapest and their pattern of settlement in a capitalist Southern European city like Barcelona, where there is also an inner-city tenement population as well as settlements and camp sites on the urban fringe. Hungarian state socialism had created inequalities in the housing system similar to those in the capitalist West and the opportunities for Roma settlement in Budapest were similarly limited to the poorest districts. If the status of the Roma under state socialism is assessed solely on the basis of quantifiable material indicators, there is no doubt that they were at the bottom of the social heap, however much they were incorporated into the wage labour force. If the elusive informal economy is considered, however, together with the ways in which the Roma manifest their wealth (often in forms other than housing, such as jewellery or horses) the picture becomes less clear. A separate Gypsy economy functioned in marginal spaces under communism but the opportunities for the Roma to accumulate wealth outside the formal economy varied considerably over space and time.

Economy and settlement in the 1990s

Since the end of the communist regime in Hungary, it is clear that there has been an acceleration of processes already under way in the socialist state. Inequalities have become more marked and statistical evidence suggests that the Roma have been particularly badly affected by the rapid shift to a market economy.

The number of jobs in Hungary decreased by about one-third between 1989 and 1993 but a large proportion of these jobs were accounted for by early retirement. Ladányi and Szelényi (1996) argue that, while there were marked regional variations in unemployment levels in the early 1990s, with lower rates in Budapest and the west and the highest rates in the north-east, all strata of society were more or less equally affected. The exception was the Roma. By 1995, Roma levels of unemployment were twice the national average, and the unemployment rate for Roma women was 83 per cent. Roma in the rural periphery were worse affected than elsewhere. Thus, in the village of Csenyéte in northern Hungary, Ladányi and Szelényi (1996) found that Roma had replaced 'Hungarians' entirely by the early 1990s and the unemployment rate stood at 100 per cent. This was consistent with a general pattern of change in the rural periphery where there was a high level of out-migration of 'Hungarians' and an increase in the proportion of Roma, with the effect that some declining villages began to expand because of the high rate of increase of the Roma population – the people Ladányi and Szelényi (1996) refer to as 'the pauper Roma'. In smaller settlements closer to Budapest, the poverty of some Gypsy quarters is still severe and housing conditions can be very bad, but it is on the rural periphery, close to the borders with Slovakia, the Ukraine and Romania, that the problem is most acute. In Budapest, the Gypsy population increased from 25,000 in 1971 to about 44,000 in 1994 and this increase has consolidated the 'Gypsy ghetto', characterised by generally poor housing. In some districts of Budapest, the problem has been exacerbated by gentrification (in the ninth district, for example), which has pushed former Roma residents into remaining poor-quality tenements (János Ladányi, personal communication).

The prospects for reintegrating the Roma in the dominant economy do not look good, particularly if we consider their experience of formal education and the skills required by the expanding tertiary sector. Statistics reported by Laki (1992), mostly relating to 1990, present a familiar picture of parents with no education beyond primary level or no formal education at all and children with not much more experience of school than their parents. However, this reflects the lack of importance attached to formal education by the Roma, who pass on their skills within the community, and racism in schools, rather than the academic failure of Gypsy children.

Similarly, we have to be cautious about employment data. The unemployment statistics cited above refer only to the formal economy. The Roma were the first to lose their jobs in the post-communist period, but this has increased the importance of the informal economy for their survival. It is very difficult to gauge the significance of traditional 'Gypsy occupations' for Roma families since the transition began – there are no ethnographies of urban Gypsies comparable to Stewart and Szegö's study of the Vlach in socialist rural Hungary. However, some activities like begging, mostly by women and young children, and market trading are flourishing, although restrictions have recently been put on street-trading in Budapest. My own observations since 1992 suggest that there is no particular niche for the Roma in market- and street-trading, but opportunities are utilised wherever possible – selling cheap, smuggled brandy from the former Yugoslavia in one large Budapest market in 1992, and doing currency deals around the main railway stations, as well as trading in more typical market goods like clothing and vegetables. A less traditional feature of the post-communist Roma economy is prostitution. Gypsy pimps and prostitutes work mostly in the Rákóczi Square district of Budapest, where most of the clients are also poor. As in Western Europe, much of the revenue-earning work is increasingly being done by women. They are now largely without work in the formal economy, but they are heavily involved in market-trading, begging and prostitution.

Like all those individuals affected negatively by the economic impacts of transition, these informal occupations indicate an alertness to money-making opportunities as a mechanism of survival. Here, the Roma are operating rather precariously in the interstitial spaces of the urban economy. Increasingly, they face competition from other traders, particularly the Chinese and Vietnamese in Budapest, and from other Central and Eastern European migrants – Poles, Ukrainians, Russians, Romanians, as well as elderly and impoverished Hungarians – in the more desperate areas of the informal economy. Judgements based on the official figures for Roma unemployment need to be qualified with observations on the informal economy, but it is unlikely that much wealth is generated by the latter, at least in the cities. For many Roma, begging and trading are work, however, and these occupations are important markers of Gypsy ethnicity. They define a relationship with the *gaje* in which the Roma retain a degree of control and this feeling of domination in their transactions is fundamental to their identity. Okely (1979) has emphasised the importance of performance, that is, playing on particular perceptions of Gypsies – as frightening or romantic – in order to get the maximum benefit from a transaction with the *gaje*. The latter are seen as gullible and subject to manipulation. Despite their dependence on the larger society and their clear economic vulnerability, both from competition and from the actions of the state, like curtailing street-trading, they retain their separateness partly by defining the terms of their transactions in the informal sector. It is thus inappropriate to judge their status solely on the basis of quantifiable indicators of economic well-being.

Nationalism, racism and the Roma

The image of Roma culture as a 'culture of survival', which carries with it a conception of power relations rather different from that suggested by theories of domination or assimilation, has to be qualified further. The clear evidence of economic dislocation and impoverishment in late socialist and capitalist Hungary, where the market economy is further advanced than in many other Central and East European states, is one cause for pessimism. The other is racism. In establishing a sense of national identity, the Roma are dispensable in the sense that they are not a part of the imagined communities of Czechs or Hungarians or Slovaks, but their presence is at the same time instrumental. For the majority, they constitute an 'other' which conveys a sense of national belonging through its exclusion. However, this is not something which is peculiarly associated with the new sense of nationalism in the region. In Central and Eastern Europe, the Roma have for a long time been the victims of what Iris Young (1990: 125) terms cultural imperialism, which renders a group 'invisible at the same time that it is marked out and stereotyped'. To put the current problem of racism into perspective, it should be noted that the Roma suffered under Fascist regimes in Hungary, Romania and in the Balkans, where client states contributed to Nazi extermination programmes during the Second World War (Fraser 1992: 264). Between 1939 and 1945, 28,000 Roma were deported from Hungary, 36,000 from Romania, 28,000 from Croatia and 12,000 from Serbia. They were sent to death camps in Germany or Poland or, in the case of most of the Romanian Roma, to concentration camps on the Black Sea at Tiraspol and Nikolayev.[4] These were the ultimate 'non-persons', invisible in the sense that they had no place in the Fascist vision of the nation, but highly visible in that they were represented by entirely negative racist stereotypes.

In a more muted way, the tradition continued under communist governments, the invisibility of the Roma was signalled by their absence from censuses, but they were visible as a stigmatised and criminalised minority although, officially, not distinguished from other workers. State racism, manifest in police and other security agency reports, is well documented for Romania and Czechoslovakia in the relevant Helsinki Watch reports (for Romania in 1991 and Czechoslovakia in 1992). This continuity should not be surprising because, under any regime, the dominant and conforming majority needs to confirm its boundaries with reference to a group which represents its antithesis. The boundaries are marked by binary opposites which reflect the fantasies of the excluding population about threats to its security as a homogeneous national community; hence, the recurrent use of clean/dirty; white/black; law-abiding/criminal, etc., in all racist discourse (Salecl 1993). One dimension of the problem which has received less attention than racist discourse, but which is particularly relevant to current problems of racism in Central and Eastern Europe, is the question of settlement and, particularly, forced relocation.

In Romania, but also in other Central and Eastern European countries, the state initiated a dispersal policy for Roma in the 1950s. Gypsy settlements were destroyed and families were relocated in villages or new housing estates. Dispersal is a common response to a minority recognised by the state as a 'problem'. Under the Ceauçescu regime, this dispersal policy did not apparently result in violence, but in the post-communist period feelings about who belongs and who does not belong have come to the surface as nationalistic feelings have become unconstrained.[5]

In a large number of well-documented cases in Romania (Helsinki Watch 1991; Fonseca 1996), the rural Roma have suffered because of their unwelcome proximity to the non-Gypsy villagers. These cases seem to demonstrate that small, closed communities turn on outsiders during periods of political and economic uncertainty (Douglas 1966; Sibley 1988), but this is particularly the case if they are given encouragement by the police or other state agencies. In one reported case in Hadreni in northern Romania in 1993 – a village with a mixed population of Romanians, ethnic Hungarians and Roma – Roma houses were set on fire, resulting in several deaths. One villager allegedly commented: 'We did not commit murder – how could you call killing Gypsies murder? . . . Gypsies are not really people, you see. They are always killing each other. They are criminals, sub-human vermin. And they are certainly not wanted here' (Bridge 1993: 12). According to a Romanian television report, the Hadreni pogrom came after 'a long period of tension between the villagers and the Gypsy community caused by robberies and aggressive actions taken by the Gypsies'. There were at least thirty recorded attacks on Gypsy communities in Romania between 1989 and 1993, but few convictions and light sentences for those who were convicted. In another incident at Ogrezeni in May 1991, twenty-one Roma houses were burned following the stabbing of a Romanian by a Gypsy teenager. According to a Helsinki Watch investigation (1991: 37–73), police were present during the attack on Gypsy property but did not intervene. A Police Department/ Ministry of the Interior report on the incident blamed the Roma:

> The atmosphere created by this minority angered many people who decided to take the law into their own hands and to bring justice where they thought it was needed. As a result, a group of Romanians set fire to and burned down many Gypsy dwellings.
>
> (Helsinki Watch 1991: 121)

Extreme violence against the Roma is not the only way in which racism has manifested itself in the post-communist period. Ladányi and Szelényi (1996) have demonstrated that the Roma have been particularly discriminated against in the rationalisation of industry, and there are worrying patterns of discrimination in schools, which may represent a continuation of practices established under the communists. Apart from discrimination in mainstream schools, remarkably high proportions of Roma children have been found in special schools for children with mental disabilities. In the Czech and Slovak Republics in 1991, for example, 41 per cent of children in special schools were identified as of 'Gypsy origin' (Helsinki Watch 1992: 39). This suggests a negative judgement of Roma culture rather than a diagnosis of mental disability. The value of mainstream education for the Roma can be exaggerated, but discrimination in education as well as in the job-market does suggest an attempt at systematic exclusion of the Roma as well as their control through such instruments as special education. There has been some organised resistance, for example, through the European Roma Parliament, based in Budapest,[6] but the prospects are not good. A recent survey of people's prejudices in Hungary (Böhm 1994) demonstrated that Gypsies were the most disliked ethnic or racial minority (presumably, for 'native' Hungarians – the opinions of the Roma were apparently not sought). They were the group respondents wanted least as neighbours, ranking above Arabs, 'coloured people', 'immigrants' and Jews – a disparate and rather confused set of categories, but the pariah status of the Roma is clear. I would emphasise, however, that racist attitudes and practices directed at the Roma are not a novel feature of the new nationalisms emerging in post-communist Central and Eastern Europe. What is happening now is a continuation of several centuries of oppression. However, racism against the Roma may be less restrained and is certainly more widely reported than it was during the communist period.

Conclusion

I started this chapter on a fairly optimistic note. Drawing on John Berger's admittedly idealised representation of peasant culture, my own ethnographic experience with Gypsies, ethnographies of indigenous peoples, and of the

Roma in Europe, I suggested that the enterprise, flexibility and opportunism of the Roma gave them a cultural endurance of which ethnocentric models of change which emphasised the transformative power of capitalism or socialism failed to take account. According to the survival model, new spaces are always there to be negotiated, whether from a base in a Budapest apartment or a remote Roma village. Changes in the dominant economy and the political system present new opportunities, invariably in interstitial spaces, the left-over bits of the dominant space economy, the invisible parts of the topography of capitalism.

A part of the problem in assessing progressive models of transition of the kind specified earlier in this chapter, in relation to minorities like the Roma, is the problem of knowledge. We might assume that knowledge of the Roma economy is accessible, if only through proxy measures like housing conditions, levels of education or unemployment rates. However, this fails to take account of cultural difference in two important respects. First, material goods, like housing, are valued differently and may not compare in importance with social 'goods', like maintaining extended family or kin networks. Second, survival for the Roma may entail deception. When begging, it is important to look poor; when selling second-hand cars, it is important to look respectable. Thus, the visible and measurable signs of economic well-being or poverty may be misleading.

Recognising that there is a problem of cultural difference and a danger of ethnocentrism, it is evident that the association between the Roma and poverty in increasingly polarised societies in Central and Eastern Europe is strengthening. In the Hungarian case, for example, this is happening because of the growing poverty of the inner city, particularly in Budapest, and the economic decline of the rural periphery, the two areas where the Roma population is increasing most rapidly. They are thus distanced from the more affluent consumers of Gypsy services, and they have to compete increasingly with other non-Gypsy poor for the meagre resources offered by their localities. This is happening at a time when factory work and other state employment, which provided security against fluctuations in the informal economy under socialism, are no longer available and when private employers refuse to take on Roma. This set of economic problems cannot be separated from racism, which has put the Roma at a disadvantageous position in the job market. It may also have contributed to their isolation in countries like Romania, where racist violence has received most publicity. The Roma survive, but there is a need for effective civil rights legislation to tackle job discrimination, discrimination in education and racism.

Acknowledgement

I am very grateful to János Ladányi, Department of Sociology, Budapest University of Economics, for providing me with unpublished material on the Hungarian Roma.

Notes

1 'Roma', meaning 'the people' in Romany, is a self-ascription widely used by continental European Gypsies. In this chapter, I use 'Roma' and 'Gypsies' interchangeably, with the occasional use of other names which indicate a more specific identity, for example, Kalderas, Vlach and Sinti.
2 Judith Okely (1979) discusses the presentation of the Gypsy self in economic transactions with the *gaje* in some detail. See also Williams (1982).
3 There are close parallels here with the important role of the 'second economy' in mainstream Hungarian society (Gábor 1979).
4 It was not until 1991 that the murder of Roma in the Sachsenhausen concentration camp near Berlin was recognised. In that year, a commemorative plaque was erected by the Sinti community on the building formerly used for surgical experiments.
5 A long history of oppression suggests that Roma have been racialised since their arrival in Europe in the fifteenth century. Racist attitudes towards the Roma were certainly a problem during the communist period when the state represented them as a criminal and generally deviant minority. I would argue that conditions in the post-communist period, when questions of national identity are being re-examined, makes the Roma more vulnerable to racist abuse. Popular racism is compounded by the negative attitudes of the state.
6 This is one of several Roma organisations across Europe that lobbies on behalf of Gypsies. Others include the Union Romani in Barcelona and the Ethnic Federation of Roma in Bucharest.

Bibliography

Berger, J. (1979) *Pig Earth*, London: Writers and Readers Publishing Co-operative.
Bhabha, H. (1994) *The Location of Culture*, London: Routledge.
Böhm, A. (1994) *Systemic Change and Local Government in Hungary*, Budapest: Institute for Political Science, Hungarian Academy of Sciences.
Bridge, A. (1993) 'Romanians vent old hatreds against Gypsies', *The Independent* 19 October: 13.
Brody, H. (1973) 'Eskimo politics: the threat from the South', *New Left Review* 79: 60–70.
Brody, H. (1975) *The People's Land: Eskimos and Whites in the Eastern Arctic*, Harmondsworth: Penguin.
Douglas, M. (1966) *Purity and Danger*, London: Routledge and Kegan Paul.
Fonseca, I. (1996) *Bury Me Standing: the Gypsies and their Journey*, London, Vintage.
Fraser, A. (1992) *The Gypsies*, Oxford: Blackwell.
Gábor, I. (1979) 'The second (secondary) economy', *Acta Oeconomica* 22: 91–111.

Giddens, A. (1979) *Central Problems in Social Theory*, London: Macmillan.

Gurvich, I. (1978) 'Contemporary ethnic processes in Siberia', in R. Holloman and S. Arutiunov (eds) *Perspectives on Ethnicity*, The Hague: Mouton.

Helsinki Watch (1991) *Destroying Ethnic Identity: the Persecution of Gypsies in Romania*, New York: Human Rights Watch.

Helsinki Watch (1992) *Struggling for Ethnic Identity: Czechoslovakia's Endangered Gypsies*, New York: Human Rights Watch.

Kemény, I. (1975) 'A Budapest cigányokrol' (Gypsies in Budapest), *Budapest* 5.

Kovács, Z. (1990) 'Rich and poor in the Budapest housing market', in C. Hann (ed.) *Market Economy and Civil Society in Hungary*, London: Frank Cass.

Ladányi, J. (1993) 'Patterns of residential segregation and the Gypsy minority in Budapest', *International Journal of Urban and Regional Research* 17, 1: 30–41.

Ladányi, J. and Szelényi, I. (1996) 'Class, ethnicity and ecological change in post-communist Hungary and at the turn of the millenium Budapest', unpublished ms.

Laki, L. (1992) 'Young people on the margin', in F. Gazsó and I. Stumpf (eds) *Youth and the Change of Regime*, Budapest: Institute for Political Sciences, Hungarian Academy of Sciences.

Okely. J. (1979) 'Trading stereotypes: the case of English Gypsies', in S. Wallman (ed.) *Ethnicity at Work*, London: Macmillan

Salecl, R. (1993) 'National identity and socialist moral majority', in E. Carter, J. Donald and J. Squires (eds) *Space and Place: Theories of Identity and Location*, London: Lawrence and Wishart.

Sibley, D. (1988) 'Purification of space', *Environment and Planning D: Society and Space* 6: 409–21.

Spivak, G. (1987) *In Other Worlds: Essays in Cutural Politics*, London: Routledge.

Tipps, D. (1973) 'Modernization theory and the study of national societies', *Comparative Studies in Society and History* 15: 199–206.

Williams, P. (1982) 'The invisibility of the Kalderas in Paris', *Urban Anthropology* 11, 3–4: 315–46.

Yates, A. (1970) 'Housing programmes for Eskimos in northern Canada', *Polar Record* 15: 45–50.

Young, I . (1990) *Justice and the Politics of Difference*, Princeton: Princeton University Press.

Part V

FROM THE DEVELOPMENTAL STATE TO HYBRID CAPITALISMS

Comparative transitions

17

DENATIONALISATION OF THE MAURITANIAN STATE

Mohameden Ould-Mey

Introduction

Since the early 1980s development policy and research have focused on processes of economic and social restructuring in the industrialised nations, structural adjustment programmes in developing countries, and transitions to market economies in the former communist bloc (Hirst and Thompson 1996; Harvey 1995; Ould-Mey 1997; Swatuk and Show 1994; Gowen 1995). One of the salient features of all these processes is the redefinition of the role and scope of the state in the light of the increasing integration of the world economy (Holloway 1994; Jessop 1991). Another is the problematic nature of the state as both an instrument and a target for the various processes of restructuring, adjustment and transition. Within the context of developing countries, I refer to this redefinition of the role and scope of the state as a process of *multilateralisation* and *denationalisation* of the state (Ould-Mey 1996). In this process, multilateralisation and denationalisation occur as national economic policy is subjugated to the supervision of multilateral organisations. These twinned processes have received much attention in the recent restructuring of state socialist societies. In this chapter I draw on the specific case of Mauritania (Figure 17.1) and argue that such a restructuring of the state is essentially designed to make it more congruent with the growing power of multilateral institutions and more suitable to the fundamental shift from national to international forms of capital accumulation. The quantitative dimension of this shift has been summed up by the 1996 Annual Report of the World Trade Organisation when it noted that: (1) during the past two decades the annual value of merchandise exports multiplied eight and a half times and the annual foreign direct investment outflows multiplied more than twelve times, and (2) in 1995 sales by overseas Japanese manufacturers exceeded Japan's total merchandise exports for the first time.

389

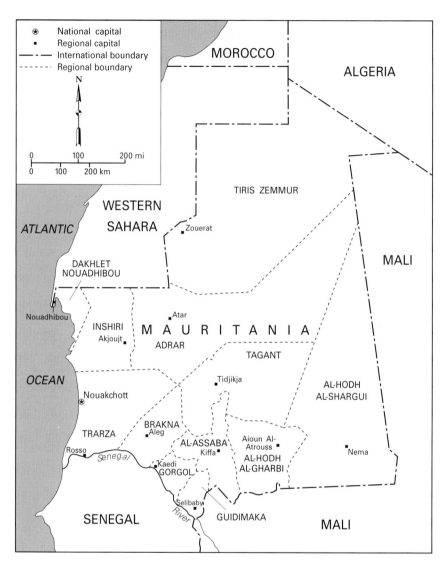

Figure 17.1 Mauritania

Between 1963 and 1984 Mauritania pursued an overall economic policy inspired by planning. Such a development model was based on: (1) unification and control of the national market; (2) protection and promotion of local industries and production (import substitution-industrialisation); (3) strict regulation of foreign trade and the maintenance of a rigid exchange rate system; (4) nationalisation of foreign capital; and (5) international co-ordination to protect the price of raw material exports (cartels). During

390

this period, a process of nation-state building was taking place in Mauritania amid the ideology of national independence and economic development. This attempt at nation building was seriously challenged by a chronic fiscal crisis of the state in the second half of the 1970s, a crisis that was exacerbated by the burden of Mauritania's involvement in the war in the Western Sahara from 1975 to 1978. The crisis was so severe that the civilian government was overthrown in a bloodless coup by the military in 1978, and this was followed by a series of nine coups and counter-coups until 12 December 1984, when a new military government adopted a structural adjustment programme in close collaboration with the IMF and the World Bank.

This chapter examines the process of multilateralisation and denationalisation of the Mauritanian state within the context of these structural adjustment programmes (SAPs). It is based on intensive fieldwork in Mauritania and extensive collection and analysis of primary documents such as the texts of Mauritania's economic and financial programmes, World Bank economic memoranda and mission reports, IMF/World Bank policy framework papers, IMF standby arrangements, Paris Club accords, Consultative Group summary of debates, loan agreements, and various minutes of proceedings of meetings between negotiation teams from the Mauritanian government and the multilateral institutions. The central argument made here draws on reading, discussion and analysis of these primary sources. Such an approach does not focus on the rhetoric of either the borrowing government or the lending multilateral institutions. By sticking to the documents (what the government and the multilateral institutions have agreed upon) one can unravel and grasp the full scope of the interplay between the state apparatus and multilateral institutions and therefore provide the key to understanding the elements of the emerging denationalised and multilateral state.

First, the shift from government to multilateral policy-making is demonstrated through United States Agency for International Development (USAID) and the World Bank's formulation of economic policy in Mauritania. Second, the retreat from the use of national plans for economic and social development is contrasted with the emergence of policy framework papers (PFPs) which are endorsed by the state but conceived and designed by the World Bank and the IMF. Third, the debt rescheduling schemes of the Paris Club are examined as an illustration of the transition from bilateral to multilateral management of foreign debt. Fourth, the role of the Consultative Group in defining and funding public investment programmes is analysed as an example of how the state is losing control over the allocation of resources in the country. Finally, the increasing role and influence of lender missions and non-governmental organisations (NGOs) within the state is highlighted as a further illustration of how deep multilateral institutions have penetrated the national space economy.

USAID and the World Bank's formulation of economic policy

Formulation of economic policy is a primary function of the state and an important element of planning. Policies are often solutions to some problems that are singled out following certain diagnostic studies that are themselves based on a great deal of data collection and analysis. The Mauritanian state began to lose control over this process of policy formulation in 1978 when the government allowed the USAID to carry out a study of the rural sector and human resources in Mauritania that was to serve as the basis for preparation of the Fourth Plan of Economic and Social Development (1981–5). The project, known as Rural Assessment and Manpower Surveys (RAMS) engaged some fifty-five specialists for varied periods of time over twenty-eight months and came up with over forty-four formal studies and reports (Ministry of Planning 1981: 23). The RAMS study sowed the seeds for the strategy of SAPs and its market-oriented philosophy of development. At the time, the RAMS' findings constituted the only comprehensive investigation covering a wide range of development issues (from environmental problems to government and politics) and the only one providing important data and market-oriented analysis.

RAMS reports were rapidly disseminated within influential circles of planners, politicians, academics and researchers. In these reports, an American functionalist approach to sociology (emphasising descriptive ethnography rather than class analysis) was applied to the study of Mauritanian society, which was accordingly divided into an ethnic mosaic. The two major studies of sociological profiles were revealing in this regard with their carefully selected titles: *Les Maures* and *La Mauritanie négro-africaine*. It is difficult to measure with precision the impact of these studies on the ethnic consciousness and ethnic strife that later plagued Mauritania. But these studies were the first to emphasise directly post-independence sub-national identities, which the nationalist state had never acknowledged because of their potential danger for national unity and inter-ethnic cohesiveness and coexistence. It should be noted that the RAMS study followed the guidelines of the Country Development Strategy Statement (CDSS) for Mauritania prepared by the USAID in January 1979. The CDSS document attacked the development strategies of the previous regime as essentially based on an ideological flirtation with socialist-type state control of the economy and a French colonial legacy that favoured centralised direction of economic activity (Ministry of Planning 1981: 23). In other words, a direct lineage in the formulation of economic policy can be traced from CDSS to RAMS to World Bank studies.

While the CDSS guidelines inspired the RAMS study, the latter paved the road for World Bank studies in Mauritania. The sequence and timing of World Bank missions and studies reveal how systematic was multilateral planning. In June 1979 the World Bank released a major report entitled *Islamic Republic of Mauritania: Recent Economic Trends and Needs for Foreign Capital*. This study constituted the basis upon which a first stabilisation programme was adopted by Mauritania in 1979 in collaboration with the World Bank and the IMF. Although the Bank carried out several other studies dealing mainly with economic sectors, the major study upon which the first structural adjustment programme was initially conceived and carefully tailored was a comprehensive one entitled *République Islamique de Mauritanie: Memorandum Économique*, released in July 1985. This study was initiated by a World Bank mission whose members stayed in Mauritania during April and May of 1984. A second mission came in December 1984 to update the findings of the first mission in the light of the military coup of December 1984. The findings, conclusions and recommendations of the study were discussed with the Mauritanian government in June 1985. The problems diagnosed and the recovery programme proposed in this study constituted the guidelines for subsequent economic policies in Mauritania and the profound denationalisation of policy formulation through the IMF and World Bank's policy framework papers.

The IMF and World Bank's policy framework papers

The multilateral design of development policy in the denationalised state is not limited to the formulation of economic policy. It also involves the design of detailed development strategies laid down in the policy framework paper (PFP). The idea of PFP was endorsed and developed by the executive directors of the World Bank and the IMF following US Secretary of State James Baker's initiative on the debt problem in 1985. The IMF describes the PFP as follows:

> The PFP is a forward-looking document, updated annually on a three-year rolling basis, that identifies the country's macroeconomic and structural policy objectives, the strategy of the authorities to achieve these objectives, and the associated financing requirements. The specific policy undertakings for the first year and general indications of policies to be pursued in the second and third years of the program are described in the PFP. The document also discusses the country's public investment program and financing requirements and outlines the likely social and environmental impact of policy changes, along with steps that can be taken to cushion the poorest segment of the population from any adverse effects. Potential creditors and donors, that is, members and multilateral organisations that may provide external assistance in support of adjustment programs, may be consulted informally in the preparation of PFPs and often draw on completed PFPs as a reference in making their

own decisions about financial assistance. The PFPs are reviewed by the Fund's Executive Board and by the Committee of the Whole of the Executive Board of the World Bank.

(IMF 1991: 79)

This IMF description of the PFP shows that it is actually a very rigid plan that guides the national economy and anatomises it for potential creditors and donors. The first policy framework paper for Mauritania (covering the period from July 1986 to July 1989) was reviewed by the executive directors of the Bank and the Fund in September 1986. This PFP represents a political commitment of the government to implement a development strategy based on a careful management of demand (stabilisation) and an increase of supply (structural adjustment) through a wide range of specific financial and economic policies. The strategy of development in the first PFP represents a departure from the previous plans of economic and social development. The public investment programme was reduced from 32 per cent of GDP in 1984 to around 20 per cent of GDP over the period covered by the first PFP, and about 42.1 per cent of total investment was allocated to rural development. The main reforms focused on economic policies in four areas: (1) management of public finance and administrative reform; (2) rehabilitation of public enterprises; (3) restructuring of the banking system; and (4) management of the external sector. The second PFP was adopted in 1987 as a document that lays out and reviews the first PFP 1986–9 by presenting development and progress accomplished during the first year and identifying actions of economic and financial policy to be pursued during the period from July 1987 to July 1990. The main areas of reform continued to be public finance management, public enterprises, banking, agriculture, fishing, and the legal and regulatory frameworks. The main factors of growth were identified in agriculture, fishing, and small and medium enterprises. Over 150 measures of economic policy were originally scheduled in the first and second PFP, but a record of 176 measures had already been implemented by 1989 (Ministry of Planning 1989a). The third PFP (1989–91) was reviewed by the executive board of the Fund and the Bank in May 1989. It was preceded by ten days of negotiations between a seventeen-member delegation of the Mauritanian government headed by the Minister of Economy and Finance, and a twenty-one-member joint delegation from the IMF and the World Bank in Washington in February 1989. Ninety measures of economic policy were specified and annexed to the thirty-seven-page PFP for implementation between 1989 and 1991. They fall within ten major policy areas that encompass all aspects of political economy in the country: (1) agricultural policy; (2) management of public sector; (3) industries, fisheries and energy; (4) public finance policy; (5) public enterprises; (6) monetary and credit policy; (7) banking sector; (8) external sector; (9) price policy; and (10) human resources.

The negotiation of the fourth PFP (1990–3) revealed even more leverage of multilateral institutions over the design of development strategies. It was drafted in early 1990, but the Governor of the Mauritanian Central Bank sent a letter to the IMF in July 1990 requesting certain modifications of some economic policy measures agreed upon in the PFP document. His request was rejected by the IMF and the World Bank in a conjoint reply, which reads:

> we have already informed our services of the positive conclusion of our consultations with the Mauritanian authorities on all points in the PFP except the issues related to the *Banque Nationale de Mauritanie* (BNM) and the *Union des Banques de Développement* (UBD). We would like to draw your attention to the fact that some of the modifications you are suggesting tend to reduce the significance of the reforms we agreed on and run the risk of weakening the process of adjustment in which your country engaged itself.

(IMF 1990)

The invasion of Kuwait by Iraq in August 1990 and the subsequent US-led Gulf War in 1991 delayed negotiations between the government and the Bank because the United States did not accept Mauritania's sympathy with Iraq. In hearings before the US Congress, held three months after the end of the war, one member of the panel noted: 'In Mauritania, the US has really few ties. We have ended our bilateral aid program and it seems that we are on the way to ending support for the multilateral aid program to Mauritania' because of its pro-Iraq position in the Gulf War even though it is now 'seeking ways to get back into the world's good grace' (United States Congress 1991: 21). Negotiations of the fourth PFP resumed again in Nouakchott in March 1992 and continued in Washington, DC, where the PFP document was finally signed in November 1992. With each successive PFP the state lost control over the articulation of development strategies and, as we shall see, over other important aspects of economic policy, particularly fiscal policy and the control of money supply and credit allocation.

IMF's monetary surveillance and discipline

Regulation of money and allocation of credit represent important functions within any state system. Indeed, control of money and credit is the first means of regulation available to the state (Amin 1980: 24). The central bank represents the instrument through which government policies determine the amount of money to be printed, the rate of exchange with other currencies, the amount of credit to be allocated, and the priorities and procedures of such allocation in accordance with a predetermined strategy of development.

395

Mauritania created its own currency, the ouguiya (UM), and central bank in 1973, and continued to perform the above functions prior to SAPs.

But IMF interference with credit and monetary policies illustrates the far-reaching process of denationalisation, since money-issuing is perhaps the last tenet of state sovereignty. Peripheral states like Mauritania have no leverage over the IMF and must comply with its recommendations. In 1996, the IMF had a membership of 181 countries. Mauritania has been a member since 1963 and has a quota of $70 million since 1992. This represents about 0.03 per cent of the $218 billion quotas of the IMF, and gives Mauritania virtually no voting power within the institution. The value of the Mauritanian currency (the ouguiya) is determined in relation to a basket of currencies: the Belgian franc, the Deutsche Mark, the French franc, the Italian lira, the Spanish peseta and the US dollar. Under obligations of Article IV, 1(iii) of the Second Amendment of the Articles of Agreement of the IMF, members shall avoid manipulating exchange rates to gain competitive advantage over other members. Section 2(b) of the same Article IV also gives full authority to the IMF to 'exercise firm surveillance over the exchange rate policies of members'. This obligation and the firm surveillance of the IMF do not leave room for Mauritania to manipulate fluctuations in the value of currencies; rather, it must adjust its exchange rate to those fluctuations. The lack of full control over the exchange rate is aggravated by the systematic devaluation of the ouguiya (agreed upon officially from time to time by the government and the IMF) and its almost annual depreciation following the adoption of a flexible exchange rate policy. In 1981, one US dollar could purchase only UM 48.3, while in early 1997 one US dollar could purchase UM 164. IMF's control over Mauritania's fiscal, monetary and credit policies is embodied in seven IMF arrangements covering the period from 1985 to 1997 (Ould-Mey 1996: 150; and IMF 1996). Under SAPs, special binding arrangements between the central bank of Mauritania and the IMF began to challenge the ability of the state to formulate credit and monetary policies independently. The IMF became to the central bank of Mauritania what the central bank of Mauritania is to the primary banks in Mauritania. In fact, IMF political leverage over fiscal policies has always belied its claim of being an *apolitical* organisation. For example, one of the first international figures to visit Mauritania following the 1992 presidential and parliamentary elections was Michel Camdessus, managing director of the IMF. He announced that he came 'to see how the President, the Premier, and the Opposition see the future of the country' (Camdessus 1992), and in the process 'encouraged' the government to adopt a 42 per cent devaluation of the ouguiya.

A brief description of the fourth arrangement further illustrates IMF growing surveillance over the economies of peripheral states. The fourth IMF arrangement (May 1987) amounted to SDR 10 million for a period of 12 months during which the government had to implement specific economic policies (IMF 1987). The IMF arrangement was prepared by the members of an IMF mission who helped the government in the drafting of the Letter of Intent and the Memorandum of Economic and Financial Policy, which were subsequently reviewed and accepted by the executive board of the IMF. The Letter of Intent was co-signed by the Minister of Economy and Finance and the Governor of the Central Bank of Mauritania, and addressed to the IMF managing director. It documented the progress of the government in the implementation of the previous IMF arrangement of 1986 and its operational readiness and political commitment to implement further policies within the framework of a new arrangement with the IMF in 1987. Paragraph 3 made it clear that the government was ready to take additional policy measures if requested to do so by the IMF managing director. These arrangements with the IMF provided temporary cash for the payments of imports and, more important, constituted a *sine qua non* condition for rescheduling Mauritania's foreign debts within the framework of the Paris Club, another international forum that has leverage on the Mauritanian state through the centralisation and management of foreign debts. The debt rescheduling schemes of the Paris Club illustrate the transition from bilateral to multilateral management of foreign debt.

Paris Club's management of foreign debts

One leitmotif of the adjustment process in Mauritania is the debt problem, which continued to worsen even after the implementation of SAPs. In 1984 the outstanding foreign debt of the country was $1.7 billion and the debt service $88 million (28 per cent of export earnings). In 1993 (by the time of the fifth Paris Club debt rescheduling), the external debt had reached SDR 1.56 billion (the equivalent of 230 per cent of the GDP) and the annual debt service was about SDR 140 million, or 44.3 per cent of export of goods and services (IMF 1995: 25). Prior to SAPs, the state was able to do business with its creditors on an exclusively bilateral basis. Foreign aid (loans, grants and technical assistance) averaged $270 million annually between 1979 and 1984, or the equivalent of 40 per cent of GDP per capita (World Bank 1985: para. 2.15; Ministry of Planning 1989b; and United States Central Intelligence Agency 1993). Within the adjustment programme, particularly its debt rescheduling component, Mauritania deals with its creditors on a multilateral basis and according to the recommendations of the Paris Club. The policy of dealing with Third World debt on a case-by-case basis has been decided and reiterated in several G7 summits, particularly at the Economic Summit in Tokyo in 1986 when the heads of state of the G7 reaffirmed 'the continued importance of the case-by-case approach to international debt problems' (University of Toronto 1997). This debt strategy aborted attempts by debtor countries to internationalise the debt problem and in a sense nipped in the bud the potential emergence of a Third World debt cartel.

Originally the Paris Club represented an informal group of finance ministers from the ten wealthiest member states of the International Monetary Fund (the Group of Ten): Belgium, Canada, the Federal Republic of Germany, France, Italy, Japan, the Netherlands, Sweden, the United Kingdom and the United States. While the London Club (of banks) deals with non-official and non-officially guaranteed debts, the Paris Club (of governments) deals exclusively with government or government-guaranteed debts. As the debt crisis deepened in the developing nations throughout the 1980s, the Paris Club became an institutionalised forum where requests from developing nations for debt rescheduling were formally presented to a World Bank-led gathering of international creditors. The Club meets monthly in Paris under the chairmanship of the French Treasury. Debt rescheduling within the framework of the Club can take place only if the country requesting it maintains a certain arrangement with the IMF and commits itself to implement a set of carefully specified economic policies. The following is a brief description of the first of five reschedulings of Mauritania's foreign debts within the framework of the Paris Club between 1986 and 1993.

The first rescheduling of Mauritania's foreign debts took place in Paris in April 1985 following the signature of an IMF arrangement in Washington (Ministry of Finance 1985). The diversity and identity of the creditors present at this meeting is worth noting because it demonstrates the extent of multilateralisation of national economic policy. They included representatives from the Federal Republic of Germany, Austria, Spain, the United States, France, Kuwait, the Netherlands and the United Kingdom. Also present were observers from the governments of Finland and Japan, from the International Monetary Fund, the International Bank for Reconstruction and Development, the secretariat of the Conférence des Nations Unies pour le Commerce et le Développement (CNUCED), and the Organisation of Economic Co-operation and Development (OECD). The Mauritanian delegation presented the economic and financial difficulties facing Mauritania and expressed the firm determination of the government to solve the financial and economic disequilibria and to reach the goals set out within the framework of the latest IMF arrangement. The IMF delegation presented a description of the economic situation in Mauritania and the main points of the adjustment programme that was implemented in the country and supported by an IMF arrangement covering the period 1985 to 1986. The representatives of the community of lenders noted with satisfaction the policy measures implemented by the government within the adjustment process and decided to recommend to their governments or organisations a rescheduling of certain categories of Mauritania's foreign debts, particularly commercial credits and loans guaranteed by the governments or appropriate organisations of the creditor countries participating (Ministry of Finance 1985).

The government committed itself to seek rescheduling of its debts according to the Paris Club terms. The participating creditors informed the government that they would communicate to each other a copy of their bilateral agreements with Mauritania in application of the present agreement. Each participating creditor was to inform the president of the Paris Club about the date of signature of its bilateral agreement with Mauritania, the interest rates applied, and the amount of debts rescheduled. By the same token the government would inform the president of the Paris Club of the contents of its bilateral agreements with the other creditors and would pay to participating creditors and observers all its other debts not included in this agreement as soon as possible, but not later than 30 September 1985. Moreover, the government was to take all necessary administrative measures to allow private Mauritanian debtors to pay to the Central Bank of Mauritania the equivalent of their debts in the local currency. Last but not least, the provisions of this agreement were to be applicable only if the Mauritanian government continued to maintain an agreement with the IMF and made sure that all agreements to be signed with banks and other creditors would meet the conditions specified in this debt rescheduling agreement. Written reports concerning bilateral debt rescheduling agreements were also to be sent on a regular basis to the president of the Paris Club. In addition to co-ordination of the management of foreign debt within the framework of the Paris Club, the creditors equally co-ordinate the design of investment programmes in Mauritania through the Consultative Group.

The Consultative Group's investment programmes

If links with the IMF and the Paris Club have subjugated fiscal and debt policies to multilateral discipline, links with the Consultative Group have defined public investment programmes in Mauritania. Within the framework of SAPs, the public investment programme of a borrowing country is presented to the Consultative Group, which is a formal gathering of international lenders seeking to invest in a particular country. It was conceived by the World Bank as an instrument for monitoring public investment programmes and centralising loans and other forms of investment that are not provided directly by the World Bank or the IMF. It also assures creditors that their money is being invested in a stable investment climate and that procedures are in place to guarantee their loans and rates of return. By 1982, dozens of Consultative Groups had already been formed to oversee investment programmes in dozens of developing countries, and in 1988 the Multilateral Investment Guarantee Agency was created as an instrument within the World Bank Group that 'ensures private foreign investment in developing countries against non-commercial risks such as expropriation, civil strife, and inconvertibility' (United States Department of State 1990:

529). Almost all the Consultative Groups are chaired by the World Bank. The first Consultative Group for Mauritania gathered in Paris in November 1985 to examine the investment programme of Mauritania for the period 1985–8, and the second meeting took place also in Paris in July 1989 (Ministry of Planning 1989c). The following is a brief description of some of the issues raised by the various representatives attending the second meeting of the Consultative Group.

The meeting was chaired by Michael Gillette, director of the Sahel Department at the World Bank. It was attended by a twenty-five-member Mauritanian delegation (headed by the Minister of Planning and Employment) and thirty-nine representatives of bilateral and multilateral lenders from seven countries and ten banks and development organisations. The bilateral lenders included the United States, France, Italy, Japan, Morocco, Brazil and the Netherlands. The multilateral lenders included African Development Bank, Islamic Development Bank, Commission of European Communities, Arab Fund for Economic and Social Development, International Fund for Agricultural Development, Kuwait Fund for Arab Economic Development, Arab Monetary Fund, International Monetary Fund, Saudi Development Fund and United Nations Development Program. These diverse investors are the ones who have the upper hand in terms of defining and funding investment programmes in Mauritania. The role of the state has become simply to stimulate them and excite them about the prospects of their investments while ensuring them of the guarantees they need.

Opening the meeting, chairman Gillette stated that the forum offers an opportunity for Mauritania to present its development programme for 1989–91 and its related public investment programme, to indicate and explain the impacts of recent civil strife (ethnic clashes in Mauritania and Senegal in April 1989), and ask for assistance in establishing a reinsertion programme for Mauritanians repatriated from Senegal. He also underlined the significance of the recent Paris Club meeting in which Mauritania's debt was rescheduled (Ministry of Planning 1989c). In his presentation, the Mauritanian Minister of Planning summarised the objectives achieved within the framework of the Economic and Financial Recovery Program 1985–8. He then reaffirmed the commitment of the government to continue the process of adjustment through the implementation of the Economic Consolidation and Growth Program of 1989–91 as defined in the third policy framework paper. He reminded the lenders that Mauritania's foreign debts represented 180 per cent of its GDP and the investment programme under consideration represented only 16.6 per cent of GDP, compared to over 30 per cent prior to adjustment (Ministry of Planning 1989b). Several members of the group noted that, despite earlier progress, the external financial situation of the country remained fragile due mainly to the burden of foreign debts. Other members expressed reservations on issues such as budgetary discipline, social dimensions, environment, reform of the banking system and execution of public investment programmes. They expressed satisfaction with the size and composition of the public investment programme for 1989–91, but they recommended more rigour in the process of selection and execution of productive projects.

Thus, the Consultative Group for Mauritania functions as an international forum where the government has to present and explain 'its' policies, especially policies of investment. How much to invest? Where to invest it? In which sector? What are the priorities of investment? These, and others, are all important decisions in which the state is now one partner among many others. The investment programme is defined within the context of a strategy of development specified in a policy framework paper prepared in close collaboration between the government and the IMF and the World Bank. The priorities in investment among regions ('projects supporting export-related activities') and sectors ('the agricultural sector has potential for strengthening the private sector') are determined following diagnoses and recommendations of consultants and experts funded and cleared by international creditors and multilateral institutions. In all the meetings and discussions in Washington, Paris and Nouakchott, the state is only one voice among others. Moreover, the state is not only facing serious challenges from multilateral institutions and donor countries within the framework of international forums, it is also facing strong competition from lender missions working within its own national territory.

Lender missions and government committees

Lender missions and government committees represent the forums where both the political and technical measures of SAPs are negotiated in detail. Their growth reflects how far multilateral institutions have penetrated the national space economy. It is estimated that African countries conducted some 8,000 separate negotiations with international creditors between 1980 and 1992 (Smith 1994: 15). Their conjoint work and close collaboration on SAPs represent a new parameter of the emerging multilateral state, since they constitute the fundamental articulation between the state and international financial institutions. While government committees tend to focus more on the day-to-day implementation of SAPs, lender missions concentrate on the identification, design, funding, monitoring and evaluation of SAPs. The first structural adjustment loan (SAL 1) for Mauritania covered the period from June 1987 to December 1988 and emphasised policy reforms in public administration, banking, public enterprises, energy, food policy, fisheries development and private sector promotion. According to the World Bank Program Completion Report of January 1990, seventy-seven out of ninety-nine monitorable policy actions had been fully implemented in these reform areas, nineteen were under way in December 1988, and nine had not been implemented (World Bank 1990: para. 5). These sweeping reforms complemented others taken in 1985 and 1986 within the framework of the Economic and Financial Recovery Program (1985–8) and opened the way for many more in the subsequent Economic Consolidation and Growth Program (1989–91). The following is a brief description of the process of identification, preparation, negotiation, financing, monitoring and evaluation of the first structural adjustment programme by World Bank missions (World Bank 1990: Annex I) and serves to illustrate the impact of multilateral institutions working within the national territory.

The first structural adjustment programme in Mauritania was initiated by a World Bank mission of seven specialists who arrived in January 1986 for the preparation of the project, especially the initiation of a policy memorandum on the basis of which the government was to draft its Letter of Development Policy. In March, a pre-appraisal mission assisted the government in drafting the Letter of Development Policy and defining the major elements of a structural adjustment programme, particularly the establishment of conditions for tranche releases. The appraisal mission came in June and included fifteen specialists. It finalised with the government the Letter of Development Policy, which identified monitorable policy actions to serve as tranche release conditions. In March 1987, a prenegotiation mission updated the Letter of Development Policy and finalised the elements of discussion on the first structural adjustment loan (SAL 1) programme, particularly the banking-sector reform component. Finally, SAL 1 was negotiated in May 1987 and approved by the executive board of the Bank in June 1987.

Monitoring the programme began with the first supervision mission in November 1987. The mission noted progress in some areas (public expenditure management and investment programming) but reported delays in others (banking reform, energy policy and public enterprises). It pointed out the 'lack of commitment of the concerned authorities', whose dedication to SAPs was weakened by their internal struggles. The World Bank Completion Report stressed that the decision-making process was almost paralysed due to a power struggle within the military, an aborted coup, social unrest, and a financial scandal that resulted in the incarceration of the Minister of Finance, the Minister of Fisheries and the Governor of the Central Bank for their involvement in illegal financial transfers. The second supervision mission of February 1988 noted progress in economic manage-ment, energy policy and fisheries policy, but criticised delays in reforms related to food policy and the restructuring of public enterprises and the banking system. The blame for these delays was put on 'the exceptionally weak economic team which was in place in late 1987 and early 1988' (World Bank 1990: para. 24).

Such critical remarks from the World Bank mission were taken seriously by the government. Following this supervision mission, several ministerial reshuffles were undertaken and the Bank Completion Report noted that between November 1987 and December 1988, the World Bank task manager of SAL 1 met with four different Ministers of Economy and Finance and three different Governors of the Central Bank. The report of the third supervision mission of May/June 1988 expressed satisfaction about the 'new government team, which was more dynamic and more committed to the adjustment effort' (World Bank 1990: para. 24). The last mission was a completion mission that took place during the month of December 1989 and involved one specialist (Miguel Saponara) from the World Bank. The Bank mission noted that the SAL programme was a success since the Mauritanian economy had become more decentralised and market-oriented. The Bank report also noted as a problem the difficulty the government had experienced in reaching consensus, even though the success of the overall programme was explained by the 'firm commitment of the President and economic teams in place at the time' (World Bank 1990: para. 81). The first structural adjustment programme prepared the ground for the more detailed sectoral adjustment programmes where many non-governmental organisations (NGOs) have often participated in collaboration with the World Bank.

NGOs and the government

Throughout the 1970s only three international NGOs operated in Mauritania: the Catholic Church-based Organisation Caritative Inter-nationale (1972), the American-based Catholic Relief Services (1973) and the Swiss-based Lutheran Federation (1974). All came as relief organisations focusing mainly on food relief and environmental protection following the severe drought that hit the Sahel region in the late 1960s and early 1970s. Following the open-door policy of SAPs in the 1980s, the number of western NGOs operating in Mauritania increased rapidly and their development activities went well beyond food relief and environmental protection to include a variety of socio-economic projects in various sectors, particularly agriculture, health, housing, water, children and pre-natal care. Other semi-NGOs dealing with local communities also arrived. The number of US Peace Corps volunteers in the country increased from thirty-five in 1984 to sixty in 1985 (Peace Corps 1991: 17). By 1992, some 600 American Peace Corps volunteers had served in Mauritania, with an overwhelming majority coming after 1984 (*Le Temps* 1991).

Beyond their obvious humanitarian relief efforts, NGOs stimulate private enterprise and establish direct links with local communities. Through their small-scale development projects, they enhance market mechanisms, and through their assistance to the poor, they contribute to the containment of political dissent and the overall social peace necessary to the development of free enterprise. In this context, NGOs constitute one of the instrumentalities of adjustment and play a crucial role in cushioning the social shock of economic and financial austerity measures.

World Bank/NGO co-operation and co-ordination parallel the development and implementation of SAPs and increased to such an extent that, in 1994, NGOs collaborated in more than half of World Bank-supported projects (World Bank 1996). In 1980, the World Bank and NGOs held a joint workshop in Paris on small enterprise development, and in May 1981 an NGO/World Bank Committee was founded and began its work under the umbrella of the World Bank and the International Council of Voluntary Agencies (Union of International Associations 1992: entries 08287 and 1206). The NGO/World Bank Committee meets once a year. It is co-chaired by the chief of the Bank's Strategic Planning and Policy Review Department and an elected NGO member, and includes eight senior World Bank staff members and twenty-six representatives of NGOs. The NGO/World Bank Committee is financed entirely by the World Bank. Within this framework, NGOs are assigned the job of bringing their experience, local knowledge, perspective, priorities and institutional alternatives to the attention of the World Bank to help it in the formulation of country programmes and policy choices, as well as to participate in the implementation and/or monitoring of Bank-financed programmes that involve the poor.

In the 1990s, the World Bank is working on a new lending strategy designed to provide *micro-loans* to local groups and even local individuals. Such micro-loans demonstrate the ability of the Bank to circumvent the state if necessary and to penetrate (using NGOs in this case) as deeply into the denationalised state as it wants. This was illustrated in June 1995, when the newly appointed president of the World Bank noted that his future visits to countries implementing SAPs

> will not be state visits. I will, of course, see the governments in the countries but it is my expectation that I will visit projects, that I will talk with project managers, that I will walk the streets, that I will meet with the locals, that I will go and have a beer at night after dinner in good Australian fashion.
>
> (Wolfensohn 1995: 993–6)

Conclusion

SAPs are unleashing a profound process of denationalisation in which the state is restructured from a national to a multilateral state, and is virtually deprived of its sovereignty over development policies. This is the most significant change in the nature of the national state since the Peace of Westphalia in 1648, which crystallised the full and exclusive sovereignty of the nation-state over its people and its territory. Such a denationalised state corresponds to the globalisation of capital and the growth of multilateral institutions. SAPs succeeded in reversing a set of nationalist policies that aimed during the 1960s and 1970s to achieve two objectives: (1) securing a stronger position within the international system of states while constantly attempting to subjugate the external relations of the state to some impera- tives of its internal relations and social structures, i.e., adjusting the 'outside' to the 'inside', and (2) assuring maximum political centralisation and socio- economic and spatial integration through state institutions and government policies, i.e., centralisation. These two objectives constitute the backbone of the national state. In contrast, the new objectives of SAPs consist of: (1) subjugating the internal relations and social structures of the state to the imperatives of its external relations, i.e., adjusting the 'inside' to the 'outside'; and (2) pursuing a certain degree of political and economic decentralisation at the national level. Such decentralisation is taking place at a time when multilateral institutions and mechanisms such as the G7, the IMF, the World Bank, the Paris Club, the Consultative Group have almost completed the centralisation of power and wealth at the global scale. National decentral- isation and multilateral centralisation embody the shift from a national state to a multilateral state and from national to international accumulation where the state's absolute control over its national space economy is shrinking amidst the rising new global command economy (Ould-Mey 1997).

Acknowledgements

This chapter develops further an argument also presented in Mohameden Ould-Mey (1994 and 1996). Its central thesis was first presented in 'The denationalisation of the Mauritanian state', a paper presented to the 91st Annual Meeting of the Association of American Geographers, San Francisco, California, April 1994.

Bibliography

Amin, S. (1980) *Class and Nation: Historically and in the Current Crisis*, New York: Monthly Review Press.

Camdessus, M. (1992) 'Conference de Presse', *Horizons* 294: 16–17 July (also *Le Temps* 48, 19–26 July 1992).

Gowen, P. (1995) 'Neo-liberal theory and practice for Eastern Europe', *New Left Review* 213, September/October: 3–60.

Harvey, D. (1995) 'Globalization in question', *Rethinking Marxism* 8, 4: 1–17.

Hirst, P. and Thompson, G. (1996) *Globalisation in Question: The International Economy and the Possibilities of Governance*, Cambridge: Polity Press.

Holloway, J. (1994) 'Global capital and the national state', *Capital and Class* 52, Spring: 23–48.

IMF (1980–94) *International Financial Statistics*, Washington, DC: International Monetary Fund.

IMF (1987a) 'Mauritanie: demande d'accord de confirmation', Document du Fonds Monetaire International, ESB/87/73, 7 April.

IMF (1987b) IMF Press Release 5 May, Washington, DC: International Monetary Fund.

IMF (1989) IMF Press Release 29 May, Washington, DC: International Monetary Fund.

IMF (1990) 'Lettre au gouverneur de la Banque Centrale de Mauritanie', 27 July.

IMF (1991) Treasurer's Department *Financial Organization and Operations of the IMF*, Pamphlet Series No. 45, Second Edition, Washington, DC: International Monetary Fund: 79.

IMF (1993) *IMF Survey* 5 April, Washington, DC: International Monetary Fund

IMF (1995) *Mauritania – Recent Economic Developments*, IMF Staff Country Report No. 95/20, March, Washington, DC: International Monetary Fund: 25.

IMF (1996) 'IMF approves ESAF for Mauritania', [On-line] IMF Press Release No. 95/5, 25 January. Available: http://www.imf.org/external/np/sec/pr/1995/pr 9505.

Jessop, B. (1991) *State Theory: Putting the Capitalist State in Its Place*, University Park, PA: Pennsylvania State University Press.

Ministry of Finance (1985) 'Procès verbal agrée relatif à la consolidation de la dette de la RIM', Paris, 26–27 April.

Ministry of Planning (1981) *Synthèse du Projet RAMS S1*, August: 23.

Ministry of Planning (1989a) 'Bilan d'execution du PREF (1985–88)', prepared for the 2nd Groupe Consultatif pour la Mauritanie, Paris, 25, 26, 27 July.

Ministry of Planning (1989b) 'Déclaration du gouvernement mauritanien', presented by the Ministre du Plan et de l'Emploi, Réunion Gouvernement Mauritanien – Donateurs, Paris, 26, 27 July.

Ministry of Planning (1989c) 'Réunion des bailleurs de fonds pour la Mauritanie, résumé des débats', 12 August.

Ministry of Planning (1990) 'Government policy letter to the World Bank', restructuring of the public enterprise sector, 25 May: para. 2.

Ould-Mey, M. (1994) 'Global adjustment: implications for peripheral states', *Third World Quarterly* 15, 2: 319–36.

Ould-Mey, M. (1996) *Global Restructuring and Peripheral States: The Carrot and the Stick in Mauritania*, Lanham, MD: Littlefield Adams Books.

Ould-Mey, M. (1997) 'The new global command economy', paper presented at the Annual Meeting of the Association of American Geographers, Fort Worth, Texas, 1–5 April.

Peace Corps (1991) *Mauritania*, Washington, DC: US Peace Corps: 17.

Smith, P. (1993–4) 'Aid conditionality is "swamping" Africa', *Africa Recovery* December 1993–March 1994: 15.

Swatuk, L. A. and Show, T. M. (1994) *The South at the End of the Twentieth Century: Rethinking the Political Economy of Foreign Policy in Africa, Asia, the Caribbean and Latin America*, New York: St. Martin's Press.

Le Temps (1991) 11, 22–28 September.

Union of International Associations (1992) *Yearbook of International Organizations, Volume 1, 1992/93*, New York: K. G. Saur Munchen: entry no. 08287, 1206.

United States Agency for International Development (1979) *Country Development Strategy Statement*, FY 1981, Mauritania, January 1979, unabridged version: 23.

United States Central Intelligence Agency (1993) 'Vital statistics, Mauritania', *Factbook 1993*, Washington, DC: Central Intelligence Agency.

United States Congress, House Committee on Foreign Affairs (1991) *Human Rights in the Maghreb and Mauritania: Hearing, June 19, 1991, before the Subcommittee on Human Rights and International Organizations and on Africa*, Washington, DC: US Government Printing Office: 21.

United States Department of State (1990) Dispatch 2, 37: 529.

University of Toronto (1997) G-7 Information Centre [On-line], 14 March. Available: http://utl2.library.utoronto.ca/www/g7/index.html.

Wolfensohn, J. D. (1995) 'Dilettante or visionary?', *West Africa* 26 June–2 July: 993–6.

World Bank (1985) *RIM: Memorandum Economique*, no. 5535–MAU, 10 July, Washington, DC: World Bank: para. 2.15, and tables 7.1 and 7.4.

World Bank (1990) *Islamic Republic of Mauritania, Structural Adjustment Program, Program Completion Report*, Washington, DC: World Bank, Country Operations Division, Sahelian Department, Africa Region, 12 January: paras 5, 24, 25 and 81; and Annex I.

World Bank (1996) 'The World Bank and NGOs' [On-line], 7 October. Available: http://www.globalpolicy.org/ngobank.

World Trade Organization (1997) *The World Trade Organization Annual Report 1996* [On-line], 6 March. Available: http://www.wto.org/wto/whats_new/anrep.htm.

18

'SEX AND VIOLENCE ON THE WILD FRONTIERS'

The aftermath of state socialism in the periphery

James Derrick Sidaway and Marcus Power

> Under Stalin the story of Bolshevism came to be rewritten in
> terms of sorcery and magic, with Lenin and then Stalin as the
> chief totems. . . . Stalinism is a complex phenomenon, which
> needs to be viewed from many angles. But when it is seen
> from the angle which we are now viewing it, it appears as the
> mongrel offspring of Marxism and primitive magic. Marxism
> has its inner logic and consistency: and its logic is modern
> through and through. Primitive magic has its own integrity
> and its particular poetic beauty. But the combination of
> Marxism and primitive magic was bound to be as incoherent
> and incongruous as is Stalinism itself.
>
> (Deutscher 1984: 115–16)

> *¡Socialismo o Muerte!/Socialismo ou Morte!* [Socialism or
> Death!]
>
> (Communist Party of Cuba/Popular Movement for
> the Liberation of Angola slogans)

Introduction: beyond socialism or death

Writing in 1984, Tariq Ali could note how '[t]he legacy of Stalinism is most
visible today in the political systems that have been established in the USSR
and Eastern Europe, China and Indo-China' (Ali 1984: 7). In recent years,
however, new totems and fetishisms have arisen in many of these states and
in descriptions of their post-communist trajectory and prospects. 'Transition'
and the 'market' have been elevated to transcendental status. In this context,
and seeking potential customers amongst the readership of a conservative
Sunday newspaper, the manager of a new *emerging markets* unit trust
informs us that '[t]he sex and violence in emerging markets is in the wild
frontiers – places such as Russia and Peru' (Hinde 1996). Clearly the fantasies

408

of such fund-managers include the notion that capitalism has some sexy-violent frontiers. Yet, to the extent that it can be said to have identifiable frontiers at all,[1] some of them might be identified in the cities of that part of the 'Third' World which was (and in some cases still is) governed by regimes that are identified as 'socialist'.

By the mid-1970s, following several decades of revolutionary and anti-colonial struggle, an array of peripheral state-elites codified rhetorics of national development, legitimacy and administration in the language of Marxism-Leninism and/or a variety of 'socialisms' (African, Arab, etc.). There is a good deal about which to be sceptical in such rhetorics and strategies of power. With notable and rare exceptions – and like the longer established 'People's Republics' implanted in Eastern and Central Europe by Stalin and his local agents – the experiences of democracy and 'worker-peasant' participation in Third World socialist states were highly ritualised, formalised and, frequently, dangerously arbitrary affairs. Malign combinations of under-development, the Stalinist tradition and the apparatus inherited from the colonial or pre-revolutionary states reinforced the limits to 'democracy' and the *instrumental* character of much Third World socialism (see Laïdi 1988). In many cases too, a discourse of socialism was no obstacle to quite extreme forms of hierarchy and elite domination. Indeed, frequently it was a strange expression of such domination. In virtually all cases, 'peripheral socialism' was a discourse and practice of statism. In many cases, such as Nasserism and Baathism, it disguised (or obscured) a system of bureaucratic petit-bourgeois or military-rentier domination in a context in which capitalist and even quasi-feudal social relations remained strong. In other words, Marxism-Leninism or state socialism in the South might be read as a particular set of elite strategies of modernisation. Such strategies offered the prospect of constructing a hegemonic state-identity around a claim of difference from 'capitalist', 'reactionary' states and/or a colonial/bourgeois past. Secure in this appealing set of binary oppositions, Third World socialist regimes also had an established model of development, rationality and planning to impose. Based in part on the Soviet experience, this reference point contained a recipe for development: comprehensive plans, the ascendancy of the party, the building of an anti-imperialist national identity. As Laïdi (1988: 4–6) notes:

> The capacity of the Bolsheviks to force their way through unfavourable 'objective conditions', thanks to a militarized organi-zation and an unquenchable voluntarism, was not without its attractions. . . . [Yet] . . . Rather than talking of a model, which evokes the idea of passive replication of Soviet reality, it would perhaps be more accurate to suggest the concept of a *Soviet reference point*. This could then be defined as the process of replication of certain realities of the Soviet system under different historical conditions and for purposes that are not necessarily congruent,

409

either with the meaning given them by the Soviets, or with the diplomatic or ideological objectives of the USSR.

At times the effects of the 'reference point' in the Third World took on tragi-comic forms that exceeded even the horror and hypocrisy of the original Stalinist mould, as the arbitrary violence and barbarism of the 'cultural revolution' in China, the 'Democratic' People's Republic of Korea and 'Democratic' Kampuchea demonstrated. More generally, the 'reference point' was disseminated through and consumed in the ideology of Third Worldism. In this respect

> the Soviet model of forced development, based on centralisation, planning and the use of the modern sector (industry) to lead the traditional sector (agriculture) has largely permeated the developmentalist ideology of Third World elites *haunted* by the idea of accumulation and catching-up.
>
> (Laïdi 1988: 12, emphasis added)

Yet, whilst it remains vital to point to the profound limits to social transformation in many self-avowed 'socialist' states, and to the frequently brutal structures of domination associated with them, it is important to recognise that the rhetorics of Marxism-Leninism and state socialism were associated with *some* distinctive forms of administration and governance (see Smith and Swain in this volume).

There is no scope here to explore these in all their historical and theoretical complexity (see Post and Wright 1989 for an attempt to unravel the 'laws of motion' of peripheral state socialisms). However, it must be noted that beyond the claims of their governments, the distinctiveness of some Third World socialisms was also *partly* based on a particular negotiation with the world capitalist system (sometimes amounting to a period of partial disengagement), combined with the pursuit of relative autonomy and sometimes a degree of popular mobilisation. In tandem with this, a number of states (most notably Cuba) sought integration into the trading system centred on the USSR. This did mean that some forms of private capital circulation were arrested. In particular, land and much of the built environment were substantially de-commodified in many peripheral socialist regimes: in China, Cuba, North Korea, Indochina, Mozambique and elsewhere, processes of bureaucratic allocation partially displaced capitalist market mechanisms in the production and allocation of urban space, as well as in industrial enterprises and social relations in the countryside.

Today, few peripheral state elites still assert a formal commitment to Marxism-Leninism. Those that do, notably North Korea and Cuba, have been where marketisation and commodification have been most resisted until relatively recently. However, even in Cuba and more markedly in China

410

and Vietnam (see Watts in this volume; Smart in this volume), the rhetoric of difference has not prevented quite dramatic and far-reaching market mechanisms and private capital. This has included not just the enhanced role of foreign joint-venture capital, but also the establishment of lively private markets and large-scale speculative investments in land and the construction of houses and offices, most notably in parts of China.[2] The discursive accommodation of such material shifts is an important and fascinating topic – one whose study enables a better understanding of how the rhetorics and strategies of communism-Marxism-Leninism-socialism operated.[3] This, however, is not our main aim here. In this chapter, we focus on what an examination of the post-socialist periphery can indicate concerning reconfigurations of hegemony and authority in the contemporary world. In doing this, we draw upon the case of Mozambique (see Sidaway and Power 1995 for a more detailed and historically grounded interpretation).

Mozambique provides a particularly useful entrée. The World Bank (1996a, 1996b) has declared that it is simultaneously the poorest (in terms of per capita domestic product) and the fastest growing economy in the world. This conjuncture comes after a long and complex crisis, registered in a civil war and the deaths of over 1 million Mozambicans, the unravelling of the socialist strategy put in place after independence from Portugal in 1975, and profound economic and social disarticulation.

The Mozambican case

A Luta Continua! (The Struggle Continues!)
(Front for the Liberation of Mozambique (Frelimo) slogan)

One is invited to steer a path between such-and-such a minister who is corrupt to the marrow, such-and-such a remarried widow of a national hero, such-and-such an ambassador or MP who has joined the cocktail set and well and truly forgotten about socialism. The reader, astonished by the abundance of information, moves between politics, religion, tribalism and the end of illusions.

(Pelissier 1997: 154)

Independence – revolution – adjustment

Mozambique gained independence from Portugal on 25 June 1975 (Figure 18.1). This event was precipitated by a coup in Portugal which took place the preceding year. In turn, the unwillingness of the Portuguese armed forces to continue anti-insurgent wars against nationalist movements in Portugal's African colonies had played an important part in prompting the coup which had brought down the authoritarian conservative regime. This regime was

Figure 18.1 Mozambique

first established as the Estado Novo (New State) under the rule of Antonio Salazar in the 1930s, and turned out to be Europe's most durable right-wing dictatorship as well as its last sizeable overseas empire.

Since independence and the coming to power of the radical Frente de Libertação de Moçambique (Front for the Liberation of Mozambique or Frelimo), Mozambique endured an externally instigated counter-revolution,

412

mediated through the rebel Resistência Nacional Moçambicana (Renamo).[4] By the mid-1980s, Renamo had become deeply embedded in large swathes of the countryside (particularly in the central and northern provinces) and the conflict had reduced much of Mozambique to chaos, greatly weakening the Mozambican state and devastating Mozambican society (Abrahamsson and Nilsson 1995; Minter 1994; Sidaway 1992).

Following more than a decade of conflict, a comprehensive cease-fire agreement, the Acordo Geral de Paz (General Peace Accord or GPA), was finally signed by Joaquim Chissano (the leader of Frelimo and head of state) and Afonso Dhlakama (the Renamo leader) in Rome in October 1992 (Alden and Simpson 1993). As part of the GPA, both sides agreed to trim down and integrate their armed forces into a new national army (Frente para a defesa de Moçambique or FADM) and contest UN-supervised multi-party elections, together with any new parties which would emerge. Nearly 8,000 foreign troops and 1,000 civilian advisers and monitors arrived to police demobilisation and the elections. Frelimo duly won the election in October 1994, with a healthy, but not overwhelming majority (for a good, critical account see Harrison 1996). With a mixture of bribes, enticements and cajolements, running into tens of millions of US dollars, Renamo were converted into a loyal opposition (McCreal 1993).

In the years of negotiation leading up to the signing of the GPA, Frelimo had altered most of its earlier proclaimed strategy. It replaced claims of 'socialist transition' with 'social progress' and its status as 'vanguard of worker-peasants' with a 'party of all the Mozambican people'. Whilst the Frelimo of 1994 contained most of the same core leadership as twenty years before,[5] its declared policies had shifted very substantially. In the window between independence and the spread of the conflict, Frelimo had proclaimed itself in pursuit of transition to socialism. Formally constituting itself as a 'Marxist-Leninist vanguard party' of the 'worker–peasant alliance' in 1977, Frelimo had put in place a policy of collectivisation of agriculture, nationalisation of large enterprises and indicative planning. All land, health and education had been nationalised at independence and private legal and medical services were abolished. Strategic sectors of the economy, such as banks, insurance, main industrial units and large-scale agriculture were also brought under the control of the state. Progressively, other abandoned and seriously mis-managed production units were added, until by 1983 at least three-quarters of all production outside the peasant sector was formally subject to state plans (Egerö 1987). A system of public education, health care and food distribution was set up to provide the core of basic needs development (O'Laughlin 1996). Though characterised by enormous and ultimately fatal contra-dictions (Mackintosh 1986; Wuyts 1989; Mackintosh and Wuyts 1988),[6] the strategy had some profound, if not entirely anticipated, impacts.

These impacts were sharply registered in respect of capital flows through the built environment. All land was nationalised at independence, as was subsequently much of the urban housing stock in 1976, thereby breaking the dynamic speculative market of the pre-independence era (Sidaway and Power 1995). The result was that the state became the only legal landlord in the downtown and suburban areas of Mozambican cities. The vast majority of Mozambicans who lived beyond the *cidades de cimento*[7] continued to inhabit housing that was self-constructed. Frelimo's plans initially included the socialisation of the countryside and the extension of 'urban'-style infra-structure throughout Mozambique (Friedmann 1980). Limited resources, capability and will (as a focus on grand projects came to the fore) and, not least, the impact of the war and economic crisis soon derailed this strategy (Sidaway 1993).

However, the relatively small proportion of the population residing in state-owned property in the cement cities belied the significance of this fraction of the population and urban property to the political economy of Mozambique. In the first place, the appearance of the cement cities, in terms of the quality of maintenance and the density of occupation, was transformed by the nationalisations. Most Portuguese settlers left in the period around independence and were rapidly replaced by indigenous Mozambicans.[8] Since then, administration of the property has been undertaken by the specially created Administração do Parque Imobiliário do Estado (APIE or State Property Agency). Second, a large, new urban constituency for the govern-ment was generated. Amongst those tens of thousands of people who were able to acquire abandoned cement city housing were many of the skilled, professional and party members who were to be the basis of Mozambique's socialist transformation and development. Notwithstanding a rhetorical government commitment to the peasantry throughout the country, these urban dwellers constituted a distinct (though heterogeneous) fraction of the Mozambican population that had directly and materially gained from independence and its aftermath, able to live in the cities for nominal rents.

A small market in private property did remain after the nationalisations, since unrented dwellings remained in private hands (a total of around 5,000). However, this market was highly regulated by the state, which fixed prices to minimise the scope for speculation. Subsequently, this small market was to evolve into a much larger and semi-legal market, as detailed below. And although Frelimo's initial radicalism and the commitment to socialist transformation eased around 1984, Mozambique was not a domain for large-scale speculative capital flows in the realm of land and property. However, from the mid-1980s, Frelimo's revolutionary strategy began at first to unravel and then to implode under the combined pressures of the spreading insurgency of Renamo in the countryside and the broader economic and social crisis and the evident inability of the Eastern bloc to provide hoped-for aid.[9]

It is clear that even before the Fifth Party Congress of Frelimo in 1989, at which the commitment to 'socialist transition' was replaced with the more modest objectives of 'social progress', illegal deals over land and property had become more common. Such deals became much more widespread after the start of the ongoing Economic Rehabilitation Programme (Programa de Rehabilitação Economica or PRE)[10] carried out since 1987 increasingly under the direction of the World Bank and the IMF.[11] In an important critical analysis of the PRE, Hanlon (1991, 1996) argued that the capitalisation and commercialisation pushed by these agencies not only came to have a determining role in Mozambique's political economy (cf. Plank 1993), but also had the effect of considerably increasing corruption. However, in tracing this, it is necessary to specify carefully the *relationships* between the Bretton Woods institutions, the party-state and other agents in moulding the context within which a new strategy of accumulation developed.

New strategies of accumulation?

In 'post-socialist' Mozambique, many state officials have charted new 'entrepreneurial' paths. In addition to commercial farming, small 'service sector' businesses, such as shops, cafés, restaurants and nightclubs, have been amongst the most favoured for those with access to capital. In other cases, though, such 'entrepreneurial channels' include quasi-legal and wholly illegal activities such as payments for import licences, for the award of contracts and, most notably, for the receipt of land and property. From the mid-1980s onwards, the formal allocation procedures for land and especially property have increasingly been subverted by illegal payments (a *de facto* revalorisation). By 1990, a 'parallel' system of property allocation had become the norm with illegal payments between approximately $200 to $2,000 routinely demanded (depending on the size, location and condition of the housing) by those in the state administrative machinery who are able to extract such 'rents'. This is in addition to 'private' subletting arrangements. Against these flows, the legal rents (now being converted into mortgages)[12] pale into relative insignificance. There are also frequent exchanges of land for large sums of hard currency (for evidence from rural areas see Myers 1994; Pitcher 1996; West and Myers 1996). Revalorisations of land and property have been generated by a combination of chronic shortages and the collapse in 'established' values, specifically the authority, patronage and provision given by Frelimo. However, in general terms such developments in Mozambique must also be understood as necessary consequences of the forms of accumulation strategy sponsored by the IMF and World Bank. In the context of public-sector cuts, the collapse in the state- and party-enforced 'socialist' values and the increased presence of foreign commodities and patterns of consumption,[13] corruption becomes an integral element of the shifting relations of political and economic power.

In the context of the crisis that accompanied the war, Mozambique became profoundly dependent on foreign aid. The IMF and World Bank mediate a good deal of this aid (about 60 per cent) – and have come to bear a central role in facilitating the redirection away from state socialism. In addition to the large flows of the emergency and development aid, dozens of foreign non-governmental organisations operate in an increasingly weakened Mozambique, mostly bypassing and subverting an already skewed and dilapidated state apparatus. By the late 1980s, Mozambique was the most aid-dependent state in the world, with over 70 per cent of its domestic product derived from aid. In turn, this generated a pattern of class formation in which powerful groups and individuals within the state apparatus sought to locate themselves as intermediaries, as close to these flows as possible, whereby they might intercept a fraction. As in other documented cases, this pattern of class (re)formation has emerged from the prior stratification of the party-state apparatus, and it is notable that a number of ministers and high state officials are some of the most prominent agents here.[14]

The emergence of a polity involving extractions from aid-led capital flows has parallels with comprador–merchant[15] capitalism. Examples of this – in which prior party-state and enterprise/managerial strata play a key role – can be found in cases as diverse as the former state socialisms of the USSR-Russia (Sautman 1995; Kryshtanovskaya and White 1996), Eastern Europe (Baylis 1994; Smith 1994) and China (Duckett 1996). But neat formulations cloak much of the complexity of the transformations that are occurring, which, as Bayart (1993) has indicated in writing about African states in more general terms, involves the crystallisation of internal social stratification astride the international system. In particular, the question of the fluid relations between the 'domestic' and the 'international' is more complex and tangled than notions of compradorism permit. As Holloway (1994: 36) suggests,

> [a]ll national states manipulate the internal/external distinction as a crucial element of practical politics. All states which have dealings with the IMF, for example, present the result of such dealings as being externally imposed, whereas in reality they are part of the seamless integration of 'national' and global political conflict.

However, it was by working with such inscriptions of inside–outside (and before–after) that Frelimo sought to articulate a very specific vision of post-colonial modernity. In this vision of a new socialist Mozambique, the manifest contradictions of colonial capitalism would be superseded. Instead, and despite significant early achievements in domains such as the establishment of basic national health and educational systems, the result after two decades of independence and conflict has been widespread devastation and the establishment of a new form of aid dependency which has a decidedly neo-colonial flavour. This is particularly registered in the capital city of Maputo and other significant urban centres. Part of this concerns the relative power of trans-national institutions to shape the Mozambican cultural and political economy. There is an intricate *nexus* of close relationships between the relative failure of the state-socialist project (itself a reflection of the war of destabilisation, inherited limits and some fundamental contradictions evident in its operation), negotiations to end the war, the macroeconomic consequences of structural adjustment led by the IMF and World Bank, the creation of receptive conditions for the emergence of private capital flows, and transformations in the social use of land and the built environment.

Phantom states?

Whilst not all of the nexus can be unpacked here, the role of foreign aid and in particular the *oversight* of the IMF and World Bank indicate how the Bretton Woods institutions might now need to be thought of in terms of their operation in the periphery and much of the post-socialist world as species of what Thrift and Leyshon (1994) have referred to as a *'phantom state'*. Thrift and Leyshon's account is one of a number of attempts to re-theorise the quasi-state systems of governance constituted by trans-national financial capital (for example, Corbridge 1994; Popke 1994). Their specific focus is on the 'discursive authority which is based in electronic networks and particular "world cities"' (Thrift and Leyshon 1994: 324). It might seem trite to note that places such as the Mozambican capital city of Maputo are at some stretch from being such a 'world city'. Nor are the impacts of powerful 'external' forces a novel phenomenon in Mozambique. However, what is new and salient is the way in which the experience and scale of aid-dependency and IMF–World Bank powers structure Maputo's political economy.

All of this is, of course, also mediated through and registered clearly in the social and cultural life of downtown Maputo. Therefore by 1990, Mozambique could already be represented – under the title 'Sex, prawns, sun and forex [foreign exchange]' – in a South African magazine as

> gradually dragging itself from the depths of misguided Third World Marxism and economic decay, as President Chissano offers a measure of freedom to private business and goes out of his way to woo foreign investment. South Africans are *welcome* here – especially when their pockets bulge with rands – whether they're businessmen looking for opportunities or holiday-makers looking for some sun, prawns and transplanted Continental culture.
>
> (*City Late* 1990: 9)

Since then, such depictions and images of exotic-sensual-decadent Mozambique have grown into a substantial genre amongst a tourist and business-investment literature. Just as elite Mozambican fractions are locating themselves in favourable positions to mediate and skim off flows of foreign capital, so the urban poor and lumpen elements seek to do the same, by whatever means possible. Such means include prostitution, begging, theft, smuggling, embezzlements, drug-dealing and a host of petty corruptions. All these forms of entrepreneurship and enterprise are now very evident in Mozambican cities. In the 1970s and 1980s, under the authoritarian Marxism-Leninism and party-led discipline of Frelimo, the cities of Mozambique were probably amongst the most secure in the world, at least in terms of crime rates and the threat of criminal violence. Today, however, they are starting to resemble places such as Los Angeles, Johannesburg or Rio de Janeiro in terms of containing areas of marginality, violence, crime and danger.

Conclusion: fantasies of transition

The head of an IMF delegation attended a major round table of Mozambican economists in early 1996. One participant commented: 'He was like a missionary priest lecturing the heathen natives. He never listened to what we said. But speaking slowly and carefully to ensure we understood, he told us: "If you ask the poor, you will see they cannot bear inflation. So to tackle poverty, we must tackle inflation first".

(Hanlon 1996: 110)

'One of the most attractive things about Mozambique to potential investors is that they are getting in on the ground floor,' says World Bank resident representative Roberto Chavez. 'This', he adds, looking out of the window of the bank's newly built Maputo office, 'is virgin territory.'

(*Institutional Investor* 1994: 2)

Fantasies and fixations such as those of IMF delegation chiefs and of publications such as *City Late* and *Institutional Investor* take us back to the fund-manager mentioned at the start of this chapter. In this respect we would argue that today the post-socialist periphery finds itself an object of fantasies of transition as lurid as those projected during an earlier age of fantastic 'visions': that of the Cold War. Recall, for example, the arbitrary scriptings by which a revolution in Central America could become a threat to America (O'Tuathail 1986) – a power that in an earlier moment of 'containing communism' (in effect, defeating revolutionary insurgency in the periphery) declared that places had to be destroyed in order to be saved. In the peculiar context of the Portuguese Empire, an elaborate colonial vision of an *anti-communist*, harmonious, multiracial and Catholic realm was set out. Therefore, a Portuguese official at the height of the Cold War could explain how '[t]he spiritual power that irradiates the Portuguese universal history through the Lusitanian community represents a competitive current of major importance in the adhesion and cooperation that must exist between the people of the Free World' (Comprido 1956: 268, our translation). Furthermore, he declared how Portugal's 'Eurafrican' dimension not only gave the West access to all kinds of strategic minerals and scope for defence in depth should Europe, in a worst case scenario, be occupied by the Soviets, 'but the possibility to mobilize around 1.5 million blacks in our African territories, who once organised by white officials and commanders constitute a human potential of high value for military ends' (Comprido 1956: 241–2, our translation). Though fortified by the Soviet 'reference point', Frelimo's radicalism, anti-imperialism and voluntarism arose from a contestation of such racist visions.

Amidst the ruins of that radicalism, today's fantasies, however, are perhaps as insidious as those of the colonial and Cold War strategists. This is particularly the case when, as in Mozambique, they sanction the elimination of public health and social provision, against backdrops of conspicuous consumption and corruption, all conducted in the names of 'adjustment', 'stabilisation', 'transition' and the 'market' (Hanlon 1991, 1996). Whilst the legacy of Frelimo's state socialism remains inscribed on the lands and subjects of the post-socialist Mozambique, it is now increasingly overlaid by the kinds of emergent ersatz-capitalist social forms whose parameters have been sketched here. Representations of Mozambique accordingly shift from that of Marxist or revolutionary realm to land of investment opportunity and touristic potential.

This chapter has pointed to the presence of phantom states – both in an abstract spectral-ideological sense (how the IMF and World Bank cast shadows over the entire Mozambican polity) and in more direct material senses as agents of consumption and governance. The 'primitive magic' that found expression in Stalinism finds new forms and adherents amongst the cults of economists and directors at the IMF and World Bank. Indeed, for George and Sabelli (1994: 3–6), the World Bank is

> very much like the Church, in fact the medieval Church. It has a doctrine, a rigidly structured hierarchy preaching and imposing this doctrine and a quasi-religious mode of self justification. Or, to borrow from a wholly different tradition, the Bank is reminiscent of a centralised political party characterised by opacity, authoritarianism and successive party lines. Could the World Bank be the last of the Leninists?

In this context, however, it should be noted that the institutional 'wizards' and 'priests' of the IMF and World Bank are heterogeneous and often in some tension with each other and with other authorities.[16] For example, in

Mozambique, 1996 saw heightened discord between the IMF and the wider donor community. A recent account notes how

> [i]n October last year [1995], the Fund refused to allow the Mozambican government to increase the minimum wage by as much as it wanted, and the Fund let it be known that it was considering pulling out of its existing arrangement with the country, because its monetary targets weren't being met quickly enough. The Maputo representatives of the main donors – with the sole exception of the UK – were so concerned at this that they broke with normal diplomatic protocol and sent an urgent letter directly to IMF headquarters in Washington. . . . The letter, which was leaked to the press, revealed a serious rift, and it was even rumoured that the World Bank's man in Maputo had a hand in drafting it.
>
> (BBC 1996: 8)

Therefore, in further developing the notion of the IMF and World Bank as modes of phantom authority, any notion that their power is unified in any *simple* sense will need to be dispensed with. Furthermore, as violent crime spirals in Maputo and in other Mozambican cities, whilst the country becomes the major conduit of narcotics, stolen cars and weapons for South Africa and beyond, we might also note Derrida's (1994: 83) heading of other phantoms that haunt the New World Order:

> How can one ignore the growing and undelimitable, that is, worldwide power of those super-efficient and properly capitalist *phantom-States* that are the mafia and drug cartels on every continent, including in the former so-called socialist States of Eastern Europe? These phantom-States have infiltrated and banalized themselves everywhere, to the point that they can no longer be strictly identified. Nor even sometimes clearly dissociated from the processes of democratization. . . . All these infiltrations are going through a 'critical' phase, as one says, which is no doubt what allows us to talk about them or to begin their analysis. These phantom-States invade not only the socioeconomic fabric, the general circulation of capital, but also statist or inter-statist institutions [emphasis added].

More specifically, Mozambican society now registers the kinds of flow that Ellis (1996: 196) describes in the following terms:

> In much of tropical Africa, the decline of state power in recent years has resulted in the emergence of networks of long-distance trade in high-value commodities including gems, weapons and drugs which

are both sources of wealth and vectors of political-military conflict. . . . This is an important development in which anti-corruption or good governance campaigns are of little relevance, since either these trades are outside the control of collapsed states, or the political powers which are emerging depend on them for their own finances.

Therefore, in much of Africa today, liberalism, democracy and good governance assume a mostly rhetorical or (new) instrumental status sometimes (as in Mozambique) replacing the old Soviet 'reference point' (Young 1995).

In the context of other accounts stressing the diversity and path dependence of post-socialist transformations (Stark 1992; see also Grabher and Stark in this volume; Smith and Swain in this volume), we must be wary of making generalisations or abstracting 'simple' histories from Mozambique. The Mozambican case has been presented here not in a representative sense, but as portent. Frelimo is now just a shadow of its former self. But, if a more general statement is allowed on the post-socialist periphery, we might conclude that after so much was gained and lost, after so much *violence*, so many words and so many attempted transformations that claimed to supersede it, capitalism has not reached its limits. This is ever more clearly the case, even as capitalist modernity overflows and thereby provincialises its historical core: the geopolitical domain known as *the West*.

Acknowledgements

We would like to thank John Pickles, Adrian Smith and Alison Stenning for comments on an earlier draft. We wish to acknowledge and thank the ESRC for financial support.

Notes

1 A moment's reflection will throw up a number of objections to the idea that the frontiers of contemporary capital are identifiable only with particular 'peripheral' places. The combined and uneven nature of capitalism means that the frontier of commodification can be as much in the deepening of capitalist social relations in its cores (Habermas (1987) speaks of the 'colonization of the lifeworld' by capitalist social relations) as the physical extension of them into peripheries. For example, in a recent report the World Bank (1991: vii) explains 'how good loans can be made to individuals and firms at the "frontier". This frontier is not geographic, but market based.' Devalorisations, de-industrialisations and redevelopments also confirm that this is the case.

We should also add that the association of sex and violence is nothing new for the periphery. Colonialism was in essence about military domination and as Said (1978, 1993) and Fanon (1968) have argued, it required the possession of land, minds and bodies.

421

2 The Chinese case is instructive, given first the historic symbolism of an autonomous (i.e. not Soviet-imposed or backed) revolution in the world's most populous state in 1949, second the subsequent revolutionary extremism and propaganda efforts of the Maoist regime, and third the enormous shift to marketisation and capitalist social relations. On the despotic nature of Maoism, see Chang (1992) and Becker (1996) on its devastating human consequences. On the scale of the shift towards capitalism, see Smith (1993), and on its mediation through the built environment see Chiu (1995), Gar-on Yeh and Wu (1996), Olds (1995), Wu (1995) and Zhou and Logan (1996). One of the best of a number of accounts of the role of *huaqiao* (overseas Chinese) in this process is Ong (1996). Two incisive journalistic accounts are Sudjic (1995) and Sayle (1992). On theorising corruption in China, see the notes by White (1996).

3 Whilst a key element of what state socialism signified (non-market allocation) is now lost, other elements (the political hegemony of the party and 'national' economic development) remain. Therefore, a study of the *contingency* of the rhetoric of socialism becomes more possible.

4 Renamo was set up by the Rhodesian security forces in 1976–7 and, after Zimbabwean independence in 1980, continued to be supplied and trained by the South African military and backed by a shadowy network which included right-wing American financiers, Portuguese former settlers and Saudi royals (see Minter 1994; Vines 1991).

5 With the notable exception of Samora Machel, the president of Mozambique from independence until his death in a suspicious air-crash over Transvaal in 1986. Machel was succeeded by Joaquim Chissano, who remains president.

6 In addition to specifying the enduring legacy of the bureaucratic and authoritarian Portuguese colonial state, these authors indicate how the contradictions in the state-led accumulation strategy in Mozambique parallel those in other peripheral socialisms. For good accounts of some others, see Ryan (1995) on Nicaragua. Post and Wright (1989) provide a good general model, building upon the pioneering work of Kornai (1980) and Kalecki (1969). On contrasts between the contradictions of state socialist strategies in Eastern Europe and the periphery, see Littlejohn's (1988) suggestive paper. For a thought-provoking consideration of the anthropological accounts of the Mozambican war, see Dinerman (1994).

7 In Mozambican cities a broad physical distinction is customarily made between the central *cidade de cimento* (cement city) and the surrounding *cidade de caniço* (cane city). The basis of the distinction is that all the buildings in the former area are of cement and brick, whilst the vast majority in the latter areas are of cane or other material (such as corrugated iron) that is relatively cheap and portable. In turn, this relates to the distinction between a 'formal' property sector of private or state-owned housing and an informal sector of largely unregistered private housing. After independence a number of houses in *caniço* areas were nationalised. However the (nominal) rents for these were rarely, if ever, collected and in 1990 it was announced that they would be 'denationalised'. In 1985, it was estimated that only about 8 per cent of the Mozambican population lived in permanent houses built of brick or cement, and although 38 per cent of housing in defined urban areas was cement or brick, only 4 per cent of housing in rural

areas fell into this category (Forjaz 1985). We should also note that in tandem with the *cimento/caniço* divide, colonial discourse had established a *cidade/campo* (city/countryside) distinction, which Frelimo sought to transcend.

8 Following the collapse of colonial authority, the initial process of reallocation was chaotic. However, Frelimo rapidly sought to assert authority over housing reallocations (and related transformations), through establishing local *grupos dinamizadores* (dynamising groups) in Mozambican cities and towns.

9 Back in 1981, Mozambique's bid to join the CMEA (Council for Mutual Economic Assistance) was turned down by the USSR, who made it clear that they were unable to subsidise Mozambique in the way they assisted Vietnam and Cuba. This rebuff lured Frelimo into looking to the West.

10 The programme was subsequently renamed the PRES (Programa de Rehabilitação Economica e Social) in belated acknowledgement of a 'social' component. Ironically, the impact of the PRES has been more negative than its predecessor in terms of increased social inequality (Hanlon 1996).

11 Though it was greatly accelerated by the PRE, the move away from the Marxist-Leninist strategy began some time earlier, following Frelimo's Fourth Party Congress in 1984. At first it was most evident within macroeconomic pricing strategy and agricultural development policies. Detailed consideration of these are beyond the remit of this chapter (see Myers 1992; World Bank 1992). However, it is worth noting that divestiture of state-held land in rural areas began in 1985, and has been accelerated since a divestiture law in 1989 (Decree 21/89). This provided a legal framework for a shift from direct state exploitation of land to its use by mixed, private, or family enterprises/farms. Although the ownership *remains* with the state, the right to use the land and fixed investment upon it is transferred and payment received for it. The details of the procedure are described in USAID/Mozambique (1989) which also indicates the foreign (in particular USAID, World Bank and IMF) role in generating the policy shift (cf. Hanlon 1991; Myers 1992).

12 By July 1996, some 41 per cent of APIE tenants (23,096) had submitted applications to buy (Summary of World Broadcasts 1996).

13 The temporary presence of nearly 10,000 well-paid UN staff and soldiers in Mozambique during the months prior to the elections added significantly to this already well-developed phenomenon.

14 It has now become clearer that even during the phase of greatest revolutionary rhetoric and transformation in Mozambique (i.e. the late 1970s) a number of senior Frelimo officials did retain and develop commercial interests (Mackintosh 1986). Although the reforms of the 1980s greatly broadened and deepened the scope for these to develop, it is important to register that they did not emerge from a vacuum.

15 The concept of *comprador* (buyer) capitalism has been formulated in Latin American theorisations of dependency. It refers to a peripheral social formation where the local (usually petit) bourgeoisie constitutes an intermediary 'broker' or type of merchant class.

16 In addition to many critical reviews of the impacts of IMF and World Bank programmes (e.g. Leftwich 1994; Leys 1994; Ould-Mey 1994; Riddell 1992), there are a number of good accounts of such *tensions* at a global level (e.g. Mosley, Harrigan and Toye 1991; George and Sabelli 1994), including a recent

study of competition between Japan and the US as mediated through the World Bank (Wade 1996).

Bibliography

Abrahamsson, H. and Nilsson, A. (1995) *Mozambique: the Troubled Transition. From Socialist Construction to Free Market Capitalism*, London: Zed.

Alden, C. and Simpson, M. (1993) 'Mozambique, a delicate peace', *The Journal of Modern African Studies* 31, 1: 109–30.

Ali, T. (1984) 'Preface', in T. Ali (ed.) *The Stalinist Legacy: Its Impact on Twentieth Century World Politics*, Harmondsworth: Penguin.

Bayart, J. F. (1993) *The State in Africa: the Politics of the Belly*, London: Longman.

Baylis, T. A. (1994) 'Plus ça change? Transformation and continuity among East European elites', *Communist and Post-Communist Studies* 27, 3: 315–28.

BBC (1996) Transcript of File on Four, 22 October 1996.

Becker, J. (1996) *Hungry Ghosts: China's Secret Famine*, London: John Murray.

Chang, J. (1992) *Wild Swans. Three Daughters of China*, London: Harper Collins.

Chiu, R. L. (1995) 'Commodification in Guangzhou's housing system', *Third World Planning Review* 17, 3: 295–312.

City Late (1990) 'Sex, prawns and forex', May 1990: 9.

Comprido, J. B. (1956) 'Importância geopolítica de Portugal para a estratégia do Mundo Livre', *Anais do Club Militar Naval* 86, 7–9: 223–68.

Corbridge, S. (1994) 'Maximizing entropy? New geopolitical orders and the internationalization of business', in G. Demko and W. Wood (eds) *New World Order: Geopolitical Perspectives*, Boulder, CO: Westview.

Derrida, J. (1994) *Spectres of Marx: the State of the Debt, the Work of Mourning, and the New International*, London and New York: Routledge.

Deutscher, I. (1984) 'Marxism and primitive magic', in T. Ali (ed.) *The Stalinist Legacy: Its Impact on Twentieth Century Politics*, Harmondsworth: Penguin.

Dinerman, A. (1994) 'In search of Mozambique: the imaginings of Christian Geffray in *La Cause d'armes au Mozambique. Antropologie d'une guerre civile*', *Journal of Southern African Studies* 20, 4: 569–86.

Duckett, J. (1996) 'The emergence of the entrepreneurial state in contemporary China', *The Pacific Review* 9, 2: 180–98.

Egerö, B. (1987) *Mozambique: a Dream Undone. The Political Economy of Democracy 1975–1984*, Uppsala: Nordiska Afrikainstitutet.

Ellis, S. (1996) 'Africa and international corruption: the strange case of South Africa and the Seychelles', *African Affairs* 95: 165–96.

Fanon, F. (1968) *Black Skin, White Masks*, London: MacGibbon and Kee.

Forjaz, J. (1985) 'Housing and planning issues in independent Mozambique', unpublished paper.

Friedmann, J. (1980) 'The territorial approach to rural development in the People's Republic of Mozambique: six discussion papers', *International Journal of Urban and Regional Development* 4: 97–115.

Gar-on Yeh, A. and Wu, F. (1996) 'The new land development process and urban development in Chinese cities', *International Journal of Urban and Regional Research* 20, 2: 330–53.

George, S. and Sabelli, F. (1994) *Faith and Credit: the World Bank's Secular Empire*, Harmondsworth: Penguin Books.

Habermas, J. (1987) *Theory of Communicative Action, vol. 2*, Boston: Beacon Press.

Halliday, F. (1989) *Cold War, Third World*, London: Hutchinson.

Hanlon, J. (1991) *Mozambique: Who Calls the Shots?*, London: James Currey.

Hanlon, J. (1996) *Peace without Profit: How the IMF Blocks Rebuilding in Mozambique*, Oxford and Portsmouth, NH: James Currey and Heinemann.

Harrison, G. (1996) 'Democracy in Mozambique: the significance of multi-party elections', *Review of African Political Economy* 67: 19–35.

Hettne, B. (1994) 'The political economy of post-communist development', *The European Journal of Development Research* 6, 1: 39–60.

Hinde, S. (1996) 'Schroders joins trek to adventure', *The Sunday Express*, 9 June.

Holloway, J. (1994) 'Global capital and the national state', *Capital and Class* 52: 23–49.

Institutional Investor (1994) 'Special edition supplement: Mozambique', June: 15–27.

Kalecki, M. (1969) *Introduction to the Theory of Growth in a Socialist Economy*, Oxford: Blackwell.

Kornai, J. (1980) *The Economics of Shortage*, Amsterdam: Elsevier.

Kryshtanovskaya, O. and White, S. (1996) 'From Soviet *nomenklatura* to Russian elite', *Europe–Asia Studies* 48, 5: 711–33.

Laïdi, Z. (1988) 'Introduction. What use is the Soviet Union?', in Z. Laïdi (ed.) *The Third World and the Soviet Union*, London: Zed.

Leftwich, A. (1994) 'Governance, democracy and development in the Third World', *Third World Quarterly* 14: 603–24.

Leys, C. (1994) 'Confronting the African tragedy', *New Left Review* 209: 33–47.

Littlejohn, G. (1988) 'Central planning and market relations in socialist societies', *The Journal of Development Studies* 24, 4: 75–101.

McCreal, C. (1993) 'Renamo puts £66 million price on keeping peace', *The Guardian*, 10 June.

Mackintosh, M. (1986) 'Economic policy context and adjustment options in Mozambique', *Development and Change* 17: 557–81.

Mackintosh, M. and Wuyts, M. (1988) 'Accumulation, social services and socialist transition in the Third World: reflections on decentralised planning based on the Mozambican experience', *Journal of Development Studies* 24, 4: 136–79.

Mathey, K. (ed.) (1990) *Housing Policy in the Socialist Third World*, London: Mansell.

Minter, W. (1994) *Apartheid's Contras: an Inquiry into the Roots of War in Angola and Mozambique*, London, Johannesburg and New Jersey: Witwatersrand University Press and Zed Books.

Mosley, P., Harrigan, J. and Toye, J. (1991) *Aid and Power. The World Bank and Policy Based Lending* (2 volumes), London: Routledge.

Myers, G. (1992) 'Agricultura e desintervençao amento das empresas agrarias estatais em Moçambique', *Extra* 9: 9–16.

Myers, G. (1994) 'Competitive rights, competitive claims: land access in post-war Mozambique', *Journal of Southern African Studies* 20, 4: 603–32.

O'Laughlin, B. (1996) 'From basic needs to safety nets: the rise and fall of urban food-rationing in Mozambique', *The European Journal of Development Research* 8, 1: 200–23

Olds, K. (1995) 'Globalization and the production of new urban spaces: Pacific Rim megaprojects in the late twentieth century', *Environment and Planning* A 27: 1713–43.

Ong, A. (1996) 'Anthropology, China and modernities. The geopolitics of cultural knowledge', in H. Moore (ed.) *The Future of Anthropological Knowledge*, London: Routledge.

O'Tuathail, G. (1986) 'The language and nature of the "new" geopolitics: the case of US–El Salvador relations', *Political Geography Quarterly* 5: 73–85.

Ould-Mey, M. (1994) 'Global adjustment: implications for peripheral states', *Third World Quarterly* 15: 319–36.

Pelissier, R. (1997) Review of Isabelle Verdier (ed.), *Mozambique: 100 Men in Power, Journal of Southern African Studies* 23, 1: 154–5.

Pitcher, M. A. (1996) 'Recreating colonialism or reconstructing the state? Privatisation and politics in Mozambique', *Journal of Southern African Studies* 22, 1: 49–74.

Plank, D. N. (1993) 'Aid, debt and the end of sovereignty: Mozambique and its donors', *Journal of Modern African Studies* 31, 3: 407–30.

Popke, E. J. (1994) 'Recasting geopolitics: the discursive scripting of the International Monetary Fund', *Political Geography* 13: 255–69.

Post, K and Wright, P. (1989) *Socialism and Underdevelopment*, London: Routledge.

Riddell, J. B. (1992) 'Things fall apart again: structural adjustment programmes in sub-Saharan Africa', *The Journal of Modern African Studies* 30, 1: 53–68.

Ryan, R. (1995) *The Fall and Rise of the Market in Sandinista Nicaragua*, Montreal and Kingston: McGill-Queens University Press.

Said, E. (1978) *Orientalism*, Harmondsworth: Penguin.

Said, E. (1993) *Culture and Imperialism*, London: Vintage.

Sautman, B. (1995) 'The devil to pay: the 1989 debate and the intellectual origins of Yeltsin's "soft authoritarianism"', *Communist and Post-Communist Studies* 28, 1: 131–51.

Sayle, M. (1992) 'A tale of two cities', *The Independent Magazine*, 28 November, Issue 221: 34–41.

Sidaway, J. D. (1992) 'Mozambique: destabilization, state, society and space', *Political Geography* 11, 3: 239–58.

Sidaway, J. (1993) 'Urban and regional planning in post-independence Mozambique', *International Journal of Urban and Regional Research* 17, 2: 241–59.

Sidaway, J. and Power, M. (1995) 'Sociospatial transformations in the "post-socialist" periphery: the case of Maputo, Mozambique', *Environment and Planning A* 27: 1463–91.

Sidaway, J. and Simon, D. (1990) 'Spatial policies and uneven development in the "Marxist-Leninist" states of the Third World', in D. Simon (ed.) *Third World Regional Development: a Reappraisal*, London: Paul Chapman.

Simon, D. (1995) 'The demise of "socialist" state forms in Africa: an overview', *Journal of International Development* 7, 5: 707–39.

Smith, A. (1994) 'Uneven development and armament industry restructuring in Slovakia', *Transactions of the Institute of British Geographers* 19: 404–24.

Smith, R. (1993) 'The Chinese road to capitalism', *New Left Review* 199: 55–99.

Stark, D. (1992) 'Path dependence and privatization strategies in East Central Europe', *East European Politics and Societies* 6, 1: 17–54.

Sudjic, D. (1995) 'Megalopolis now', *The Guardian Weekend*, 24 June: 26–35.

Summary of World Broadcasts (1996) 'Paper reports earnings from sale of houses nationalized in 1975', SWB ACW/044 WA/5, 16 July: 18.

Thrift, N. and Leyshon, A. (1994) 'A phantom state? The de-traditionalisation of money: the international financial system and international financial centres', *Political Geography* 13: 299–327.

USAID/Mozambique (1989) 'Options for state farm divestiture and the creation of secure tenure', unpublished report, Maputo: US Agency for International Development.

Vines, A. (1991) *Renamo: Terrorism in Mozambique*, London: James Currey.

Wade, R. (1996) 'Japan, the World Bank, and the art of paradigm maintenance: the East Asian miracle in political perspective', *New Left Review* 217: 3–36.

West, H. G. and Myers, G. W. (1996) 'A piece of land in a land of peace. State farm divesture in Mozambique', *Journal of Modern African Studies* 34, 1: 22–57.

White, G. (1996) 'Corruption and market reform in China', *IDS Bulletin* 27, 2: 40–7.

World Bank (1991) *Finance at the Frontier: Debt Capacity and the Role of Credit in the Private Economy*, Washington, DC: World Bank.

World Bank (1992) *Mozambique, Rural Rehabilitation Project*, Maputo: World Bank Resident Mission.

World Bank (1996a) *Global Economic Prospects and the Developing Countries*, Washington, DC: World Bank.

World Bank (1996b) *World Bank Annual Report*, Washington, DC: World Bank.

Wu, F. (1995) 'Urban processes in the face of China's transition to a socialist market economy', *Environment and Planning C: Government and Policy* 13: 159–77.

Wuyts, M. (1989) *Money and Planning in Socialist Transition*, Aldershot: Gower.

Young, T. (1995) 'A project to be realised: global liberalism and contemporary Africa', *Millennium: Journal of International Relations* 24, 3: 527–46.

Zhou, M. and Logan, J. R. (1996) 'Market transition and the commodification of housing in urban China', *International Journal of Urban and Regional Research* 20, 3: 400–21.

19

ECONOMIC TRANSFORMATION IN CHINA

Property regimes and social relations

Alan Smart

Introduction

The collapse of the Soviet bloc nations was taken by many as vindication of capitalism's success, proof of socialism's inevitable failure, and evidence for the desirability of a transition to capitalist institutions and practices as quickly as possible. For this view, China poses a problem, since the fastest growing economy in the 1990s is located in a Marxist fossil. This awkward situation can be explained away, though, by appealing to China's reforms, and the vitality of markets and capitalist relations of production within the shell of the planned economy. Thus, it is still capitalism that accounts for dynamism and efficiency.

There are some problems with this analysis, however, and in order to understand better the transitions to post-socialism we should take stock of the nature of China's successes since 1978. Nolan (1994: 12–13) notes that 'China's post-Mao economic policies and performance are a puzzle: why did it perform so well, despite the apparent deficiencies both with economic institutions and policies?' Furthermore, in 'almost all key aspects of institutional arrangements and policy China's post-reform economy in the 1980s appears as the kind of interventionist halfway house that most economists would predict would perform very badly' (Nolan 1994: 10). China has contravened most of the orthodox transformation policies proposed for Central and Eastern Europe (CEE) by the World Bank and the IMF (Angresano 1994). Yet, as well as significantly outperforming the post-socialist economies of Central and Eastern Europe (CEE),[1] China's real GDP growth since 1978 has been amongst the highest ever recorded, comparable to Japan from 1960 to 1974, and 'somewhat superior to that of South Korea from 1965–1978' (Yusuf 1994: 74). Like the examples of Japan and South Korea, much of the growth in China has been based on manufacture exports to developed economies, a feat that the CEE economies have yet to accomplish (Perkins 1994: 32).

Like other Pacific Asia 'miracle' economies, the experience of China reinforces the lesson that success at the capitalist game need not retrace the steps of western capitalism.[2] In this chapter, I examine the impact of regimes of property rights in China's transformation, and explore why the ambiguous, uncertain and politically dependent character of property rights in China has not crippled its economic expansion, but may have played an important part in enabling it.

Walder (1995a) has argued that it has not been the increase of marketisation or privatisation but the clarification of property rights that made possible the dynamism of township and village collective industries (the fastest growing sector since 1978). He rejects the tendency to emphasise the 'hybrid' nature of township and village enterprises (TVEs), and asserts that the key to their performance has been that 'governments at the lower levels are able to exercise more effective control over their assets than are governments at higher levels' (Walder 1995a: 270). While this may be valid in many cases, it is too sweeping to account, however, for the remarkable local diversity that is found in emerging property regimes. The claim that TVEs are owned and controlled by local governments, and that the clarification of local government property rights explains China's successes, cannot be generalised to all the dynamic sectors of China. Instead, there is a great diversity of organisational forms, many of which are characterised by a high level of uncertainty of property rights. Patterns of development under the Chinese variety of socialism have left different regions with widely varying sets of economic, spatial and social resources, and taking advantage of these can encourage distinct organisational forms and practices.

Transition, social economics and regimes of property

Transformation policies proposed for CEE were based on the assumption that 'the reduction of state influence over the economy would automatically and quickly stimulate private sector recovery' as long as an appropriate 'regulatory and incentive framework' to support a market economy was put in place (Angresano 1994: 82). The need for 'big bangs' and 'shock therapy' is justified by fears of inefficiencies resulting from undesirable responses to market signals in the absence of private property rights. Prybyla claimed that:

> The sad chronicle of China's post-Mao attempt to introduce a modern economic system contains a useful lesson which others, notably the East Europeans are taking to heart. The lesson is that to address the economic problem in a modern way in the context of a low calibre, inefficient, slothful, wasteful, cronified socialist system,

one must go all the way to the market system, do it quickly, and not stop anywhere on the way.

(quoted in Nolan 1994: 5–6)

The argument is that political distortions create market inefficiencies, and the most radical reforms are those that are most likely to succeed.

In China, the many areas of incompleteness in the reforms are seen as obstacles which distort efficient economic behaviour, leading to rent-seeking, corruption, and the preservation of 'soft-budget constraints' for state-sector enterprises. All of these impede the generation of more efficient production and distribution systems (Saich 1992: 31). The successes are ascribed to the market-based reforms, and problems are attributed to their incompleteness.

Outcomes do not yet demonstrate the superiority of transitions involving rapid adoption of western capitalist institutions. If anything, so far, China's approach seems to be yielding a more competitive economy. Published comparisons between China and Eastern Europe carry echoes of the criticisms of neo-liberal faith in market governance based on interpretations of the economic miracles of the Asian Newly Industrialising Countries (NICs). For example, Amsden (1990) and others emphasise the role played by 'the developmental state', and argue that late developing economies can succeed only with effective state intervention, whereas the introduction of free-market reforms on their own will lead only to chaos. It is clear, though, that state intervention can also fail (Evans 1996) and it is important to move beyond dualistic claims over the effectiveness of state intervention versus market forms of governance (see Smith and Swain in this volume). Consequently, some of the answers to the important but contentious questions concerning the sources of economic success can be found in recent work in the interdisciplinary field of social economics. This approach sees economies as inherently socially and culturally embedded, and argues that free markets cannot by themselves generate economically efficient behaviour, especially because of the problem of trust and the accompanying transaction costs (Granovetter and Swedberg 1992).

Issues regarding the embeddedness of economic activity have, however, been ignored in neo-classical analyses which have been based on a range of unrealistic assumptions, such as perfect information (Stiglitz 1993) and cost-free transactions. Such assumptions are justified as necessary in order to model the fundamental processes, thereby separating the most important factors from those less central. Neo-classical economics also depends on 'a denial of the need for built-in', or endogenous, enforcement (Bowles and Gintis 1993: 85). Market exchanges are assumed to be enforceable without cost to the participants. This enforcement is treated as exogenous (a variable that arises from outside the model and is thus taken as given) in the argument, as one of the few legitimate functions of the state. Adam Smith, along with Karl Marx, realised that the world was rather more complicated than this, and that real economies included 'issues of opportunism, strategic action, changes in tastes, norms, and sentiments, collusion among agents, and reciprocity and altruism as well' (Bowles and Gintis 1993: 83). It is extremely difficult and costly to police to enforce the actual processes of market exchange.

There is an irony at the centre of the neo-classical model which is that it assumes 'a world where a handshake is a handshake, in which individuals *irrationally* did not take advantage of each other' (Stiglitz 1993: 110). People characteristically *under-invest* in opportunism: employees work hard when no one will notice; business associates stick to their deals even when they might gain by breaking their word, and so on. In general, 'the enforcement costs of a society without trust would be monumental' (Bowles and Gintis 1993: 95). A world populated by amoral economic agents would not be a more efficient one because inefficiently high levels of resources would need to be devoted to monitoring agreements, and desirable exchanges would not be undertaken. The 'under-socialised' firm, in which everyone is 'out for number one', is likely to be as ineffective in the real world as is the 'over-socialised' firm where obligations to kin and other personal responsibilities outweigh competitive demands (Granovetter 1990). The result is that, where enforcement is endogenous,

> the value of the exchange depends on the commitments of the parties to the exchange. Because the exchange is durable and personal, the exchanging parties have an interest in shaping the structure of the transaction to mold the personalities, objectives, and other characteristics of the other parties to the exchange.
>
> (Bowles and Gintis 1993: 89)

Similarly, Block asserts that the theories of both market self-regulation and planning assume that efficient outcomes can be produced by the operation of only one of the three levels of market, government and social regulation, while in practice 'the efficiency of a particular economy will depend on how the three fit together' (Block 1990: 42). Social economists argue that the reliance of real-world economic agents on shared cultural understandings and ideas about propriety and justice, and the way in which such ideas can reduce certain economic costs, means that giving free rein to markets may not necessarily increase efficiency.

It would seem that purism is a dangerous indulgence in the world of real economies. Recent developments in evolutionary economics provide support for the idea that there may not be a single optimum solution for economic efficiency, and that instead there may be multiple stable equilibria (see Grabher and Stark in this volume). Furthermore, path-dependent development (Stark 1994) means that you might not easily be able to get to another solution from the place you are already at. Which organisational and institutional compromises are feasible and effective in a particular context may depend on social and cultural resources and expectations. Economic 'big bangs' may be like designing efficient technical solutions which cause chaos when implemented. The real challenge is not to design something that is technically efficient, but is the task of getting to that solution without killing the patient in the process (Stark 1994).

Ironically, 'big bang' economic transformations assume the kind of top-down planning and control in implementation that economists have criticised as doomed to failure in running centrally planned economies (Stark 1994: 64). The *ad hoc* chaotic process of economic reform in China, where some localities move far in advance of what the central authorities have expected or even what they might be prepared to accept (Young 1994; Smart 1995), actually fits better with market ideology. The difference from 'big bang' scenarios is that there is competition between different localities experimenting with the institutional foundations and property rights, rather than the adoption of national rules and centralised regulatory authorities. This type of radical decentralisation, implemented in a context where hard-budget constraints apply to localities, seems to set up a kind of natural laboratory for developing solutions to the economic challenges, building on the local resources available.

What is efficient will depend partially on opportunities provided by social and economic resources within historically specific local regimes,[3] and there is no single approach that can be applied in all contexts. Valuable as such insights are, however, it seems desirable to go further by identifying common conditions of viability in these varying patterns.

In post-socialist transition, the potential costs of negotiating and enforcing agreement may be magnified since routines and expectations are disrupted, and multiple legitimating principles are available to agents (Stark 1996). The constraints under which successful capitalist investment takes place in China are similar but even more fundamentally uncertain, since the legal and administrative framework is less developed or established against such practices. The concrete relations of production and distribution for capitalist enterprises in China are, in a sense, between state and market: agents can depend on neither to provide a reliable and supportive framework within which to get on with business. This does not mean that they are separate from or isolated from state and market, since both institutions are pervasive. But the construction of boundaries between state and market, and the subordination of activities to predominantly one or the other institution, are much less routinised and predictable than in the developed capitalist countries. Instead, both state and market institutions are selectively utilised, and this is usually accomplished through personalised claims. Operating underneath the principles of market and planned economy are the practices of *guanxi* (connections or relationships). Personal relationships and obligations are pervasive in market activities in contemporary China, but they were also important during the hardline days of the Cultural Revolution (Yan 1996; Kipnis 1997), and in both cases they have served to help deal with conflicts and contradictions between rules and practical requirements. The outcome of intersections between personal relations and the theoretically impersonal institutions of market and state can also vary dramatically, from unproductive rent-seeking, on the one hand, to collusive manoeuvring to avoid the 'structural impediments' imposed by the central state. Personal relationships more generally can serve as a way to get things done when the rules seem to be acting as impediments to practical require-ments: such processes can be seen in all societies although they may be of greater significance in contexts of transformation.

From a neo-classical point of view, these processes are directly un-productive and should be minimised through more complete specification of property rights and reduction of state intervention. But are such activities always unproductive in periods of transition, or can they be seen as ways in which the absent or underdeveloped state and societal infrastructure for capitalist practices are substituted for in local contexts? More generally, Aglietta (1979) has reminded us that concrete economies, or 'regimes of accumulation', are always dependent on the entanglement of power with markets. Cross-national variation results from this entanglement and whether the resultant mode of social regulation is capable of producing sustained growth. In examining transitional societies and economies, we must pay attention to the conditions of existence which underpin accumulation even as they are being transformed.

Stark (1996: 1016) has suggested that the transformation of enterprise relations in Hungary can be seen as involving what he calls 'recombinant property', with the three features of 'blurring of public and private, blurring of enterprise boundaries, and blurring the boundedness of legitimating principles'. He suggests that the blurring of enterprise boundaries might be a 'viable strategy to promote organizational flexibility' and that ambiguity of property claims might provide flexible forms for adaptation to the market (Stark 1996: 1021). Property rights are not clear in this case, or in China, and many practices would seem unproductive, if not corrupt, from the perspective of the developed capitalist economies. Yet, in both cases, ambiguous boundaries and property rights facilitate tactical alliances that cope with a myriad of obstacles facing the conduct of business in transitional contexts.

Neither can the mode of regulation that structures and supports a regime of accumulation be limited to national-level state policies, but must incorporate other regulatory sites including modes of enterprise regulation (Smith 1995: 761; see also Smith and Swain in this volume). In periods of rapid and largely chaotic transition, networks of individuals and organisa-tions may be able to support certain forms of accumulation more effectively than can central state institutions. Circumstances like those that Stark calls recombinant property, and I have called local capitalisms (Smart 1995), may reflect a shift downwards of the sites of effective regulation.

The path-dependent nature of development means that transformation must always build on what already exists, regardless of how much is rejected. For example, Smith (1995: 763) has described the situation in socialist Slovakia as involving an 'extensive regime of accumulation' based on forced industrialisation. This regime produced convergence towards less uneven development, but the resultant industrial structure had differential capacities to cope with industrial restructuring. The outcome has been fragmentation and strongly divergent pathways which are the product of the intersection of past regional economic structures and 'the way in which these structures interrelate with the particular restructuring strategies pursued by firms' (Smith 1996: 153). Changing conditions can therefore turn advantages into obstacles, or convert backwardness into opportunity (Grabher and Stark in this volume).

When we recognise that the mode(s) of regulation associated with a regime of accumulation can exist at a variety of levels, it becomes clear that the complex interactions of real people must also be brought into the model, and that political economy must also include a moral economy. Personal networks can provide security that facilitates investment where purely economic self-interest might not. Commitment to a place or a firm can produce a 'stickiness' to capital that can provide the seeds for stable expansion in one location, while distrust may result in negative-sum outcomes in another. The entities created in the pre-reform era in China offer many advantages (but also obstacles) which can be adapted to a newly market-oriented era. But these are not the only resources that can be adopted and reinterpreted. While the clarification of property rights can make important contributions to increasing efficiency, in certain circumstances the retention of their ambiguity can also help get things done. Further, it is critical to realise that neither claim is mutually exclusive: both clarification and obfuscation can be used to promote local development and to increase prosperity.

The character of China's reforms

In considering China's reforms we must keep clear what they are not. They have not involved a substitution of the market for central planning as the integrative mechanism for the economy. Central planning was never as fully developed in China as in Eastern Europe, so a large proportion of China's economy before 1978 was integrated by neither plan nor market but through local collective production and through inter-regional patterns of relational contracting (Meaney 1989). In much of the country today, local governments control enterprises and the local cadres supervise the process (Oi 1994; Selden 1994).

While government expenditure fell from 41.1 per cent of national income in 1978 to 23 per cent in 1989 (Goodman 1994: xi), the private sector is still relatively small in most parts of China, accounting for about 9 per cent of industrial production (Nolan 1994: 6). The fastest growing part of the economy is the rural 'collective' sector, an ill-defined category which includes enterprises owned by local governments, by workers, by shareholders, and in some cases private enterprises disguising themselves as collectives (Odgaard 1992).

The Chinese reforms have involved 'a devolving of property rights downward in political or administrative hierarchies, or reassigning and clarifying property rights among institutions and households' (Walder 1994: 6). These shifts in property rights do not necessarily mean the construction of private property in the capitalist sense. Instead, some of the most important sets of changes 'have given local governments . . . greater rights over the increased revenues they generate' (Walder 1994: 10), with the result that extra-budgetary funds (those held outside financial plans sent down from above) grew from 31 per cent of national budgetary revenues in 1978 to 90 per cent in 1987 (Walder 1994: 10). One result of this has been a weakening of control by central authorities over implementation of policy in provinces and localities. Related to this has been

> the rise of local corporatism and cadre entrepreneurship in China's townships and villages. Fiscal reforms have given these local governments greater rights to retain revenues they generate. . . . The rapidly growing rural industrial sector has been centered on enterprises collectively owned by villages and townships.
>
> (Walder 1994: 10)

However, this is not necessarily the only outcome. In Guangdong (the province with the fastest growing economy since 1978), capitalist enterprises with foreign investment have been the most important product, while in Wenzhou, in central coastal China, private domestic capitalist enterprises have been the vibrant part of the economy (Liu 1992) (Figure 19.1).

One thing that all these diverse enterprise forms have in common is that they do not seem to have the same 'soft-budget constraints' as does the state sector (Perkins 1994: 37). Even in the state sector, the reforms have been partially successful in creating more efficient enterprises, as a result in part of competition with the collective sector.

Fundamental to understanding the diversity of local economic forms in China is consideration of how the reforms were implemented. The reform process has not been one in which the reforms were

> planned at the center and imposed from the top down in conformity with a predetermined scheme. On the contrary, the reform process

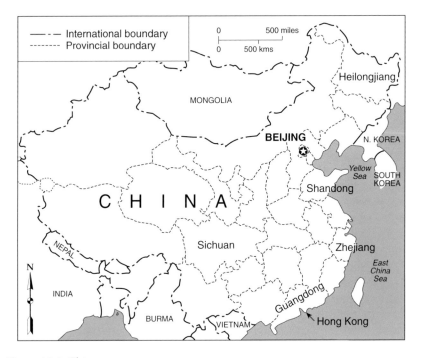

Figure 19.1 China

has been highly pragmatic, experimental, and sequential. . . . [O]ne of the lessons from China is the advantage of proceeding in a flexible and incrementalist way. One consequence of such an approach is that the reforms are not uniform across time and space.

(Griffin and Khan 1994: 86)

The adoption of this incrementalist reform approach is hardly one that was reasoned and theoretically based. Rather it was grounded in the constraints of Chinese politics. The strategy was that:

Instead of attacking the perquisites and powers of the central bureaucracies head-on, Deng Xiaoping decided to encircle the bureaucracies by creating new forms of business exempt from normal state rules, such as private and collective firms and Special Economic Zones. . . . The dynamic growth of this nonstate sector put competitive pressure on state-owned firms and the government bureaucrats responsible for them; soon state managers and bureaucrats were demanding the same market freedom for their state-owned firms.

(Shirk 1993: 15–16)

Reforms that have encouraged standardisation, such as tax reforms, have been regularly and often successfully resisted, whereas 'reform proposals that applied different rules to each enterprise and locality were widely popular' (Shirk 1993: 16). Such particularism allowed political benefits to both sides with potential efficiency costs, but ones that seem not to have as yet derailed the reforms.

The strategy of expanding the market sector gradually while maintaining the planned sector created a

> transitional two-track economy, with numerous accompanying economic problems, but it was politically very successful. Managers and industrial bureaucrats were protected by the security of the old system . . . while gaining access to the profitable opportunities of the market.
>
> (Shirk 1993: 130)

It also created potential for corruption and rent-seeking. This potential could easily have derailed the reforms, and has undoubtedly done so in some counties,[4] rather than leading to such surprisingly vibrant macroeconomic performance.

Another side of the reforms has received less attention than that of rapid growth – the way in which they have resulted in a dramatic reduction in the state welfare system (Vohra 1994: 55). Goodman (1994: xvii) considers the 'most remarkable indicator of change' in China to be that the proportion of the working population which received housing, health care and pensions from the state dropped from 19 per cent of the workforce in 1980 to 9 per cent in 1990. Although Gini coefficients of inequality have not increased as sharply as might have been expected, there has been a rise from 0.21 in 1978 to 0.3 in 1988 in the countryside (Ma 1994: 25). In addition, the disappearance of collective agriculture has resulted in massive interregional and rural–urban migrations (Fan 1995). While real rural incomes increased substantially between 1978 and 1984, since then they have been stagnant in most regions, with exceptions limited to areas with rapid industrialisation (Rozelle 1996). In combination with the household registration system, these migrations have resulted in an increasing proportion of the population being completely without any social welfare 'safety net'. In the countryside, welfare provisions and state redistribution more generally are provided on the basis of the place of permanent household registration. While rural dwellers can now move to cities or other provinces in a way that was not possible before 1978, they receive only temporary household registration which does not make them eligible for government services other than subsidised health care. Even education for their children is priced at prohibitive levels for those who do not have permanent registration. Migrants must rely on the market or *guanxi* ties to obtain many of the necessities of life, and in the rapidly growing factory towns of Guangdong the resulting patterns are remarkably similar to those found in the capitalist Third World.

The welfare outcomes cannot be seen in isolation from the nature of the transformation of the economy. They vary quite considerably depending on the kind of local economy that has been produced. The way in which the reforms have been instituted has resulted in a great degree of local diversity in rates of growth, the kinds of state institutions and regimes of property that have been constructed and the distribution of the new forms of wealth that are being produced. For example, the successful reforms in agriculture, which largely laid the groundwork for subsequent light manufacturing successes by producing effective demand among the rural population for their products, did not result from the implementation of a top-down reform but from the acceptance of local-level experimentation. Many of these experiments were initially beyond the limits of what was envisaged as acceptable, but local successes resulted in their adoption elsewhere.

There are, however, lively disagreements among observers about how and why China's economic expansion since 1978 was accomplished. Leaving aside the foreign-invested sector for the moment, a critical clue to these accomplishments lies in the TVE sector, with annual growth above 20 per cent for over a decade. Walder (1995a: 264) identifies three main explanations for this success: first, TVEs represent an ownership form distinct from that of the state sector, exposed to strong market competition and seen as semi-private or hybrid forms; second, the spread of market mechanisms has created incentives for firms, and budget constraints are harder for TVEs; and third, incentives for government officials have been changed, producing entrepreneurial local governments. All three explanations revolve around issues of regimes of property rights and their impact on market incentives.

Acknowledging that each explanation offers some valuable insights, Walder argues that none of them cope adequately with the relationship between government and enterprises. To correct for this, Walder (1995a) has produced an elegant revision of Kornai, demonstrating that Kornai's economic analysis of the inevitable softening of budget constraints under socialism (e.g. Kornai 1986) assumes a number of characteristics that do not apply or are variably applicable in different circumstances for China. He counters the argument that the dynamism of township and village government industries can be attributed to their 'hybrid' form. Instead he asserts that, contrary to common assumptions, 'governments at the lower levels are able to exercise more effective control over their assets than are governments at higher levels' (Walder 1995a: 270). The reforms have not privatised TVEs. Instead, they have decentralised ownership to local governments and removed the attenuation of property rights characteristic of centrally planned economies. They have also produced clearer financial incentives for local governments as enterprise owners, and expanded their capacity to monitor their activities. As a result, the fastest growth in output and productivity has occurred where government ownership rights are clearest (Walder 1995a: 271).

Walder offers an 'unequivocal portrayal' of the TVEs as government-owned and operated in the 'same sense as the urban state sector' (Walder 1995a: 271; 1995b). The definitional issue is important here if the validity and implications of the argument are to be understood. For Walder, to say that a government jurisdiction owns an enterprise means that

> the government holds all rights to control income flows, and sale or liquidation except for those rights it chooses to transfer to agents who are either hired to manage those assets or who obtain these rights in lease contracts. Less abstractly, with regard to control, this means that the government hires and replaces managers or allocates contracts to lease assets.
>
> (Walder 1995a: 270)

It should be noted that this definition describes a wide range of relationships between government jurisdiction and enterprises/assets. It is the kind of rights that are transferred and retained that distinguishes between public and private enterprise. When the government jurisdiction extracts surplus only in the form of rent or tax, the enterprise that appropriates the surplus in the form of profit should be seen as private. Only if profits in the narrow sense accrue to the government does it make sense to talk about a public enterprise. Therefore, while Walder may be correct given his definition of government ownership, I would argue that this definition is far too broad to be useful. Furthermore, the practical interpretation of his conclusions involves conceptual slippage between the broader and narrower definitions of government ownership.

I would agree with Walder that many TVEs fit within the narrower definition of government ownership (the accrual of profits to the government), and furthermore accept the claim that the impressive results of this kind of enterprise are at least partially accounted for by clarification of rights and the attendant hardening of budget constraints. However, to extend this analysis to include all local variants of TVEs is analytically confusing. For example, shareholder-controlled collectives can have a character very different from others that are creatures of governments. We need to distinguish between rights of property that extract profit and rights that extract rent (Burawoy 1996). Nowhere is this distinction made in Walder's discussion of property rights, but I believe it to be a central one, and elaborating it may allow us to produce a typology or model of property rights in China which can aspire to the elegance of Walder's solution while explaining a greater fraction of local diversity and variance.

Such a model is beyond the scope of this chapter. Instead, I will support my criticism of Walder by describing some of the local variations that exist within the legal category of TVE, the variety of relations of production and distribution of property rights that are found among them, and some of the social and organisational implications of the varieties. First, I present cases that fit Walder's account most closely: those that have been presented as local corporatism or local market socialism. Second, I turn to shareholder-controlled collectives. Finally, I discuss private- and foreign-controlled enterprises operating under the legal fiction of collective or TVE status.

Local economic regimes in China
Local corporatism

Observers of rural China have suggested that less has changed in China's countryside than is apparent on the surface (Oi 1989), and that rather than extensive privatisation, what has resulted can be referred to as 'local corporatism'. Lin (1995: 340) defines local corporatism as 'an institutional arrangement that consists of a hierarchically-ordered set of organizations, a central authority, a functional unity, with local (territorial and network) imperatives and the duality of internal (co-ordination)–external (competitiveness) dependence'. The concept of local corporatism allows for the extensive decentralisation of property rights simultaneous with the preservation of large amounts of (local) governmental control of the economy, and for responsiveness to external market incentives and constraints despite the absence of substantial numbers of enterprises controlled by agents other than governmental ones.[5] Judd's (1994) book on Qianrulin village in Shandong provides an example. This village is one of a very small percentage that still have not de-collectivised agriculture. Villagers who are assigned to work in the rural factories that have been developed receive work-points rather than cash wages, which reduces income disparities between agriculture and industry. Yet, despite such continuities with pre-1978 practices, 'Qianrulin has demonstrated a high level of competence in managing sustained and diversified growth' in rural industry (Judd 1994: 69).

Lin (1995) has taken this line of argument a step further by proposing that the dynamism and local diversity of rural China can be encompassed within what he calls 'local market socialism'. This notion 'makes explicit that the reform economy system should be analyzed in terms of three components: (socialist) bureaucratic coordination, market coordination, and local coordination, with the last playing the pivotal role' (Lin 1995: 310). Local social networks, particularly those based on patrilineal kinship, provide a stable basis for a locality using non-market structures (particularly bureaucratic units derived from the pre-reform era) to engage in external market transactions in order to accumulate collective capital. In the extreme case, the locality retains 'the centralized functions of setting plans and directions for the entire economy' and 'owns all properties and resources' (Lin 1995: 310), except for individual earnings and properties.

It is clear that instances of local corporatism or local market socialism are common in China. It is also undeniable that in many cases (but certainly not always) they have been capable of producing remarkable rates of growth, are an important, if not the most important, element of China's accomplishments, and appear to be a stable organisational form worthy of attention. However, the whole range of diverse local patterns found in China cannot be encompassed within these concepts without stretching them beyond the bounds of utility. Local corporatism or local market socialism exists where the majority of productive property is held by a local governmental agency or agencies, where all local citizens (which usually excludes migrant workers, who may outnumber citizens) have rights to some revenues or benefits from the collective property, and where few significant alternative channels to prosperity exist. This restricted definition does not fit the kind of TVE that is discussed in the next section, much less the disguised capitalist enterprises oriented towards export-based industrialisation.

Shareholder-controlled collectives

Shenquan village in Sichuan was formerly the poorest village in the area, but now is among the most prosperous (Yang 1994). This economic expansion has been made possible by the development of a collective rural enterprise specialising in health-care products. Unlike the corporatist enterprises discussed in the last section, to be a member of this collective a family had to purchase at least one share at a price of RMB 2,000. Each share entitled a family to one job. The collective expanded rapidly, but eventually stopped selling shares and began hiring 'temporary' workers who were not shareholders and who received lower wages and less security (Yang 1994: 166). The management of this factory's operations is 'independent from the village government. It is, however, advantageous for the factory to retain its image as a village collective enterprise' (Yang 1994: 168–9). Certain tax exemptions and other advantages are available that would not be if the enterprise registered as a private-sector company.

The success of the collective has led to social differentiation within the village, with the wealthiest having annual household incomes up to RMB 50,000 and the poorest as little as RMB 1,000. The new elite produced their wealth through marketing the factory's products by selling them at wholesale prices to managers who sell them in the cities and keep the profits, essentially creating private companies attached to the factory (Yang 1994: 167). It is clear that the relations of production in this factory differ considerably from those in Qianrulin despite both being officially labelled as part of the collective sector. The disguised capitalist enterprises of Guangdong take us even further from local corporatism.

Export-oriented capitalist industrialisation

In some parts of China, unambiguously capitalist relations of production have become common. Most of these capitalist enterprises are in the foreign-investment sector. Although many of these enterprises are officially joint ventures and thus distinct from TVEs, many others operate on paper as entities owned by local governments or governmental agencies. In one township, the standard procedure for Hong Kong investment is to register as an export-processing contract with a newly registered entity that has little more than a paper existence. To all intents and purposes these are foreign-owned and managed enterprises with the local counterpart receiving revenue streams primarily composed of rent (Smart and Smart 1991).

During 1985–91, Guangdong averaged 43.7 per cent of the total direct foreign investment in China (Chen 1994: 166). 'Guangdong's dynamic economic growth in the 1980s was based on highly labor-intensive and export-oriented industrialization, similar to what Hong Kong and Taiwan experienced in the 1960s' (Chen 1994: 169). With only 6 per cent of China's population, Guangdong accounts for 21 per cent of the country's total exports (Chen 1994: 169).

Josephine Smart and I have been conducting research since 1987 in Henggang township near Hong Kong. In 1993, it was incorporated into an expanded Shenzhen Special Economic Zone. The township has grown from about 10,000 people in 1985 to perhaps 140,000 today, largely on the basis of over 600 Hong Kong-run factories. A shoe factory that we have studied intensively (Smart and Smart 1993), although on paper only an export-processing subcontract with a local enterprise counterpart, is in reality exclusively Hong Kong operated. It operates in ways comparable to small, labour-intensive factories in Hong Kong. Most workers are hired on piece-rates, and are hired and fired without restrictions. Capitalist practices in this factory have consistently run ahead of what is allowed by foreign investment policies and the resulting obstacles are mitigated with the assistance of locals and their connections in the local government (Smart 1993a).

The local residents, unlike the migrant workers who have temporary registration and are over nine-tenths of the population, have become increasingly prosperous. Luxurious homes of over 2,000 square feet, furnished with the fashionable conveniences of colour television, cordless phones, electric fans, and sometimes even laser-disc karaoke machines, are increasingly common. This rapidly increasing prosperity is usually based on various forms of rent: shares of village dividends produced by the leasing of land, real-estate investments, renting of business permits to migrants, access to well-paying jobs as factory managers, and renting of land to migrant agricultural workers (Smart 1993b). The locals are becoming an elite rentier class benefiting from the export-generated profits resulting from foreign investment and the poorly remunerated efforts of migrant workers.

I have suggested elsewhere (Smart 1995) that situations like this can be seen as involving 'local capitalisms' where the conditions of existence for capitalist relations and practices are accomplished locally, rather than through national laws and property rights. Unlike 'routine capitalism' (i.e. once developed and legally accepted), where property rights and economic governance are constructed and guaranteed by national governments, these ground rules are locally accomplished, primarily through *ad hoc* and pragmatic arrangements between outside investors and local officials,[6] often through the intermediation of kinship and other social connections (Smart and Smart 1992).

Therefore, although organisational outcomes of reform experimentation differ significantly between places, the processes by which they have been accomplished share strong similarities. Stark's (1996) account of recombinant property in Hungary suggests that the comparison could be stretched even further.

Despite apparent differences between these versions of TVEs there are also similarities that span the divide between local market socialism and disguised capitalist enterprises. We can see in local varieties the impact of the process involved in higher-level reforms and legislation, but also the diversity of ways local governments have responded to new opportunities. As Lin (1995: 307) points out, 'each model seems to have emerged from local roots and each seems to be thriving on its own terms'. Local governments have everywhere been important in providing the social infrastructure for new organisational forms which are instituted in order to take advantage of the opportunities opened up by reforms. They have often gone beyond what has been officially accepted by the central authorities, with local experiments sometimes subsequently acknowledged as models for other localities, leading to their further dissemination. In a study of four counties in Guangdong, Fitzgerald (1996: 21–2) concludes that when the central government has presented short-term obstacles to reform and development

> local government has tried to walk around the obstacles. Central policy, like everything else, is adapted to 'suit local condition'. Devising strategies for coping with central policy is thus a major task of local government in the Delta. 'Superiors have policies,' runs a local ditty, 'and subordinates have strategies for coping with them.'

Conclusions

China's experience does not provide much comfort for those with an inclination to economic purism. An important question is whether the messy character of the reforms is part of the problem requiring further market-oriented reform, or a source of some of the remarkable accomplishments of the last fifteen years. An appreciation of the ways in which the reforms have developed, and the tenuous legal basis for the many local experiments and innovations may offer part of the answer, or perhaps only further questions. At this point we cannot make any definitive statements about the results of this historic experiment. Still, it does seem that it offers opportunities for exploring the processes of transformation of political and moral economies.

There is a danger of being too impressed with the variety of local systems that seem to be able to deliver the goods in China. We might be tempted to conclude that we can say little more than that we can identify a situation of 'path-dependent development' requiring different solutions building from local conditions, so that 'where you end up depends on where you've been, and whatever optimality properties may be claimed for the equilibria are at most local rather than global' (Bowles and Gintis 1993: 97). One possible terminus, but one that I believe erroneous, of such an argument is that economically 'anything goes', as long as local actors are convinced of the viability of such arrangements. Still, it is salutary to realise that more variability is possible, that even in dealing with a capitalist world economy a variety of approaches can facilitate competitive edges (Mahon 1991). Clearly, a task for the future is to identify what the minimal conditions are, and what might account for the variety of successful patterns.

One commonality might be, for example, the need to keep labour costs down and productivity up, given that China's main competitive advantage in the world economy is its large supply of cheap labour.[7] Another factor, although one that may derive from the competitive disadvantages of a socialist state, is the need to avoid potentially heavy transaction costs. An attendant factor, given the socialist heritage and the ability of Chinese citizens to make effective discursive use of this heritage to criticise the emerging managerial elite, is being able to keep labour costs down without setting off local dissatisfaction heavy enough to derail development. There is certainly scope within the system for dissatisfied groups to impede development. The main factor in accounting for the differential growth strategies may lie in what solutions exist to the twin problems of keeping down labour costs while avoiding effective resentment. In our field sites in Guangdong, this is accomplished partly through the legitimacy of 'the open door' policy of developing links with the outside world, partly through the buying off of local residents through rent, and through placing the costs of the strategy on migrant workers, who have little capacity to use the local government to disrupt production. In other places, it seems to be accomplished through corporate or quasi-corporate means. Oi (1994: 77) suggests that redistribution of some of the wealth of entrepreneurs is common and may serve to 'mediate possible class conflicts that might otherwise develop amidst the increasing stratification'. Bowles and Dong (n.d.) found that workers in TVEs in Heilongjiang had high levels of commitment to their enterprise, being 49 per cent more willing to take a cut in pay to help their firm than workers in private enterprises. Resolving these twin challenges successfully means not only that labour costs can be kept down and productivity up but that the transaction costs of doing this can be minimised.

Given the temper of the times, it is necessary to balance recognition of the potential competitive disadvantages of China's socialist heritage with the positive elements of the legacy (Weller 1994). There is the remarkably high life expectancy, generally good standards of public health (by contrast to comparable-income countries), a low level of inequality so that an increase in inequality does not necessarily yet mean dramatic polarisation, fairly equal access to land in the countryside and the absence of the entrenched interests of foreign or private capitalists (Pearson 1991: 14). The outcome of similar reforms in countries like the Philippines would probably be very different and less positive, as powerful rural elites utilised their monopolisation of resources to control opportunities. While China's cadres are better positioned than ordinary villagers to benefit, and are among the main benefactors of local growth in the reform era (Oi 1994), their capacity to dominate is limited by ideology and the lesser degree of concentration of these advantages. Again, we see the necessity to attend to the path-dependent character of development and the danger of assuming that what works in one place can be successfully transferred elsewhere.

The question that will arise in the future is whether strategies that are effective now will continue to work later. In our field site, the creation of a non-productive rentier elite may mean that, when investors move on to greener pastures, local entrepreneurs will not be able to take up the slack, while collectives that produce a managerial elite may be creating a growing sense of resentment that immobilises future developments through internecine fights. The outcome will be influenced by whether and when the period of experimentation will be succeeded by standardisation and the possibilities and dangers that such a stage will pose. Yet, can expansion continue indefinitely without such standardisation? As companies grow, the cost of doing business in places with different ways of doing things and local protectionism may undermine continued growth.

Notes

1 Jefferson and Rawski (1994: 66) argue that the performance difference is too large to be explained by differences in initial conditions.
2 In emphasising here China's economic accomplishments, it should not be forgotten that they have come with a full set of accompanying problems, such as environmental degradation, increasing inequality (Rozelle 1996), regional disparities between a rapidly growing coast and much slower growth in much of the interior, dislocation through massive levels of redundancy in agriculture, and high rates of inflation. This chapter will not concentrate on most of these problems, since they are less surprising than the fact of any success at all in producing a competitive socialist state. Furthermore, all these problems have been

observed in the periods of early development of all the advanced capitalist economies and the Newly Industrialising Economies. They will be considered, of course, where they are relevant to the argument, or to the likelihood of continued success.

3 Smith's (1995) suggestion that regulation theory needs to be expanded by the recognition that there are many sites of regulation, including those within enterprises, is consistent with the argument here.

4 Peng (1996: 68–9) suggests that in economically backward districts in the interior where it is hard to develop local industry, local governments act like predatory states, 'extracting revenue from the public without necessarily promoting economic development'.

5 Space does not permit discussion of the relationship between such local corporatism and national-level corporatism, which Unger and Chan (1995) suggest is broadly operating in China.

6 Certainly, the willingness of the party-state to countenance such activities, and its provision of certain basic legal and political ground rules, is an important enabling factor in this process, but it is one that seems to be necessary, but not sufficient, for the capitalist investors.

7 Here, Zhao and Nichols's (1996) fascinating discussion of the introduction of primitive accumulationist techniques of labour control in state sector enterprises, and how institutions and routines are reinterpreted in maintaining labour discipline, is very relevant.

Bibliography

Aglietta, M. (1979) *A Theory of Capitalist Regulation*, London: NLB.

Amsden, A.H. (1990) 'Third world industrialization: "global Fordism" or a new model?' *New Left Review* 182: 5–31.

Angresano, J. (1994) 'Institutional change in Bulgaria: a socioeconomic approach' *Journal of Socio-Economics* 23, 1/2: 79–100.

Block, F. (1990) *Postindustrial Possibilities*, Berkeley: University of California Press.

Bowles, P. and X. Dong (no date) 'Enterprise ownership and worker attitudes in Chinese rural industry', unpublished manuscript.

Bowles, S. and H. Gintis (1993) 'The revenge of Homo economicus: contested exchange and the revival of political economy' *Journal of Economic Perspectives* 7, 1: 83–102.

Burawoy, M. (1996) 'The state and economic involution: Russia through a China lens' *World Development* 24, 6: 1105–17.

Chen, X.M. (1994) 'The new spatial division of labor and commodity chains in the Greater South China Economic Region', in G. Gereffi and M. Korzeniewicz (eds) *Commodity Chains and Global Capitalism*, Westport: Greenwood Press.

Evans, P. (1996) 'Embedded autonomy and industrial transformation', in D. Davis (ed.) *Political Power and Social Theory*, Greenwich: JAI Press.

Fan, C. (1995) 'Of belts and ladders: state policy and uneven regional development in post-Mao China' *Annals of the Association of American Geographers* 85, 3: 421–49.

Fitzgerald, J. (1996) 'Autonomy and growth in China: county experience in Guangdong Province' *Journal of Contemporary China* 5, 11: 7–22.

Goodman, D.S.G. (1994) 'Introduction: the political economy of change', in D. Goodman and B. Hooper (eds) *China's Quiet Revolution*, Melbourne: Longman Cheshire.

Granovetter, M. (1990) 'The old and the new economic sociology', in R. Friedland and A. Robertson (eds) *Beyond the Marketplace*, New York: Aldine de Gruyter.

Granovetter, M. and R. Swedberg (eds) (1992) *The Sociology of Economic Life*, Boulder: Westview Press.

Griffin, K. and A.R. Khan (1994) 'The Chinese transition to a market-guided economy: the contrast with Russia and Eastern Europe' *Contention* 3, 2: 85–107.

Jefferson, G.H. and T.G. Rawski (1994) 'Enterprise reform in Chinese industry' *Journal of Economic Perspectives* 8, 2: 47–70.

Judd, E.R. (1994) *Gender and Power in Rural North China*, Stanford: Stanford University Press.

Kipnis, A.B. (1997) *Producing Guanxi*, Durham, NC: Duke University Press.

Kornai, J. (1986) *Contradictions and Dilemmas*, Cambridge, MA: MIT Press.

Lin, N. (1995) 'Local market socialism: local corporatism in action in rural China' *Theory and Society* 24: 301–54.

Liu, Y. (1992) 'Reform from below: the private economy and local politics in rural industrialization' *The China Quarterly* 130: 293–316.

Ma, G. (1994) 'Income distribution in the 1980s', in D. Goodman and B. Hooper (eds) *China's Quiet Revolution*, New York: Longman Cheshire.

Mahon, R. (1991) 'Post-Fordism: some issues for labour', in D. Drache and M. Gertler (eds) *The New Era of Global Competition*, Montreal: McGill-Queen's University Press.

Meaney, C. (1989) 'Market reform in a Leninist system' *Studies in Comparative Communism* 22: 203–20.

Nolan, P. (1994) 'Introduction: the Chinese puzzle', in Q. Fan and P. Nolan (eds) *China's Economic Reforms*, New York: St. Martin's Press.

Odgaard, O. (1992) 'Entrepreneurs and elite formation in rural China' *Australian Journal of Chinese Affairs* 28: 89–108.

Oi, J. (1989) *State and Peasant in Contemporary China*, Berkeley: University of California Press.

—— (1994) 'Rational choices and attainment of wealth and power in the countryside', in D. Goodman and B. Hooper (eds) *China's Quiet Revolution*, New York: Longman Cheshire.

Pearson, M. (1991) *Joint Ventures in the People's Republic of China*, Princeton: Princeton University Press.

Peng, Y. (1996) 'The politics of tobacco: relations between farmers and local governments in China's southwest' *The China Journal* 36: 67–82.

Perkins, D. (1994) 'Completing China's move to the market' *Journal of Economic Perspectives* 8, 2: 23–46.

Rozelle, S. (1996) 'Stagnation without equity: patterns of growth and inequality in China's rural economy' *The China Journal* 35: 63–92.

Saich, T. (1992) 'The reform decade in China', in M. Dassu and T. Saich (eds) *The Reform Decade in China*, London: Kegan Paul International.

Selden, M. (1994) 'Russia, China and the transformation of collective agriculture' *Contention* 3, 3: 73–93.

Shirk, S. (1993) *The Political Logic of Economic Reform in China*, Berkeley: University of California Press.

Smart, A. (1993a) 'Gifts, bribes and *guanxi*: a reconsideration of Bourdieu's social capital' *Cultural Anthropology* 8, 3: 388–408.

—— (1993b) 'The political economy of rent-seeking in a Chinese factory town' *Anthropology of Work Review* 14, 2–3: 15–19.

—— (1995) *Local Capitalisms: Situated Social Support for Capitalist Production in China*, Department of Geography, Chinese University of Hong Kong Occasional Paper Series No. 121, Hong Kong: Chinese University of Hong Kong.

Smart, A. and J. Smart (1992) 'Capitalist production in a socialist society: the transfer of manufacturing from Hong Kong to China', in F. Rothstein and M. Blim (eds) *Anthropology and the Global Factory*, New York: Bergin & Garvey.

Smart, J. and A. Smart (1991) 'Personal relations and divergent economies: a case study of Hong Kong investment in South China' *International Journal of Urban and Regional Research* 15: 216–33.

—— (1993) 'Obligation and control: employment of kin in capitalist labour management in China' *Critique of Anthropology* 13, 1: 7–31.

Smith, A. (1995) 'Regulation theory, strategies of enterprise integration and the political economy of regional economic restructuring in Central and Eastern Europe: the case of Slovakia' *Regional Studies* 29, 8: 761–72.

—— (1996) 'From convergence to fragmentation: uneven regional development, industrial restructuring, and the "transition to capitalism" in Slovakia' *Environment and Planning A* 28: 135–56.

Stark, D. (1994) 'Path dependence and privatization strategies in East Central Europe', in J. Kovacs (ed.) *Transition to Capitalism?*, New Brunswick: Transaction Publishers.

—— (1996) 'Recombinant property in East European capitalism' *American Journal of Sociology* 101, 4: 993–1027.

Stiglitz, J. (1993) 'Post Walrasian and Post Marxian economics' *Journal of Economic Perspectives* 7, 1: 109–14.

Unger, J. and A. Chan (1995) 'China, corporatism, and the East Asian model' *The Australian Journal of Chinese Affairs* 33: 29–53.

Vohra, R. (1994) 'Deng Xiaoping's modernization: capitalism with Chinese characteristics' *Developing Societies* 10: 46–58.

Walder, A. (1994) 'Evolving property rights and their political consequences', in D. Goodman and B. Hooper (eds) *China's Quiet Revolution*, New York: Longman Cheshire.

—— (1995a) 'Local governments as industrial firms: an organizational analysis of China's transitional economy' *American Journal of Sociology* 101, 2: 263–301.

—— (1995b) 'China's transitional economy: interpreting its significance' *The China Quarterly* 144: 963–79.

Weller, R. (1994) 'Cultural legacies and development: a view from East Asia', in ~~J. Kovacs (ed.) Transition to Capitalism?, New Brunswick: Transaction~~ Publishers.

Yan, Y. (1996) *The Flow of Gifts*, Stanford: Stanford University Press.

Yang, M. (1994) 'Reshaping peasant culture and community' *Modern China* 20, 2: 157–79.

Young, S. (1994) 'Private entrepreneurs and evolutionary change in China', in

D. Goodman and B. Hooper (eds) *China's Quiet Revolution*, New York: Longman Cheshire.

Yusuf, S. (1994) 'China's macroeconomic performance and management during transition' *Journal of Economic Perspectives* 8, 2: 71–92.

Zhao, M. and T. Nichols (1996) 'Management control of labour in state-owned enterprises: cases from the textile industry' *The China Journal* 36: 1–21.

20

RECOMBINANT CAPITALISM

State, de-collectivisation and the agrarian question in Vietnam

Michael J. Watts

There can be no comprehensive and consistently radical transformation in other spheres while the key feature of the old classical structure, the Communist Party's power, remains.
(Janos Kornai, *The Socialist System*, 1992)

In the Third World, the degree to which a successful decollectivization depends on a significant political liberalization does not appear great.
(Frederic Pryor, *The Red and the Green*, 1992)

If the Ho Chi Minh trail still existed, says Thomas Friedmann of the *New York Times*, it would almost certainly be a toll road (*New York Times*, 15 January 1995: 17). In comparison to other post-socialist transitional states, Vietnam has liberalised relatively quickly and, in many respects, has embarked upon a reform programme that has moved faster and further than other comparable state socialist economies with a large agrarian base (for example China, Cuba or Laos). Vietnam remains, of course, rural and overwhelmingly impoverished – according to the latest World Bank report (1995: 7) 51 per cent of the Vietnamese population is classified as poor – lagging behind other Asian economies, in a distant league from the Asian newly industrialising states. Vietnam is a case of striking socialist reform, and according to some very possibly an Asian 'tiger' in the making (Reidel 1993; EIU 1994). As if to ratify these predictions, the Vietnamese leadership has openly acknowledged the South Korean and Taiwanese economies as provisional sorts of models for its curious blend of free markets and Leninism, a political economy described by the Communist Party in its extraordinary linguistic copulations as 'socialist-oriented multi-sectoral economy driven by the state-regulated market mechanism', what the *Far Eastern Economic Review* has dubbed 'Vietnamonomics' (*FEER*, 26 October

1995: 7). Whether one interprets this hybrid economy, as ranking Politburo member Dao Duy Tung does, as a confirmation that Vietnam has 'skipped capitalism' (*FEER*, 26 October 1995: 52) or more cynically as a case of 'the party plus capitalism equals socialism' (Kolko 1995), is perhaps of less relevance than the fact that the model has unequivocally produced tiger-like economic growth rates (Figure 20.1).[1]

Prime Minister Vo Van Kiet told the national assembly at the end of November 1995 that the economy had been growing at 9.5 per cent (up from 8.8 per cent) in 1994; industrial output expanded by 14 per cent and agriculture by almost 5 per cent. Exports are expected to rise by 30 per cent (up from 24 per cent in 1994 and 15 per cent in 1993).

The purpose of this chapter is to account for the particular character and trajectory of Vietnam's agrarian transition – what I shall call the Vietnamese agrarian question – in the period of reform (*doi moi*) and market socialism. While I shall outline the *general* contours of Vietnamese agrarian policy in the socialist (1947–85) and *doi moi* (1986–present) periods, I shall focus *regionally* on North Vietnam, and on field research conducted in one province

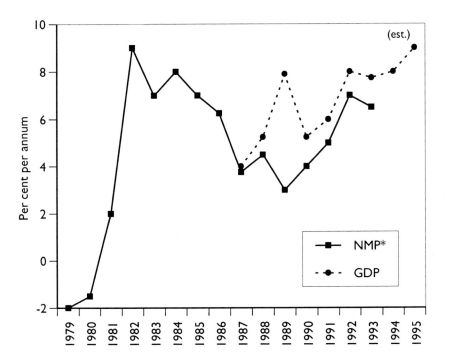

Sources: International Labour Office 1994: 8; *Far Eastern Economic Review* 1994–5
Figure 20.1 Vietnam: growth rates, 1979–95 (*calculated on the basis of the Soviet material product system)

(*tinh*) in the northern and western reaches of the Red River delta (Vinh Phu Province (see Figure 20.2)). There is no presumption, of course, that the Red River delta, or the Midland region in which Vinh Phu is located, are in any sense representative of Vietnamese agriculture *in toto* (the latter being, in any case, something of a statistical fiction). Rather this chapter accounts for the morphology and the dynamics of rural accumulation (and secondarily rural differentiation) in a number of northern (i.e. inland) Deltaic villages (*thon*) in several communes (*xa*) south-east of Viet Tri city insofar as they seem to exemplify certain tendencies within the agricultural sector as a whole. Land redistribution, peasant commercialisation, off-farm income and new forms of agrarian accumulation, and revenue demands by increasingly decentralised provincial and local governments, are the central planks of the agrarian transition but, as I seek to show, each can only be grasped in regard to concrete political forces, specifically the enduring centrality of a refigured party-state. This intersection of politics, a nascent civil society and agrarian accumulation raises obvious parallels with Stark's (1996) account of recombinant development in Eastern Europe building, as he felicitously says, *with* the ruins of communism. 'Ruins' perhaps exaggerates the decay of the old and the birth of a new political order in the Vietnamese case, but Stark highlights the new spaces within the interstices of a reformed system, what Luong has properly referred to as 'the interplay of national reform policy and local sociocultural dynamics' (1992a: 26). Indeed, it is precisely these tensions – and sometimes the contradictions – between national, regional and local levels that are encased in the oft-quoted metaphor that rural Vietnam resembles a 'tiger on a bicycle', that is to say an assertive class of farmers located atop a weak and ineffective institutional base (see Fforde 1993, the work cited in Long 1993, and Luong 1994, and also Tuan 1995; INSA 1995; CIRAD 1995).

A secondary concern in this chapter is to locate Vietnamese agrarian reform on a larger comparative canvas of de-collectivisation within formerly socialist agricultures. An obvious comparison in view of the timing and the shape of de-collectivisation is China, which affords an opportunity to reflect upon property rights, networks, markets and political power in the agrarian roads from socialism to capitalism. Three processes seem striking in the Vietnam case, however. First is the fact that the reforms were incubated in the agrarian sector (beginning in the late 1970s) and without these changes, driven in large measure by the production crisis within the sector, the entire reform package would surely not have proceeded as far, and as quickly, as it has to date. In this sense, *agriculture was the incubus for de-collectivisation*, and after 1989 in the wake of a rapid and irrevocable de-collectivisation and a stunning collapse of the co-operative structure, agriculture stands at the thermidor of the Vietnamese return to capitalism. Second, Vietnam is, I suggest, a compelling refutation of Kornai's claim that *fundamental reform is not possible without the destruction of the one Communist Party and the*

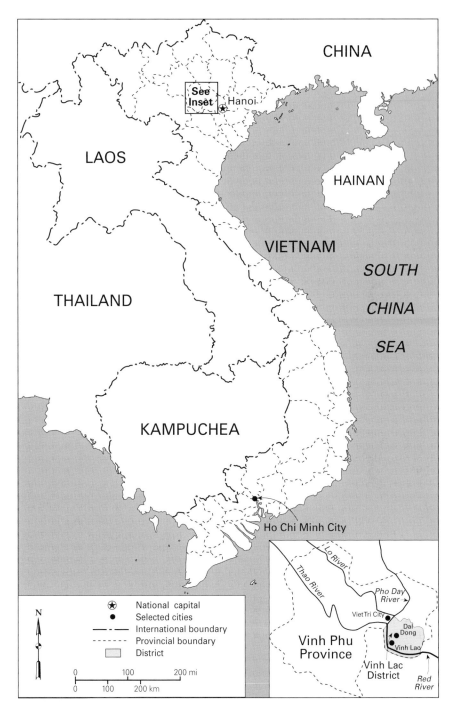

Figure 20.2 Vietnam

453

centralised and authoritarian party-state apparatuses. The Vietnam case suggests a greater flexibility to adapt and evolve, certainly greater than Russia or Bulgaria or Romania and at least as adaptable as most of the more successful East European regimes (Poland or Hungary). Vietnam confirms the so-called neo-liberal paradox – namely, that the transition to capitalism perhaps requires a strong and often a party-state to sail through the choppy waters of property rights reform, market liberalisation and global competition. And third, Vietnam is an instance of *macroeconomic stabilisation which reverses the standard Bretton Woods package*; the IMF-type stabilisation came in the late 1980s after almost a decade in which, to employ Fforde and de Vylder's (1989) felicitous phrase, institutional reforms laid the basis for making prices matter (see Irwin 1995: 739). The microeconomic crucible in which the macroeconomic reforms were made possible (if not necessarily desirable) was, in fact, Vietnamese agriculture and specifically the relatively rapid creation, especially in the old north, of an egalitarian peasantry of a distinctly Chayanovian sort.

De-collectivisation and the agrarian question

Aristocratic and bourgeois writers set out in their cars and speed through the countryside; they see the great and fragrant fields, the thick smoke rising from the thatched roofs in the evening and immediately invent a picture of 'poetic flavour', but they don't know. . . . The peasants want to improve their lot; they have suffered too long already. The hour has come. No more hesitation; it is impossible to deceive them any longer.
(Truong Chinh and Vo Nguyen Giap, *The Peasant Question*, 1937–8)

As the millennial clock ticks towards the end of the twentieth century, there is a profound sense in which the hour of the Vietnamese peasantry has come yet again. In a curious historical irony, the revolutionary peasantry, which Truong Chinh and Vo Nguyen Giap praised over half a century ago in their classic political tract *The Peasant Question* and which essentially disappeared as such in the wake of the collectivisation efforts of the 1950s, has re-emerged fully formed a generation later. Now the peasant occupies centre stage in the theatre of *doi moi* reform launched by the Vietnamese government in 1986. In late 1987 the rationing of all agricultural produce except rice was ended. While the government retained some influence over pricing by selling inputs for rice at a fixed barter rate via the co-operatives, subsequently reforms in 1988 allowed farmers to decide which inputs and services they purchased from local government (EIU 1994: 43). Restrictions on inter-provincial grain trade were lifted and prices doubled in real terms in 1988–9. In fact, without the radical changes in ownership, price and market relations that began to surface formally in 1981 in the form of household product contracting (*khoan san pham*) in agriculture, 'the impressive degree of openness' much praised by the World Bank (1993: 4) would not have been possible. Sustained growth rates and relatively low inflation since 1989 – GDP in Vietnam during 1993 grew by 8 per cent (exports by 20 per cent) (*Economist*, 7 January 1995: 27) – were built on the back of a largely agrarian, and by the early 1990s an essentially peasant, economy accounting for one-third of GDP and three-quarters of employment.[2] Notwithstanding the persistence of rural poverty (51 per cent of the population resides below the UN poverty line)[3] and sharp regional inequalities (per capita income in the Red River delta is only half of that in the Mekong), the results of a wide-ranging agrarian reform have been, according to Ljunggren (1994: 19), 'spectacular'. Prime Minister Vo Van Kiet's central objective – to 'unfetter the productive forces' (Vietnam 1993) – was, in other words, built around the recreation of an aggressively commercial and relatively autonomous peasantry which converted a 1.1 million ton rice deficit in 1980 into 1.4 million tons of rice exports nine years later (*Far Eastern Economic Review*, 10 May 1990: 32). Vietnamese rice output reached new records of 23.5 million tons in 1994, and 25.5 million tons in 1995, of which perhaps 300,000 tons was smuggled illegally into China (The Rice Paper 1995). Furthermore, as Fforde (1993: 52) and others have pointed out, this dynamism within the agrarian sector occurred in spite of the long-standing state priorities granted to industrialisation and urbanisation. But what sort of agrarian transition – what sort of agrarian question – is under way in Vietnam in the mid-1990s?

In his classic text on the nineteenth-century rural economy in Europe, Karl Kautsky (1899: 12) posed the agrarian question with admirable clarity: 'whether and how is capital seizing hold of agriculture, revolutionising it, making old forms of production and property untenable and creating the necessity for new ones?' Kautsky was of course largely concerned with Prussian modernisation of German agriculture, and more specifically with the consequences of the world grain market on agrarian structure. His anomalous findings that a deepening of market relations had underwritten the growth of a middle peasantry rather than a class polarisation fuelled by economies of scale in land-based production suggested both the capacity of small farmers to maintain their competitive edge through self-exploitation and their simultaneous subordination to large capital through ostensibly new forms of finance, credit and institutional integration. Kautsky was not simply concerned with the scale of production, but also with rural social structure and more generally with the *politics of transition*. Kautsky sought out those aspects of agriculture – agriculture's exceptionalism one might say – which sufficiently distinguished it from industrial manufacturing to suggest different tendencies and trajectories under capitalism.

Kautsky addressed the growing commercialisation of an estate and peasant-based agriculture long embedded in commodity relations and wage labour. But the collapse of the former Soviet Union and more generally the post-1989 reintegration of the socialist bloc into the world market have bathed Kautsky's agrarian question in a somewhat different light. How was socialist agriculture in its panoply of forms – state farms and centralised marketing institutions, peasant co-operatives, collective enterprises, private plots – to be subject to privatisation, de-collectivisation and the introduction of new forms of property rights? How was capital seizing hold of putatively socialist forms of production and property and creating new ones? Posed in this way, the agrarian question is framed by the particular structures of socialist economy – and the diverse forms of socialist agriculture – that evolved in the post-revolutionary period: agrarian reform from below versus collectivisation from above, or state farms versus Maoist collectives. Hence, in the 1960s, Hungary and China represented two rather contrary modalities of socialist agriculture in which the forms of production, the role of the market, and the significance of the household economy were sharply dissimilar. Not only was the *starting point* of the transition to market agriculture substantially different in these two cases, but the post-collective *trajectory* of each has diverged markedly. Selden (1993: 209) refers to a 'European model' characterised by land restitution and compensation to pre-revolutionary owners, rapid privatisation of land ownership and marketing rights, and the conversion of collectives into corporations and decentralised family farms. In Szelenyi's account of Hungary, this model of transition produces the reappearance of pre-socialist families and enterprises, the so-called 're-embourgeoisement' of Hungarian agriculture (in press). China, conversely, represents a case of gradualism in embracing private ownership and the market economy – in contradistinction to the 'big bang' or 'shock therapy' approach of Eastern Europe – and a more complex evolution of property rights and commodity relations which by the 1990s came to resemble a two-tier ownership system.[4] In 1992, 45 per cent of the total income of China's rural economy was generated by collective and co-operative organisations (Bowles and Dong 1994: 65).

These discussions of agrarian de-collectivisation, which in orthodox neo-liberal thinking rest on the presumption that private ownership and free prices are necessary preconditions for productivity enhancement, typically turn on four key processes:

1   *Property rights reform*: the clarification and reassignment of a panoply of ownership rights between a varied constituency of actors whether they are 'government jurisdictions, agencies, public or private corporations, households or individuals' (Walder 1994: 53).
2   *Privatisation*: a subset of property reform referring to the reallocation of rights from the public and state sphere to private firms and enter-prises, and to 'legal and administrative reforms . . . designed to provide

guarantees of the rights of firms against state manipulation or abrogation of such rights' (Walder 1994a: 53).

3 *Price and fiscal reform*: sweeping liberalisation of prices removed from virtually all state controls and fiscal reform to consolidate stabilisation, i.e. the reduction or removal of budgetary subsidies (the soft-budget constraint) and a restructuring of the public sector.

4 *Temporalities of liberalisation*: the distinction between *gradualism* (reform cautiously implemented and often without a co-ordinated overall plan), typically seen as a function of the ideological commitment of Communist Parties or of unbridgeable splits between conservative and pragmatic reformers (see Sachs and Woo 1993) and the '*big bang*' approach most closely associated with Eastern Europe, simultaneously involving radical price liberalisation, devaluation to unify the exchange market, and sharply tightened credit markets, i.e. radical stabilisation.

Each of the four broad characteristics is usually attached to a broader model of transition. In Burawoy's (1994) imaginative typology, models tend to privilege either the weight of the past (the pre-existing political order) or the virtues of capitalism ('capitalism by design' as Burawoy calls it). Hence the *totalitarian* approach emphasises the absolutist nature of the party-state and its precipitous collapse into liminality, confusion and anarchy, while the *society-centred* model points to the incompleteness of party control which leaves a residual civil society within socialism. In the two capitalist models, the *neo-classical* variety sees shock therapy as a necessary catalyst to smash the old order and create a space which is 'naturally' occupied by new market institutions, while the *evolutionary* model locates stable external institutions for dynamic capitalism and privileges careful, piecemeal social engineering to create the institutional preconditions of the market.

All of these analytics are relatively silent on the question of agrarian trajectories – that is to say the social relations of production and accumulation appearing in recently de-collectivised agriculture – which was the heart of Kautsky's concern. Both Kautsky (and later Lenin) highlighted the heterogeneity of ways in which capitalism takes hold of agriculture, and both pointed to the various paths by which such transitions might be affected. For example, the conversion of semi-feudal estates into capitalist enterprises (the so-called Junker path) contrasted with the more 'revolutionary' American road in which a commercially oriented and socially differentiated family farm sector (so-called simple commodity producers) provided the propulsive energy within a capitalist agriculture. Of course the recent departures from collective agriculture are hardly comparable to the intensification of commodity relations associated with the long arch of capitalist growth in nineteenth-century Europe. Rather, the starting point for the analysis of the agrarian question in formerly socialist states is the legacy of the revolutionary period, the abruptness of its collapse, and *the key role of the party-state* – for which there is obviously no historical parallel in the cases described by Kautsky and Lenin – in engineering the socialist transition to a market economy. Indeed, it is precisely the role of the party-state which strikes to the heart of both the debates over the likelihood of successful de-collectivisation, and the distinctiveness of the agrarian question in circumstances of socialist market transition. In this sense, the 'success' of the Chinese case and the 'failure' of Russia turn less on the relative merits of gradualism versus the radical break (shock therapy), as on the necessity and capacity of the party-state to shape the transition through the decentralisation of property rights, the promotion of competition and the construction of 'socialist markets' (Walder 1994a; Oi forthcoming; Shirk 1993; Burawoy 1994). It is against this backdrop of the particularities of socialist legacies and models of transition, the distinctive role of the party-state, and the classical agrarian question debates that the Vietnamese case can, in my view, be productively located.

Reinventing Vietnamese socialism

The state develops a commodity economy system composed of many sectors along a market economy model under the management of the state according to socialist principles. The multi-sector economy with its varied production and investment organizational structures is based on the principle of ownership by the entire people, collective ownership and private ownership.

(Article 15, 1992 *Constitution of the Socialist Republic of Vietnam*)

Perhaps the most compelling illustration of Vietnam's New Times is the recent removal of Ho Chi Minh's portrait from the roof of the State Bank in Hanoi for 'structural reasons', a loss which coincides with the disappearance of statues from parks and other public places because 'prostitutes and drug addicts were leaning on them' (Greenfield 1994: 203). Such changes in political symbolism are not of course the result of popular protest or crowds celebrating the collapse of communist regimes. Nevertheless, the refashioning of the culture of Vietnamese socialism marks a watershed of sorts '[in which] the ruling class is consolidating itself in a new social order' (Greenfield 1994: 203). Systematic undermining of a sclerotic Vietnamese system of bureaucratic centralism – a Soviet-style central planning system – was certainly assisted by the ascendance of Mikhail Gorbachev (and its prelude in the cessation of Soviet and Chinese aid between 1979 and 1980). But it was driven primarily by developments in Vietnam itself (Porter 1993: 151). In particular it was the onset of worsening budget deficits, high inflation and the crisis of productivity within agriculture and industry which prompted and sustained the reformist initiatives begun in the late 1970s.[5]

Revisionist scholarship on the pre-unification period in North Vietnam has contributed to a more refined understanding of the timing and genesis of the market transition in Vietnam. On the one hand, there is the question of the extent to which a Soviet–type political economy was ever *fully* instituted (Kerkvliet 1993, 1995); or put differently, what transpired in the name of collectivisation did not necessarily approximate the models of the central planners or indeed have the practical, moral or ideological consequences envisioned by the cadres and the *nomenklatura*. In the North, which underwent a relatively successful collectivisation after 1959, the period up to 1975 was as much a war economy as a mature bureaucratic centralism. Fforde (1989) and Beresford (1990) have further emphasised both the resistance to, and the uneven performance of, collectivisation in the North, while van Arkadie (1993: 439) is right to note that

> in contrast to the industrial sector where large scale industrial enterprise became a dominant mode (though with much less central planning control than the Soviet case) effective control was never established over rural production throughout the country, even before the reform process started.

On the other, the sporadic and uneven process of collectivisation in the South after 1975 proved to be an unmitigated disaster, a crisis reflected in a sharp drop in rice output, peasant foot-dragging and the mass slaughtering of buffaloes in the Mekong where the collectivisation drive was met with fierce resistance. While officials claimed that collectivisation goals had been achieved in the South by the early 1980s, in practice they had resigned them-selves to failure (Quang Truong 1987). It is precisely the incompleteness of and the resistance to the collectivisation project between 1960 and 1980 – pressure from below as Kerkvliet (1993) sees it – which undergirds the changes of the late 1970s and subsequently the more radical opening initiated by the *doi moi* reforms in 1986.

In the period between 1976 and 1980, the economy grew at barely 2 per cent and none of the state industrial or agricultural targets were achieved. The year 1978 proved to be a crisis year because production in the rice sector fell sharply – compelling 1.4 million tons of grain imports – while simul-taneously the government inflicted a 'procurement crisis' on itself (White 1985). In August 1979 the Sixth Plenum decided to soften the system of directive planning through the contract system in agriculture (the devolution of production decisions and some land rights to peasant households)[6] and the so-called 'fence-breaking system' among state-owned enterprises (i.e. the allocation of self-accounting and self-financing autonomy reminiscent of the Soviet *khozrastchet* form of enterprise management, Andreff 1993: 518).[7] These reforms were, in form and timing, comparable in many respects to the post-Mao new responsibility system in China but they failed to stabilise the economy except in agriculture. Economic growth accelerated after 1981, but the economy was marred by recession and massive inflation in 1985. Once more it was a grave fiscal crisis that precipitated the second, and more radical, phase of reform initiated by the Eighth Plenum in June 1985, what was to become the *doi moi* (renovation) reforms.

It was in December 1986 that the Sixth Party Congress formally launched *doi moi*, marking in many respects a more profound shift from a 'subsidised centrally administered state economy' to what was referred to in party-speak as a 'multi-sector economy'. Led by reformists such as Nguyen Van Linh with substantial experience in the South, the reformist wing of the party gained further credibility from the nascent reforms of Gorbachev. However, fiscal and banking reform, the reduction of consumer price subsidies and the recognition and guarantee of private industry (including the lifting of all limitations on its ability to hire labour) in 1988 all proved to be inadequate to the task. Monetary reform squeezed the liquidity of state-run enterprises which could only be sustained by large government deficit spending. Hyperinflation followed and in 1989 Vietnam adopted more radical *doi moi* surgery:

> During 1985–1988, Vietnam implemented a gradual reform strategy that failed to address serious macroeconomic imbalances. The program failed: inflation accelerated while growth and trade performance remained unchanged. . . . In 1989 Vietnam enacted a Eastern-Europe style 'big bang' including price liberalization, a 450 per cent devaluation to unify the exchange market and sharply tightened credit policy. The collective farms were returned to family farms with long term leases. Growth accelerated, inflation ended, agricultural productivity soared and small, non-state enterprises proliferated.
>
> (Sachs and Woo 1993: 2)

Whether one agrees with Sachs and Woo's diagnosis, it is incontestable that the growth performance in Vietnam since 1989 has been quite impressive. Initially the surge in output – at a time when Vietnam was still receiving substantial assistance from the Soviet Union – was substantial, GDP grew at 8 per cent per year and the performance of all sectors has been solid. Agricultural output grew sharply and rice production – already stimulated by the 1981 reforms – was further enhanced by price liberalisation and property rights reforms (Table 20.1). Average growth rates in agriculture (value-added) in 1984–8, and 1988–92 were 2.5 per cent and 3.9 per cent, comparable to Thailand, Taiwan and the Punjab (Mellor 1993: 16–17). The growth of rural incomes – on an admittedly low base – was sufficient to have spill-over effects in construction and services; average per capita income of rural households increased by almost 30 per cent between

Table 20.1 Vietnam: food production, 1976–95

	1976	1980	1984	1988	1989	1990	1991	1992	1994	1995
Food production paddy equivalent (million tons)[1]	13.5	14.4	17.8	19.6	21.4	22.0	23.0	21.0	26.4	27.4
Food production per capita (kg)	274.4	265.2	303.2	307.3	332.9	332.3	339.5	303.5	322.2	369.0
Percentage of requirement (100% = 365 kg)	75	73	83	82	91	91	93	83	87	100
Population (millions)	49.2	53.7	58.7	63.7	64.4	66.2	67.7	69.2	72.5	75.0

Sources: UNDP 1990 and The Rice Paper 1996.
Note
1 Food production is in 'paddy-equivalent terms'. For practical purposes, 90 per cent consists of paddy itself. ADB source used for 1990–2 gives paddy production only; topping up was done based on coefficients for previous years.

1989 and 1992 (Tuan 1995: 142) while the proportion of households under the poverty line fell from 25.3 to 12.1 per cent over the same period according to Vietnamese government statistics. In spite of the collapse of the Soviet Union and the reduction in foreign assistance, the recession of the early 1990s proved to be mild – GDP grew by 5.1 per cent in 1990 and 6.0 per cent in 1992 – and by 1992 the growth rate was in excess of 8 per cent. By this time, industry had come to play a leading role (World Bank 1993). Since 1989 exports have been growing at an average annual rate of over 30 per cent, but in 1992 rice production also hit a historic high. GDP growth in 1993 was 8 per cent, and exports boomed by 20 per cent. In 1994 and 1995 both of these figures were exceeded. By 1993, the private sector accounted for two-thirds of GDP. In sum, the post-1989 *doi moi* reforms – however one assesses their 'shock' quality – had placed Vietnam ahead of most reforming economies in Eastern Europe and Russia. Vietnam remains, of course, a poor country: 70 million people, a population density (900 persons per square km) higher than China, a land per capita ratio similar to Bangladesh, and a GNP per capita of less than US$200.00 per year. But the robustness of the reform period is impressive, not least because it was achieved during a trade embargo and without substantial foreign assistance.

The role of the party during *doi moi* has been complex and contradictory. On the one hand the economy was under less central government control than, say, the former Soviet Union and to this degree has been capable of greater political flexibility. On the other, economic liberalisation was both a cause and effect of political opening, and underwrote the gradual emergence of what has been called a 'nascent' civil society outside the control of the party-state (see Thayer 1992).[8] As Fforde and Porter (1994: 16–17) put it, the party is simply no longer sure of what the correct ideological or institutional line should be:

> Assertions of the absolute superiority of the socialist sectors . . . have been replaced by the use of a language of constraint . . . the 1992 Constitution asserts the right of all Vietnamese citizens to carry out legal business; the June 1993 Plenum . . . asserts the positive role to be played by farmers' own organizations; workers are given the right to strike even if they do not belong to an approved trade union.

In addition to the important, if occasionally ambiguous, changes inscribed in the 1992 Constitution, the Vietnamese government unashamedly claimed in its 1993 presentation to the donors' conference in Hanoi that a 'bureaucratic centralised mechanism based on subsidies is not appropriate' and that 'a market oriented economy is best'. Of course, in a one-party communist state, the CPV has been singularly concerned with mass organisations under the umbrella of the Vietnam Fatherland Front and their compatibility with the prevailing vision of a normative socialist order, a vision largely incompatible with a fully developed sense of civil society.[9] But it is also true that people no longer organise their lives *solely* with respect to the party and mass organisations, and to this extent Pike (1994: 78) is surely correct in concluding that 'the top leadership is committed – or resigned – to moving Vietnam from a hard to a soft authoritarian system, more or less following the pattern of South Korea and Taiwan'. The growth of new farmer associations and networks, the reform debates within the party, or the emergence of the so-called umbrella system in which new networks of political and economic power emerged outside the party-state during the 1980s, should not imply an ineluctable political decentralisation or the attenuation of central power. Indeed, the party central committee has responded to the obvious proliferation of state corruption and inefficiency in the reform period with a fierce internal critique (Goodman 1995), and new efforts at centralisation and 'strengthening of management by the state' (Reuters, 11 November 1995). In late 1995 the National Assembly centralised a number of state agencies into three 'mega ministries' which included two new state enterprises which virtually monopolise rice production, distribution and export. At the 1996 Plenary Congress of the party, the old guard built upon fears of corruption, social degeneration and loss of sovereignty to reaffirm the centrality of the state in the management of the Vietnamese political economy.

The agrarian question and rural de-collectivisation in Vietnam, 1975–95: a tale of two deltas

Currently, 2.1 million hectares are under controlled (flood) irrigation in Vietnam, roughly two-thirds of which are located in the two vast deltaic areas in the Mekong and Red River systems. Average cropping intensity across these low-lying alluvial lands is high – typically 220 per cent (i.e. 2.2 crops per year per plot) – but per capita cultivated area tends to be extraordinarily small, approximately 0.25 hectares per household in the Red River and 1 hectares per household in the Mekong. Thirty-seven per cent of Vietnam's flood irrigated lands lie within the Red River delta and 27 per cent in the Mekong delta. These two vast, but quite different, irrigation systems account for almost half of the population of Vietnam and the majority of domestic grain production. But insofar as they stand at the epicentre of Vietnam's agrarian transition, they also encapsulate in their differences the very deep regional bifurcation and contrasting geographical trajectories which have marked the last two centuries of Vietnamese history (Tables 20.2 and 20.3).

Prior to the unification of Vietnam in 1975, the division between the North and the South of the country corresponded to regionally distinctive patterns of agrarian political economy that had been fundamentally shaped by French colonialism. Land in the northern Tonkin delta was the most minutely divided in all of Indochina. According to colonial data collected in

Table 20.2 The evolution of agrarian systems in the Mekong and Red River deltas, 1930–90

	Rural population (millions)	Arable land & permanent crops (thousand ha)	Land density (m²/cap)	Food crops production (thousands)	Food yield (t/ha)	Food per capita (kg)
Red River delta						
1930	6.5	1.2	1846	1.8	1.5	277
1990	11.9	0.8	689	4.9	5.9	411
Growth rate (%)	1.0	−0.6	−1.6	1.7	2.3	0.6
Mekong delta						
1930	3.2	2.0	6250	2.6	1.8	812
1990	11.8	2.3	1949	9.6	4.2	816
Growth rate (%)	2.2	0.2	−2.0	2.2	2.0	−0.01

Source: Tuan 1994; 1997.

Table 20.3 Peasant living standards and social equity in the Red River and Mekong deltas, 1930–90

	Year	Average net income (kg paddy/cap)[1]	Gini coefficient
Red River delta	1930	584	0.43
	1945	370	0.59
	1954	501	0.35
	1957	568	0.07
	1965	596	0.15
	1970	570	0.26
	1978	680	0.25
	1990	692	0.25
Mekong delta	1930	782	0.87[2]
	1955	600	0.84[2]
	1966	866	0.80[2]
	1972	863	0.55[2]
	1981	1009	0.30
	1990	1259	0.35

Source: Tuan 1994, 1997.
Notes
1 The paddy price in the Mekong delta is about 80% of that of the Red River delta.
2 Gini coefficient of land ownership.

the 1930s, 90 per cent of all land-holders held less than 1.8 hectares, and an estimated 36 per cent of rural households were landless. Colonial rule had certainly stimulated the process of social differentiation – Murray (1980: 414) refers to increasing contrasts in landed wealth and the emergence of huge land-holdings – but in the densely populated Red River delta and the coastal lowlands (unlike Cochinchina), agrarian class structure remained dominated by fragmented holdings, small-scale petty commodity production and households increasingly compelled to sell wage labour in order to survive. Colonial policies deliberately attempted to preserve some aspects of the traditional moral economy of rural communities as 'one means of ensuring the survival of the migrant labor system' (Murray 1980: 8) which was central to the agro-mineral concessions in the South. Ironically, it was precisely the pre-capitalist traditions of the North – most especially collective worship of tutelary deities, village endogamy and the institution of communal land[10] – which nurtured values 'conducive to the growth of nationalism . . . collectivism . . . and to an hierarchical organizational framework' (Luong 1992a: 228–9).

Conversely, in Cochinchina in the South, and in the Mekong delta in particular, the process of frontier occupation from the fifteenth century had produced a quite different settlement pattern and an average land-holding substantially in excess of the North's. French conquest intensified the rate at which rice lands were brought under cultivation in the Mekong – two-thirds of Cochinchina's rice lands were opened as newly reclaimed lands after 1880 – but more critically it was the presence of metropolitan capital in conjunction with a colonial administration anxious to promote export-oriented rice production which facilitated the concentration of rice lands under the control of large landlords. Land-holding concentration, precipitated by credit policy to landlords and a draconian colonial tax policy, was accompanied by exploitative forms of tenancy and serious pauperisation through land dispossession. By the 1930s Pierre Gourou estimated that 2.5 per cent of the rural households owned 45 per cent of the total rice area under cultivation and presided over thousands of sharecroppers and tenants.[11] Twenty years later, 79 per cent of the rural households in Cochinchina were landless. The trajectory of agrarian change in Cochinchina, in other words, turned on the role of merchant capital linking export-oriented rice-growing regions to the world market through landlords relying upon 'labor squeezing practices' (Murray 1980: 449), processes which were ultimately hemmed in by the closing of the land frontier (by the 1930s) and the definite limits on the quantity of rent which landlords could extract.

This pattern of uneven regional development – the Red River (Tonkin) approximating a sort of 'American road' involving the evisceration of owner-occupancy and the growth of petty commodity production within the context of a long-standing and robust pre-capitalist tradition, and the Mekong (Cochinchina) representing an 'English road' through the rapid dissolution of non-capitalist relations, the emergence of an export-oriented landlord class reproduced through tenancy and sharecropping relations, and the creation of a large rural proletariat – provides a legacy of enormous historical and contemporary significance. It proved to be relevant not only to the success of the revolutionary communist movement in the North but also, as we shall see, to the character of post-socialist reforms during the 1980s and 1990s.

When the Democratic Republic of Vietnam (DRV) came to power in 1945, land-poor and landless peasants represented roughly 60 per cent of the Tonkin population and controlled 10 per cent of the land (conversely, 2.5 per cent of wealthy rural households controlled 24 per cent of the cultivated area). This rural social structure was, of course, shattered by the land reform and subsequent collectivisation implemented after 1954 by the DRV. Households were classified in terms of land-holding size and the degree of exploitation (usury, rental and so on), and landlords were required to reimburse sharecroppers for the difference between actual and government-determined rents (Luong 1992a: 188; Kerkvliet 1993). Subsequently, rich and 'brutal' landlords were publicly denounced by local peasant associations and land was confiscated, appropriated, coercively purchased and redistributed to landless, poor and middle peasants. If Luong's

(1992a: 192) reconstruction of a community in Vinh Phu Province is at all illustrative, during the 1954 reform average land-holding of landless, poor and middle peasants increased by 1400 per cent, 250 per cent and 45 per cent respectively.[12] In 1958, the DRV launched the co-operative programme aimed at resolving the peasant differentiation problem and in a relatively short period of time the collectivism principle gained ascendancy through control of non-land assets, credit and output markets. Labour deployment was, at face value, quite simple; co-operatives were divided into work brigades which performed all production tasks (Long 1993: 166) and work-points (based on duration not quality or skill) were allocated. At harvest after the deductions of taxes, production costs and welfare expenses, members received shares in proportion to work-points. By 1960 there were 414,000 co-operatives (hop tac xa) in the North containing 2.4 million families accounting for 85 per cent of the total farm population and 76 per cent of the cultivated area.[13] In hindsight, this early period in fact proved to be, in spite of the technical problems of assessing labour performance, the golden age of collectivisation: the cultivated area increased by almost 50 per cent and worker productivity doubled (Tong Cuc Thong Ke, cited in Long 1993: 167; Kerkvliet 1993). Subsequently, the US invasion compelled the unification of hamlet co-operatives into 'higher level' multi-village co-operatives, an aggregation which demanded new divisions of labour in which specialised units (based on a system of three contracts[14] (ba khoan)) existed side by side with the older basic production brigades. By 1975 the average size of the North Vietnamese co-operative was roughly 100 acres and contained roughly 220 families (Lam 1993: 153). However, efforts to capture economies of scale through aggregation resolved few of the seemingly intractable problems associated with loose work-point standards, conflicts over payment in kind versus cash, and the fact that brigades rather than individuals were tied to final products, all of which contributed to a dramatic decrease in production after 1966 (see Fforde 1989; Kerkvliet 1993). By the late 1960s and early 1970s living standards had stagnated, communal property had deteriorated, administrative burdens within the communes became ever more onerous, and individual production (linked in some cases to deliberate experimentations and refashionings of the co-operative system) accounted for larger proportions of co-operative land and household income (perhaps 13 per cent and 60 per cent respectively (Quang Truong 1987: 91; Kerkvliet 1993: 11)).

Agrarian structure in the South stands in sharp contrast to Tonkin. During the Diem period, 40 per cent of the land in the south-central region was rented out by landlords, and in the Mekong 6,300 landlords (0.25 per cent of the rural population) owned 45 per cent of the rice land; 6,000,000 tenants cultivated two-thirds of the land (Porter 1993: 59). Southern agrarian structure was radically reshaped after unification in 1975, but in the wake of the defeat of the French at Dien Bien Phu, the Delta region was restructured by two agrarian land laws in 1956 and 1970. Neither of these reforms can be understood outside the circumference of the DRV agrarian reform in the North and the growing presence of Viet Cong forces throughout the South. The first, Government Ordinance No. 57, limited paddy ownership to a maximum of 100 hectares (the surplus being sold to the government and redistributed to former tenants, who could acquire up to 5 hectares), and a cadastral survey limited rents to 25 per cent of the expected crop to be paid in kind (see Hickey 1964: 46). In the case of Khanh Hau village, studied by Hendrey (1964: 33–9), the impact of the reform was to increase the proportion of households holding less than 2 hectares from 46.2 per cent to 62.1 per cent and to reduce the number of the 2-plus hectare land-holders from 53.8 per cent to 37.9 per cent. The Saigon government also carried out a far-reaching land reform in 1970, during which 1 million hectares of rice land was distributed to 858,000 households over a two-year period (Luong 1992b: 115). In Khanh Hau village, for example, a landowner was limited to 5 hectares of patrimony fields and 15 hectares of rice land if already under direct cultivation (Luong 1994: 89). After the reform (and prior to the redistribution of land in 1977 following unification),[15] 6.3 per cent of households were classified as landlord/rich, 63 per cent as middle peasants and 19.8 per cent were poor or landless (the corresponding figures for Long An Province as a whole were 4 per cent, 63.5 per cent and 31.8 per cent respectively). While land inequalities remained more pronounced than in the North (prior to 1954), the two southern reforms had nonetheless reshaped the agrarian structure of the Mekong, reducing the landless class by half and substantially reducing the concentration of land in the hands of very large landowners.[16] During the Vietnam conflict the Delta was transformed by war, US intervention and by the introduction of new rice technologies. According to Porter (1993: 60), 'wartime labor shortages, loss of buffaloes, reduced crop areas and the introduction of new rice varieties and cheap [food aid] . . . caused . . . South Vietnamese farmers to become dependent on expensive agricultural inputs and machinery to grow rice.' On the eve of unification, agrarian differentiation in the South seemed to resemble Lenin's description of capitalism in late nineteenth-century Russia, resting less on land – the reforms having undercut the colonial pattern – than on the ownership of farm machinery and on commercial and mercantile capital.

In theory the distinctive regional agrarian structures were to be finally obliterated in 1975 with the historic victory of the DRV, which brought within a collectivised northern economy suffering from land scarcity, a fragile local ecology and declining productivity, the vast agricultural potential of the Mekong River delta. In practice the picture is much more complex. In the North, the collectivisation process had been far from smooth and took place in the context of the Indochina War and US bombing and in the absence of a systematic effort to industrialise (Selden 1993: 217). Luong's account of an area in which the party had been active and could pull upon substantial poor peasant sympathy reveals how co-operative formation proceeded 'fairly smoothly', yet even here 15 per cent of households withdrew at various points in the process (1994: 86). Fforde, Vickerman, Beresford and others have argued that collectivisation was never fully implemented[17] and have highlighted its economic failures, encapsulated in rice yields that did not reach the historic high of 1958 until the early 1970s. Per capita grain production has only recently reached the 1959 figure in the Red River delta; indeed, North Vietnam remained import-dependent for rice throughout the war years. In the context of war and popular mobilisation, and the unquestionable energies unleashed by collectivisation, some of these failures should be carefully located in relation to the stringencies and rigidities imposed by the war economy. But it is also clear that by the early 1970s the collective system was in crisis:[18] 'The division of labor in Vietnamese cooperatives was, therefore an artificial one which, combined with the workpoints system, actually helped to stifle productivity increases. It encouraged peasants to concentrate . . . on subsistence . . . [which was] reinforced by macroeconomic policies' (Beresford 1990: 479–80). Matters were compounded after 1974 as efforts to consolidate and expand the co-operatives – the average acreage and numbers of households at least doubled between 1975 and 1980[19] – were implemented with devastating consequences. As wartime rhetoric lost its appeal after 1975 and a new system of management introduced between 1974 and 1978 made matters worse by re-centralising management under state control, more and more households turned to their own '5 per cent plots',[20] which contributed to the steady decline in yield per labour input. A study in 1977 of 21 district co-operatives revealed that families took in between 178 and 323 kilos of paddy per crop *less* than in the 1970–4 period. Another survey in 1979 in the Red River delta showed that increasing co-operative size was correlated with lower productivity and lower income (Long 1993: 171–2).[21] In the North rice output fell by 2.1 per cent per annum (Beresford 1990: 470), which in conjunction with low state procurement prices, high taxes, and efforts by peasants to evade taxation through falsifying yields, manufactured a crisis of sufficient gravity to push the party towards the reforms of 1981.[22] In practice, however, the system had been reconfigured from within (by farmers) and from without (by some local cadres) by the late 1960s. The rise of so-called 'sneaky contracts' (*khoan chui*) in which households were contracted for particular tasks for a share of earnings suggests, in a sense, that the first phase of Vietnamese collectivisation was in its death throes before the fall of Saigon.

The situation in the South after unification proved to be quite different. In November 1975, the DRV agreed upon a gradualist approach to collectivisation in all sectors of the southern economy. In agriculture peasants were to join labour-exchange teams and low-level co-operatives which would be subsequently upgraded to higher level collectives. In its Second Plenum in

1977, the Communist Party hastened the pace of collectivisation and by 1978 rice land was, in theory, fully collectivised (in Khanh Hau, for example, thirty-five solidarity teams were converted into seventeen co-operative teams during 1978 (Luong 1994: 89)). Unlike the northern collectivisation drive during the 1950s, however, there were no landlord denunciations, arrests or executions, and villagers were permitted to *retain* non-land means of production that could be contracted to co-operative teams (which had become during the war years *the* primary form of social and economic differentiation within the peasantry). But the effort to integrate a highly marketised and relatively mechanised free peasantry into a collective division of labour proved to be impossible. Peasants, particularly in the Mekong and the south-east, proved to be ornery and recalcitrant; farmers refused to engage in collective labour and devoted increasing time and energy to private gardens and fishponds. Rice yields in the South fell by 25 per cent between 1976 and 1980, and rice yields in the Mekong languished at 2.25 tons per hectare in spite of huge increases in fertiliser availability (Tran 1991). Real agricultural output per capita and output per investment dong fell by 1.5 per cent and 5.7 per cent per annum between 1976 and 1979.[23]

The proportion of families in co-operatives in the South stood at 25 per cent in the late 1970s (compared to 65 per cent for the entire country) and by 1980 at least 8,514 of the 12,246 co-operative teams in the southern third of Vietnam had collapsed altogether (Lam 1993: 115). A compulsory collectivisation drive began in 1981 but there was no real increase in peasant families joining between 1981 and 1983. The proportion of peasant families in co-operatives in 1985 was 3.8 per cent in the Mekong (a rise from 2 per cent for the previous year) compared to 28.5 per cent for the South as a whole (Tran 1991: 12–13). A survey of eighty rural areas in southern Vietnam in 1981 revealed nothing like a collectivised agriculture but an agrarian structure not radically dissimilar from the post-1970 reform period. One-quarter of all households were effectively landless, 56 per cent were middle peasants who owned 60 per cent of the land, 12 per cent were upper middle peasants controlling 27 per cent of the land who regularly rented out equipment and hired in labour, and 2 per cent of the households were rural capitalists(!) who owned 7 per cent of the land and more than half of all agricultural machinery and livestock (Tran Quoc Khai cited in Porter 1993: 60). While collectivisation was in other words in crisis in *general*, the state confronted two very different agrarian universes (see the data on population density, land availability and gini coefficients in Tables 20.2 and 20.3).

Driven by a domestic production crisis and a two-front war with Kampuchea and China (to say nothing of the US trade embargo), the Vietnamese government was compelled to make concessions to agricultural producers which, like the new responsibility system in China, took the form of a household contract system (Selden 1993; Lam 1993). Anxious to improve collective performance, the Sixth Plenum in August 1979

(Resolution No. 6) ordered that distribution according to labour replace distribution by quota. While the reform was incapable of resolving all problems – particularly those associated with subsidies for veterans – it initiated, or perhaps more properly provided a space for, local experimentation which generated the so-called new contract that was formally ratified in 1981. Co-operative officials still assigned tasks to collective units, families and so on, labour arrangements remained essentially the same, and the power of the cadres went unchallenged. What the contract system conferred, however, was direct payments to households – in effect groups of labourers – on the basis of specific yields on specific plots of land contracted to them (Long 1993: 174). In other words, households, through two- to five-year contracts which stabilised the production quota, took responsibility for key production tasks (transplanting, tending, harvesting) while the co-operative maintained its role in ploughing and irrigation. Collective land was allocated to households on the basis of their adult workforce and each family earned work-points (as before), but now households could retain the entire surplus of the crop above the co-operative quota while being directly responsible for all deficits. The contracts placed compulsory quotas on households but did not either provide for security of land tenure or make provisions for private marketing. In essence, the reformed collectives were slowly *producing sharecroppers*, the institution that had drawn so much criticism in Ho Chi Minh's attack on landlording and kulak exploitation during the 1950s. This innovation replaced the system by which peasants were allocated to specialised teams and accumulated work-points which entitled them to a harvest share less taxes, obligations and welfare contributions. As a consequence, output increased sharply (from 11.6 million tons in 1975 to 18.4 million tons in 1986); Luong's data from Vinh Phu show that rice output increased by 20 per cent over three years, incomes doubled, and rice yields reached 4.2 tons per hectare (1992a: 208–9).

In the Mekong delta, where only 5.9 per cent of rural households were members of co-operatives in 1988 (see Table 20.4),[24] contracting was not implemented until 1983 and had a rather different impact. Indeed, the reforms of 1981 – and the more radical reforms in 1988 known as Resolution No. 10 in which the full contract (*khoan trang*) leased paddy-fields to households without any responsibility to the co-operative – were obviously directed mainly to the collectivised northern and central provinces. Local and provincial authorities in the South had already confronted peasant opposition to collectivisation, and political demonstrations had been made in favour of land restitution. In Khanh Hau in An Long Province, for example, within several years of the reforms the provincial government returned to the former owners of rice land most of what they had directly cultivated during the mid-1970s prior to collectivisation (Luong 1994: 102). Former landlords were restituted (often to the tune of 9 hectares) and as a consequence about 400 village households became landless though each was offered about 1.5

Table 20.4 The collective agricultural production system in 1986

Region	No. of provinces	No. of co-operatives	No. of co-operative farmers (thousands)	% of co-operative farmers to total farmers	Average size of the co-operative (ha)	Average no. of farmers per co-operative
Northern provinces						
Northern region	10	7,724	1,212.6	92.8	93	157
Red River delta	6	2,768	2,216.8	99.4	250	800
Former fourth	3	3,673	1,284.6	98.0	155	345
Southern provinces						
Central coastal	4	1,284	778.0	89.5	300	606
Western plateau	3	509	140.6	51.8	180	276
South-eastern region	5	411	95.5	20.0	152	232
Mekong delta	9	374	117.1	5.9	235	313

Source: *Statistical Yearbook of Vietnam* (Hanoi, various issues).

hectares of land in a recently reclaimed region call the Plain of Reeds.[25] In short, southern communities saw the rapid *dissolution* of a compromised and weakened collective system and the emergence of an aggressively commercialised peasantry which in some cases reconstituted pre-revolutionary land-holding patterns (i.e. the 're-embourgeoisement' described by Ivan Szelenyi (1988) for Hungary but in this case re-emerging after less than a decade!). Insofar as the 1970 reforms represented the baseline for any restitution, the privatisation of land, which had proceeded much more quickly in the South[26] because it was never fully collectivised, had not (at least by the early 1980s) produced a pattern of marked land-holding inequality replicating the late colonial period. In the Mekong it appears that about 90 per cent of households held less than 1 hectare and rarely did more than 1 per cent of peasants hold in excess of 2 hectares (Luong 1994: 117).

If the 1981 reforms produced short-term output increases in the North, they were, as Long (1993: 176) properly notes, 'one time effects' and as a consequence incentive problems returned to haunt the co-operative system. Furthermore, one of the major innovations – the ability to retain surpluses over and above quota – was undercut by the demands of local cadres. Porter's comment (1993: 135) that local exactions often accounted for a staggering 60–70 per cent of total grain (85 per cent in some northern villages) explains why surplus retention was so trivial (rarely more than 20 per cent), why 60 per cent of co-operative members' total income derived from the 5 per cent plots, and not least why a survey of a co-operative in Thai Binh Province in 1985 revealed that almost two-thirds of households suffered from severe food shortage (Long 1993: 176).

The reformist initiatives of the early 1980s were, not surprisingly, deepened and extended in 1988 with Resolution No. 10, which, like the Chinese contracting system, offered fifteen-year contracts (named 'contract 10' or *khoan muoi*) to all households with target yields and tax obligations fixed for the first five years (Selden 1993: 239).[27] The share of the crop appropriated by households accordingly doubled (from 20 per cent in 1981 to 40 per cent in 1988) since contractors who exceeded the contracted yield could retain the *entire* surplus. In addition, the long-term leases on land also nominally included inheritance and transfer rights, for the first time some collective assets such as buffaloes were sold to co-op members, and private marketing was officially sanctioned.[28] Resolution No. 10 maintained state ownership of land but private use rights granted by the state were also recognised (rights which could be recorded on a cadastral survey and evidenced by a certificate of use). Land rights conferred to households – which at a stroke substantially undermined the existing co-operatives and in effect *created* a full-blown peasantry – were not leases from the state but were rights in property subject to a variety of controls, regulations and taxes. These rights could not be fully transferred or alienated, and hence not pledged as collateral (World Bank 1993: 27). By 1992, however, 6 out of 7 million hectares of cropland were farmed under direct household use rights (either leased from state farms, allocated by collectives or dissolved state farms, or in the case of the South as restituted land claims). While the 1988 law did not permit formal transfers or inheritance, the growing commercialisation of a peasant economy underwrote a number of *de facto* sales and rentals. In order to regularise (and tax) these transactions and to further the marketisation process, a major revision of the 1988 law was introduced in 1992, and formalised in July 1993 as the Land Law. While land 'belongs to the people', land use rights can now be leased, sold or transferred and inherited and mortgaged.[29] Land allocation was granted for twenty years for annual crops and fifty years for perennial croplands.[30]

In sum, what began in the late 1970s and early 1980s as an effort to reform but bolster and partially re-collectivise a sick co-operative system had, fifteen years later, converted Vietnam into a nation of family farms linked through kin, long-term usufructory rights over land, and the market. The number of various forms of collectives fell from 26,073 in 1989 to 7,413 in 1992, and those that remained were Ptemkin villages,[31] so circumscribed in form and function that they did not warrant the term co-operative. A survey in 1992, prior to the 1993 Land Law, estimated that 80–90 per cent of rural income in Vietnam was derived from the 'household economy' (Ljunggren 1994: 19). In the same year, labour and capital had become increasingly commoditised, and it was commonplace throughout the country for farmers to be able to quote daily and piece wage rates for a panoply of productive tasks and services.[32] A survey in Ha Bach Province in 1993 revealed that only 1.7 per cent of peasants sold produce to the state, the remainder was fully privatised; in the same survey, 26 per cent of local households hired workers and 68 per cent of households polled said they had worked as hired labourers (Huong 1994: 71).[33] The contraction of the co-operative sphere was, if not complete, very close to it. In many cases the role of the co-operative institutions had simply evaporated and former co-op buildings crumbled under the weight of their obsolescence.[34] Between 1988 and 1991, the number of low-level co-operatives shrank by 60 per cent and the number of cadres by half. A 1990 survey established that 5 per cent had successfully shifted to new service functions but the remainder were 'confused and disorganised' (EIU 1994: 42). In the South, 80 per cent of the low-level co-operatives had ceased to function (EIU 1994: 42).

The overall impacts of the reforms on both agrarian output and rural income seem unequivocally positive. A 26 per cent increase in paddy production between 1987 and 1989 converted Vietnam from a 750,000 ton grain deficit in 1986 to the world's third largest exporter in 1992. Estimates of regional food balances reveal that exports have been driven by huge marketable surpluses from the Mekong delta and modest surpluses (except in 1991) from the Red River. Four regions – the northern mountains, north central/central coastal, central highlands and the 'north-east' of the South –

have had persistent and non-decreasing deficits, and the North is probably (for the first time in several decades) self-sufficient in staple foods (World Bank 1993: 127). The exact income effects are not readily discerned but suggest marked regional differences. In 1989 the average national net monthly income per capita (constant dong) was 21,428. The corresponding figures for the regions are: Northern Highlands 18,789, Red River 19,203, Central Highlands 21,104, and Mekong 27,285 (Long 1993: 185). These differences are clearly rooted in land availability and the size of land-holdings.[35] Compared to the Red River delta household, the Mekong delta household owned over four times more cultivated land and six times more paddy land, growing two crops per year (Long 1993: 184). A 1990 Survey in the Red River delta determined that an average household held less than 0.1 hectares of land consisting of seven to seventeen highly fragmented and geographically dispersed plots; in contrast Mekong holdings were 1.5 hectares and typically unfragmented. The market opportunities in agriculture accordingly conferred substantial advantages upon Mekong peasants who owned larger areas, possessed on average greater quantities of domestic adult labour (3.4 persons versus 2.1 in the Red River), were less bound by the rigidities of collectivisation, and not least who saw their per capita income at least double between 1989 and 1990 (see Tuan 1995, 1994).[36] If the collectivist period retarded in some respects the different regional dynamics which have for so long marked the Vietnamese agrarian sphere, the period of de-collectivisation has seemingly unleashed new forces of polarisation within and between the two deltas.

'Tiger on a bicycle': land, labour and rural accumulation along the Red River (Vinh Lac District, Vinh Phu Province in the 1990s)[37]

The Red River delta contains the oldest man-made irrigation and drainage in Vietnam and currently over thirty main systems serve about 600,000 hectares, roughly 78 per cent of the total irrigated area. While the Red River has long been diked to protect paddy fields from the annual floods, inundation still persists due to local rainfall and runoff from the higher lands whose drainage is prevented by the levees. The pumping systems which are required to facilitate drainage during the monsoons are employed for water distribution during the dry season. About 1,700 power plants with 7,600 electric pumps feed a system of primary and secondary canals from which water flows by gravity to tertiary systems. Criss-crossed by three major rivers and tributaries and a maze of dikes and canals, the Red River is one of the world's great rice-growing regions. When Pierre Gourou (1936/1955) completed his famous study of land use in the Tonkin (Red River) delta over fifty years ago, the population stood at 6.5 million; today the figure is close to 20 million.

As Gourou made clear, the Delta is a spatially restricted but morphologically quite varied region. Vinh Phu Province (4,824 sq. km), the research site for the fieldwork conducted in 1994, was included in Gourou's discussion of Tonkin because it is cross-cut by the Red River (and its tributaries the Lo and Thao), even though the province is often seen as a part of the Vietnamese Midlands (Cuc *et al.* 1990: 45), a region distinguished by low rounded hills and narrow river valleys. Properly speaking, Vinh Phu is a *transition zone* ecologically between the delta proper and the highlands to the north and west. It forms an arc around the Red River delta, but Gourou rightly included some of the alluvial villages of what is now Vinh Phu Province in the northern periphery of his Tonkin region. Vinh Lac District, in which several village studies were conducted, is located some 15 km to the south-east of the provincial capital of Viet Tri and occupies an area of 250 sq. km immediately adjacent to the Red River (Figure 20.2). Table 20.5 provides a schematic socio-economic and demographic profile of both province and district in relation to Vietnam as a whole. While the average cultivated land devoted to staples is almost identical across all three units (national, provincial and district), the percentage of land cultivated is three times higher in Vinh Lac than the national average and twice as high as Vinh Phu Province. The percentage of cultivated land devoted to rice and the population density are also substantially higher in Vinh Lac, confirming Gourou's picture of the district as a modal Red River rice economy. What is particularly striking, however, is my calculation of the nutritional population density (i.e., population per sq. km of *cultivated* area); in 1993, the Vinh Lac figure is close to 2,000 per sq. km, roughly *four times* the figure that Pierre Gourou found so impressive and disconcerting in the 1930s, and substantially higher than China.

Vinh Phu Province – formed in early 1968 from the unification of Phu Tho and Vinh Phu Provinces[38] and in 1997 returned to its pre-1968 status – provides the political and administrative structure for Vinh Lac District. The province was a Vietminh stronghold but has an unusual political history with respect to the experience of collectivised agriculture. During the US bombing between 1965 and 1968, local cadres and People's Committees experimented with decentralised production contracts to inter-household 'task' brigades. Party Secretary of Vinh Phu Province Kim Ngoc was in fact forced to resign because of his de-collectivisation innovations during the late 1960s (Fforde and Porter 1994: 23). By 1977 a number of Vinh Phu co-operatives had further experimented with the household contract system that was adopted finally by the Sixth Plenum in 1979. Yet party delegates from the province – with some backing from rural dwellers – were some of the most vociferous opponents of the dismantling of the co-operative structure and the privatisation of land ownership (Long 1993: 192). Whatever the claims of the local cadres and party members, they were unable – and perhaps in the final analysis unwilling – to halt the rising tide of

Table 20.5 Vinh Lac District in provincial and national context, 1993

Variable	Vietnam	Vinh Phu Province	Vinh Lac District
Land area (km²)	331,040	4,824	250.66
Population (millions) 1989	69.405	2.236	0.316[1]
Agricultural pop. (mill.)	49.574	1.897	0.30 (est.)
No. of households	15 million (est.)	405,300	65,366
No. of farming households	10,281,000	65,366	65,366
Population density/sq. km	209	463.5	1,259.5
Nutritional population density			
(pop. per km² of cultivated area)	1,068	1,354	1,848
Population growth 1979–89	2.1	2.85	2.0 (est.)
% population female	51.3	51.7	51.7
% population rural	80.9	92.6	100.0
Land use			
Total agricultural (ha.)	6,493,200	165,077	17,094
% cultivated (excluding forests)	21	34	68
% of cultivated land devoted to rice	50 (est.)	60.9	79.3
Average paddy/capita (ha per person)	0.059	0.064	0.039
% of cultivated land devoted to:			
Staples	88.0	87.9	87.8
Vegetables	5.2	6.6	6.1
Industrial crops	6.8	5.5	6.1

Source: Vinh Lac District Government Statistics 1993; Statistical Yearbook of Vietnam, Hanoi, 1994.
Note
1 1993 data.

liberalisation and de-collectivisation of the 1980s. By 1994, what was most striking was the rapidity with which the co-operative structure had collapsed and how residual administrative responsibilities were transferred to village people's committees. As a consequence, within the province itself the state is sometimes difficult to identify and is made palpable only through 'an incredible melange of deals ... struck between representatives of people's councils, Party bosses, state [and] private enterprises and occasional foreign investors' (Marr 1994: 10).

Population growth in Vinh Phu Province is high – it was 2.87 per cent per annum over the period 1979–89 – and average population density in 1993 was in excess of 463.5 persons per square kilometre. Within Vinh Lac District the figures are even more striking (Table 20.6). The population density is 1,082 per sq. km[39] but if the ratio of population to cultivable land is calculated, the population density is well over 1,800 per sq. km; each hectare of cultivated land is supporting almost twenty people. Indeed, the average in the three villages in which surveys were conducted was slightly

Table 20.6 Vinh Lac District: land, population and production, 1990–3

Variable	Unit	1990	1991	1992	1993	%increase 1990–3
Land area	km²	250.66	250.66	250.66	250.66	—
Population	nos.	291,214	303,475	307,610	312,546	7.3
Cultivable area	ha	15,699	15,699	15,699	15,699	—
Sown area	ha crops	34,700	36,319	35,778	36,816	3.2
Cropping density	per year	2.17	2.25	2.27	2.28	—
Cereal production	tons	81,314	60,693	99,431	108,185	33
Cereal production per cap	kg (mill.)	279	201	324	350	26
Gross production	dong (mill.)	nd	166,122	208,797	221,324	33
Agricultural production	dong	nd	155,904	203,678	214,684	37
Population growth	% per annum	2.13	2.05	2.01	1.93	—
Total rice production	tons	68,553	47,795	81,835	89,358	30
Average rice yield	tons/ha	2.99	2.09	3.54	3.53	18
Spring rice	ha	12,012	12,674	12,528	12,452	4
Winter rice	ha	10,910	10,167	10,055	10,872	—
Maize	ha	4,768	8,425	6,084	6,354	33
Sweet potatoes	ha	1,403	1,927	2,382	2,472	71
Soy	ha	290	219	228	247	—
Sugar	ha	1,500	1,122	829	1,210	—
Cattle	no. of head	20,919	21,272	21,855	21,500	2
Pigs	no. of head	61,979	60,845	68,056	68,000	10
Fish production	tons	800	634	689	830	4
Brick production	millions	42	41	48	48	13

Source: Vinh Lac District Government Statistics 1990–3.

higher still. At present population growth in the district is, according to village data, just over 2 per cent per annum, though the actual rate of growth is almost certainly higher. While the cropping intensity is high – for example, 2.8 crops per year in Dai Dông village[40] over the period 1990–3 – there is no land available to be brought under cultivation, and the potential increases in cereal productivity from new high-yielding varieties are, for the foreseeable future, quite limited. Land–labour ratios are, in the light of Gourou's earlier analysis, unthinkably high, and the prospects for radical improvements in productivity appear to be, in the short term, constrained.

The Vinh Lac District agrarian economy faces a low-level equilibrium trap, and it is in relation to this conundrum that the economic reforms and their local consequences have to be assessed. Table 20.6 assembles some aggregate data on production trends in Vinh Lac between 1990 and 1993 (i.e. after the second phase of radical reforms in 1988 and 1989) and reveals quite impressive growth rates. Total rice output, total cereal production and the value of agricultural production increased respectively by 30 per cent, 33 per cent and 37 per cent over a three-year period. Cropping intensity stood at 2.28 by 1993. A somewhat clearer and fuller picture emerges from Table 20.7, a statistical summary of one of the villages in Vinh Lac District in which research was conducted. Dai Dông is a village of nearly 8,000 (1,631 households), located within the commercial sphere of Viet Tri city, on a major road linking the provincial capital with Hanoi, and in close proximity to a number of flourishing market towns. A picture of an increasingly prosperous and dynamic rural economy appears with great clarity, suggesting a quite radical unfettering of the social relations of production and as a consequence a number of key loci of agrarian change.

The intensification of land-based production is contained in the growth of cropping intensity from 2.5 in 1990 to 3.0 in 1993, and a high degree of market integration (the proportions of farm, livestock and off-farm output that are sold are respectively 35 per cent, 61 per cent and 86 per cent, all of which have grown by 10–40 per cent over three years). Village GDP has increased by 87 per cent since 1990, and its composition (35 per cent tilled, 35 per cent livestock, 30 per cent off-farm) reflects a gradual diminution of the role of agriculture (by 4–5 per cent) and an 8 per cent increase in the share of off-farm production. While the official statistics are incomplete, all sources of income (agrarian, livestock, off-farm) have probably doubled in four years, and average household income is currently 6 million dong, a 196 per cent increase over 1990.[41] Whatever one makes of the income estimates, it is incontestable that the output of most major crops and livestock has grown by a minimum of 50 per cent since 1990, and the prosperity of the village is clear in the proliferation of brick houses, a vibrant market, and investments in all manner of private enterprises.[42] A deepening and widening of market relations, an *intensification* and *diversification* of land-based activities at the household level (in which soy and vegetables seem to

Table 20.7 Dai Đông; village statistics, 1993

Variable	Unit	1990	1991	1992	1993	% increase/decrease
Uncultivated	ha	485.81	485.81	485.81	485.81	—
Cultivated	ha	377.28	377.28	377.28	377.28	—
Population	no.	7,066	7,169	7,399	7,892	11.6
Total no. households	h/h	1,385	1,485	1,502	1,631	17.7
Total labour force	workers	3,184	3,209	3,349	3,465	8.8
Total sown area	ha	967.9	1,011.6	1,123.4	1,132.0	16.9
Cropping intensity	crops/year	2.56	2.68	2.97	3.00	17.0
Village GDP (perm. price)	million dong	3,377.8	3,832.2	4,757.3	6,331.3	87
Agriculture	million dong	2,068.5	2,884.8	3,553.3	3,446	67
Industry	million dong	420	480	529	580	38.1
All services	million dong	289.3	467.4	669	1,286.1	344
Cereal production (1)	tons	2,070.3	1,924.0	3,025.0	3,090.0	49
Non-rice cereal (2)	tons	96.3	173	212	180.2	87
(2) as a % of (1)	%	4.6	8.9	7.0	5.8	1.2
Average cereals/cap.	kg	293	268	409	392	34
Village GDP (composition)						
Tilled land	%	40.7	39.3	41.6	35.4	-5.3
Livestock	%	38.2	36.0	33.1	35.1	-3.1
Off-farm	%	21.1	24.7	25.3	29.5	8.4
Value village GDP sold	million dong	2,870	3,447.9	5,641.7	3,923.7	276.1
% of farm prod. sold	% val.	25 662	24 833	31 1,807	35 2,135	322
% output sold off-farm	% mill.	54 1,626	57 1,870	63 2,717	61 4,058	249
Product sold	% dong	75 582	72 744	85 1,117	86 1,729	297
Crops						
Area	ha	747.3	750.0	754.4	750.6	0.4
Rice: output	tons	1,974	1,751	2,942	2,910	47.6
	ha	30.6	30.6	28.8	46.8	53.1

Table 20.7 continued

Corn: annual	tons	59.3	55.0	63.9	97.3	64.1
	ha	67	54	150	166	167.0
Soy: output	tons	70.3	59.4	187.5	192.5	157.0
	ha	105	112	118	132	25.7
Vegetables: annual	tons	2,184	2,419	2,678	3,036	39.0
	ha	—	11	11	11	
Sugar: output	tons	—	470	480	480	2.0
Pigs	no.	1,936	2,016	2,895	2,920	51.0
Labour force	nos.	1,846	—	—	2,035	—
Agriculture	%	57.9	—	—	58.7	
	nos.	284	—	—	462	
Livestock	%	9	—	—	13.3	
	nos.	310	—	—	775	
Off-farm	%	9.7	—	—	22.3	
	nos.	744	—	—	193	
Unemployed	%	23.3	—	—	5.7	
Income						
Average income/cap.	1000 dong	608.9	747	1,074	1,262	207
Agriculture	"	281.2	363.2	591	579.9	206
Livestock	"	252.2	275.2	349.9	473.3	187
Off-farm	"	75.5	108.6	134	209	276
Average h/h income	"	3,106	3,606	5,295	6,108	196
No. of h/h[1]	no.	1,385	1,485	1,502	1,631	17
Rich	%/no.	10/138	—	—	20/326	
Middle	%/no.	70/969	—	—	70/1,142	
Poor	%/no.	20/278	—	—	10/163	

Source: Vinh Lac District Statistics Office 1993.
Note
1 Total number of households increased from 1,385 in 1990 to 1,631 in 1993.

be prominent), a steady growth of the livestock sector, and off-farm income (trade, services) as the major source of local economic growth define the co-ordinates of Dai Dông's post-reform economy.

I have chosen to focus on three aspects of the Dai Dông research: *land and labour, taxation* and *rural accumulation*, particularly off-farm sources of income. Necessarily, this picture is incomplete and awaits further ethnographic and careful local community studies, but these three foci have, I suspect, general relevance for many villages across the Red River delta (see INSA 1995; CDM 1995).

The land and labour question

The land is the most important thing after water. When the people fought for independence, they were fighting for the land.
(Cam Ngoan, Vice-Chairman, Vietnamese Union of Peasants, 1993)

In the light of the land scarcity and high population densities in the Delta and in Vinh Lac District, the questions of land access and social distribution of landed property, and the extent to which land provides the basis for some measure of cereal self-sufficiency and economic security for village households, assume a singular importance. As a result of the equitable privatisation of the agricultural co-operatives beginning in 1991, however, Vinh Lac District does not as yet suffer from the debilitating economic consequences of high population densities coupled with a high degree of landlessness and rural poverty typically associated with, for example, Java, Bangladesh and parts of India. In most of our study villages, land distribution began in April 1992 and was completed by October 1993; in Dai Dông village, there was a second land distribution which allocated surplus, and for the most part poorer quality, land on the basis of competitive bids. The majority of commune land was, however, distributed through twenty-year usufructory rights on the basis of household demography: each person received 1.6 sao,[43] and a half measure was granted to every person over and above the state-mandated limit of two children. In practice this meant that each household obtained a number of geographically dispersed plots – it is not unusual to have twelve or thirteen per family with the most distant parcel 1.5 kilometres from the village – the quality of which should reflect the village-wide pattern of land qualities.

The 1993 Land Law formalised in Dai Dông, as elsewhere, the transfer, inheritance and sale of land use rights that had been broached in the 1988 Resolution No. 10, and subsequently in the 1992 Constitution. There is evidence to suggest that even in Dai Dông – and certainly in the southern communes – land transactions of various sorts had continued with varying degrees of secrecy after 1980 and, since 1986, were increasingly above-ground. But the final assignation of land rights – a move which in effect not only allocated private land rights but also shifted co-operative functions to private enterprises and the People's Committees – seems to have been effected with little malfeasance and with an impressive accountability. Oversight of the land management functions of the state lies with the General Department of Land Management, but these functions (recording of rights, land classification, issue of land certificates) and the land allocation itself are performed through branches of the local government (Departments of Land Management) in conjunction with People's Committees and the co-operative. As the World Bank (1993: 28) points out, this system produces a high degree of local autonomy, flexibility and an air of *ad hoc* local practice. Land allocation has as a consequence been extremely varied in form and character throughout Vietnam (CDM 1995).

The consequence of what appears to be an effective and well-implemented land redistribution is the creation of a relatively equitable land-holding peasantry based on a modified Chayanovian (household demographic composition) principle. To give some sense of land distribution, a 15 per cent random sample of households was taken from the tax records of Dai Dông village. The average household land-holding is 6 sao and the average household size is 4.75 persons. Only 11 per cent of households possess more than 10 sao (the largest holding in the village appears to be about 20 sao) and 19 per cent have less than 3 sao. The 5 per cent of households that have roughly 1 sao of land are all single-person families (some of whom are old and infirm). The average land-holding per active worker is 3.1 sao.

Dai Dông's experience paints a picture of quite striking (if fragmented) equality[44] – though it needs to be said that the principles of land distribution benefited those who paid no heed to the state family planning policy! – and of average land-holdings capable of producing 2,100 kg of rice equivalents per year. The gini coefficients for Dai Dông are roughly 0.18, comparable to the astonishingly low coefficients for agricultural households computed by Tuan (1995: 147) for three Vinh Phu villages (0.14, 0.13 and 0.17). The Land Law of 1993 has recreated a Red River delta that in one sense looks back to Gourou's model of Tonkin; parcelised, fragmented land-holdings dominated by a densely settled, relatively commercialised peasantry with a tenacity for access to and control over paddy. What fifty years and a de-collectivisation produced, however, was a much more *equal* land-holding distribution than 1936, a population density four times higher and a modal per capita land-holding of 0.05 hectares in 1993, only one-quarter of the average holding size fifty years earlier. Notwithstanding these differences, land-holdings at present seem capable of providing even the poorest with a modicum of economic security and a sort of subsistence safety net. It appears that there is no landless class to speak of[45] – although the question of whether and how land might be transferred to the next generation of post-co-operative peasant farmers in Vinh Lac will be of immense significance in terms of the possible fragmentation and diminution of family land-holdings.

The distribution of land use rights has not precipitated, in other words, the beginnings of rapid land accumulation as such (or for that matter dis-possession through collateralised land). In Vinh Son, Dai Dông and Tho Thang there is a brisk market in land 'rental' (in effect the purchase of long-term use rights). Land is rarely rented for less than 5–10 years and the renter typically pays 30,000–50,000 dong per sao per year and the taxes appropriate to the quality of land rented. On balance, the rental market seems relatively sluggish and at this point is not a source of social or economic differentiation in either land-holding or income for several reasons. First, plots of land available for rent are small in area and appear on the local market quite irregularly, and many households are in any case reluctant to forgo their own subsistence rice production unless compelled to do so. Second, in view of the relative ease with which land can be taxed (in contradistinction to livestock, which is exempt, and industry and trade, which are difficult, if not impossi-ble, to monitor and assess) and the greater profitability of other non-farm sectors, land is not necessarily seen as a source of accumulation. Indeed it was noted by several farmers in the course of interviews that the daily return to self-cultivation of rice was roughly 2,000 dong per day whereas the prevailing peak wage rate for transplanting and watering was 8,000–10,000 dong per day. And third, land is attractive for that *limited number* of families for whom some agricultural output may be significant for the production of livestock feedstuffs (soy, maize and rice) as part of an expanding off-farm enterprise. For all these reasons the land question seems characterised by a relative equality in holdings and, at this point at least, land-holding transactions, while certainly increasing, remain sluggish and are not seen as either a profit-able investment or a source of speculation. Nevertheless, it is clear that the government is sensitive to the question of the genesis of land concentration – there is a statutory limit of 2 hectares in the North and 4 hectares in the South, though this is widely understood to be rarely enforced[46] – and Vietnamese government policy, fully aware of the experience of China with regard to land speculation in Guandong and elsewhere, holds up the cases of Taiwan and South Korea as potential models of small-holder-based agrarian development which Vietnam seeks to replicate.

As regards the question of labour, it appears that the very fact of widely held land in most communes has had the effect of *stimulating* the growth and development of a local wage labour market for two clear reasons. First, the 20 per cent of households with three or fewer persons (and whose producer–consumer ratios accord with the demographic structure of young families) typically need to hire labour at key periods during the labour-intensive periods of rice production to complement their domestic labour force. A Vinh Son household, for example, consisting of a married couple in their mid-twenties and their two young daughters, needed to employ 10–15 man-days per cropping season in order to cultivate their 5.3 sao of land. Other survey research in the Red River shows that a majority of households are both buying and selling labour,[47] and this must be understood in relation to the labour demands of paddy cultivation (even on small-holdings) for households that are immature in Chayanovian terms (i.e. a high consumer–producer ratio due to the presence of young children) and to the need for off-farm income among those households for whom Lilliputian rice-holdings are incapable of reproducing the household economy. Second, the possibilities for commercial development in the non-farm sector – especially in trade, industry and pig and fish production – draw family labour away from agricultural pursuits even in relatively large (and prosperous) families, creating a need for hired farm labour. The fact that all households have access to land, coupled with the necessity to maintain land-based production for commercial households actively engaged in off-farm production, means that the Vietnamese reforms have *contributed to and stimulated the development of a labour market* alongside a robust peasant economy.

The tax question and the politics of state decentralisation

Agricultural tax collection in 1990 was a mere 1.5 percent of the sector's value added.
(World Bank, *Vietnam: Transition to the Market*, 1993)

Whether surpluses are created and reinvested in the rural economy is of central significance in understanding the opportunities and constraints for rural growth and dynamism within Vinh Lac District and within the regional economy more generally. To this extent the role of the state in appropriating rural surpluses as district and village tax revenues must be analysed as one of several key entry points for our study of the linkages between agricultural and non-farm production and between town and country. The tax question and the politics of local government budgets in Vietnam are, however, exceedingly complex and require much more serious and systematic research than our field studies to date can provide (see Porter 1995). This study can only offer the most superficial analysis of taxation in the rural economy based in large measure on a detailed examination of tax loadings in Dai Dông village (Table 20.8), but in our view it is a key component of future policy reforms and development strategies.

In 1983 a new land tax system based on area and quality of land was introduced. Cultivated land was subject to a sevenfold classification according to yield, and five fixed tax rates were levied on three types of land (delta, midland and mountain) for all of the seven categories (Tran 1991: 17). Private gardens (the so-called 5 per cent plots) were subject to an additional levy ranging from 1 to 7 per cent of total paddy output, equivalent to 10–60 per cent of the total household tax obligation. In April 1987, the tax rate on

Table 20.8 Tax loadings in Dai Dông village, Vinh Lac District, 1993

Dai Dông Village	Ward A	Ward B	Total (15% sample)
No. of households	90	92	182[1]
No. of people who received full land allotment[2]	399	466	865
No. of people who received reduced land allotment[3]	394	457	851
No. of able-bodied workers	5	9	14
Total land in sao	299	356	655
(i) Land tax kg/crop[4]	526.01	573.10	1,099.11
(ii) Command tax kg/crop	5,926.2	6,479.8	12,406.0
(iii) Social fund (village) kg/crop	855	1,080	1,935
(iv) Health tax kg/crop	725	875	1,600
I Total tax (i), (ii), (iii), & (iv) kg/crop	199.5	283	432.5
(a) Irrigation tax kg/crop	4,014.1	8,659.8	16,365.5
(b) Rice protection fees kg/crop	4,014.1	4,397.9	8,412.0
(c) Canal repair kg/crop	253.4	287.1	540.5
(d) Debt kg/crop	253.4	287.1	540.5
II Total tax (a), (b), (c) + (d)	5,281.9	6,675.5	11,957.4
Tax grand total I + II Kg/crop	12,987.6	14,735.3	27,722.9
Average household tax burden: kg/crop	144.30	160.16	152.32

Source: Village tax records, Dai Dông Co-operative, 1993.
Notes
1 Average household size is 4.75 persons.
2 Each household member received 1.6 sao.
3 The aged and infirm received 1 sao.
4 All taxes are calculated in kg of rice equivalent per crop.

private plots was raised by the Politburo to an incredible 60 per cent of paddy output but it is clear that the tax rates – which in effect subsumed a dizzyingly complex array of charges – for agriculture, for industry and trade and for other social obligations varied widely from locality to locality. Tax exemptions in particular were used as punitive means both to maintain collective production and service provision and to stimulate industrial and craft enterprises. It is also important to recognise that after the 1981 reforms, peasants borrowed heavily either because of the low returns from contract production or because they ran up debts through a failure to meet the co-operative production quotas. Either way, peasants faced debt repayments *in addition* to tax obligations, with the result that (i) by 1988 20–40 per cent of co-operative members had their access to co-operative assets such as land and buffaloes withdrawn to cover debts (Fforde 1993: 60), and (ii) in some co-operatives households are continuing to repay debts incurred during the 1981–9 period in addition to the standard [*sic*] tax obligations (see Table 20.8, Id).

Either way, tax revenues levied on rural communes have represented, and continue to represent, a substantial flow of resources out of the countryside, and suggest a somewhat different picture from the conventional macro-level view that agricultural taxes are declining and represent only 1.6 per cent of value added in agriculture and only 7 per cent of government revenue (Mellor 1993: 37; see also Riedel 1993: 407). According to aggregate tax data at the district level, Vinh Lac District raises 11 billion dong per annum from its forty-five administrative units (i.e. roughly 35,000 dong per capita); 9 billion (81 per cent) derives from the agriculture tax, 270 million dong from garden taxes, 560 million from house taxes and 800 million (8 per cent) from industrial and trade taxes.[48] In order to gain a better appreciation of the tax question and its developmental implications, we have examined data furnished for 192 households in Dai Dông village. There are four broad categories of taxation at the village level: the slaughter tax (levied on butchered animals), the agricultural tax (embracing land and taxes for irrigation, protection and infrastructural services), a series of communal taxes (the social fund, health), and taxes on trade and industry. In the case of Dai Dông village, the agricultural and communal taxes account for almost 90 per cent of total tax revenues generated by the village (and this is probably quite typical for most communes in the district). The slaughter tax is totally insignificant, amounting to 5,000 dong per household per year (livestock production is tax exempt), while the trade tax represents a very small proportion of incomes actually generated in that sector. The calculation of industrial and other taxes is quite complex (if not baroque) and, to the degree that it is quantitatively important, only falls on a small number of larger household enterprises. A seven-worker furniture operation in Dai Dông pays 170,000 dong per month, and a restaurant in the same village pays 60,000 dong per month. Notwithstanding the limited number of households on which these sorts of tax burdens fall, it is clear that the real tax loadings on trade and industry are trivial as a percentage of turnover or profit within the sector. The total tax in Dai Dông village for 1993 was 12 million dong (roughly 10,000 dong per household).

For the purposes of this discussion, the land taxes and the irrigation taxes are taken together. The land tax is defined by the quality of land: the respective tax levels for land level I (richest) to land level V (poorest) are as follows: 12, 9, 7, 6 and 5 kg per sao per crop. There are also three taxes levied by the village for water, for irrigation repair and protection, and for other infrastructural services. The irrigation/protection/infrastructure taxes are by far the most important, amounting to roughly 10 kg rice equivalent per sao per crop. Finally, there are a series of taxes that I have termed 'communal': these are taxes levied by the district for services provided, a social fund levied by the village, and a health tax also levied by the village. On average these communal taxes amount to roughly 25 per cent of the land tax.

For purposes of calculation I have aggregated all of these taxes, recognising that the land tax and the irrigation taxes constitute the lion's share (on average) of a household's tax burden (they are on average at least 80 per cent of total taxes paid each year). In 1993 the total tax burden for Dai Dông village was 202,000 kg rice equivalent per crop (i.e. over 600 tons per annum assuming a cropping intensity of 3.0), plus 18 million dong for trade/industry and the slaughter tax which are *not* included for the purposes of calculating average tax loadings. The average household tax loading therefore is 144 kg per crop. Assuming a cropping intensity of three, the total annual loading is 432 kg per household. For the average household with 4.75 persons and 6 sao of land, this translates into roughly 1 sao of output per cropping season, that is a 17 per cent tax burden on land.

These tax loadings are substantially in excess of both the calculations by Alagre (1995), who argues that in Thai Ninh co-operative the tax burdens (at roughly 10 per cent) have at least fallen since the mid-1980s, and Fforde (1993), who argues that they remain onerous (roughly 300 kg of rice equivalent per household per year) in Thai Binh. Luong's 1989 calculation in Vinh Phu and Ha Bac Provinces (in close proximity to Dai Dông village) seems at first blush to present a somewhat different picture. His estimates are that respectively 46 per cent and 49 per cent of the rice crop on contracted land was appropriated (1992a: 27). However, in these calculations only 8 per cent is accounted for by agricultural taxes; the remainder is 7 per cent from co-operative and local administration, and 31.85 per cent by production costs (i.e. in effect a share of harvest paid to cover all production inputs). If the panoply of other local and state taxes (and production costs) were included in my calculations, the figures would obviously be much higher; similarly, Fforde's (1993) estimates also include levies that do not fall within the categories I have chosen to calculate the agricultural (land and irrigation) taxes. The central point is that Dai Dông's agricultural tax loading exceeds by a substantial amount the oft-quoted 5–10 per cent figure that is traditionally assumed to hold in Vietnamese communes and raises the question of whether (and for whom) this loading might represent a structural constraint to economic well-being and security. Insofar as livestock activities are tax exempt, the burden of the general tax system seems to fall disproportionately on those with land and perhaps with limited capacity to generate off-farm income. There is good reason to presume that taxes are especially onerous for poorer families with limited economic sidelines. Conversely, the district and the village are losing very substantial tax revenues not simply from the most dynamic sectors in the rural economy (livestock and industry in Dai Dông accounts for 65 per cent of village GDP but only 12 per cent of taxation) but also from those prosperous families that can, and should, draw a heavier tax burden. Given the low-level equilibrium trap which we have described in our account of the post-co-operative rural economy, it is entirely possible that the tax burden is acting as a constraint on the material prosperity of some rural constituencies.

These tax burdens must, however, be situated on the larger contemporary political economic canvas of government fiscal reforms on the one hand and the reduction of state revenues for local development on the other. The neo-liberal reforms have necessarily undercut state revenues from the collective sector and reduced central expenditures and subsidies,[49] with the result that local governments (i.e., the provincial level) – if they are to maintain any semblance of a local revenue base for both developmental and social purposes – are necessarily thrown back on their ability to extract surpluses from the peasantry (Porter 1995). Much of the commune and district level has been effectively undermined by the fiscal centralisation which has enhanced the power of central and provincial governments. District resources are increasingly dependent on tax revenues (see Porter 1995: 242) but their budget is increasingly dependent upon a strapped provincial state which provides grants and levies. The agricultural taxes in particular have seemingly become the centrepiece of local government revenue systems, which suggests that the Dai Dông tax loadings are probably indicative of a widespread underestimation of the current fiscal burden imposed on rural peasants. More poignantly, the taxation questions give reason to think carefully about the beneficiaries of Dai Dông's economic boom during the reform period (indeed in Vietnam generally) and the possible burdens that taxes may impose upon a rural poor now largely beyond the security granted by the collectivist safety net of the 1960s and 1970s.

Rural accumulation: recombinant strategies and social networks in the Red River delta

The dynamism unleashed by *doi moi* reforms has commercialised the Vietnamese countryside and ushered in a new phase of rural growth. But what exactly is the relation between agrarian reform and accumulation? Is this a sort of capitalism without capitalists? While private property of sorts, and markets of sorts, are in place, the extent to which either has precipitated systematic accumulation in the Red River remains an open question. What is clear is that access to and control over key resources – among them markets, credit, labour and what one might call social capital – are fundamental for those rural and agrarian accumulation strategies (all diverse and recombinant in character) that are emerging in the interstices of the party-state. At the heart of the accumulation question is the political, cultural and social capital which is required to prise open the possibilities for systematic accumulation – to use the language of Sara Berry (1989) how networks and social relations, how culture and power shape the prospects for growth. This is a complex story even within the confines of Vinh Lac District, and all I can do here is highlight a number of key strategies – and their frailties and structural limits – in the heart of the old North. For expository purposes I simply want briefly to discuss three strategies, all of which highlight that accumulation is most dynamic in the non-rice sector, in agro-industrial pursuits (pig, livestock, fish), and in the mercantile sphere.

Nomenklatura entrepreneurs

One strategy turns on party connections – political capital – and the ability of both local government officials and influential party members (former co-operative Chairs, notables within the People's Committees) to move into entrepreneurial pursuits. Intensive (and large-scale) pig production is key, often predicated on access to state credit typically beyond the reach of most peasants, and guaranteed markets via government contracts (some export-oriented) at the provincial level. Such accumulators are often active in the rice trade – indeed in mercantile activities generally – which is a prerequisite for pig feed, and may also be central figures in the informal credit 'market' insofar as they become financial middlemen between local entrepreneurs and the state banks.

Social networks and mercantile traditions

A second strategy relates to social networks and mercantile traditions. A striking instance of this sort of strategy is the groundnut sector in Vinh Lac. Here families with pre-revolutionary involvement in trade (particularly into China) are now reactivating the social networks as a way both of taking command of the peanut economy and more generally of deepening commercial ties in a panoply of exchanges with southern China. One such individual employed social networks – long-standing familial connections with ethnic Chinese traders in the region – to create a now flourishing export-oriented fishery (turtles, in particular to Hong Kong). The gradual re-emergence of some of the artisanal lineages described by Gourou a half-century earlier, represents another nascent 'reinvention of tradition' fitted to the purposes of market liberalisation. In these cases, there is both a reactivation of historical mercantile and cultural connections, and the deployment of political connections to ensure market access and lack of surveillance (much of the trade is illicit across the Chinese border). Many of these families also use their local standing – often increasingly legitimated in the 1990s by their active participation in flourishing religious networks – as a way of recruiting local labour. In much of this, however, *pace* China, local and provincial governments have been largely passive.

Remittance accumulation

While remittances (and foreign flows in particular) are less significant than in the South, the limited availability of rural credit represents – in the context of a marked agrarian egalitarianism – a major barrier to investment and innovation. Remittance income – typically through family connections and the larger Vietnamese diaspora – is central, and it is often the case that many of the emerging petty enterprises (restaurants, shops) have been built upon the backs of what one can legitimately call social (kin) capital.

* * *

If the Vinh Lac experience is at all suggestive for the Red River delta as a whole – and the new survey work by INSA (1995) and others suggests that it is – then the broad contours of the agrarian transition in the North are quite clear. De-collectivisation began in a sense organically from below in the 1970s through various forms of illegal co-operative household contracting arrangements, and subsequently picked up momentum through the reforms from above in 1981 and 1988. In the course of a decade, long-term usufructory land rights were carved out of collective (commune) property to create, quite literally, an egalitarian Chayanovian peasantry (land was allocated on the basis of household demography). The relative solidity of the co-operative structure in the North, coupled with the power and commitment of the party-state, ensured a remarkably egalitarian (more equitable than in Taiwan, South Korea or China) and accountable redistribution of property rights. With the Land Law of 1993, and the ninety-nine year leases issued in 1994, the basis was laid for the emergence of a land market (rental, lease and sale are already part of the peasant lexicon). In fact the relatively rapid disappearance of many co-operative functions after 1986 facilitated the emergence of relatively free markets in both inputs and outputs. Real increases in producer prices for staples, and new market opportunities for livestock and vegetable products, propelled the emergence of a commercially oriented 'free peasantry'. Is this picture, however, of a prosperous commercial peasantry – a small-holder growth strategy with equity – an adequate account of the Red River delta in the 1990s?

The peculiarities of the Red River delta – agro-ecological and demographic – suggest that this picture is only partially accurate. On the one hand, the very success of the land distribution in a land-scarce environment has produced a situation not only of physical land fragmentation, but of barely sufficient paddy land for household reproduction. In other words, for some households land reform created a semi-proletarianised rural workforce which in some ways approximates Lenin's notion of the 'propertied worker'. On the other, the insufficiency of average land assets compels households to engage in off-farm employment, encouraging among other things the growth of nascent labour markets throughout the North. But it is precisely access to off-farm income sources which is less equitably distributed and capable of accounting for new forms of social and economic differentiation which the low gini coefficients in land tend to obscure.

In this sense, the significance of the Vinh Lac case is to reveal how the centre of gravity in the local economy now lies, as a result of Vietnam's *doi moi* reforms, not so much in land-based agricultural production as in the off-farm sector. The constraints on paddy productivity, coupled with the disincentives created by the current tax structure and falling producer prices for rice, all hamper the rice sector as a source of serious profitability. It is true that the winter crops such as soy and maize (for pig feed) and vegetables for local markets are likely to be more profitable than paddy, but the real dynamism in the district seems to reside in livestock, trade and a limited number of small-scale industries. The paddy sector will retain local significance simply because widely held access to some rice-land confers a measure of subsistence security for the rural poor. But the Red River is not able to compete with the Mekong delta as a dynamic rice-growing region (in part because of natural factors and in part because of the structure of land-holdings). As sources of accumulation and social differentiation, it is not so much land as livestock and trade that command centre stage. I have emphasised, however, the extent to which these nascent enterprises and circuits of accumulation are both constrained and facilitated by two broad sets of forces: one is the continuing role of local politics (i.e., bureaucratic centralism) associated with what van Arkadie (1993) calls the 'double subordination' of local government to People's Councils and central government and the party. The other is what one might call 'recalcitrant tradition', that is the resurfacing in new and not so new ways of social processes, cultural relations and social networks which facilitate access to markets, capital and credit. It is in this respect that the reports of the reappearance of 'traditional' kinship and patrilineages – the pagoda, dinh renovation and other rituals – may be of lasting consequence, not simply as the renewal of traditions long suppressed, but as *the ways in which the market and accumulation strategies are being constituted through a new capitalist nexus.*

The significance of the rural dynamics in the North are portrayed in Tuan's (1994, 1995) regional analysis of socio-economic differentiation in Vietnam. On the one hand, his work shows how the southern agrarian trajectory is characterised by a larger percentage of commercialised farmers, by the presence of relatively large holdings, and higher rates of growth of rural income. On the other, the gini coefficients of household income distribution suggest that the tendencies towards greater inequality which hold throughout the rural sector (from 0.258 in 1989 to 0.298 in 1992) may in fact be *greater* in the North than the South. If these trends are indeed accurate, the differential access to and control over differing forms of non-farm employment in the North seem to be of particular salience. The equity conferred by land reform, in other words, is contradicted by the polarising tendencies within the non-farm sphere. In the North, where large numbers of relatively poor – if increasingly commercialised – peasants depend in large measure on irrigated paddy cultivation, the decline in agricultural infrastructure associated with the collapse of the communal institutional structure (and the diminution of state investments in agriculture) may further exacerbate the economic prospects of the poor and semi-proletarian households that populate one of the most densely settled places on earth.

Vietnamese de-collectivisation through a Chinese lens: some comparative speculations

> Whilst de-collectivisation of farming practices in post-reform China has received considerable attention, the part played by the continued collective ownership of land and the role of collective organisations in promoting growth . . . have been less appreciated. . . . China's economic success can be interpreted as the success of an economy which gives significant scope to market forces but which also retains a critical role for the state and widely uses the institution of social ownership of the means of production.
>
> (Paul Bowles and Xao-yuan Dong, 'Currents Successes and Future Challenges in China's Reforms', 1994)

Does Vietnam's 'successful' renovation have anything at all to do with gradualism, flexibility, or the pacing and timing of macroeconomic 'shock therapy' after a period of initial microeconomic reform? According to Sachs and Woo (1993), Vietnam's record rests on the fact that its economic structure ('a peasant agricultural society') is much easier to reform than an urban, 'over-industrialised' socialism of the Soviet or East European sort. It possesses, in short, Gerschenkron's (1962) advantages of backwardness. These properties have permitted communes – in general less hampered by subsidies and soft-budget constraints than heavy industry – to be dismantled and to 'spur enormous flows of workers out of subsistence into new sectors of the economy' (Sachs and Woo 1993: 4–5). Sachs and Woo are centrally concerned with an account of the 'success' of China, but in so doing they raise the intriguing question of whether China and Vietnam are comparable forms of post-socialist transition.

Both China and Vietnam were largely agrarian economies in 1978, in which agriculture accounted for roughly 70–75 per cent of employment, and collective agriculture (distinct from state enterprises, urban collectives and the private sector) was the dominant type of productive organisation. Each country had similar sorts of pre-revolutionary agrarian structure. In China 10 per cent of the landlords and rich peasantry accounted for 56 per cent of the land, while 68 per cent of the rural poor owned 14 per cent of the land area; in Vietnam between 60 and 70 per cent of landowners owned just 12 per cent of the cultivated area. China and Vietnam initiated a *first phase* of land reform in the 1940s and 1950s – a sort of land-to-the-tiller programme – in which similar amounts of land per capita were redistributed (0.39 acres per capita in China, 0.25 acres per capita in Vietnam), followed by a *second phase* of higher order collectivisation (China rapidly after 1956, Vietnam more tardily in the 1960s) orchestrated by 'mobilizational authoritarian states' (Kerkvliet 1995: 66). Vietnam has no equivalent of the Great Leap Forward but the exigencies of war provided a certain legitimacy for the Vietnamese collectivisation drive and for the subsequent enlargement of the co-operatives. The tripartite productive structure – commune, brigade and team – required, in China and Vietnam alike, the maintenance of a private plot sector, the tight regulation of markets, and coercive local regulation through party and state apparatuses. As Kerkvliet and Selden (1995: 6) note, the collective production system was more shallowly rooted in Vietnam and, in contrast to China, its productivity was radically compromised by the 1970s. China increased per capita grain output between 1957 and 1978 by 2.1 per cent per annum; in Vietnam, conversely, rice productivity was much lower than China and per capita grain production fell steadily from the early 1960s up to 1980.

Similarly, the *third phase* of de-collectivisation also suggests intriguing temporal and institutional parallels. Reforms in China initiated after Mao's death referred to as the household responsibility system are in many respects identical to the Vietnamese 'contract 100' of 1979. Land was gradually privatised in the sense that households were granted various forms of usufructory rights over plots, producing in both cases a quite strikingly equal land distribution (the gini for land in China is roughly 0.54 and for Vietnam 0.30); in both cases landlessness is around 4–5 per cent (McKinley 1995; World Bank 1995). This reformist move seems to have been driven both from below and from above. Rapid acceleration in agricultural output, especially of grain, occurred in China between 1979 and 1984 (grain output increased by a third between 1980 and 1984, and rural incomes by 11 per cent per annum between 1978 and 1981), largely as a result of a 40 per cent increase in state procurement prices coupled with large increases in the application of fertiliser. Vietnamese rice productivity lags behind China, but the rural and agrarian growth rates of the late 1980s are comparable. Jean Oi (forthcoming) and others have shown that agricultural incomes in China have *fallen* since 1985, largely because of the continued role of state manipulation of prices and the profitability squeeze imposed by the inflation of input costs. But the confluences between the reformist phases are striking nonetheless.

Vietnam and China seem in short to have adopted de-collectivisation strategies which involve four common attributes: the creation of a relatively egalitarian peasantry with some form of private property rights; a complex hybrid of new forms of state and market regulation; a 'Bukharinist' trajectory in which a substantial degree of local state intervention steered the reform process; and the transfer of the burden of absorbing and supporting the enormous rural surplus labour onto individual peasant families for whom the state is not directly responsible (Chang 1993: 444). In a curious way peasantisation entailed a continuity between the collectivist and reform periods:

> family farming can be seen as the same type of non-modernised or non-rationalised economies in which work and morality . . . are inextricably intertwined and in which the balance between production and consumption rules over any reified efficiency criteria such as productivity or profit.
>
> (Chang 1993: 439)

But it is here that the Vietnamese and Chinese roads part company. The increase in grain output after 1981 in Vietnam has been similar to that of China but the speed and pace of de-collectivisation has proceeded much faster and further. Land is now almost fully privatised,[50] state marketing has been abolished and the co-operative structure has in large measure been liquidated. This has unleashed the energies of some 10 million Vietnamese farming households which have lent to the rural sector a visible dynamism (as indeed did the creation of 200 million peasant contractors in China in 1980). But rural dynamism in China has been sustained since the mid-1980s not by agriculture – which remains in significant measure collective (Bowles and Dong 1994) – but by so-called sidelines and especially by rural township enterprises (TVEs) (Oi forthcoming). Building upon a prior Maoist tradition of rural industrialisation, the Chinese model drew labour from agriculture and from the collectives into non-farm activity without attempting a radical introduction of private property and market liberalisation. China has in this sense a *two-tiered property system* consisting of private plots *and* the preservation of (and in some cases the reconstitution of) village collectives which have a key role in managing land contracts and service provisions. In a 1990 survey, village collectives in China accounted for 35 per cent of ploughing, 75 per cent of irrigation; 62 per cent provided more than a third of inputs and on average owned 44 per cent of non-land capital assets. In 1992, roughly half of total rural income still came from collective sources (Bowles and Dong 1994: 65). Distributive corporatism associated with the hybrid privatisation from below – the genesis of township and village enterprises currently employing close to 100 million workers – provides continued investment funds for the continuity of the collective operations at the local level. By 1988 TVE industry accounted for 23 per cent of rural employment, and over 50 per cent of rural output. The growth of rural industrialisation, coupled with the termination of the growth phase of agricultural liberalisation in the 1980s, has seen China shift from a net food exporter in 1990 to a major importer (close to 20 million tons) in 1997. Vietnam, conversely, remains a major food exporter.

In the China case, then, it is the maintenance of a two-tiered system coupled with rural industrialisation by township enterprises in which the local state is a residual claimant with an incentive to maximise local economic growth rather than bargain with the centre for more resources, which contrasts so sharply with Vietnam. Incomplete collectivisation in Vietnam – the North–South divide – ensured that collective institutions were in many cases less robust in large parts of Vietnam, and hence the possibility of a two-tiered system, had an effort been made to replicate the Chinese experience, would have been unlikely. In the North at least, I have argued that it was precisely the strength of the collectives and People's Committees which ensured the equity and accountability of a radical land distribution and privatisation, a strength which is, *pace* China, legislating its own extinction. The continued centrality of the party in China – in spite of the 'quiet revolution from within' that Walder (1995) describes in the erosion of enforced discipline within the centralised party apparatus – has ensured a much stronger hand in maintaining this co-operative and collective set of institutions, some of which are distinctly new, if not hybrid.

As a consequence, to return to Sachs and Woo, the agrarian and rural trajectories in China and Vietnam by the early 1990s have important structural similarities and contrasts. Furthermore, these differences cannot be accounted for by their identical starting points, i.e., their 'peasant' character. Vietnamese agrarian structure is in some respects quite similar to other parts of South-east Asia and to the land-to-the-tiller programmes of Taiwan and South Korea in the 1950s. Conversely, the Chinese case is a hybrid, two-tiered system in which household and collective are articulated in complex ways with dynamic circuits of rural industrialisation. What seems to be key in both cases is, in a strange way, a partial confirmation of Kornai; not so much that reform cannot happen without the abolition of the party (in this Kornai is demonstrably wrong) but rather that the party is fundamental in shaping (engineering perhaps) the agrarian question broadly understood. In this sense Verdery (1992) is perhaps right to talk of a transition from socialism to 'feudalism', understood as the parcelisation of sovereignty. In China it is the new sovereignty associated with distributive corporatism and in Vietnam through socio-economic differentiation and the local despotisms of economically enfranchised notables. This parcelisation is naturally happening in different ways with different consequences in Vietnam and China but it alerts us to the complex and differentiated ways in which politics and economics – rather than some undifferentiated capitalism – are being decomposed, reconstituted and refigured in the post-socialist order.

Acknowledgements

The fieldwork upon which this chapter is based was conducted with a small interdisciplinary team of University of Hanoi and University of California, Berkeley faculty and students in Vinh Lac District (see Figure 20.2) during August 1994. I am grateful for the support of Dr Pham Van Phe and his colleagues at the University of Hanoi for their support and assistance. I am especially indebted to Hy van Luong, Mark Selden, Carmen Diana Deere and Ivan Szelenyi for critical comments, though I remain solely responsible for the content of the chapter. This research was supported by the MacArthur Foundation.

Notes

1 'Amid a sea of unapologetic capitalism, Vietnam's socialist market economy is poised this year to grow faster than just about all of its neighbours' (*Far Eastern Economic Review*, January 1995: 23).

2 According to the World Bank Report (1993) private employment in agriculture (i.e. peasants) accounts for 32 per cent of total employment in the sector, and co-operatives 60 per cent; the official Vietnamese government statistics in 1992 report 12 per cent and 87 per cent respectively. Both of these figures are ridiculous underestimations of the actual preponderance of independent family farms which, in view of the near *total collapse* of the co-operatives across the country, currently dominate the agrarian landscape.

3 According to Fforde (1993: 58), 30 per cent of the Vietnamese rural population are 'hungry and in poverty'; Tran Lan Houng cites a figure of 35.9 per cent of the rural Vietnamese population as 'absolutely poor' (1994: 45). A survey in Quang Tri Province reported 61 per cent of families short of food for 3–7 months of the year.

4 According to Bowles and Dong (1994: 65), because of their economic potential 'village collectives have in fact been on a steady *increase* in China since the early 1980's. More than 2.05 million villages . . . and sub-villages . . . *reestablished* collective organizations' (emphasis added).

5 There is a large literature on the process and politics of reform during the 1980s, most especially the early initiatives in 1979, the *doi moi* reforms of 1986, and the 'big bang' reform of 1989 (see Andreff 1993; Ljunggren 1994; Fforde and de Vylder 1996; Turley and Selden 1993). No effort is made to review this literature here.

6 In 1979, one year after the Chinese reforms, the Vietnamese agrarian reforms were intended to be 'inner-systemic', that is to say they limited material incentives for peasants using market-type measures to enhance local autonomy and responsibility. In 1981 the party explicitly sanctioned the contract system in which households received land on multi-year contracts for which they were made responsible for significant labour input. Price reform was coupled with the retention of surplus above the co-operative quota by participating households.

7 More critically, fence breaking was a sort of spontaneous bottom-up innovation

by state-owned enterprises in which they swapped or sold goods on the open market in order to raise cash and buy materials or pay bonuses to workers (see Fforde and de Vylder 1989, and Ljunggren 1994: 20).

8 Good accounts of the political content of *doi moi* appear in Porter (1993), Pike (1992, 1994), Tonnesson (1992) Marr (1994) and Fforde and de Vylder (1996).

9 According to the party organ, the CPV was itself by 1990 in crisis; 30 per cent of the party cells were on the point of collapse and 60 per cent of party members 'failed to develop their vanguard role' (cited in Joiner 1990: 1060). By 1991 there were only 2 million party members according to Elliot (1993: 82).

10 During the colonial period the percentage of communal land in total cultivated acreage was roughly 25 per cent compared to 3 per cent in Cochinchina (Luong 1992a: 229).

11 According to Gourou (1936/1955), 34 per cent of all land-holding households owned less than 1 hectare of rice land and held roughly 12.5 per cent of the total rice area. While there is some variability in the surveys conducted on land distribution issues during the 1930s, there is a general consensus that large land-holdings and tenancy predominated and small-scale owner-occupancy was relatively insignificant, in sharp contrast to the North.

12 Luong's (1992a: 192) estimates are that average landless, poor and middle peasant holdings after reform were: 0.4, 0.48 and 0.57 mau respectively (1 mau = 0.36 hectares).

13 According to Selden (1993: 221), 74 per cent of the co-operatives in 1960 were 'low co-operatives' but by 1969, 90 per cent were 'high level co-ops' accounting for 84 per cent of cultivated land. This development from elementary forms of co-operation based on mutual and small-scale co-operation to fully collectivised co-operatives working on a labour point system was shaped profoundly by the Chinese experience of course.

14 *Product contracts* prescribed that a brigade should deliver a certain quantity and quality of grain from an assigned land area of specific fertility; *production cost contracts* provided seed, fertilisers and fuel to grow certain quantities of food; and *piece-work contracts* fixed work days for specific tasks (Long 1993: 167–8).

15 In 1977 the local government redistributed 70 hectares of landlord land to 50 poor village households, and 'encouraged' 117 middle peasant households to transfer rice land to 150 landless families. In other parts of the Mekong the proportion of households involved in such transfers was much higher (66 per cent or more) in contrast to the 21 per cent of households in Khanh Hau (Luong 1994: 89, 115).

16 Pingali and Xuan (1992: 703) argue that, when completed in 1974, the southern reform compared favourably to the North; land to the tiller redistributed 1.3 million hectares to 1 million farmers at an average of 1.3 hectares per farmer while by 1957 in the North 810,00 hectares were granted to 2.1 million farmers at 0.4 hectares per farmer.

17 Fforde (1989) has shown how collective property was in any case 'privatised' through negotiations between co-operative leadership and local households. In this sense the history of experimentation with the Maoist model of collectivisation long pre-dated the 1981 reforms (at least to the 1960s) since the degree of autonomy of the co-operatives from the state permitted party figures and cadres to work the system to their own advantage.

18 'Not only were output and yields rising too slowly to offset the increase in population, cooperatives were experiencing difficulties in mobilizing workers for collective labor' (Beresford 1990: 478). According to Lam (1993: 163), by 1974 one-quarter of all machines had broken down, rice output per capita had fallen 17 per cent since 1960, the cattle herd by 10 per cent, the cultivated area by 3.65 per cent and the costs of production were increasing three times as fast as personal income.

19 By 1979 the North had 4,150 co-operatives cultivating areas from 500 to 1,000 plus hectares. Over the same period the number of specialised brigades grew from 3,182 to 18,041.

20 Like the Chinese system, Vietnamese households were granted small private plots – typically gardens – the product being taxable but for personal use (and sale).

21 According to Kerkvliet (1993: 13), by the late 1970s 150,000 hectares of collectivised land lay fallow due to farmers' lack of enthusiasm for production.

22 Luong (1992a: 204) attributes this decline in Son-Duong village in Vinh Phu Province to (i) a decline in cultivated area per capita, (ii) declining state purchase price for paddy, which was only 12 per cent of market value, and (iii) a series of bad harvests in 1978–80.

23 The implications of resistance to collectivisation in the South were profoundly devastating for state procurements, which fell from 2 million tons in 1976 to 1.4 million tons in 1979. According to Beresford (1990: 473), the most serious falls occurred in the Mekong River delta (on the order of 60 per cent).

24 Furthermore, in Mekong half of the co-operatives were located in the relatively poor province of Tien Giang where 40 per cent of the households were members. In the most aggressively commercial rice provinces, co-op membership was barely 2 per cent.

25 In the case of Khanh Hau, the allocation of recently reclaimed land reduced the landless class to 9.8 per cent of all households compared to a figure of 83 per cent in 1957 and 58 per cent in 1959.

26 Land rental and land sale were, if not widespread, then at the very least on record in the Mekong region by the mid-1980s (10 per cent of households had purchased or rented land according to Luong 1994: 117).

27 The process by which Resolution No. 10 was instituted – like the subsequent Land Law in 1993 – was extremely varied. In some cases the collective maintained responsibility for tractor ploughing, irrigation, pesticide application and so on. In other instances, some of these services were privatised (though irrigation has been for the most part a non-private service on which taxes are paid either to the co-operative, local government or both). Luong (1992a: 212–13) describes how contracted land in Vinh Phu was classified by fertility and was distributed to different sorts of household contractors (poor, average and good), only the better contractors being permitted to *bid* for high fertility land. Poor quality land, conversely, was distributed to each household on the basis of the number of consumers.

28 By 1989, 95 per cent of all co-operatives in northern and central provinces had implemented Resolution No. 10. As a consequence of the reforms the number of co-operative personnel fell by 40 per cent between 1988 and 1990 in the North.

29 Article 3:2 of the Land Law says that 'any household or individual shall be entitled to exchange, assign, rent, inherit and put the right of land use in the pledge toward the land allocated by the state.'

30 These leases were extended to 99 years in 1994. While the Land Law extends to state farms and to forest lands, the privatisation of the former has proceeded very slowly (see Barker 1994).

31 As one might anticipate there are clear regional variations in the attitudes of local leaders – indeed of rural householders – to the central planks of the reform programme. If private ownership was rarely compromised in the South, there was continuing *support* for state ownership in much of the North. In 1991, for example, the provincial delegates of Thai Binh voted overwhelmingly to retain 'unified state management' of rural land even though private ownership was seen to have 'some advantages' (Elliot 1993: 89).

32 In Vinh Lac District in late 1994, wage rates for transplanting were 5,000 dong per day and 8,000 dong per day for irrigating and harvesting.

33 In the World Bank survey (1995: 168), 32.6 per cent of rural households hired labour; the figures for the Red River and Mekong were respectively 14.5 per cent and 70 per cent. For wealthier households, the proportion hiring labour was 30 per cent and 85 per cent respectively.

34 A survey in 1990 showed that in 26 districts across the country, the majority of co-operatives had no funds to work with and provided little more than farming advice and administered certain state policies (Kerkvliet 1995: 71).

35 Cultivated land per household is graduated from North to South: the average is 2,643 sq. m in the northern highlands, 3,231 in the Red River, 3,678 in the Panhandle delta, 5,468 in the Central Region and 13,814 in the Mekong (Department of Agriculture cited in Long 1993: 187).

36 The percentage of households with annual income above 800,000 in the Red River in 1990 was 2 per cent, compared to 16 per cent in Mekong.

37 Since the study was conducted Vinh Phu Province has been split into two provinces: Phu Tho and Vinh Phuc Districts have also been reorganised. Strictly speaking the study is located now in Phu Tho Province. I have, however, retained the nomenclature of the period in which the research was conducted.

38 The province currently consists of 11 districts, 413 communes, 2.3 million persons and an average population density of 463 per sq. km.

39 More recent data collected by John Gallup reveals that Vinh Lac's population density in 1993 was 1,250 per sq. km; the figure for population in relation to cultivated area rises to 2,000 per sq. km.

40 Cropping intensities actually vary quite substantially between villages, depending in large measure on the ability to grow winter vegetables and other crops. Cropping densities in Vinh Thinh village (an area adjacent to the Red River dikes in Vinh Lac District), for example, are roughly half that of Dai Dông village (1.6 versus 3.0 over the period 1990–3). Whether and how cropping densities can be increased is an issue that requires systematic field research.

41 Inflation over this period has been quite high, somewhere between 14–20 per cent over the last three years, and it is unclear from the village data whether income figures have employed real or nominal prices in their calculation. Calculations by John Gallup suggest that real GDP per capita in two Vinh Lac villages (Dai Dông and Te Lo) increased respectively by 18 per cent and 12 per

cent annually between 1990 and 1993. These figures are comparable with the Mien *et al.* (1995) study of Thanh Liem District in Nam Ha Province in the Red River delta.

42 Tuan *et al.*'s excellent study (1995: 146) in Hai Hung Province near Hanoi shows that net revenue in two villages increased by 80 per cent between 1988 and 1993; the growth sectors, like Dai Dông, were off-farm incomes and gardening, and secondarily livestock. Paddy land per capita fell by roughly 1.5 per cent in Hai Hung and increased very slightly in Dai Dông.

43 1 sao is 360 square metres; 1 thuoc is 24 square metres.

44 According to Long (1993) the rural gini coefficient in the Red River delta is 0.11 (compared with 0.26 in Taiwan, 0.42 in Indonesia and 0.40 in Bangladesh). My preliminary estimate of the land-holding gini coefficients in Dai Dông is 0.1.

45 When Gourou conducted his study he estimated a landless population of 36 per cent; in Dai Dông the figure is probably no more than 3 per cent. According to the World Bank (1995) the degree of landless nationally among all rural households is roughly 4 per cent.

46 Interview with Prof. Dr Chu Huu Quy, Chairman, Vietnam Integrated Rural Development Program, Berkeley, 29 November 1994.

47 In three villages in Ha Bac Province, the proportions of households hiring in and hiring out labour were as follows: 26 per cent and 68 per cent; 16 per cent and 15 per cent; and 34 per cent and 24 per cent (Huong 1994: 71–2).

48 I am grateful to Dr Phe, who provided me with this information.

49 Since 1989 transfers from state enterprises fell 66 per cent, subsidies (amounting to 4.9 per cent of GDP) have been eliminated, and fiscal revenues fell by almost half between 1989 and 1991 (Leipziger 1992: 8). This has necessitated 'forced reductions in the levels of current and capital public spending' (Lipworth and Spitaller 1993: 14).

50 In China ownership rights in land are vested in villages but private ownership rights are very much restricted (McKinley 1995: 15). In Vietnam the ninety-nine year leases introduced as an extension of the Land Law have *de facto* produced a land market in which authority is not vested in villages or communes but in district and provincial government.

Bibliography

Andreff, W. (1993) 'The Double Transition from Underdevelopment and from Socialism in Vietnam', *Journal of Contemporary Asia* 23, 4: 515–531.

van Arkadie, B. (1993) 'Mapping the Renewal Process: The Case of Vietnam', *Public Administration and Development* 13: 433–451.

Barker, R. (ed.) (1994) *Agricultural Policy Analysis for Transition to a Market Oriented Economy in Viet Nam*, Rome: FAO.

Beaulieu, C. (1993) *Whose Land Is It Anyway?* Hanover: Institute of Current World Affairs.

Beresford, M. (1990) 'Vietnam: Socialist Agriculture in Transition', *Journal of Contemporary Asia* 20, 4: 466–485.

Beresford, M. and McFarlane, B. (1995) 'Regional Inequality in Vietnam and China', *Journal of Contemporary Asia* 25, 1: 51–72.

Berry, S. (1989) 'Social Institutions and Access to Resources', *Africa* 59, 1: 41–55.

Bowles, P. and Dong, X. (1994) 'Current Successes and Future Challenges in China's Economic Reform', *New Left Review* 208: 49–77.

Burawoy, M. (1994) 'Industrial Involution: The Dynamics of a Transition to a Market Economy in Russia', paper presented to the SSRC Workshop on Rational Choice Theory and Post Soviet Studies, Harriman Institute, New York, 9 December 1994.

CDM (1995) 'Fleuve Rouge', *Les Cahiers d'Outre Mer* (Special Issue) April–June.

Chang, K. (1992) 'China's Rural Reform', *Economy and Society* 21, 4: 430–452.

—— (1993) 'The Peasant Family in the Transition from Maoist to Lewisian Rural Industrialization', *Journal of Development Studies* 29, 2: 220–244.

CIRAD (1995), *Durabilité du développement agricole au nord Vietnam*, Hanoi: Maison d'Edition de l'Agriculture.

Cuc, N. *et al.* (1990) *Ecology and Agroecosystems in Vinh Phu Province*, Honolulu: East–West Center.

Dinh, Q. (1993) 'Vietnam's Policy Reforms and Its Future', *Journal of Contemporary Asia* 23, 4: 532–553.

EIU (1994) *Indochina: Country Profile*, London: The Economist Intelligence Unit.

Elliot, D. (1993) 'Dilemmas of Reform in Vietnam', in W. Turley and M. Selden (eds) *Reinventing Vietnamese Socialism: Doi Moi in Comparative Perspective*, Boulder: Westview.

Fforde, A. (1989) *The Agrarian Question in North Vietnam 1974–1979*, New York: M. E. Sharpe.

—— (1993) 'Vietnam: Economic Commentary and Analysis', Canberra: ADUKI.

Fforde, A. and Porter, D. (1994) 'Public Goods, the State and Civil Society and Development Assistance in Vietnam', paper presented to the Doi Moi, the State and Civil Society Conference, Canberra, Australia, 10–11 November 1994.

Fforde, A. and de Vylder, S. (1989) *Vietnam: An Economy in Transition*, Stockholm: SIDA.

Fforde, A. and de Vylder, S. (1996) *From Plan to Market: The Economic Transition in Vietnam*, Boulder: Westview.

Gerschenkron, A. (1962) *The History of Economic Backwardness*, Cambridge, MA: Harvard University Press.

Gourou, P. (1936) *Les Paysans du delta tonkinois*, Paris: Ecole française d'Extrême-Orient (reprinted 1955, Human Relations Area Files, Yale University).

Greenfield, G. (1994) 'The Development of Capitalism in Vietnam', *Socialist Register* 30: 202–234.

Hart, G. (forthcoming) *Interstitial Spaces*, Berkeley: University of California Press.

Hendrey, J. (1964) *The Small World of Khanh Hau*, Chicago: Aldine.

Hickey, G. (1964) *A Village in Vietnam*, New Haven: Yale University Press.

Hunt, D. (1995) 'Prefigurations of the Vietnamese revolution', in J. Schneider and R. Rapp (eds) *Articulating Hidden Histories*, Berkeley: University of California Press.

Huong, T.(1994) 'Effect of Broadening Democracy in the Transition to Market Reform in the Countryside', Ho Chi Minh National Political Institute.

INSA (1995) *L'Agriculture du delta du Fleuve Rouge*, Hanoi: Maison d'Edition de l'Agriculture.

502

Irwin, G. (1995) 'Vietnam: Assessing the Achievements of Doi Moi', *Journal of Development Studies* 31, 5: 725–750.

Joiner, C. (1990) 'The Vietnam Communist Party Strives to Remain the Only Force', *Asia Survey* 30: 1053–1065.

Kautsky, K. (1899/1906) *La Question agraire*, Paris: Maspero.

Kerkvliet, B. (1993) 'State–Village Relations in Vietnam', Working Paper, Centre for Southeast Asian Studies, Monash University, Clayton, Australia.

—— (1995) 'Rural Society and State Relations', in B. Kerkvliet and D. Porter (eds) *Vietnam's Rural Transformation*, Boulder: Westview.

Kerkvliet, B. and Selden, M. (1995) 'Cycles of Agrarian Transformation in China and Vietnam', unpublished manuscript, State University of New York, Binghamton.

Kolko, G. (1995) 'Vietnam since 1975', *Journal of Contemporary Asia* 25, 1: 1–49.

Kornai, J. (1992) *The Socialist System: The Political Economy of Communism*, Princeton: Princeton University Press.

Lagrée, S. (1995) 'Evolution de l'agriculture vietnamienne dans un district du delta du Fleuve Rouge', *Les Cahiers d'Outre Mer*: 139–156.

Lam, C. (1993) 'Doi Moi in Vietnamese Agriculture', in W. Turley and M. Selden (eds) *Reinventing Vietnamese Socialism: Doi Moi in Comparative Perspective*, Boulder: Westview.

Lampland, M. (1995) *The Object of Labor*, Chicago: University of Chicago Press.

Leipziger, D. (1992) 'Awakening the Market', World Bank Discussion Paper No. 157. Washington, DC: The World Bank.

Lipworth, G. and Spitaller, C. (1993) 'Vietnam Stabilization and Reform 1986–1992', Washington, DC: IMF Working Paper.

Ljunggren, B. (1994) 'Beyond Reform: On the dynamics between Economic and Political Change in Vietnam', paper presented to Doi Moi, the State and Civil Society Conference, Canberra, Australia, 10–11 November 1994.

Long, N. (1993) 'Reform and Rural Development', in W. Turley and M. Selden (eds) *Reinventing Vietnamese Socialism: Doi Moi in Comparative Perspective*, Boulder: Westview.

Luong, H. van (1992a) *Revolution in the Village: Tradition and Transformation in North Vietnam*, Honolulu: University of Hawaii Press.

—— (1992b) 'Local Community and Economic reform: a microscopic perspective from two northern villages', in N. Jameison *et al.* (eds) *The Challenges of Vietnam*, Honolulu and Fairfax: East–West Center and George Mason University.

—— (1994) 'The Marxist State and the Dialogic Restructuration of Culture in Rural Vietnam', in D. Elliot *et al.* (eds) *Indochina: Social and Cultural Change*, Monograph Series No. 7, Keck Center for International and Strategic Studies. Claremont: Claremont McKenna College.

McKinley, T. (1995) *The Distribution of Wealth in Rural China*, Boulder: Westview.

Marr, D. (1994) 'The Vietnamese Communist Party and Civil Society', paper delivered to the Doi Moi, State and Civil Society Conference, Canberra, Australia, 10–11 November.

Mellor, J. (1993) *An Agriculture-Led Strategy for the Economic Transformation of Vietnam*, Rome: FAO.

Murray, C. (1980) *The Development of Capitalism in Colonial Indochina (1870–1940)*, Berkeley: University of California Press.

Oi, J. (forthcoming) *Rural China Takes Off*, Berkeley: University of California Press.

Pike, D. (1992) 'Vietnam in 1991: A Turning Point', *Asian Survey* 32: 1.

—— (1994) 'Vietnam in 1993: Uncertainty Closes In', *Asian Survey* 34: 1.

Pingali, P. and Xuan, V. (1992) 'Vietnam: Decollectivization and Rice Productivity Growth', *Economic Development and Cultural Change* 16: 697–718.

Porter, D. (1995) 'Economic Liberalization, Marginality and the Local State', in B. Kerkvliet and D. Porter (eds) *Vietnam's Rural Transformation*, Boulder: Westview.

Porter, G. (1993) *Vietnam: The Politics of Bureaucratic Socialism*, Ithaca: Cornell University Press.

Pryor, F. (1991) 'Third World De-collectivisation', *Problems of Communism* 12: 97–108.

—— (1992) *The Red and the Green*, Princeton: Princeton University Press.

Quang Truong (1987) *Agricultural Collectivisation and Rural Development in Vietnam*, Amsterdam: Vrije Universiteit te Amsterdam.

Quy, C. (1994) 'Personal Interview' (Chairman, Vietnam Integrated Rural Development Program), Berkeley, 29 November.

Riedel, J. (1993) 'Vietnam: On the Trail of the Tigers', *World Economy* 16, 4: 401–422.

Sachs, J. and Woo, M. (1993) 'Structural Factors in the Economic Reforms of China, Eastern Europe and the Former Soviet Union', paper presented to the Economic Policy Panel, Brussels, Belgium, 22–23 October 1993.

Selden, M. (1993) 'Agrarian Development Strategies in China and Vietnam', in W. Turley and M. Selden (eds) *Reinventing Vietnamese Socialism: Doi Moi in Comparative Perspective*, Boulder: Westview.

Shirk, S. (1993) *The Political Logic of Economic Reform*, Berkeley: University of California Press.

Stark, D. (1996) 'Recombinant Property in Eastern European Capitalism', *American Journal of Sociology* 101, 4: 146–153.

Szelenyi, I. (1988) *Socialist Entrepreneurs*, Madison: University of Wisconsin Press.

—— (ed.) (in press) *Reforming Collectivized Agriculture: Failures and Successes*, Boulder: Westview.

Thayer, A.C. (1992) 'Political reform in Vietnam', in R. Miller (ed.) *The Development of Civil Society in Communist Systems*, Sydney: Allen and Unwin.

Tonnesson, S. (1992) 'Democracy in Vietnam', NIAS Report No. 16, Copenhagen.

Tran, D. (1991) 'Socialist Economic Development and the Prospects for Economic Reform in Vietnam', Working Paper No. 2, East–West Center, University of Hawaii.

Tuan, D.T. (1994) 'The Agrarian Transition Process in Vietnam as Institutional Change', paper delivered to the Agrarian Question Conference in Wageningen, May 1994.

—— (1995) ' The Peasant Household Economy and Social Change', in B. Kerkvliet and D. Porter (eds) *Vietnam's Rural Transformation*, Boulder: Westview.

Tuan, D.T. (1997) 'The agrarian transition in Vietnam', in M. Spoor (ed.) *The 'Market' Panacea: Agrarian Transformation in Developing Countries and Former Socialist Econnomies*, London: Intermediate Technology Publications.

Truong Chinh and Vo Nguyen Giap (1937/8 [1974]) *The Peasant Question* (trans. C. White), Data Paper No. 94, Ithaca: Cornell University, Asian Studies.

Turley, W. and Selden, M. (eds) (1993) *Reinventing Vietnamese Socialism: Doi Moi in Comparative Perspective*, Boulder: Westview.

Verdery, K. (1992) 'Transition from Feudalism to Socialism', Lewis Henry Morgan Memorial Lecture.

Vietnam (1992) *Constitution of the Socialist Republic of Vietnam*, Hanoi: Government Press.

—— (1993) 'Vietnam: A Development Perspective', prepared for the Donor Conference, Hanoi, Vietnam.

Walder, A. (1994a) 'Corporate Organization and Local Property Rights in China', in V. Milor (ed.) *Changing Political Economies*, Boulder: Westview.

—— (1994b) 'The Varieties of Public Enterprise in China: An Institutional Analysis', unpublished paper, Department of Sociology, Harvard University.

—— (ed.) (1995) *The Waning of the Communist State*, Stanford: Stanford University Press.

White, C. (1985) 'Agricultural Planning, Pricing Policy and Cooperatives in Vietnam', *World Development* 13: 97–114.

World Bank (1993) *Vietnam: Transition to the Market*. Washington, DC: The World Bank.

—— (1995) *Viet Nam: Poverty Assessment and Strategy*, Washington, DC: The World Bank.

INDEX

Abrahams, R., on national identity
298, 303
accountability 36, 63, 66, 188
accumulation: Bulgaria 182; capitalism
82; China 433; Czech Republic
218–19, 234–5; land and labour
475–82; Mozambique 415–17; rate
of 83; regulation 27–8, 32–4, 47;
remittance 491–3; restructuring and
7, 389; second economy 61; Soviet
Union 39; Third World 410;
Vietnam 452, 484, 489–90
adaptability 42, 54, 58, 62–3, 66–7
Aggar, Ben, on fast capitalism 355
Aglietta, M., on concrete economies
433
Agricultural Circles (KZKiOR)
(Poland) 270, 274, 276–7, 279–80
agriculture: Bulgaria 123, 181, 332–4,
337, 347; Bulgarian path
dependency 243–59; China 437–8;
Estonia 293, 298–301; GDR 100–1;
Mozambique 415; Russia 150–1;
Soviet Union 83–4, 89; Vietnam
450–96
aid, foreign 397, 403, 415–17
Ali, Tariq, on legacy of Stalinism 408
Allen, P.M., on organisational analysis
55
Altvater, E.: on Soviet socialism and
Fordism 33–4; on transformation of
modernism 174
ambiguity and evolutionary theory 63,
65–6, 68, 71

Amin, A.: on globalisation 284, 288–9;
on socio-economic practices 31
Amsden, A.H., on developmental state
430
van Arkadie, B.: on agrarian reform
459; on 'double subordination' of
local and central government in
Vietnam 492
Arthur, W. Brian, on technology and
development 244
asset ambiguity 65–6, 68, 71
assimilation, of Roma 373–4, 381
autonomy 175, 351, 353, 363, 366–8,
374, 377, 410, 483

Balcerowicz-Sachs Plan (Poland) 270,
280
banking: bankruptcy and 204; Bulgaria
246, 250–1, 355; GDR 101;
Mauritania 395–6; Poland 273–4,
276, 278; Russia 164; Soviet Union
89, 99; Vietnam 460
bankruptcy 45, 91, 98, 204
bargaining, collective 35–6, 131,
179–80, 198, 200, 203–4, 206–8,
234, 252
barter 46, 85, 126, 131, 139, 186–9,
454
Bauman, Zygmunt, on social
transformation 351
Bayart, J.F., on African transformations
416
Beresford, M., on collectivisation 459,
469

Berger, John, on survival culture 375, 383

Berov, Lyuben 249, 353

Berry, Sara, on accumulation in Vietnam 489

'big bang' 429, 432, 456–7

Block, F.: on economic and social practices in Russia 167; on market regulation and planning 431

Bohemia, coal mining in 218–35

Borrowed Nature (Bulgaria) 360, 365–6

Bowles, Paul: on collectivisation in Vietnam 493; on worker commitment in China 444

Boyer, R., on implementation of capitalism 99, 105–6

branch plants: Bulgaria 116, 119, 124–5, 127, 130–42, 181–3, 186–7, 342, 361; GDR 44, 59; Soviet Union 37

Bristow, J.A., on wage setting in Bulgaria 133

Brody, H., on marginalisation of indigenous peoples 374

Buckwalter, D.W., on investment and economy in Bulgaria 117

Budapest, Roma in 377–80

budget balancing in Estonia 292, 295

budget constraints 60, 91, 126, 166, 432, 438–9, 458; soft 35, 179, 430, 435, 457, 493

built environment: Mozambique 413, 417

Bulgaria: democratisation and water politics in 347–68; GDP 77; imagined and imaginary equality in 330–43; industrial change in 115–42; path dependency in agriculture 243–59; production and politics 177–9; reform and flexibility 454; trade unions 201, 205, 208; transition 1–2, 5

Bulgarian Socialist Party (BSP) 2, 243, 247, 253–4, 339–40, 353–4, 359, 363–4, 366

Burawoy, Michael: on capitalism by design 457; on economic reproduction 172; on merchant capitalism 187; on privatisation and production in Russia 224; on regional integration in Russia 46; on Russian economy 91, 96; on social transformation 174; on 'transitology' 4, 173

Burbach, R., et al. on Russian regional restructuring 97

Cam Nguan, on land in Vietnam 482

capital: Bulgarian reserves 181–2, 189; exchange rates in Russia 149; flow in Mozambique 413, 417; foreign flows 42, 177, 205; investment in Bulgaria 118–19; nationalisation in Mauritania 390; nationalisation in Soviet Union 83, 85; political 116–17; 'social' 64

'capital hover' 149–50

capitalism: Bulgaria 140, 351, 355–7; China 435, 442–4; Czech Republic 219–20, 224, 226, 234; East Europe 106–7, 172–3; Estonia 290–2, 293, 297, 300, 302; 'frontiers' of 409–11, 421; globalisation and 284–9; Marxism 6, 8–9; regulating and institutionalising 25–47; Russia 166–7; Soviet Union 85, 89–99, 428; trade unions and 198–9, 204, 210; transition and 4–5, 184–7; western model of 429–30, 433

Ceauçescu, Nicolae 374, 382

central planning: breakdown 4, 39, 43; Bulgaria 123, 126, 132, 140–2, 181–2, 186–7, 190, 246–7, 250, 252, 255, 257, 334; China 432–3, 434–5, 437–8, 443, 496; Czech Republic 221, 223–4; modernisation and 172–4, 179; Soviet Union 32–8, 44, 46, 83, 89, 96; state and market 25; trade unions and 197–8, 200, 206; Vietnam 458–60, 462

centralisation 66, 202–3, 392, 403, 410, 452, 456, 469

de Certeau, M., on representational spaces 352–3

chemicals industry 122–4, 218–19

China: agriculture 258, 456, 458;
cultural revolution 410–11;
de-collectivisation 452; economic
reforms in 428–45; household
contract 473; industrialisation
495–6; land 476, 484; reform 2,
459; trade 490
Chissano, Joaquim, of Mozambique
413
citizenship 291–2, 297, 310, 312, 322,
342
Civil Initiative Committee for the
Protection of Rila Waters (Bulgaria)
360, 364, 366, 368
civil rights 198, 384
Clarke, S. *et al.* on enterprise
restructuring 224
class: difference in Bulgaria 142, 182,
351, 355; difference in Mozambique
416; difference and power in Estonia
297–8, 302; forms of struggle in
Czech coal mining 220, 222–3,
233–4; interest and Polish peasantry
262–81; rentier in China 442, 445;
trade unions and working 209–10
clean air legislation, Czech Republic
221, 227, 233
co-operatives: Bulgaria 130, 243,
245–9, 251–2, 253, 255, 257–8;
networks and 69; Roma 376; Soviet
Union 89; Vietnam 452, 454, 456,
467, 469–74, 476–7, 482–3, 487,
489, 494–6
Coalition Party (Estonia) 294–5
Cold War 418–19
collective action 36–8, 43, 45, 313–14,
319, 322–4, 338
collective farms 243, 246–7, 253–8,
262, 264–6, 270–1, 277, 300, 343,
376
collectivisation: Bulgaria 137, 179,
186, 188; China 434–7, 439, 441,
445, 493–4, 496; Mozambique 413;
Poland 265; Soviet Union 83, 86–7,
90; Vietnam 456, 459, 466–73,
475–6, 486, 489, 493–6
colonialism 392, 409, 411, 416,
418–19, 463, 465–6, 473

combinats 59, 100, 116, 121, 123–4,
126, 132, 162, 176–7, 180, 184–5,
361
command economy: Bulgaria 115–17,
125–7, 130–1, 133–4, 137–8,
140–1; control systems 175–8;
global 405; modernisation and 174;
Soviet Union 34, 83–4, 85
commercialisation 39, 452, 455, 474,
489, 492
Commonwealth of Independent States
(CIS) 92, 95, 100, 148
communism: Bulgaria 130, 134, 177,
246–7, 253, 255; collapse of 5,
76–9, 82, 284, 286; East Europe
107, 452; environment and 366;
Poland 262–70, 272–3, 275–9, 281,
314; Roma and 373, 379, 381, 383;
Soviet Union 83–9; trade unions
and 199–201
'company towns' 37
compartmentalism 58–61, 68, 71
competition: Bulgaria 119, 133, 142,
178, 252, 259; China 430, 432,
435–6, 438, 444–5; Czech coal
mining 221–4, 226–8; enterprise 60;
Estonia 293, 300; free market 56–7;
GDR 104; Novosibirsk 163, 165–6:
Poland 271, 277; Roma and 380–1,
384; Russia 149, 160; socialism 458;
Soviet Union 85–6, 89, 93, 96, 97,
98; transition and 187–8; Vietnam
454; wages and trade unions 203
conflict: trade union 205, 207–9,
223–5; war in Mozambique 415–20;
war in Vietnam 468–70, 494
construction industry 85, 91, 101, 333,
335–6, 460
consumer goods 85–6, 90–2, 100, 102,
131, 149–50, 161, 206, 251, 460
corporatism 65–6, 202, 312, 435,
440–1, 444, 458, 495–6
corruption 131, 415, 418–20, 430,
437, 463
Council of Europe and Estonia
299–301
Council for Mutual Economic
Assistance (CMEA or COMECON)

4, 45, 91, 104, 115, 119, 148–9, 172, 181, 183, 251
coupling: institutional 38–9, 47; loose 62–4, 71
Creed, G., on agricultural liquidation in Bulgaria 253
crime 96, 160, 418, 420
cropping intensity in Vietnam 463, 479–80, 488
culture: assimilation 373–4, 384; capitalism and 167; hierarchy and women in Bulgaria 331; identity in Bulgaria 351, 364; Polish women's political 314–16, 318; reintegration into capitalism 287; resources and political organisation 340; sub-, and networks 62; survival 375–6, 380–1, 384; Vietnam 458, 489, 492
cumulative causation 115, 117, 119
Czech Republic 1–2, 5, 92, 95, 205, 208, 336, 338, 341, 383; privatisation and coal mining 218–35
Czechoslovakia 32, 65, 77, 134, 172, 201, 382

Dai Dông village, Vietnam 479–89
de-collectivisation 249, 253–5, 258, 337, 440; Vietnamese agriculture 450–96 de-industrialisation 154, 161 174, 220, 343
de-monopolisation 2, 133, 178
de-regulation 38, 165, 310
debt 82, 91, 98, 107, 125, 181, 183, 391, 397–400, 486–7
decentralisation: Bulgaria 125, 132–3, 138, 141, 178, 188–9, 243, 258–9, 341–2; China 432, 438, 440; and multilateralism 405; networks 45; and Polish women 311, 324; Russia 147, 162; Soviet Union 34, 98–9; trade unions and 206; Vietnam 458, 463, 476, 485–9
democracy: African 421; Bulgaria 132, 140, 341, 343, 364; civil society and 197; East Europe 4–5; Estonia 290, 291, 300–2; Poland 262, 266, 322; revolution and liberal 284–9; Third

World socialism 409; trade unions and 206–7
democratisation: breakdown 39; Bulgaria 188–9; Bulgarian water politics 347–68; Czech Republic coal mining 227–8; East Europe 1–2, 4; Estonia 297; modernisation and 173, 178; role of state in 309–12; Soviet Union 36, 88, 90
denationalisation of the Mauritania state 389–405
Deng Xiao Ping, of China 436
Derrida, Jacques: on deconstruction and Marxism 6; on neo-liberalism 8; on 'phantom states' 420
Deutscher, I., on Stalinism 408
development, economic 8–10, 31, 39–40, 57–8, 69, 200, 205–6, 210
Dhlakama, Afonso, of Renamo 413
differential development and comparative transitions to capitalism 76–108
discrimination against Roma 383–4
discursive constitution of social change 26–7, 31
disinvestment 246, 257, 361
distribution systems in China 430, 432, 440
division of labour 104, 172, 331, 336
Djerman-Skakavitsa water project (Bulgaria) 347, 350–1, 352–9, 359–66
doi moi reform in Vietnam 451, 454, 459–60, 462, 489, 492
domination: command economy 175–7; enterprise 123–4, 126, 157; of Estonia by foreign powers 298, 302; of ethnic minorities 375, 381–2; gendered power in Poland 311, 313–14, 317, 31; of state 30, 184; state and socialism 9
Dong, Xao-Yuan: on collectivisation in Vietnam 493; on worker commitment in China 444
drought 353, 357, 359, 403
dualism 25–6, 199
Dupre, J.A., on evolutionary models 57

Ecoglasnost-National Movement (Bulgaria) 360, 364–6
ecology, locality as 66–8, 70–1
'economic involution' 96
Economic Social Council of the Basin Region (ESCBR) (Czech Republic) 233–4
economy: denationalisation of Mauritania 392–5; transformations in China 428–45
education 61, 101, 331–2, 334–5, 380, 383–4, 413, 416, 437, 440
efficiency 34–5, 54, 56–60, 62–3, 67, 244, 431
Eisenstadt, S.N., on collapse of communism 284
elections: Bulgaria 337, 339, 342, 353, 358–9, 363–4, 367; Estonia 290–2, 294–7, 301; Mozambique 413; Poland 316–17, 319
elitism 61, 85, 96–7, 134, 409–10, 417–18, 441–2, 444
Ellis, S., on long-distance trade 420–1
embeddedness 28, 30–1, 41, 44–5, 167, 173, 178, 180, 430
employment: Bulgaria 181, 183, 331–2; China 493, 495; Czech Republic 218, 223, 225; East Europe 78–9, 330; Estonia 300; GDR 100–4; in Novosibirsk 162; of Roma 376–9, 384; Russia 153–60; Soviet Union 37, 83, 85–6; Vietnam 455, 493
empowerment 188, 310, 312, 319, 324, 368
energy industry 149, 160, 218, 226
enforcement of market exchange 430–2
'enterprise desertion' 186
enterprises: in Bulgaria 116, 123–30, 140, 186–90, 334, 336, 355; Chinese town and village 429–30, 433–5, 438–45, 495–6; in Czech Republic 221–5, 234; domination 157; in Estonia 293–5, 294; in GDR 103; in Mozambique 413; network analysis and 58–61, 65–6; networks and FDI 42–3; nomenklatura 211; Novosibirsk 162;

restructuring state 172–90; Roma 378; in Russia 150, 157–60; in Soviet Union 34–5, 37, 40, 45–6, 85–7, 89, 91, 93, 96–7, 99; trade unions and 205–7; in Vietnam 459–60, 463, 479, 483, 486–7, 491
entrepreneurialism: Bulgaria 139, 178, 188, 190, 246, 257; China 435, 438, 444–5; Czech Republic 234; Mozambique 415, 418; network analysis and 56, 60–1, 63–5, 68–71; nomenklatura 490; Poland 268, 281; Russia 157; Soviet Union 90, 96; trade unions and 206
environment: despoliation in Estonia 297; harshness in Russia 154, 158, 160; management in Bulgaria 353, 363–6; mining devastation of Bohemia 219–20, 222, 224, 226–33, 235; Neftochim and reconstruction of Bulgarian 183, 185, 189–90
Environmental Impact Assessment 363–6
equality: economic spatial 76–82; imagined and imaginary in Bulgaria 330–43; in Poland 267–8, 278
Estonia, rurality and construction of nation in 284–303
ethnic minorities: imagined and imaginary equality in Bulgaria 330–43; Russian-speaking in Estonia 290–2, 299, 303; social exclusion of Roma 373–84; Turks in Bulgaria 116, 130, 134–7, 135–9, 142, 249, 254
European Union (EU): integration of Estonia 287, 289–90, 292, 295, 300–2; transformation of south 107; unification of Germany 100, 104–5
evolutionary theory 36, 54–71, 431
exchange rates 103–4, 390, 395–6
exclusion: of Bulgarian women 331, 342; of Polish women 310–12, 314, 323; of Roma in transition 373–84
expenditure, state 89, 96–8, 105, 164, 181, 221–2, 435, 489

exports 119, 153, 155, 160–1, 163, 251, 258–9, 389, 451, 455, 462, 474

Falbr, Richard, on trade union 206
family farms in Poland 262–4, 270–1, 273, 275, 277, 281
Farkasova, Etela, on socialism and gender equality 331
Farmers' Trade Union 'Self-Defence' (Samoobrona) (Poland) 274, 276–80
Fassman, H., on unemployment in East Europe 128
Federal Republic of Germany 100, 105
feminism 311, 319–20, 338, 340
Fforde, A.: on agrarian dynamism 455; on collectivisation 459, 469; on macroeconomic stabilisation in Vietnam 454; on socialism and Vietnam 462
Fitzgerald, J., on central-local reform in China 443
flexibility 35, 45, 66, 86, 98, 174, 375–6, 453, 462, 483, 493
Fordism 29, 33–4, 38, 174
foreign direct investment 42–4, 46, 116–18, 157–8, 160, 174, 293, 300, 302, 389, 435, 442
Fossum, J.A., on role of trade unions 203
Foucault, Michel: on 'governmentality' 351; on liberalisation 2; on political modernity 368
fragmentation of land in Bulgaria 245–8, 255, 257
Fraser, N., on subaltern counter-publics 311
freedom 286–8, 301, 331, 341
Frente de Libertação de Moçambique (Frelimo) 412–16, 418–19, 421
friction, economic 70–1
Friedman, Thomas, on Vietnam 450
Fukuyama, Francis: on end of history 8; on neo-conservatism 286; on transition 197

Gabor, I., on second economy 60–1, 69

gaje, Roma and 376–7, 381
GDP: Bulgaria 177, 355–6; China 428; CIS and East Europe 148; Estonia 293; GDR 100–2, 104–5; hierarchy by major regions 76–82; Mauritania 393, 397, 400; Russia 150–1; Soviet Union 88–9, 91–2, 94–5; Vietnam 455, 460, 462, 479–80
geography: Bulgarian agriculture 253–5; industrial and transition 172–90; political interest in Poland 267–8; trade unions and 198, 201, 204–5, 210; voting in Estonia and 295–6; water and power in Bulgaria 347–52
geopolitics of transition 1–6, 9–10
George, S., on World Bank 419
German Democratic Republic 44, 59–60, 76, 82, 201, 338, 342, 343; transition to capitalism 99–107
Germany, trade unions 202, 206–7
Gheyselinck, T.O.J., plans for coal industry in Czech Republic 221–2
Glenny, Misha, on patronage 253
globalisation: cultural difference and 351; East Europe 39–40; Estonia 284, 286–8, 289, 292, 297, 301; European Union 107; locality and 67; and multilateralism 404–5; Novosibirsk 165–6; regulation theory 29; Soviet Union 43; transition and 4, 7; Vietnam 456, 466
Goodman, D.S.G., on welfare in China 437
Gorbachev, Mikhail, of Soviet Union 220, 341, 458, 460
Gourou, Pierre, on Red River Delta, Vietnam 466, 475–6, 479, 483, 490
governance 25, 27–32, 68, 149, 332–3, 335, 410, 419–21, 443
'governmentality' 351, 358, 365
Gowan, P.: on communism to capitalism 90–1; on global political economy 285, 289, 302
Grabher, Gernot: on GDR 44; on Soviet system geography 37

gradualism 436, 456–8, 469, 493
Graham, A., on gendered politics 315
Green Patrols (Bulgaria) 360, 365–6
Guangdong, China 435, 438, 441–4
Gurvich, I., on ethnic minorities in Siberia 374

Habermas, Jürgen: on discursive democracy 367; on modernisation and political freedom 286–7, 302; on social movements 352
Hachey, Jr, G.A., on Estonian independence 293
Hanlon, J., on Economic Rehabilitation Programme in Mozambique 415
Harrison, M., on labour and wage increase 87
Hausner, J., *et al.* on accumulation and transformation 39; on regional development in Poland 70
Havel, Vaclav, of Czechoslovakia 341
Havelkova, Hana, on political organisation of Czech women 341
Hayenko, F.S., on role of Soviet trade unions 200, 201
heavy industry 150, 161, 162, 165–6, 177
Held, David, on democracy and global order 287, 288, 301
Hendrey, J., on Khanh Hau reform 468
heritage 140, 272, 297, 299, 444–5
'hidden unemployment' 150, 157
Hirst, P., on nation-state and globalisation 288
history of Bulgarian agriculture and legacies 244–7, 257
Ho Chi Minh, of Vietnam 458, 471
Holloway, J., on internal-external politics 416
'hollowing out' of nation-state 288, 301
household: contracts in Vietnam 455, 459, 470–1, 473, 476, 486, 491, 494; registration in China 437, 494; survival strategy 7

housing 91, 294, 377–9, 384, 411, 414, 437
Hungary: agriculture 458; corporatism 59–61; GDP 77, 92, 94; privatisation 96; property and corporatism 65; recombinant property 443; reform 131–2, 134; reform and flexibility 454; Roma in 376–81; second economy 32; as specialist industrial region 174; trade unions 201–2, 244

identity: in Bulgaria 332–3, 340–1, 351, 353, 368; collective 62, 69; in Estonia 287–92, 294, 297–303; in Mauritania 392; in Poland 263–4, 266, 278–9, 310–11, 314, 317–18, 320, 322, 324; of Roma 373, 375–7, 381; in Third World 409
Illner, M., on enterprises and cities in Czech Republic 231
incentives 66, 87, 90, 116, 131, 176, 178–9, 184, 186, 438, 440
income: Bulgaria 251, 335, 337, 342, 361–2; China 437, 440–1; Estonia 294; GDR 102, 104–5; Poland 268, 270–2, 271, 275; of Roma 377–8, 378, 384; Soviet Union 92, 98; Vietnam 460, 465, 467, 469, 473–5, 479, 481, 484–5, 488
independence: Estonia from Soviet Union 290–7, 299, 303; Mauritania 391; Mozambique 411–12, 414, 416; trade unions 200–2
Independent Self-Governing Union for Individual Farmers Solidarity (NSZZ RI) (Poland) 266, 268–9, 273–81
Indjova, Renata, of Bulgaria 353, 358
industrial districts, Italy 67–9
industrialisation: Bulgaria 138, 141, 331; China 437, 440–3, 495–6; East Europe 10, 78; Estonia 298–9; Russia 162; Soviet Union 33, 83; Vietnam 455, 468
industry: Bulgaria 177, 181–5, 332–3, 335–7, 347; change in Bulgaria 115–42; decline and restructuring 173–4; Russia 150–1; Soviet Union

83, 86, 88–9, 93; Vietnam 460, 462, 486–78, 492–3
inequality: in China 445; gender in Poland 311, 313, 315, 318–21; regional in Bulgaria 117, 120; Roma and economic 378–9, 384; in Soviet Union 92–3; transition and 5, 7
inflation 1, 85, 89, 91, 98, 104, 149, 226, 355–6, 455, 458, 460
innovation 58, 66, 87, 90, 108, 164, 491
institutionalised capitalisms and regulation theory 25–47
institutionalisation and evolutionary theory 54–8, 65, 67, 70–1
institutions 44–5, 82, 88, 98–9, 100, 105, 107–8, 118, 454
integration: Estonia and West 287, 289–90, 297; Poland and West 274–5
interest rates 250, 257, 259, 273
International Monetary Fund 5, 98, 286, 292, 301, 428; and Mauritania 391, 393–9, 401, 405
internationalism 103, 160, 163, 165–6, 184, 207, 289, 390, 416
investment: Bulgaria 116–20, 132, 138, 181, 183, 185, 187, 189, 246, 257–8; China 432, 495; Czech Republic 223; GDR 44, 104–5; loose coupled networks 63; Mauritania 391, 393, 399–401; modernisation and 172; in Novosibirsk 164–5; Russia 149–50, 160; Soviet Union 35, 83–7, 89, 90, 93, 96, 98, 99; Third World 411; Vietnam 479, 491, 493
invisibility: of gender in Poland 313–23; of Roma 381–2
irrigation 256–7, 463, 475, 487–8

Jalusic, V., on 'political' and Slovenia 314
Janowski, Gabriel, Polish agrarian leader 275
Japan 202, 389, 428
Jaruzelski, General Wojcieich, of Poland 269

Judd, E.R., on Qianlrulin village Shandong 440

Kagarlitsky, B., on modernism in Russia 9
Kahk, J., on national identity 298
Katz, H.C., on trade unions 197
Kautsky, Karl, on nineteenth-century European rural economy 455
Kemény, I., on Roma in Hungary 377–8
Kerkvliet, B., on collectivisation 459, 494
Khanh Hau, Vietnam 468, 470, 471
Kim Ngoc, party secretary, Vietnam 476
Kisczkova, Zuzana, on socialism and gender equality 331
Klaus, Vaclav, of Czech Republic 2, 65, 206
Knight, C.G., et al. on water shortage in Bulgaria 353
Kochan, T.A., on trade unions 197
Konakchiev, Doncho, Bulgarian minister 358
Korbonski, A., on peasant dissent and land reform in Poland 269–70
Kornai, Janos: on budget constraints and socialism 438; on central planning and budgets in Bulgaria 179; on communist power 450, 452, 454; on party and reform 496
Kovács, Z., on Roma in Budapest 378
Krastev, Amadeus, of Green Patrols 365
Kremikovtsi, Bulgaria 124–6, 358
Krotov, P.: on economic reproduction 172; on privatisation and production in Russia 224; on regional integration in Russia 46
Kuczi, T., on network linkages 68–9, 70
Kuratowska, Zofia, on women in Polish political parties 316

labour: Bulgaria 126, 136–7, 177; collective 43; Czech Republic 224–5, 235, 336; Estonia 299;

Fordism and 33–4; German 103;
  hoarding 87, 131, 179, 200, 206;
  intensity 262, 442, 444–5, 484–5;
  land and accumulation 475–82;
  relations and transition 197, 201,
  203–4, 209–10; Russia 149, 159,
  163; shedding 115–19, 127, 137,
  140–1, 157, 181–3, 188; Soviet
  Union 35, 37, 83, 85, 87, 173
Ladányi, János, on Roma in Hungary
  376, 377, 378, 379, 383
Laïdi, Z., on Soviet model of socialism
  409–10
Laki, L., on education of Roma 380
land: access in China 445;
  de-commodification 410; holding in
  Bulgaria 245–50, 257; holding in
  Poland 262, 266; holding by Roma
  376; holding in Vietnam 452, 458,
  465–71, 473–4, 476–80, 482–5,
  488, 491–3, 495–6; labour and
  accumulation 475–82;
  nationalisation in Mozambique
  413–15, 417; reform in Estonia 295,
  297; rights 482–3, 484, 491, 494
language and ethnic identity in
  Bulgaria 332–3, 340
legacies and evolutionary theory 56–61,
  64–5, 71
legislation: Bulgarian agriculture 243;
  Decree 56 (Bulgaria) 132; ecological
  in Czech Republic 220, 226–7, 235;
  Environmental Protection Act
  (Bulgaria) 365; Equal Status Act
  (Poland) 320–1; Estonian enterprises
  293; Land Act (Bulgaria) 247–50,
  253, 257; land laws in Vietnam 468,
  470, 473, 474, 482–3, 491; Land
  Reform Law in Estonia 297–8, 303;
  Polish elections and women 316–17;
  trade union 208
Lenin, V.I., of Russia 8, 162, 197,
  457–8, 468, 491
Leninism 33, 409–11, 413, 450
Lepper, Andrzej, Polish agrarian leader
  274, 276
Leyshon, A., on 'phantom states' 417
liberalisation: Bulgaria 123, 142, 178,

243; central planning 39; China
  495; East Europe 2, 172; Russia
  151, 160, 164; transition and 310;
  Vietnam 450, 454, 460, 462, 477,
  490
liberalism 173–4, 178, 187, 189,
  313–14, 421
Lieven, A., on Estonian politics 294
Lin, N.: on local corporatism 440; on
  local market socialism 440, 443
linkages: Bulgaria 126, 131, 139, 142,
  178–9; and evolutionary theory 56,
  62–6, 68, 71; Neftochim 185–7;
  supplies 148, 155
Lipietz, A., on regulation theory 141
liquidation of property 100, 248–9,
  253, 343
Ljunggren, B., on Vietnamese reform
  455
localism: Bulgaria 175–8, 253–5, 334,
  341–3; Bulgarian water crisis 349,
  351–2, 360, 364–7; capitalism and
  accumulation 27; China 429, 432,
  434–6, 438–44; Czech Republic
  227–35; gender and transition
  309–10, 312, 318–19, 321–4;
  Neftochim 185–6, 188–9; Poland
  268, 279; Russia 147, 150–1, 154,
  160; Soviet Union 37–8, 97;
  transition and trade unions 210;
  Vietnam 452, 483, 485, 489, 492
localities and evolutionary theory 31,
  56, 62, 66–71
Lonc, T., on farmers' organisations 270
Long, N., on reform and output in
  Vietnam 473
Lugus, 0., on Estonian independence
  293
Luong, H. van: on land-holding and
  peasantry in Vietnam 467, 469; on
  reform in Vietnam 452; on rice
  yields in Vietnam 471; on tax
  loadings in Vietnam 488

McDermott, G., on networks in
  Czechoslovakia 65, 70
McGlade, J.M., on organisational
  analysis 55

McIntyre, R., on production quotas in Bulgaria 131–2
Maddison, A., on productivity and economy 78
mafia 7, 91, 96, 358–9
Makó, C., on network linkages 68–9, 70
managers: appointment and control 175; command economy in Bulgaria 116, 131–2; Czech coal mining 222, 224, 226, 235; enterprise behaviour and 178; labour in Soviet Union 35; Neftochim 184, 186–9; power and networks in Bulgaria 134, 139; profitability and 66; trade unions and 197–9, 204
manufacturing 100–1, 124, 149–50, 157, 161, 181, 293, 428, 438
Mao Zedong, of China 494
Maputo, Mozambique 416–17, 420
marginalisation 310–11, 321, 323, 373–5, 377–9
market: Bulgarian agriculture 251–2; economic co-ordination in Russia 148, 154; exchange 430–1; free 56–7, 61, 167, 284–5, 288–9, 318; free in Estonia 291, 292–5, 299–303; orientation in China 433–4, 436–8, 443; orientation in Mauritania 392, 403; orientation in Vietnam 450–1, 459, 462, 479; property in Mozambique 414–15; state and 25–6, 29; unification in Mauritania 390
market economy: Bulgaria 115–17, 119, 127, 255, 258–9; China 432; Czech Republic 220, 224; East Europe 82, 105, 107, 108, 429; industrialisation 187, 190; institution 98–9, 107–8; modernisation 173, 178; Poland 262, 264–5, 272, 275, 277–8; regulation 42; Roma and 373–81, 384; Russia 149, 151, 158; socialist agriculture 458; Soviet Union 89–90, 93, 96–9; trade unions and 197, 199, 203–4, 206–9; transition 2, 4, 389

marketisation: brown coal mining 226–7; Bulgaria 187, 189, 352; capitalism 25; China 429–30; command economy 178; East Europe 2, 4–5, 27; efficiency 54; networks 45–6; regulation 45; Russia 154, 157, 162, 167; second economy 60; Third World resistance 410; trade unions 206, 209; transition 40; Vietnam 474
Markoff, J., on collapse of communism 285
Marx, Karl, on market exchange 430–1
Marxism 8–9, 27–8, 374, 428; –Leninism 9, 409–11, 413
Mauritania, denationalisation of 389–405
Mazowiecki, Tadeusz, of Poland 263, 273
Mekong delta region (Vietnam) 455, 459, 463–8, 470–2, 473–5, 492
merchant capital 8, 187, 416
migrant workers 139, 442, 444, 465
migration 87, 104, 155, 377–8, 379
Milanovic, B., on poverty 7
military defence industry 86–7, 150, 161, 162, 164, 166
mining 100–1, 160, 218–35
modernisation: Bulgaria 118, 182, 245; East Europe 9, 172–4, 374–5; Estonia 290, 292, 300–1; political freedom and 286–8; socialism and elitist 409; Soviet Union 83, 93; trade unions and 201; transition and 4–5; US aid and 81; western model of 197, 210
Mokrzycki, E., on communist legacy in Poland 272–3
monopoly 96, 98–9, 186, 189, 226, 252, 258–9
Mooney, Patrick H., on class interests in Poland 262–81
Most Coal Company (Czech Republic) 224, 230–2, 235–9
Most region, Czech Republic 218–20, 223–4, 227–39
Movement for Rights and Freedoms (MRF) (Bulgaria) 134–5, 247,

249, 254–5, 337, 339–40, 342–3, 353
Mozambique, aftermath of socialism 408–21
Mukha, Vitalii, governor of Novosibirsk 165
Multigroup (Bulgaria) 355, 358–9
multilateralism 2, 5, 389, 391–3, 398–402, 404–5
Murphy, A.B., on regional industrial production in East Europe 120
Murray, C., on landholding in Vietnam 465
Murray, R., on Soviet Fordism 33
Myrdal, G., on regional poverty and inequality 117

Naskapi Indians, Labrador 54–5, 68, 71
nation construction in Estonia 284–303
nation-state 287–8, 391, 404
nationalism 266–8, 285–7, 289, 291, 301–2, 381–3
Neftochim (Bulgaria) 121, 123–6, 177, 183–9
neo-classicalism 28, 55–7, 115–18, 130, 140, 430–1, 433, 457
neo-liberalism 2, 5, 7–11, 25–6, 32, 39–40, 46–7, 54–6, 65, 82, 89, 93, 97, 99, 107–8, 116, 165, 220, 258, 294, 430, 454, 456, 489
network analysis 29–31; and post-socialism 54–71
networks: Bulgaria 118, 130–1, 134–7, 138–41; East Europe 47; gendered in Poland 318, 324, *guanxi* 432–4, 437, 440, 443–4; Polish agriculture 264–5, 281; regulation and 39, 42, 45; Soviet Union 34–8, 44, 91; survival and 7; Vietnam 389–90, 463, 492
Newly Industrialising Countries 430, 450
Nguyen Van Linh, Vietnamese reformer 460
Nolan, P., on performance in China 428

nomenklatura 96, 118, 132, 134, 140, 206–8, 210, 275–6, 351, 459, 490
non-governmental organisations 309, 318, 320–2, 324, 360, 364–6, 391, 403–4, 416
North Atlantic Treaty Organisation (NATO), Estonian integration 287, 300–1
North Bohemian Brown Coal Mines (NBBCM) 223, 228
North, Douglas, on state and reform in Bulgaria 244
Novosibirsk, Siberia 149, 161–7

oblast and *obstina*, industrial unemployment in Bulgaria 128–30, 134–6, 138
off-farm enterprise 452, 479, 484–5, 488, 491–2
Offe, Claus, on transition of East Europe 263, 280–1
Oi, Jean: on incomes in China 494; on wealth distribution 444
Okely, J., on Roma and manipulation of *gaje* 381
oppression 311, 313–14, 322, 331, 373
Ost, David: on class and trade unions 209; on transition of East Europe 263, 280–1
Ottoman empire, legacy for Bulgaria 245–6, 316
output: Bulgaria 115, 118, 120–4, 131–2, 140, 181; Bulgarian agriculture 243, 246, 250, 256, 258; China 438, 441, 495; Czech coal mining 225; GDR 100–1, 103–4; Russia 148, 150–5; Soviet Union 91–2, 96; Vietnam 451, 455, 460, 470–1, 473–4, 479, 484, 494

Paris Club 391, 397–400, 405
participation, political 312–13, 315–16, 318–22, 324, 331, 337–43
paternalism 246, 255, 318
path dependency 10, 26–7, 39, 46, 57–9, 421, 431, 434, 444–5
patriarchy 311, 315, 323, 331

Peasant-Christian Party (SL-Ch)
(Poland) 274–5, 281
peasantry: in Poland 262–81; in
Vietnam 454–5, 465–71, 473–5,
483, 491–2, 494–6
Pelissier, R., on Third World socialism
411
perestroika 88–9, 160
periphery: Bulgaria 119–21, 128,
138–9, 142, 177, 181–3; collapse 5;
North Rila as 351–3, 355, 360–4;
Roma in Hungary 377–9, 384;
Russia 154, 158
phantom states 417–20
Pike, D., on Vietnamese
authoritarianism 463
Piore, C., on capitalism and
institutionalism 25–6
Piore, M.J., on market economy 204
'plan bargaining' 35
'plan capture' 36–7
plan circumvention 36–7
plant closure 115–6, 129, 138–9, 183
Poland: class interests in 262–81; GDP
77, 92, 95; locality and development
70; 'political' meaning for women in
309–24, 338; reform 134, 141, 454;
shock therapy 89; trade unions
201–2, 208–9, 244
Polevanov, head of Russian
privatisation 93
policy framework papers in Mauritania
391, 393–5, 400–1
Polish Peasant Party 'People's
Agreement' (PSL-PL) 274–5, 277,
279, 281
Polish Peasant Party (PSL) 273–5, 277,
279–80
Polish United Workers' Party (PZPR)
265, 274
politics: Bulgarian agriculture 247–50,
254–5; change in Central and East
Europe 350–2; economy of Estonia
289–97; Polish and agriculture
264–81; reorganisation in Bulgaria
337–42; rights of Polish women
310–12, 314, 318–19, 321, 324;
transitional in Poland 309–24

pollution 177–8, 185, 188, 219, 354
Popular Front (Estonia) 290, 294
population density in Vietnam 462,
464, 476–7, 482–3
Porter, D., on socialism and Vietnam
462
Porter, G., on reform and output in
Vietnam 473; on Vietnam
agriculture 468
Poulantzas, N., on international
integration 107
poverty 7, 117, 450, 455, 462, 482
power: accumulation and 489; of
Bulgarian Socialist Party and
agriculture 246–8; of capital 26;
power of command economy 175–6; of
communism 34; of Czech state and
coal mining 219; domination and
ethnic minorities 375, 381–2; of
enterprise in Bulgaria 125–6, 133;
gendered, in Poland 311, 313–14,
317, 321; of managers 141–2; of
markets in China 433; of Neftochim
185; regional in Bulgarian
agriculture 253; social and political
in Bulgaria 351, 358; of state in
Vietnam 491; of trade unions 210;
of trans-national institutions 416–17
Pravda, A., on trade unions 199
Pred, A., on cultural difference and
territorial identity 351
prices: agricultural in Poland 270–1,
274–6; GDR 102, 104; liberalisation
of Czech coal 220–1, 223, 226;
liberalisation in Soviet Union 89,
96–7; relative in Bulgarian
agriculture 250–2, 256, 258–9;
relative in Russia 151, 157; relative
in Soviet Union 90, 92–3
private farms: Bulgaria 243, 245–9,
252, 254–9, 337; Estonia 297–8,
300–1, 303; Poland 262–4, 270–1,
273, 275, 277, 281
privatisation: Bulgaria 115–16, 118,
123, 132–4, 139, 182, 184–6,
188–9, 336, 342–3, 352; Bulgarian
agriculture 243, 244, 252, 256–7;
central planning 39; China 429,

438, 440, 495; coal mining in Czech Republic 218–35; command economy 178; efficiency and 54 65; Estonia 293–5; GDR 100, 104; Hungary 244; Poland 265, 275; regulation and 42; role of state in 309; second economy 60; Soviet Union 90–1, 93, 96–8; trade unions and 204, 206–8, 210–11; transition and 2, 4, 310; Vietnam 458, 473–4, 476, 482, 494–6; western society and 197–8

production: Bulgarian agriculture 36, 61, 363; Bulgarian decline 115, 119–20, 123–4, 126, 132–3, 138–9, 141, 181–2, 188, 251; capitalism 442, 444–5; China 430, 432, 435, 440; CMEA 172; Czech coal mining 218, 223–8, 235; Estonia 293; GDP in East Europe 78–9, 82; GDR 102, 104–5; Mauritania 390; Mozambique 413; Novosibirsk 162; organisation and transition 60, 173–4; Poland 262, 265–6, 268, 270, 273; quotas and state control 175–6, 179; quotas and trade unions 197–8, 200, 203; Russia 150–4, 157; second party 186; Soviet Union 32, 33–5, 37, 84; specialisation 91, 97, 99, 105, 174, 246, 252, 255–8; systems diversity 28; Vietnam 456, 458–9, 461, 464, 467, 469–71, 473, 478–9, 486, 492, 494

profit 64, 66, 184–5, 188–9, 221–3, 225, 228, 231, 439

property rights 90, 98, 149, 245, 247, 454, 456, 460, 491, 494; in China 428–45

protection: Bulgaria 116, 181, 183–5, 187, 189–90; Czech Republic 223; enterprise 125, 131, 139–40, 177; Poland 274–5, 277, 280; Russia 154; trade unions and 199–202, 207

protest: agricultural, in Poland 263, 265–6, 268, 271, 273–80; collectivism in Vietnam 459; local and Bulgarian water crisis 347–68;

union dissent 201–2; women in Bulgaria 340–1

Pryor, Frederic, on de-collectivisation 450

Putnam, Robert, on differential development in Italy 244

racism 373, 377, 380–4, 419

rationalisation 116, 131, 139, 142, 166–7, 186–8, 224–6

rationality 36, 40, 42–3, 45, 54

recombinant property 43, 58–9, 65–6, 71, 96, 204, 433–4; capitalism and Vietnam 450–96

Red River Delta, Vietnam 452, 455, 463–6, 469, 472–85, 489, 491–2

Reform Party-Liberals (Estonia) 294–6

regionalisation: Bulgaria 115–42; China 429; East Europe 47; Russia 147, 151–61; Soviet Union 37–8, 43–6, 87–8, 91, 93, 97; Vietnam 455, 463, 466, 468, 474–5, 485, 492

regulation: accumulation and 433–4; Bulgaria 118, 141–2, 177–8, 181–5, 188; enterprise reporting 175; Estonia 289; Mauritania 390, 395; modes of 76–108, 25–47; Russia 149, 167–8; theory and institutional capitalisms 25–47; transition to capitalism and 76–108; Vietnam 494

relations: enterprises and local community in Czech Republic 228–35; social of enterprises and communities 173, 178, 180; state and peasant in Poland 269–70

religion 262, 266–7, 278, 322–3, 340, 490

rents 186, 429, 430, 433 437, 444, 466–8, 470, 484

representation: political in Bulgaria 338; 'spaces' 352–3; women in Poland 311, 315, 318–21, 323

reproduction: diversity 59; household 340–1; sexual, and Polish women's rights 317, 323; social 28, 180, 186; Soviet system 34–7

Resistência Nacional Moçambicana (Renamo) 413–14

resources: allocation in Bulgarian agriculture 257; allocation in Mauritania 391; allocation and socialism 58, 60, 64–6, 68; allocation in Soviet Union 89; allocation in Vietnam 486–7, 489; availability in Bulgaria 120, 123; Bulgarian appropriation of public 182; Bulgarian cultural 138–40, 142; concealment and trade unions 200; efficiency in China 432; exploitation in Novosibirsk 162; individual and economy 204; lack in East Europe 55; mobilisation 8; natural and cultural in Rila 361, 363, 367; natural in Soviet Union 83–4, 97; Polish women and access to 312, 316–17, 319–21; raw materials in Russia 149, 151, 154, 158, 160–1; representation of water crisis in Bulgaria 351; west–east flow in GDR 105

restitution of land in Bulgaria 246–9, 254, 256–9

restructuring: Bulgaria 1, 115–16, 128, 130, 134, 137, 140, 243–5, 249–53, 255, 330–1, 335–7, 355; corporations 65–6; Czech Republic 218–35; Estonia 292–3, 295; GDR 100, 104; networks 56, 70; Novosibirsk, Siberia 161–7; political economy and 5–12; regional in Bulgaria 116–23; regional in Russia 147–68; Soviet Union 82, 83, 89, 91, 93, 97–9; state enterprises 172–90; trade unions and 197, 204–8, 210; Vietnam 1

restructuring: capitalism 457

restructuring: transition and 310, 389

Revoluční Odborové Hnutí 205–7

revolution 409–10, 418, 457

rice production in Vietnam 454–5, 459–63, 476–7, 479, 484–5, 491–2, 494

Rila region (Bulgaria) 347, 349–50, 355, 357, 359–61, 364–6, 368

Roma, social exclusion and transition 373–84

Romania 77, 201, 205, 208, 381–2, 384, 454

Ruble, B.A., on trade unions 199

'Rural Solidarity' 268

Rural Union (Estonia) 294–6

rurality and construction of nation in Estonia 284–303

Rusnok, Jií, on Czech unemployment 207

Russia 8–9, 63–4, 107, 199–203, 205, 208, 252, 454, 458

Rutland, P., on 'transitology' 173

Sabel, C.F., on co-operative relations 69

Sabelli, F., on World Bank 419

Sachs, Jeffrey: on big bang in Vietnam 460; on enterprise reform in Bulgaria 141; on privatisation and command economy 134; on reform in Vietnam 493, 496; on shock therapy 26, 285

Salazar, Antonio, of Portugal 412

Sanders, Irwin, on Bulgarian village life 245–6

Sapareva-Bania barricade (1995) (Bulgaria) 347–51, 358–9, 361–4, 366–8

Sapir, J.: on command economy 83, 87–8; on Russian regulation 97–8

Schroeder, G., on service gap in Russia 157

Schumpter, J.A., or entrepreneurship and personality 64

Scott, J.C., on public transcripts 266

second economy 32, 35, 45, 60–1, 68–9

Selden, M.: on collectivisation in Vietnam 494; on European model of agrarian reform 456

service sector: Bulgaria 332–6; China 494; Estonia 293; GDR 101; Mozambique 415; Novosibirsk 165–6; Russia 150, 153, 155,

157–8, 157–60, 161; Soviet Union 86–7, 91; Vietnam 460, 486
settlement of Roma 376–82
'shadow plan' 35–6, 45
'shadow transition' 357
shareholdings 93, 439, 441
shock therapy 4, 26, 89–91, 107–8, 285, 302, 429, 456–8, 462, 493
shortages 35–6, 85–6, 89, 91, 98, 179–80, 186, 415
Siberia 87, 97, 374
Sik, O., on networks as resources 134
Simmel, G., on commodity exchange and Russian system 63–4
size: of Bulgarian enterprises 120, 123–4, 131, 140, 177; of Bulgarian farms 246–7; of East European enterprise 181; of Polish farms 262, 269; of Soviet enterprises 86–7; of Vietnamese co-operatives 467, 472
Slabakov, Petr, Bulgarian environmentalist 365
Slovak Republic 45–6, 205, 383, 434
Smart, Josephine, on Henggang township, Guangdong 442
Smith, Adam, on market exchange 430–1
Smith, Adrian, on Slovakia accumulation and industrialisation 434
Smith, G., on ethnic democracy 291, 302
social change, theories of, and Roma 373–6
Social Democratic Party (SdRP) (Poland) 273–4, 280
social practice: Bulgaria 142; capitalism and 167–8; economic life as 147; governance and 29–32, 39; networks and 134; plant economy as 130; Russia 46–7; Soviet Union 36–8
socialism: aftermath and the periphery 408–21; and agriculture 456–8; Bulgaria 333, 336–41, 350, 355, 360, 367; capitalism and 6, 33, 37; China 429; East Europe 1, 5, 493; liberalism and Polish women

313–14; 'local market' 440–1, 443; post- 32, 39, 47, 54–71; restructuring 389, 392; Roma and 376–9, 384; Soviet Union 428; Vietnam 450–1, 458–63
Sofia, Bulgaria, water crisis in 347, 349, 352, 352–9, 357–9
Solidarity (Poland) 202, 208–9, 244, 263–6, 268, 274–81
Somogyi, Paul, on trade unions 206
South Korea 428, 450, 463, 484, 491
Soviet Union: collapse 5, 25, 34, 38, 147, 150, 428, 456, 462; reform 131–2, 141; regulation and institutionalisation 32–8; socialism and Third World 409–10, 419, 421; state control 30; transition to capitalism 76, 82–99; Vietnam and 460
space economies 9, 43–4, 46–7
spatiality 29, 31, 67–8
stabilisation, economic 28–30
Stalin, Josef, of Russia 83, 162, 331, 374, 409
Stalinism 408–10, 419
standard of living: Bulgaria 355, 361; East Europe 78; GDR 104; Poland 268; Soviet Union 83, 90, 93, 97; trade unions and 199, 206, 208–9; Vietnam 465, 467
Stark, David: on economic transition in East Europe 244; on enterprise reform in Hungary 433; on recombinant property 96, 204, 434, 443, 452; on socialism and restructuring 264
state: Bulgarian water crisis 350–2, 357, 359, 368; Czech control of coal mining 219–23, 225, 227–30, 233–5; global integration 389; governance in Soviet Union 99; international trade in Russia 160; labour employment in Bulgaria 331; market in China 322–3; monopoly and market economy 204; non-intervention in Estonia 293–4; politics and transition 309–11; trade unions and 197–8, 200, 203, 205,

210–11; women and politics in Poland 313, 315–19

state farms 262, 270, 277, 300, 376, 456

state intervention 96, 99, 116–17, 119, 126, 131, 151, 154, 429–30, 433, 494

Stewart, Michael, on Vlach gypsies 376, 380

street-trading 380–1, 418, 420

strikes 200, 203, 205–6

structural adjustment 107–8, 389, 391–3, 396–405, 417, 419

'subaltern counter-publics' 311–12, 321–2, 324

subsidies: Bulgaria 179, 181, 186, 190, 254, 337; Czech coal mines 221; Estonia 289, 293, 295, 301; GDR 105; Russia 150, 155, 157; Soviet Union 91, 96; trade unions and 208; Vietnam 457, 460, 462, 471, 489, 493

Suchocka, Hanna, of Poland 275

surplus and Vietnamese peasants 471, 473–4, 485, 489

Szegö, Judit, on Vlach gypsies 376, 380

Szelenyi, Ivan on reform in Hungary 476, 473

Szelényi, Ivan on Roma in Hungary 376, 377, 379, 383

Szurek, J.-Ch., on Farmers' Solidarity 269

Taiwan and Vietnam 450, 463, 484, 491

taxation 97, 174, 250–1, 259, 441, 469, 484–9, 492

technology 85, 96, 138–9, 163–5, 183, 185, 187–8, 203

Théret, B., on economic, political and domestic order 288

Thompson, G., on nation-state and globalisation 288

Thrift, N., on 'phantom states' 417; on socio-economic practices 31

Tilly, C., on collapse of communism 285

Tipps, D., on assimilation 374

tobacco industry in Bulgaria 249, 252, 254–5, 335, 337, 342

trade: Bulgaria 243, 259; Estonia 292–3; GDR 100–1; high-value, long-distance 420; Poland 270; Russia 148–9, 151, 157–8, 160–1; Soviet Union 86, 91, 96; Vietnam 486–8, 490, 492

trade unions: Bulgaria 184–5, 188, 336, 339; Czech coal mining 222–3, 233–4; Poland 274, 276, 279; in transition 197–211

trans-nationalism 107, 416–17

transfers, financial, west to east Germany 105–6

transformation 4–5, 125, 140, 174–5, 373–5, 384, 410; China 428–45; East Europe 25–47

'transitional recession' Russia 148–51, 162, 166

'transitology' 4, 173

transport industry 100–1, 332–3, 336

Trotsky, Leon, on trade unions 197

Truong Chinh, on Vietnamese peasants 454

Tuan, D.T., on regional difference in Vietnam 492

unemployment: Bulgaria 115–42, 183, 186, 255, 335–7, 343, 355–6, 361; Czech trade unions 205, 207; East Europe 108; enterprise behaviour 179; GDR 102–3; Novosibirsk 162; of Roma 379–80, 384; Russia 150, 153–7: Soviet Union 87; trade unions 200, 210

unification 100, 103–5, 463, 467

Union of Democratic Forces (UDF) (Bulgaria) 247–9, 253–5, 339, 359, 363

United Peasant Party (ZSL) (Poland) 269–70

United States Agency for International Development (USAID) 391–3

universalism 10, 342, 364

urban–rural conflict and identity in Estonia 297–302

urbanisation 5, 138, 361–2, 414, 455
US 107–8, 202, 395, 403

Vaptsarov, Nikoli, Bulgarina
vuzhrazhdane 363
Verdery, K., on socialism and
feudalism 496
Videnov, Zhan, of Bulgaria 353, 355,
363
Vietnam 1–2, 5, 411; recombinant
capitalism in 450–96
Vinh Phu province, Vietnam 452, 467,
475–87, 491
VK-Rila project (Bulgaria) 357–60,
364–6
Vlach gypsies 376–7, 380
Vo Nguyen Giap, on Vietnamese
peasants 454
Vo Van Kiet, of Vietnam 451, 455
voucher privatisation 65, 93, 98, 224,
355
Vylder, S. de, on macroeconomic
stabilisation in Vietnam 454

wages: in Bulgaria 117–19, 126,
132–3, 141, 177, 182, 335–7, 356;
in Czech coal mines 223, 225;
foreign capital and 205; in GDR
102, 104, 105; in Neftochim 183,
184; in Novosibirsk 166; in Russia
160; in Soviet Union 87, 90; trade
unions and 200, 203, 206; in
Vietnam 484
Wainwright, Hillary, on democratic
governance 343
Walder, A.: on centralisation in China
496; on property rights in China
429, 438–9
water, democratisation and politics in
Bulgaria 347–68
Watts, Michael: on cultural difference
and territorial identity 351; on fast
capitalism in Nigeria 355
Weick, K.E., on loose coupling 62
welfare 37, 92, 96, 437, 438

western model of trade unions 202–4,
207–9, 211
Wilkin, J., on Polish peasant farming
264
Williamson, O., on economic
transactions 70
women: imagined and imaginary
equality in Bulgaria 330–43;
meaning of political in Poland
309–24
Woo, M., on reform in Vietnam 460,
493, 496
worker-peasant, Third World socialism
409, 413–14
workers: Bulgaria 116, 126, 134, 139,
142, 183–90; trade unions 197–201,
203, 205, 209–10, 233; Vietnamese
brigades 467, 476, 494
workpoints and agriculture in Vietnam
467, 469, 471
'works councils' 207
workshops in Bulgaria 125, 130,
132–4, 138–9, 141, 182–3
World Bank 5, 286, 391–3, 395,
399–405, 411, 415, 417, 419–20,
428, 455
Wyzan, M.: on Bulgarian restructuring
and transition 355; on labour
shedding 172

Yanchoulev, mayor of Sofia 358
Yates, A., on marginalisation of
indigenous peoples 374
Yeltsin, Boris, of Russia 165
yields, agricultural 243, 255–6, 258,
463–4, 469–70
Young, Iris, on cultural imperialism
381

Zamfirovo (Mikhailovgrad), Bulgaria
256
Zhivkov, Todor, of Bulgaria 132, 138,
140, 243, 360, 364
Žižek, S., on nationalism 289, 291,
302